SECOND EDITION

SIMULATION USING PROMODEL

Dr. Charles Harrell

Professor, Brigham Young University, Provo, Utah
Director, PROMODEL Corporation, Orem, Utah

Dr. Biman K. Ghosh, Project Leader

Professor, California State Polytechnic University,
Pomona, California

Dr. Royce O. Bowden, Jr.

Professor, Mississippi State University,
Mississippi State, Mississippi

Boston   Burr Ridge, IL   Dubuque, IA   Madison, WI   New York   San Francisco   St. Louis
Bangkok   Bogotá   Caracas   Kuala Lumpur   Lisbon   London   Madrid   Mexico City
Milan   Montreal   New Delhi   Santiago   Seoul   Singapore   Sydney   Taipei   Toronto

Higher Education

SIMULATION USING PROMODEL, SECOND EDITION

Published by McGraw-Hill, a business unit of The McGraw-Hill Companies, Inc., 1221 Avenue of the Americas, New York, NY 10020. Copyright © 2004, 2000 by The McGraw-Hill Companies, Inc. All rights reserved. No part of this publication may be reproduced or distributed in any form or by any means, or stored in a database or retrieval system, without prior written consent of The McGraw-Hill Companies, Inc., including, but not limited to, in any network or other electronic storage or transmission, or broadcast for distance learning.

Some ancillaries, including electronic and print components, may not be available to customers outside the United States.

This book is printed on acid-free paper.

2 3 4 5 6 7 8 9 0 QPF/QPF 0 9 8 7 6 5 4

ISBN 0–07–248263–X

Publisher: *Elizabeth A. Jones*
Senior sponsoring editor: *Suzanne Jeans*
Developmental editor: *Amanda J. Green*
Marketing manager: *Sarah Martin*
Project manager: *Joyce Watters*
Production supervisor: *Kara Kudronowicz*
Lead media project manager: *Judi David*
Senior coordinator of freelance design: *Michelle D. Whitaker*
Cover designer: *Jamie E. O'Neal*
Cover image: *© Paul H. McGown for ProModel Corporation*
Senior photo research coordinator: *Lori Hancock*
Compositor: *Interactive Composition Corporation*
Typeface: *10/12 Times Roman*
Printer: *Quebecor World Fairfield, PA*

Library of Congress Cataloging-in-Publication Data

Harrell, Charles, 1950–
    Simulation using ProModel / Charles Harrell, Biman K. Ghosh, Royce O. Bowden. — 2nd ed.
      p. cm.
    Includes index.
    ISBN 0–07–248263–X
    1. Computer simulation. 2. ProModel. I. Ghosh, Biman K. II. Bowden, Royce. III. Title.

QA76.9.C65H355 2004
003'.35369—dc21

2003010420
CIP

www.mhhe.com

McGRAW-HILL SERIES IN INDUSTRIAL ENGINEERING AND MANAGEMENT SCIENCE

Consulting Editors

Kenneth E. Case

Department of Industrial Engineering and Management, Oklahoma State University

Philip M. Wolfe

Department of Industrial and Management Systems Engineering, Arizona State University

Dedication

To Paula, Raja, and Pritha
B.K.G.

To Yvonne, Emily, Elizabeth, Richard, Ryan, and Polly
C.H.

To CeCelia, Taylor, and Robert
R.O.B.

BRIEF CONTENTS

CONTENTS

PART III

CASE STUDY ASSIGNMENTS

Simulation is a modeling and analysis technique used to evaluate and improve dynamic systems of all types. It has grown from a relatively obscure technology used by only a few specialists to a widely accepted tool used by decision makers at all levels in an organization. Imagine being in a highly competitive industry and managing a manufacturing or service facility that is burdened by outdated technologies and inefficient management practices. In order to stay competitive, you know that changes must be made, but you are not exactly sure what changes would work best, or if certain changes will work at all. You would like to be able to try out a few different ideas, but you recognize that this would be very time-consuming, expensive, and disruptive to the current operation. Now, suppose that there was some magical way you could make a duplicate of your system and have unlimited freedom to rearrange activities, reallocate resources, or change any operating procedures. What if you could even try out completely new technologies and radical new innovations all within just a matter of minutes or hours? Suppose, further, that all of this experimentation could be done in compressed time with automatic tracking and reporting of key performance measures. Not only would you discover ways to improve your operation, but it could all be achieved risk free—without committing any capital, wasting any time, or disrupting the current system. This is precisely the kind of capability that simulation provides. Simulation lets you experiment with a computer model of your system in compressed time, giving you decision-making capability that is unattainable in any other way.

This text is geared toward simulation courses taught at either a graduate or an undergraduate level. It contains an ideal blend of theory and practice and covers the use of simulation in both manufacturing and service systems. This makes it well suited for use in courses in either an engineering or a business curriculum. It is also suitable for simulation courses taught in statistics and computer science programs. The strong focus on the practical aspects of simulation also makes it a book that any practitioner of simulation would want to have on hand.

This text is designed to be used in conjunction with ProModel simulation software, which accompanies the book. ProModel is one of the most powerful and popular simulation packages used today for its ease of use and flexibility. ProModel was the first fully commercial, Windows-based simulation package and the first to introduce simulation optimization. ProModel is already being used in thousands of organizations and taught in hundreds of universities and colleges throughout the world. While many teaching aids have been developed to train individuals in the use of ProModel, this is the first full-fledged textbook written for teaching simulation using ProModel.

Simulation is definitely a learn-by-doing activity. The goal of this text is not simply to introduce students to the topic of simulation, but to actually develop competence in the use of simulation. To this end, the book contains plenty of real-life examples, case studies, and lab exercises to give students actual experience in the use of simulation. Simulation texts often place too much emphasis on the theory behind simulation and not enough emphasis on how it is used in actual problem-solving situations. In simulation courses we have taught over the years, the strongest feedback we have received from students is that they wish they had more hands-on time with simulation beginning from the very first week of the semester.

This text is divided into three parts: a section on the general science and practice of simulation, a lab section to train students in the use of ProModel, and a section containing cases to assign as student projects. While the book is intended for use with ProModel, the division of the book into these three distinct parts permits a modular use of the book, allowing any part to be used independently of any other part. For example, the lab section based on ProModel could be replaced with training on some other simulation product. If you already have a background in simulation and want to use the book as a primer on ProModel, the labs can be completed independently of Parts I and III.

Part I consists of study chapters covering the science and technology of simulation. The first four chapters introduce the topic of simulation, its application to system design and improvement, and how simulation works. Chapters 5 through 11 present both the practical and theoretical aspects of conducting a simulation project and applying simulation optimization. Chapters 12 through 14 cover specific applications of simulation to manufacturing, material handling, and service systems.

Part II is the lab portion of the book containing exercises for developing simulation skills using ProModel. The labs are correlated with the reading chapters in Part I so that Lab 1 should be completed along with Chapter 1 and so on. There are 14 chapters and 14 labs. The labs are designed to be self-teaching. Students are walked through the steps of modeling a particular situation and then are given exercises to complete on their own.

Part III is a series of case studies taken mostly from actual scenarios that can be assigned as simulation projects. They are intended as capstone experiences to give students an opportunity to bring together what they have learned in the course to solve a real-life problem.

This text focuses on the use of simulation to solve problems in the two most common types of systems today: manufacturing and service systems. Nearly 15 percent of the U.S. workforce is employed in manufacturing. In 1955, about one-half of the U.S. workforce worked in the service sector. Today nearly 80 percent of the American workforce can be found in service-related occupations. Manufacturing and service systems share much in common. They both consist of activities, resources, and controls for processing incoming entities. The performance objectives in both instances relate to quality, efficiency, cost reduction, process time reduction, and customer satisfaction. In addition to having common elements and objectives, they are also often interrelated. Manufacturing systems are supported by service activities such as product design, order management, or maintenance. Service systems receive support from production activities such as food production, check processing, or printing. Regardless of the industry in which one ends up, an understanding of the modeling issues underlying both systems will be helpful.

Acknowledgments

No work of this magnitude is performed in a vacuum, independently of the help and assistance of others. We are indebted to many colleagues, associates, and other individuals who had a hand in this project. John Mauer (Geer Mountain Software) provided valuable information on input modeling and the use of Stat::Fit. Dr. John D. Hall (APT Research, Inc.) and Dr. Allen G. Greenwood (Mississippi State University) helped to develop and refine the ANOVA material in Chapter 10. Kerim Tumay provided valuable input on the issues associated with service system simulation.

We are grateful to all the reviewers not only for their helpful feedback, but also for their generous contributions and insights: particularly, William Giauque (BYU), George Johnson (Idaho State University), Stephen Chick (INSEAD), Patrick Delany (U.S. Army), Ed Williams (Production Modeling Corp.), Michael E. Nowatkowski (United States Military Academy), Kellie B. Keeling (Virginia Tech), Aparna Gupta (Rensselaer Polytechnic Institute), Alok K. Verma (Old Dominion University), Allen G. Greenwood (Mississippi State University), and Christos Alexopoulos (Georgia Institute of Technology).

Many individuals were motivational and even inspirational in taking on this project: Peter Kalish, Rob Bateman, Lou Keller, Averill Law, Richard Wysk, Dennis Pegden, and Joyce Kupsh, to name a few. We would especially like to thank our families for their encouragement and for so generously tolerating the disruption of normal life caused by this project.

Thanks to all of the students who provided valuable feedback on the first edition of the text. It is for the primary purpose of making simulation interesting and worthwhile for students that we have written this book. In particular we would like to acknowledge and thank the contributions of Aaron Cheng Wee Tan (Mississippi State University).

We are especially indebted to all the wonderful people at PROMODEL Corporation who have been so cooperative in providing software and documentation. Were it not for the excellent software tools and accommodating support staff at PROMODEL, this book would not have been written.

Finally, we thank the editorial and production staff at McGraw-Hill: Amanda Green, Suzanne Jeans, and Joyce Watters. They have been great to work with.

P A R T

I STUDY CHAPTERS

1   INTRODUCTION TO SIMULATION

"Man is a tool using animal. . . . Without tools he is nothing, with tools he is all."

—Thomas Carlyle

1.1 Introduction

On March 19, 1999, the following story appeared in *The Wall Street Journal:*

> Captain Chet Rivers knew that his 747-400 was loaded to the limit. The giant plane, weighing almost 450,000 pounds by itself, was carrying a full load of passengers and baggage, plus 400,000 pounds of fuel for the long flight from San Francisco to Australia. As he revved his four engines for takeoff, Capt. Rivers noticed that San Francisco's famous fog was creeping in, obscuring the hills to the north and west of the airport.
>
> At full throttle the plane began to roll ponderously down the runway, slowly at first but building up to flight speed well within normal limits. Capt. Rivers pulled the throttle back and the airplane took to the air, heading northwest across the San Francisco peninsula towards the ocean. It looked like the start of another routine flight. Suddenly the plane began to shudder violently. Several loud explosions shook the craft and smoke and flames, easily visible in the midnight sky, illuminated the right wing. Although the plane was shaking so violently that it was hard to read the instruments, Capt. Rivers was able to tell that the right inboard engine was malfunctioning, backfiring violently. He immediately shut down the engine, stopping the explosions and shaking.
>
> However this introduced a new problem. With two engines on the left wing at full power and only one on the right, the plane was pushed into a right turn, bringing it directly towards San Bruno Mountain, located a few miles northwest of the airport. Capt. Rivers instinctively turned his control wheel to the left to bring the plane back on course. That action extended the ailerons—control surfaces on the trailing edges of the wings—to tilt the plane back to the left. However, it also extended the

spoilers—panels on the tops of the wings—increasing drag and lowering lift. With the nose still pointed up, the heavy jet began to slow. As the plane neared stall speed, the control stick began to shake to warn the pilot to bring the nose down to gain air speed. Capt. Rivers immediately did so, removing that danger, but now San Bruno Mountain was directly ahead. Capt. Rivers was unable to see the mountain due to the thick fog that had rolled in, but the plane's ground proximity sensor sounded an automatic warning, calling "terrain, terrain, pull up, pull up." Rivers frantically pulled back on the stick to clear the peak, but with the spoilers up and the plane still in a skidding right turn, it was too late. The plane and its full load of 100 tons of fuel crashed with a sickening explosion into the hillside just above a densely populated housing area.

"Hey Chet, that could ruin your whole day," said Capt. Rivers's supervisor, who was sitting beside him watching the whole thing. "Let's rewind the tape and see what you did wrong." "Sure Mel," replied Chet as the two men stood up and stepped outside the 747 cockpit simulator. "I think I know my mistake already. I should have used my rudder, not my wheel, to bring the plane back on course. Say, I need a breather after that experience. I'm just glad that this wasn't the real thing."

The incident above was never reported in the nation's newspapers, even though it would have been one of the most tragic disasters in aviation history, because it never really happened. It took place in a cockpit simulator, a device which uses computer technology to predict and recreate an airplane's behavior with gut-wrenching realism.

The relief you undoubtedly felt to discover that this disastrous incident was just a simulation gives you a sense of the impact that simulation can have in averting real-world catastrophes. This story illustrates just one of the many ways simulation is being used to help minimize the risk of making costly and sometimes fatal mistakes in real life. Simulation technology is finding its way into an increasing number of applications ranging from training for aircraft pilots to the testing of new product prototypes. The one thing that these applications have in common is that they all provide a virtual environment that helps prepare for real-life situations, resulting in significant savings in time, money, and even lives.

One area where simulation is finding increased application is in manufacturing and service system design and improvement. Its unique ability to accurately predict the performance of complex systems makes it ideally suited for systems planning. Just as a flight simulator reduces the risk of making costly errors in actual flight, system simulation reduces the risk of having systems that operate inefficiently or that fail to meet minimum performance requirements. While this may not be life-threatening to an individual, it certainly places a company (not to mention careers) in jeopardy.

In this chapter we introduce the topic of simulation and answer the following questions:

- What is simulation?
- Why is simulation used?
- How is simulation performed?
- When and where should simulation be used?

- What are the qualifications for doing simulation?
- How is simulation economically justified?

The purpose of this chapter is to create an awareness of how simulation is used to visualize, analyze, and improve the performance of manufacturing and service systems.

1.2 What Is Simulation?

The *Oxford American Dictionary* (1980) defines simulation as a way "to reproduce the conditions of a situation, as by means of a model, for study or testing or training, etc." For our purposes, we are interested in reproducing the operational behavior of dynamic systems. The model that we will be using is a computer model. Simulation in this context can be defined as the imitation of a dynamic system using a computer model in order to evaluate and improve system performance. According to Schriber (1987), simulation is "the modeling of a process or system in such a way that the model mimics the response of the actual system to events that take place over time." By studying the behavior of the model, we can gain insights about the behavior of the actual system.

Simulation is the imitation of a dynamic system using a computer model in order to evaluate and improve system performance.

In practice, simulation is usually performed using commercial simulation software like ProModel that has modeling constructs specifically designed for capturing the dynamic behavior of systems. Performance statistics are gathered during the simulation and automatically summarized for analysis. Modern simulation software provides a realistic, graphical animation of the system being modeled (see Figure 1.1). During the simulation, the user can interactively adjust the animation speed and change model parameter values to do "what if" analysis on the fly. State-of-the-art simulation technology even provides optimization capability—not that simulation itself optimizes, but scenarios that satisfy defined feasibility constraints can be automatically run and analyzed using special goal-seeking algorithms.

This book focuses primarily on discrete-event simulation, which models the effects of the events in a system as they occur over time. Discrete-event simulation employs statistical methods for generating random behavior and estimating

FIGURE 1.1

Simulation provides animation capability.

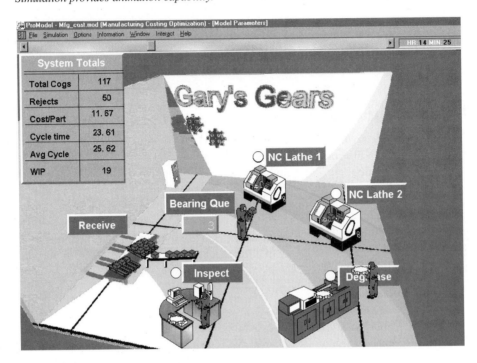

model performance. These methods are sometimes referred to as Monte Carlo methods because of their similarity to the probabilistic outcomes found in games of chance, and because Monte Carlo, a tourist resort in Monaco, was such a popular center for gambling.

1.3 Why Simulate?

Rather than leave design decisions to chance, simulation provides a way to validate whether or not the best decisions are being made. Simulation avoids the expensive, time-consuming, and disruptive nature of traditional trial-and-error techniques.

Trial-and-error approaches are expensive, time consuming, and disruptive.

With the emphasis today on time-based competition, traditional trial-and-error methods of decision making are no longer adequate. Regarding the shortcoming of trial-and-error approaches in designing manufacturing systems, Solberg (1988) notes,

> The ability to apply trial-and-error learning to tune the performance of manufacturing systems becomes almost useless in an environment in which changes occur faster than the lessons can be learned. There is now a greater need for formal predictive methodology based on understanding of cause and effect.

The power of simulation lies in the fact that it provides a method of analysis that is not only formal and predictive, but is capable of accurately predicting the performance of even the most complex systems. Deming (1989) states, "Management of a system is action based on prediction. Rational prediction requires systematic learning and comparisons of predictions of short-term and long-term results from possible alternative courses of action." The key to sound management decisions lies in the ability to accurately predict the outcomes of alternative courses of action. Simulation provides precisely that kind of foresight. By simulating alternative production schedules, operating policies, staffing levels, job priorities, decision rules, and the like, a manager can more accurately predict outcomes and therefore make more informed and effective management decisions. With the importance in today's competitive market of "getting it right the first time," the lesson is becoming clear: if at first you don't succeed, you probably should have simulated it.

By using a computer to model a system before it is built or to test operating policies before they are actually implemented, many of the pitfalls that are often encountered in the start-up of a new system or the modification of an existing system can be avoided. Improvements that traditionally took months and even years of fine-tuning to achieve can be attained in a matter of days or even hours. Because simulation runs in compressed time, weeks of system operation can be simulated in only a few minutes or even seconds. The characteristics of simulation that make it such a powerful planning and decision-making tool can be summarized as follows:

- Captures system interdependencies.
- Accounts for variability in the system.
- Is versatile enough to model any system.
- Shows behavior over time.
- Is less costly, time consuming, and disruptive than experimenting on the actual system.
- Provides information on multiple performance measures.
- Is visually appealing and engages people's interest.
- Provides results that are easy to understand and communicate.
- Runs in compressed, real, or even delayed time.
- Forces attention to detail in a design.

Because simulation accounts for interdependencies and variation, it provides insights into the complex dynamics of a system that cannot be obtained using other analysis techniques. Simulation gives systems planners unlimited freedom to try out different ideas for improvement, risk free—with virtually no cost, no waste of time, and no disruption to the current system. Furthermore, the results are both visual and quantitative with performance statistics automatically reported on all measures of interest.

Even if no problems are found when analyzing the output of simulation, the exercise of developing a model is, in itself, beneficial in that it forces one to think through the operational details of the process. Simulation can work with inaccurate information, but it can't work with incomplete information. Often solutions present themselves as the model is built—before any simulation run is made. It is a human tendency to ignore the operational details of a design or plan until the implementation phase, when it is too late for decisions to have a significant impact. As the philosopher Alfred North Whitehead observed, "We think in generalities; we live detail" (Audon 1964). System planners often gloss over the details of how a system will operate and then get tripped up during implementation by all of the loose ends. The expression "the devil is in the details" has definite application to systems planning. Simulation forces decisions on critical details so they are not left to chance or to the last minute, when it may be too late.

Simulation promotes a try-it-and-see attitude that stimulates innovation and encourages thinking "outside the box." It helps one get into the system with sticks and beat the bushes to flush out problems and find solutions. It also puts an end to fruitless debates over what solution will work best and by how much. Simulation takes the emotion out of the decision-making process by providing objective evidence that is difficult to refute.

1.4 Doing Simulation

Simulation is nearly always performed as part of a larger process of system design or process improvement. A design problem presents itself or a need for improvement exists. Alternative solutions are generated and evaluated, and the best solution is selected and implemented. Simulation comes into play during the evaluation phase. First, a model is developed for an alternative solution. As the model is *run,* it is put into operation for the period of interest. Performance statistics (utilization, processing time, and so on) are gathered and reported at the end of the run. Usually several *replications* (independent runs) of the simulation are made. Averages and variances across the replications are calculated to provide statistical estimates of model performance. Through an iterative process of modeling, simulation, and analysis, alternative configurations and operating policies can be tested to determine which solution works the best.

Simulation is essentially an experimentation tool in which a computer model of a new or existing system is created for the purpose of conducting experiments. The model acts as a surrogate for the actual or real-world system. Knowledge gained from experimenting on the model can be transferred to the real system (see Figure 1.2). When we speak of *doing* simulation, we are talking about "the

FIGURE 1.2

*Simulation provides a
virtual method for
doing system
experimentation.*

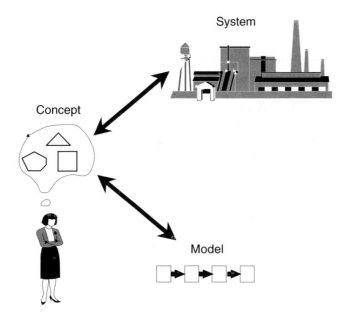

process of designing a model of a real system and conducting experiments with
this model" (Shannon 1998). Conducting experiments on a model reduces the
time, cost, and disruption of experimenting on the actual system. In this respect,
simulation can be thought of as a virtual prototyping tool for demonstrating proof
of concept.

The procedure for doing simulation follows the scientific method of (1) formulating a hypothesis, (2) setting up an experiment, (3) testing the hypothesis
through experimentation, and (4) drawing conclusions about the validity of the
hypothesis. In simulation, we formulate a hypothesis about what design or
operating policies work best. We then set up an experiment in the form of a
simulation model to test the hypothesis. With the model, we conduct multiple
replications of the experiment or simulation. Finally, we analyze the simulation
results and draw conclusions about our hypothesis. If our hypothesis was correct, we can confidently move ahead in making the design or operational
changes (assuming time and other implementation constraints are satisfied).
As shown in Figure 1.3, this process is repeated until we are satisfied with the
results.

By now it should be obvious that simulation itself is not a solution tool but
rather an evaluation tool. It describes how a defined system will behave; it does
not prescribe how it should be designed. Simulation doesn't compensate for one's
ignorance of how a system is supposed to operate. Neither does it excuse one from
being careful and responsible in the handling of input data and the interpretation
of output results. Rather than being perceived as a substitute for thinking, simulation should be viewed as an extension of the mind that enables one to understand
the complex dynamics of a system.

FIGURE 1.3

The process of simulation experimentation.

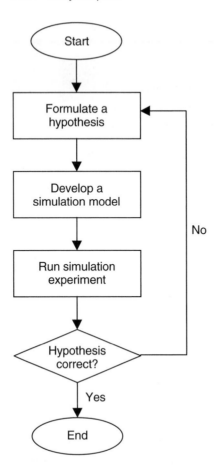

1.5 Use of Simulation

Simulation began to be used in commercial applications in the 1960s. Initial models were usually programmed in FORTRAN and often consisted of thousands of lines of code. Not only was model building an arduous task, but extensive debugging was required before models ran correctly. Models frequently took a year or more to build and debug so that, unfortunately, useful results were not obtained until after a decision and monetary commitment had already been made. Lengthy simulations were run in batch mode on expensive mainframe computers where CPU time was at a premium. Long development cycles prohibited major changes from being made once a model was built.

Only in the last couple of decades has simulation gained popularity as a decision-making tool in manufacturing and service industries. For many companies, simulation has become a standard practice when a new facility is being planned or a process change is being evaluated. It is fast becoming to systems planners what spreadsheet software has become to financial planners.

The surge in popularity of computer simulation can be attributed to the following:

- Increased awareness and understanding of simulation technology.
- Increased availability, capability, and ease of use of simulation software.
- Increased computer memory and processing speeds, especially of PCs.
- Declining computer hardware and software costs.

Simulation is no longer considered a method of "last resort," nor is it a technique reserved only for simulation "experts." The availability of easy-to-use simulation software and the ubiquity of powerful desktop computers have made simulation not only more accessible, but also more appealing to planners and managers who tend to avoid any kind of solution that appears too complicated. A solution tool is not of much use if it is more complicated than the problem that it is intended to solve. With simple data entry tables and automatic output reporting and graphing, simulation is becoming much easier to use and the reluctance to use it is disappearing.

The primary use of simulation continues to be in the area of manufacturing. Manufacturing systems, which include warehousing and distribution systems, tend to have clearly defined relationships and formalized procedures that are well suited to simulation modeling. They are also the systems that stand to benefit the most from such an analysis tool since capital investments are so high and changes are so disruptive. Recent trends to standardize and systematize other business processes such as order processing, invoicing, and customer support are boosting the application of simulation in these areas as well. It has been observed that 80 percent of all business processes are repetitive and can benefit from the same analysis techniques used to improve manufacturing systems (Harrington 1991). With this being the case, the use of simulation in designing and improving business processes of every kind will likely continue to grow.

While the primary use of simulation is in decision support, it is by no means limited to applications requiring a decision. An increasing use of simulation is in the area of communication and visualization. Modern simulation software incorporates visual animation that stimulates interest in the model and effectively communicates complex system dynamics. A proposal for a new system design can be sold much easier if it can actually be shown how it will operate.

On a smaller scale, simulation is being used to provide interactive, computer-based training in which a management trainee is given the opportunity to practice decision-making skills by interacting with the model during the simulation. It is also being used in real-time control applications where the model interacts with the real system to monitor progress and provide master control. The power of simulation to capture system dynamics both visually and functionally opens up numerous opportunities for its use in an integrated environment.

Since the primary use of simulation is in decision support, most of our discussion will focus on the use of simulation to make system design and operational decisions. As a decision support tool, simulation has been used to help plan and make improvements in many areas of both manufacturing and service industries. Typical applications of simulation include

- Work-flow planning.
- Capacity planning.
- Cycle time reduction.
- Staff and resource planning.
- Work prioritization.
- Bottleneck analysis.
- Quality improvement.
- Cost reduction.
- Inventory reduction.

- Throughput analysis.
- Productivity improvement.
- Layout analysis.
- Line balancing.
- Batch size optimization.
- Production scheduling.
- Resource scheduling.
- Maintenance scheduling.
- Control system design.

1.6 When Simulation Is Appropriate

Not all system problems that *could* be solved with the aid of simulation *should* be solved using simulation. It is important to select the right tool for the task. For some problems, simulation may be overkill—like using a shotgun to kill a fly. Simulation has certain limitations of which one should be aware before making a decision to apply it to a given situation. It is not a panacea for all system-related problems and should be used only if the shoe fits. As a general guideline, simulation is appropriate if the following criteria hold true:

- An operational (logical or quantitative) decision is being made.
- The process being analyzed is well defined and repetitive.
- Activities and events are interdependent and variable.
- The cost impact of the decision is greater than the cost of doing the simulation.
- The cost to experiment on the actual system is greater than the cost of simulation.

Decisions should be of an operational nature. Perhaps the most significant limitation of simulation is its restriction to the operational issues associated with systems planning in which a logical or quantitative solution is being sought. It is not very useful in solving qualitative problems such as those involving technical or sociological issues. For example, it can't tell you how to improve machine reliability or how to motivate workers to do a better job (although it can assess the impact that a given level of reliability or personal performance can have on overall system performance). Qualitative issues such as these are better addressed using other engineering and behavioral science techniques.

Processes should be well defined and repetitive. Simulation is useful only if the process being modeled is well structured and repetitive. If the process doesn't follow a logical sequence and adhere to defined rules, it may be difficult to model. Simulation applies only if you can describe how the process operates. This does not mean that there can be no uncertainty in the system. If random behavior can be described using probability expressions and distributions, they can be simulated. It is only when it isn't even possible to make reasonable assumptions of how a system operates (because either no information is available or behavior is totally erratic) that simulation (or any other analysis tool for that matter) becomes useless. Likewise, one-time projects or processes that are never repeated the same way twice are poor candidates for simulation. If the scenario you are modeling is likely never going to happen again, it is of little benefit to do a simulation.

Activities and events should be interdependent and variable. A system may have lots of activities, but if they never interfere with each other or are deterministic (that is, they have no variation), then using simulation is probably unnecessary. It isn't the number of activities that makes a system difficult to analyze. It is the number of interdependent, random activities. The effect of simple interdependencies is easy to predict if there is no variability in the activities. Determining the flow rate for a system consisting of 10 processing activities is very straightforward if all activity times are constant and activities are never interrupted. Likewise, random activities that operate independently of each other are usually easy to analyze. For example, 10 machines operating in isolation from each other can be expected to produce at a rate that is based on the average cycle time of each machine less any anticipated downtime. It is the combination of interdependencies and random behavior that really produces the unpredictable results. Simpler analytical methods such as mathematical calculations and spreadsheet software become less adequate as the number of activities that are both interdependent and random increases. For this reason, simulation is primarily suited to systems involving both interdependencies and variability.

The cost impact of the decision should be greater than the cost of doing the simulation. Sometimes the impact of the decision itself is so insignificant that it doesn't warrant the time and effort to conduct a simulation. Suppose, for example, you are trying to decide whether a worker should repair rejects as they occur or wait until four or five accumulate before making repairs. If you are certain that the next downstream activity is relatively insensitive to whether repairs are done sooner rather than later, the decision becomes inconsequential and simulation is a wasted effort.

The cost to experiment on the actual system should be greater than the cost of simulation. While simulation avoids the time delay and cost associated with experimenting on the real system, in some situations it may actually be quicker and more economical to experiment on the real system. For example, the decision in a customer mailing process of whether to seal envelopes before or after they are addressed can easily be made by simply trying each method and comparing the results. The rule of thumb here is that if a question can be answered through direct experimentation quickly, inexpensively, and with minimal impact to the current operation, then don't use simulation. Experimenting on the actual system also eliminates some of the drawbacks associated with simulation, such as proving model validity.

There may be other situations where simulation is appropriate independent of the criteria just listed (see Banks and Gibson 1997). This is certainly true in the case of models built purely for visualization purposes. If you are trying to sell a system design or simply communicate how a system works, a realistic animation created using simulation can be very useful, even though nonbeneficial from an analysis point of view.

1.7 Qualifications for Doing Simulation

Many individuals are reluctant to use simulation because they feel unqualified. Certainly some training is required to use simulation, but it doesn't mean that only statisticians or operations research specialists can learn how to use it. Decision support tools are always more effective when they involve the decision maker, especially when the decision maker is also the domain expert or person who is most familiar with the design and operation of the system. The process owner or manager, for example, is usually intimately familiar with the intricacies and idiosyncrasies of the system and is in the best position to know what elements to include in the model and be able to recommend alternative design solutions. When performing a simulation, often improvements suggest themselves in the very activity of building the model that the decision maker might never discover if someone else is doing the modeling. This reinforces the argument that the decision maker should be heavily involved in, if not actually conducting, the simulation project.

To make simulation more accessible to non–simulation experts, products have been developed that can be used at a basic level with very little training. Unfortunately, there is always a potential danger that a tool will be used in a way that exceeds one's skill level. While simulation continues to become more user-friendly, this does not absolve the user from acquiring the needed skills to make intelligent use of it. Many aspects of simulation will continue to require some training. Hoover and Perry (1989) note, "The subtleties and nuances of model validation and output analysis have not yet been reduced to such a level of rote that they can be completely embodied in simulation software."

Modelers should be aware of their own inabilities in dealing with the statistical issues associated with simulation. Such awareness, however, should not prevent one from using simulation within the realm of one's expertise. There are both a basic as well as an advanced level at which simulation can be beneficially used. Rough-cut modeling to gain fundamental insights, for example, can be achieved with only a rudimentary understanding of simulation. One need not have extensive simulation training to go after the low-hanging fruit. Simulation follows the 80–20 rule, where 80 percent of the benefit can be obtained from knowing only 20 percent of the science involved (just make sure you know the right 20 percent). It isn't until more precise analysis is required that additional statistical training and knowledge of experimental design are needed.

To reap the greatest benefits from simulation, a certain degree of knowledge and skill in the following areas is useful:

- Project management.
- Communication.

- Systems engineering.
- Statistical analysis and design of experiments.
- Modeling principles and concepts.
- Basic programming and computer skills.
- Training on one or more simulation products.
- Familiarity with the system being investigated.

Experience has shown that some people learn simulation more rapidly and become more adept at it than others. People who are good abstract thinkers yet also pay close attention to detail seem to be the best suited for doing simulation. Such individuals are able to see the forest while still keeping an eye on the trees (these are people who tend to be good at putting together 1,000-piece puzzles). They are able to quickly scope a project, gather the pertinent data, and get a useful model up and running without lots of starts and stops. A good modeler is somewhat of a sleuth, eager yet methodical and discriminating in piecing together all of the evidence that will help put the model pieces together.

If short on time, talent, resources, or interest, the decision maker need not despair. Plenty of consultants who are professionally trained and experienced can provide simulation services. A competitive bid will help get the best price, but one should be sure that the individual assigned to the project has good credentials. If the use of simulation is only occasional, relying on a consultant may be the preferred approach.

1.8 Economic Justification of Simulation

Cost is always an important issue when considering the use of any software tool, and simulation is no exception. Simulation should not be used if the cost exceeds the expected benefits. This means that both the costs and the benefits should be carefully assessed. The use of simulation is often prematurely dismissed due to the failure to recognize the potential benefits and savings it can produce. Much of the reluctance in using simulation stems from the mistaken notion that simulation is costly and very time consuming. This perception is shortsighted and ignores the fact that in the long run simulation usually saves much more time and cost than it consumes. It is true that the initial investment, including training and start-up costs, may be between $10,000 and $30,000 (simulation products themselves generally range between $1,000 and $20,000). However, this cost is often recovered after the first one or two projects. The ongoing expense of using simulation for individual projects is estimated to be between 1 and 3 percent of the total project cost (Glenney and Mackulak 1985). With respect to the time commitment involved in doing simulation, much of the effort that goes into building the model is in arriving at a clear definition of how the system operates, which needs to be done anyway. With the advanced modeling tools that are now available, the actual model development and running of simulations take only a small fraction (often less than 5 percent) of the overall system design time.

Savings from simulation are realized by identifying and eliminating problems and inefficiencies that would have gone unnoticed until system implementation. Cost is also reduced by eliminating overdesign and removing excessive safety factors that are added when performance projections are uncertain. By identifying and eliminating unnecessary capital investments, and discovering and correcting operating inefficiencies, it is not uncommon for companies to report hundreds of thousands of dollars in savings on a single project through the use of simulation. The return on investment (ROI) for simulation often exceeds 1,000 percent, with payback periods frequently being only a few months or the time it takes to complete a simulation project.

One of the difficulties in developing an economic justification for simulation is the fact that it is usually not known in advance how much savings will be realized until it is actually used. Most applications in which simulation has been used have resulted in savings that, had the savings been known in advance, would have looked very good in an ROI or payback analysis.

One way to assess in advance the economic benefit of simulation is to assess the risk of making poor design and operational decisions. One need only ask what the potential cost would be if a misjudgment in systems planning were to occur. Suppose, for example, that a decision is made to add another machine to solve a capacity problem in a production or service system. What are the cost and probability associated with this being the wrong decision? If the cost associated with a wrong decision is $100,000 and the decision maker is only 70 percent confident that the decision is correct, then there is a 30 percent chance of incurring a cost of $100,000. This results in a probable cost of $30,000 (.3 × $100,000). Using this approach, many decision makers recognize that they can't afford *not* to use simulation because the risk associated with making the wrong decision is too high.

Tying the benefits of simulation to management and organizational goals also provides justification for its use. For example, a company committed to continuous improvement or, more specifically, to lead time or cost reduction can be sold on simulation if it can be shown to be historically effective in these areas. Simulation has gained the reputation as a best practice for helping companies achieve organizational goals. Companies that profess to be serious about performance improvement will invest in simulation if they believe it can help them achieve their goals.

The real savings from simulation come from allowing designers to make mistakes and work out design errors on the model rather than on the actual system. The concept of reducing costs through working out problems in the design phase rather than after a system has been implemented is best illustrated by the *rule of tens*. This principle states that the cost to correct a problem increases by a factor of 10 for every design stage through which it passes without being detected (see Figure 1.4).

Simulation helps avoid many of the downstream costs associated with poor decisions that are made up front. Figure 1.5 illustrates how the cumulative cost resulting from systems designed using simulation can compare with the cost of designing and operating systems without the use of simulation. Note that while

FIGURE 1.4

Cost of making changes at subsequent stages of system development.

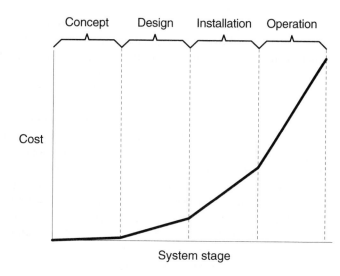

FIGURE 1.5

Comparison of cumulative system costs with and without simulation.

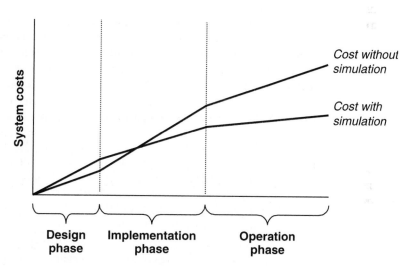

the short-term cost may be slightly higher due to the added labor and software costs associated with simulation, the long-term costs associated with capital investments and system operation are considerably lower due to better efficiencies realized through simulation. Dismissing the use of simulation on the basis of sticker price is myopic and shows a lack of understanding of the long-term savings that come from having well-designed, efficiently operating systems.

Many examples can be cited to show how simulation has been used to avoid costly errors in the start-up of a new system. Simulation prevented an unnecessary expenditure when a *Fortune* 500 company was designing a facility for producing and storing subassemblies and needed to determine the number of containers required for holding the subassemblies. It was initially felt that 3,000 containers were needed until a simulation study showed that throughput did not improve significantly when the number of containers was increased from 2,250 to 3,000. By purchasing 2,250 containers instead of 3,000, a savings of $528,375 was expected in the first year, with annual savings thereafter of over $200,000 due to the savings in floor space and storage resulting from having 750 fewer containers (Law and McComas 1988).

Even if dramatic savings are not realized each time a model is built, simulation at least inspires confidence that a particular system design is capable of meeting required performance objectives and thus minimizes the risk often associated with new start-ups. The economic benefit associated with instilling confidence was evidenced when an entrepreneur, who was attempting to secure bank financing to start a blanket factory, used a simulation model to show the feasibility of the proposed factory. Based on the processing times and equipment lists supplied by industry experts, the model showed that the output projections in the business plan were well within the capability of the proposed facility. Although unfamiliar with the blanket business, bank officials felt more secure in agreeing to support the venture (Bateman et al. 1997).

Often simulation can help improve productivity by exposing ways of making better use of existing assets. By looking at a system holistically, long-standing problems such as bottlenecks, redundancies, and inefficiencies that previously went unnoticed start to become more apparent and can be eliminated. "The trick is to find waste, or *muda*," advises Shingo; "after all, the most damaging kind of waste is the waste we don't recognize" (Shingo 1992). Consider the following actual examples where simulation helped uncover and eliminate wasteful practices:

- GE Nuclear Energy was seeking ways to improve productivity without investing large amounts of capital. Through the use of simulation, the company was able to increase the output of highly specialized reactor parts by 80 percent. The cycle time required for production of each part was reduced by an average of 50 percent. These results were obtained by running a series of models, each one solving production problems highlighted by the previous model (Bateman et al. 1997).

- A large manufacturing company with stamping plants located throughout the world produced stamped aluminum and brass parts on order according to customer specifications. Each plant had from 20 to 50 stamping presses that were utilized anywhere from 20 to 85 percent. A simulation study was conducted to experiment with possible ways of increasing capacity utilization. As a result of the study, machine utilization improved from an average of 37 to 60 percent (Hancock, Dissen, and Merten 1977).

- A diagnostic radiology department in a community hospital was modeled to evaluate patient and staff scheduling, and to assist in expansion planning over the next five years. Analysis using the simulation model enabled improvements to be discovered in operating procedures that precluded the necessity for any major expansions in department size (Perry and Baum 1976).

In each of these examples, significant productivity improvements were realized without the need for making major investments. The improvements came through finding ways to operate more efficiently and utilize existing resources more effectively. These capacity improvement opportunities were brought to light through the use of simulation.

1.9 Sources of Information on Simulation

Simulation is a rapidly growing technology. While the basic science and theory remain the same, new and better software is continually being developed to make simulation more powerful and easier to use. It will require ongoing education for those using simulation to stay abreast of these new developments. There are many sources of information to which one can turn to learn the latest developments in simulation technology. Some of the sources that are available include

- Conferences and workshops sponsored by vendors and professional societies (such as, Winter Simulation Conference and the IIE Conference).
- Professional magazines and journals (*IIE Solutions, International Journal of Modeling and Simulation,* and the like).
- Websites of vendors and professional societies (www.promodel.com, www.scs.org, and so on).
- Demonstrations and tutorials provided by vendors (like those on the ProModel CD).
- Textbooks (like this one).

1.10 How to Use This Book

This book is divided into three parts. Part I contains chapters describing the science and practice of simulation. The emphasis is deliberately oriented more toward the practice than the science. Simulation is a powerful decision support tool that has a broad range of applications. While a fundamental understanding of how simulation works is presented, the aim has been to focus more on how to use simulation to solve real-world problems.

Part II contains ProModel lab exercises that help develop simulation skills. ProModel is a simulation package designed specifically for ease of use, yet it provides the flexibility to model any discrete event or continuous flow process. It is similar to other simulation products in that it provides a set of basic modeling constructs and a language for defining the logical decisions that are made in a system. Basic modeling objects in ProModel include *entities* (the objects being processed), *locations* (the places where processing occurs), *resources* (the agents used to process the entities), and *paths* (the course of travel for entities and resources in moving between locations such as aisles or conveyors). Logical behavior such as the way entities arrive and their routings can be defined with little, if any, programming using the data entry tables that are provided. ProModel is used by thousands of professionals in manufacturing and service-related industries and is taught in hundreds of institutions of higher learning.

Part III contains case study assignments that can be used for student projects to apply the theory they have learned from Part I and to try out the skills they have acquired from doing the lab exercises (Part II). It is recommended that students be assigned at least one simulation project during the course. Preferably this is a project performed for a nearby company or institution so it will be meaningful. If such a project cannot be found, or as an additional practice exercise, the case studies provided should be useful. Student projects should be selected early in the course so that data gathering can get started and the project completed within the allotted time. The chapters in Part I are sequenced to parallel an actual simulation project.

1.11 Summary

Businesses today face the challenge of quickly designing and implementing complex production and service systems that are capable of meeting growing demands for quality, delivery, affordability, and service. With recent advances in computing and software technology, simulation tools are now available to help meet this challenge. Simulation is a powerful technology that is being used with increasing frequency to improve system performance by providing a way to make better design and management decisions. When used properly, simulation can reduce the risks associated with starting up a new operation or making improvements to existing operations.

Because simulation accounts for interdependencies and variability, it provides insights that cannot be obtained any other way. Where important system decisions are being made of an operational nature, simulation is an invaluable decision-making tool. Its usefulness increases as variability and interdependency increase and the importance of the decision becomes greater.

Lastly, simulation actually makes designing systems fun! Not only can a designer try out new design concepts to see what works best, but the visualization makes it take on a realism that is like watching an actual system in operation. Through simulation, decision makers can play what-if games with a new system or modified process before it actually gets implemented. This engaging process stimulates creative thinking and results in good design decisions.

1.12 Review Questions

1. Define simulation.
2. What reasons are there for the increased popularity of computer simulation?

3. What are two specific questions that simulation might help answer in a bank? In a manufacturing facility? In a dental office?

4. What are three advantages that simulation has over alternative approaches to systems design?

5. Does simulation itself optimize a system design? Explain.

6. How does simulation follow the scientific method?

7. A restaurant gets extremely busy during lunch (11:00 A.M. to 2:00 P.M.) and is trying to decide whether it should increase the number of waitresses from two to three. What considerations would you look at to determine whether simulation should be used to make this decision?

8. How would you develop an economic justification for using simulation?

9. Is a simulation exercise wasted if it exposes no problems in a system design? Explain.

10. A simulation run was made showing that a modeled factory could produce 130 parts per hour. What information would you want to know about the simulation study before placing any confidence in the results?

11. A PC board manufacturer has high work-in-process (WIP) inventories, yet machines and equipment seem underutilized. How could simulation help solve this problem?

12. How important is a statistical background for doing simulation?

13. How can a programming background be useful in doing simulation?

14. Why are good project management and communication skills important in simulation?

15. Why should the process owner be heavily involved in a simulation project?

16. For which of the following problems would simulation likely be useful?
    a. Increasing the throughput of a production line.
    b. Increasing the pace of a worker on an assembly line.
    c. Decreasing the time that patrons at an amusement park spend waiting in line.
    d. Determining the percentage defective from a particular machine.
    e. Determining where to place inspection points in a process.
    f. Finding the most efficient way to fill out an order form.

References

Audon, Wyston Hugh, and L. Kronenberger. *The Faber Book of Aphorisms.* London: Faber and Faber, 1964.

Banks, J., and R. Gibson. "10 Rules for Determining When Simulation Is Not Appropriate." *IIE Solutions,* September 1997, pp. 30–32.

Bateman, Robert E.; Royce O. Bowden; Thomas J. Gogg; Charles R. Harrell; and Jack R. A. Mott. *System Improvement Using Simulation.* Utah: PROMODEL Corp., 1997.

Deming, W. E. *Foundation for Management of Quality in the Western World.* Paper read at a meeting of the Institute of Management Sciences, Osaka, Japan, 24 July 1989.

Glenney, Neil E., and Gerald T. Mackulak. "Modeling & Simulation Provide Key to CIM Implementation Philosophy." *Industrial Engineering,* May 1985.

Hancock, Walton; R. Dissen; and A. Merten. "An Example of Simulation to Improve Plant Productivity." *AIIE Transactions,* March 1977, pp. 2–10.

Harrell, Charles R., and Donald Hicks. "Simulation Software Component Architecture for Simulation-Based Enterprise Applications." In *Proceedings of the 1998 Winter Simulation Conference,* ed. D. J. Medeiros, E. F. Watson, J. S. Carson, and M. S. Manivannan, pp. 1717–21. Institute of Electrical and Electronics Engineers, Piscataway, New Jersey.

Harrington, H. James. *Business Process Improvement.* New York: McGraw-Hill, 1991.

Hoover, Stewart V., and Ronald F. Perry. *Simulation: A Problem-Solving Approach.* Reading, MA: Addison-Wesley, 1989.

Law, A. M., and M. G. McComas. "How Simulation Pays Off." *Manufacturing Engineering,* February 1988, pp. 37–39.

Mott, Jack, and Kerim Tumay. "Developing a Strategy for Justifying Simulation." *Industrial Engineering,* July 1992, pp. 38–42.

Oxford American Dictionary. New York: Oxford University Press, 1980. [compiled by] Eugene Enrich et al.

Perry, R. F., and R. F. Baum. "Resource Allocation and Scheduling for a Radiology Department." In *Cost Control in Hospitals.* Ann Arbor, MI: Health Administration Press, 1976.

Rohrer, Matt, and Jerry Banks. "Required Skills of a Simulation Analyst." *IIE Solutions,* May 1998, pp. 7–23.

Schriber, T. J. "The Nature and Role of Simulation in the Design of Manufacturing Systems." *Simulation in CIM and Artificial Intelligence Techniques,* ed. J. Retti and K. E. Wichmann. S.D., CA.: Society for Computer Simulation, 1987, pp. 5–8.

Shannon, Robert E. "Introduction to the Art and Science of Simulation." In *Proceedings of the 1998 Winter Simulation Conference,* ed. D. J. Medeiros, E. F. Watson, J. S. Carson, and M. S. Manivannan, pp. 7–14. Piscataway, NJ: Institute of Electrical and Electronics Engineers.

Shingo, Shigeo. *The Shingo Production Management System—Improving Process Functions.* Trans. Andrew P. Dillon. Cambridge, MA: Productivity Press, 1992.

Solberg, James. "Design and Analysis of Integrated Manufacturing Systems." In W. Dale Compton. Washington, D.C.: National Academy Press, 1988, p. 4.

The Wall Street Journal, March 19, 1999. "United 747's Near Miss Sparks a Widespread Review of Pilot Skills," p. A1.

2   SYSTEM DYNAMICS

"A fool with a tool is still a fool."
—Unknown

2.1 Introduction

Knowing how to do simulation doesn't make someone a good systems designer any more than knowing how to use a CAD system makes one a good product designer. Simulation is a tool that is useful only if one understands the nature of the problem to be solved. It is designed to help solve systemic problems that are operational in nature. Simulation exercises fail to produce useful results more often because of a lack of understanding of system dynamics than a lack of knowing how to use the simulation software. The challenge is in understanding how the system operates, knowing what you want to achieve with the system, and being able to identify key leverage points for best achieving desired objectives. To illustrate the nature of this challenge, consider the following actual scenario:

> The pipe mill for the XYZ Steel Corporation was an important profit center, turning steel slabs selling for under $200/ton into a product with virtually unlimited demand selling for well over $450/ton. The mill took coils of steel of the proper thickness and width through a series of machines that trimmed the edges, bent the steel into a cylinder, welded the seam, and cut the resulting pipe into appropriate lengths, all on a continuously running line. The line was even designed to weld the end of one coil to the beginning of the next one "on the fly," allowing the line to run continually for days on end.
>
> Unfortunately the mill was able to run only about 50 percent of its theoretical capacity over the long term, costing the company tens of millions of dollars a year in lost revenue. In an effort to improve the mill's productivity, management studied each step in the process. It was fairly easy to find the slowest step in the line, but additional study showed that only a small percentage of lost production was due to problems at this "bottleneck" operation. Sometimes a step upstream from the bottleneck would

have a problem, causing the bottleneck to run out of work, or a downstream step would go down temporarily, causing work to back up and stop the bottleneck. Sometimes the bottleneck would get so far behind that there was no place to put incoming, newly made pipe. In this case the workers would stop the entire pipe-making process until the bottleneck was able to catch up. Often the bottleneck would then be idle waiting until the newly started line was functioning properly again and the new pipe had a chance to reach it. Sometimes problems at the bottleneck were actually caused by improper work at a previous location.

In short, there was no single cause for the poor productivity seen at this plant. Rather, several separate causes all contributed to the problem in complex ways. Management was at a loss to know which of several possible improvements (additional or faster capacity at the bottleneck operation, additional storage space between stations, better rules for when to shut down and start up the pipe-forming section of the mill, better quality control, or better training at certain critical locations) would have the most impact for the least cost. Yet the poor performance of the mill was costing enormous amounts of money. Management was under pressure to do something, but what should it be?

This example illustrates the nature and difficulty of the decisions that an operations manager faces. Managers need to make decisions that are the "best" in some sense. To do so, however, requires that they have clearly defined goals and understand the system well enough to identify cause-and-effect relationships.

While every system is different, just as every product design is different, the basic elements and types of relationships are the same. Knowing how the elements of a system interact and how overall performance can be improved are essential to the effective use of simulation. This chapter reviews basic system dynamics and answers the following questions:

- What is a system?
- What are the elements of a system?
- What makes systems so complex?
- What are useful system metrics?
- What is a systems approach to systems planning?
- How do traditional systems analysis techniques compare with simulation?

2.2 System Definition

We live in a society that is composed of complex, human-made systems that we depend on for our safety, convenience, and livelihood. Routinely we rely on transportation, health care, production, and distribution systems to provide needed goods and services. Furthermore, we place high demands on the quality, convenience, timeliness, and cost of the goods and services that are provided by these systems. Remember the last time you were caught in a traffic jam, or waited for what seemed like an eternity in a restaurant or doctor's office? Contrast that experience with the satisfaction that comes when you find a store that sells quality merchandise at discount prices or when you locate a health care organization that provides prompt and professional service. The difference is between a system that has been well designed and operates smoothly, and one that is poorly planned and managed.

A *system,* as used here, is defined as a collection of elements that function together to achieve a desired goal (Blanchard 1991). Key points in this definition include the fact that (1) a system consists of multiple elements, (2) these elements are interrelated and work in cooperation, and (3) a system exists for the purpose of achieving specific objectives. Examples of systems are traffic systems, political systems, economic systems, manufacturing systems, and service systems. Our main focus will be on manufacturing and service systems that process materials, information, and people.

Manufacturing systems can be small job shops and machining cells or large production facilities and assembly lines. Warehousing and distribution as well as entire supply chain systems will be included in our discussions of manufacturing systems. Service systems cover a wide variety of systems including health care facilities, call centers, amusement parks, public transportation systems, restaurants, banks, and so forth.

Both manufacturing and service systems may be termed *processing systems* because they process items through a series of activities. In a manufacturing system, raw materials are transformed into finished products. For example, a bicycle manufacturer starts with tube stock that is then cut, welded, and painted to produce bicycle frames. In service systems, customers enter with some service need and depart as serviced (and, we hope, satisfied) customers. In a hospital emergency room, for example, nurses, doctors, and other staff personnel admit and treat incoming patients who may undergo tests and possibly even surgical procedures before finally being released. Processing systems are artificial (they are human-made), dynamic (elements interact over time), and usually stochastic (they exhibit random behavior).

2.3 System Elements

From a simulation perspective, a system can be said to consist of entities, activities, resources, and controls (see Figure 2.1). These elements define the *who, what, where, when,* and *how* of entity processing. This model for describing a

FIGURE 2.1

Elements of a system.

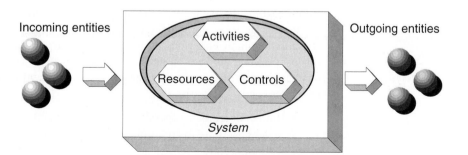

system corresponds closely to the well-established ICAM definition (IDEF) process model developed by the defense industry (ICAM stands for an early Air Force program in integrated computer-aided manufacturing). The IDEF modeling paradigm views a system as consisting of inputs and outputs (that is, entities), activities, mechanisms (that is, resources), and controls.

2.3.1 Entities

Entities are the items processed through the system such as products, customers, and documents. Different entities may have unique characteristics such as cost, shape, priority, quality, or condition. Entities may be further subdivided into the following types:

- Human or animate (customers, patients, etc.).
- Inanimate (parts, documents, bins, etc.).
- Intangible (calls, electronic mail, etc.).

For most manufacturing and service systems, the entities are discrete items. This is the case for discrete part manufacturing and is certainly the case for nearly all service systems that process customers, documents, and others. For some production systems, called *continuous systems,* a nondiscrete substance is processed rather than discrete entities. Examples of continuous systems are oil refineries and paper mills.

2.3.2 Activities

Activities are the tasks performed in the system that are either directly or indirectly involved in the processing of entities. Examples of activities include servicing a customer, cutting a part on a machine, or repairing a piece of equipment. Activities usually consume time and often involve the use of resources. Activities may be classified as

- Entity processing (check-in, treatment, inspection, fabrication, etc.).
- Entity and resource movement (forklift travel, riding in an elevator, etc.).
- Resource adjustments, maintenance, and repairs (machine setups, copy machine repair, etc.).

2.3.3 Resources

Resources are the means by which activities are performed. They provide the supporting facilities, equipment, and personnel for carrying out activities. While resources facilitate entity processing, inadequate resources can constrain processing by limiting the rate at which processing can take place. Resources have characteristics such as capacity, speed, cycle time, and reliability. Like entities, resources can be categorized as

- Human or animate (operators, doctors, maintenance personnel, etc.).
- Inanimate (equipment, tooling, floor space, etc.).
- Intangible (information, electrical power, etc.).

Resources can also be classified as being dedicated or shared, permanent or consumable, and mobile or stationary.

2.3.4 Controls

Controls dictate how, when, and where activities are performed. Controls impose order on the system. At the highest level, controls consist of schedules, plans, and policies. At the lowest level, controls take the form of written procedures and machine control logic. At all levels, controls provide the information and decision logic for how the system should operate. Examples of controls include

- Routing sequences.
- Production plans.
- Work schedules.
- Task prioritization.
- Control software.
- Instruction sheets.

2.4 System Complexity

Elements of a system operate in concert with one another in ways that often result in complex interactions. The word *complex* comes from the Latin *complexus,* meaning entwined or connected together. Unfortunately, unaided human intuition is not very good at analyzing and understanding complex systems. Economist Herbert Simon called this inability of the human mind to grasp real-world complexity "the principle of bounded rationality." This principle states that "the capacity of the human mind for formulating and solving complex problems is very small compared with the size of the problem whose solution is required for objectively rational behavior in the real world, or even for a reasonable approximation to such objective rationality" (Simon 1957).

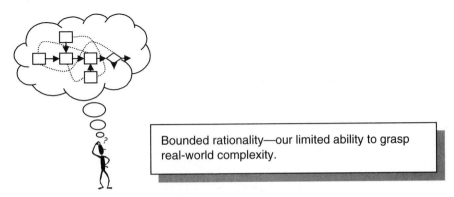

Bounded rationality—our limited ability to grasp real-world complexity.

While the sheer number of elements in a system can stagger the mind (the number of different entities, activities, resources, and controls can easily exceed 100), the interactions of these elements are what make systems so complex and difficult to analyze. System complexity is primarily a function of the following two factors:

1. *Interdependencies* between elements so that each element affects other elements.
2. *Variability* in element behavior that produces uncertainty.

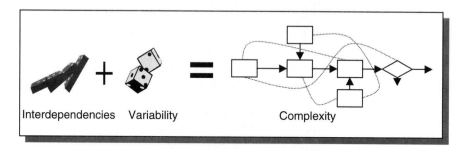

These two factors characterize virtually all human-made systems and make system behavior difficult to analyze and predict. As shown in Figure 2.2, the degree of analytical difficulty increases exponentially as the number of interdependencies and random variables increases.

2.4.1 Interdependencies

Interdependencies cause the behavior of one element to affect other elements in the system. For example, if a machine breaks down, repair personnel are put into action while downstream operations become idle for lack of parts. Upstream operations may even be forced to shut down due to a logjam in the entity flow causing a blockage of activities. Another place where this chain reaction or domino effect manifests itself is in situations where resources are shared between

FIGURE 2.2

Analytical difficulty as a function of the number of interdependencies and random variables.

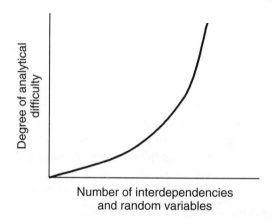

two or more activities. A doctor treating one patient, for example, may be unable to immediately respond to another patient needing his or her attention. This delay in response may also set other forces in motion.

It should be clear that the complexity of a system has less to do with the number of elements in the system than with the number of interdependent relationships. Even interdependent relationships can vary in degree, causing more or less impact on overall system behavior. System interdependency may be either tight or loose depending on how closely elements are linked. Elements that are tightly coupled have a greater impact on system operation and performance than elements that are only loosely connected. When an element such as a worker or machine is delayed in a tightly coupled system, the impact is immediately felt by other elements in the system and the entire process may be brought to a screeching halt.

In a loosely coupled system, activities have only a minor, and often delayed, impact on other elements in the system. Systems guru Peter Senge (1990) notes that for many systems, "Cause and effect are not closely related in time and space." Sometimes the distance in time and space between cause-and-effect relationships becomes quite sizable. If enough reserve inventory has been stockpiled, a truckers' strike cutting off the delivery of raw materials to a transmission plant in one part of the world may not affect automobile assembly in another part of the world for weeks. Cause-and-effect relationships are like a ripple of water that diminishes in impact as the distance in time and space increases.

Obviously, the preferred approach to dealing with interdependencies is to eliminate them altogether. Unfortunately, this is not entirely possible for most situations and actually defeats the purpose of having systems in the first place. The whole idea of a system is to achieve a synergy that otherwise would be unattainable if every component were to function in complete isolation. Several methods are used to decouple system elements or at least isolate their influence so disruptions are not felt so easily. These include providing buffer inventories, implementing redundant or backup measures, and dedicating resources to single tasks. The downside to these mitigating techniques is that they often lead to excessive inventories and underutilized resources. The point to be made here is that interdependencies, though they may be minimized somewhat, are simply a fact of life and are best dealt with through effective coordination and management.

2.4.2 Variability

Variability is a characteristic inherent in any system involving humans and machinery. Uncertainty in supplier deliveries, random equipment failures, unpredictable absenteeism, and fluctuating demand all combine to create havoc in planning system operations. Variability compounds the already unpredictable effect of interdependencies, making systems even more complex and unpredictable. Variability propagates in a system so that "highly variable outputs from one workstation become highly variable inputs to another" (Hopp and Spearman 2000).

TABLE 2.1   **Examples of System Variability**

Type of Variability	Examples
Activity times	Operation times, repair times, setup times, move times
Decisions	To accept or reject a part, where to direct a particular customer, which task to perform next
Quantities	Lot sizes, arrival quantities, number of workers absent
Event intervals	Time between arrivals, time between equipment failures
Attributes	Customer preference, part size, skill level

Table 2.1 identifies the types of random variability that are typical of most manufacturing and service systems.

The tendency in systems planning is to ignore variability and calculate system capacity and performance based on average values. Many commercial scheduling packages such as MRP (material requirements planning) software work this way. Ignoring variability distorts the true picture and leads to inaccurate performance predictions. Designing systems based on average requirements is like deciding whether to wear a coat based on the average annual temperature or prescribing the same eyeglasses for everyone based on average eyesight. Adults have been known to drown in water that was only four feet deep—on the average! Wherever variability occurs, an attempt should be made to describe the nature or pattern of the variability and assess the range of the impact that variability might have on system performance.

Perhaps the most illustrative example of the impact that variability can have on system behavior is the simple situation where entities enter into a single queue to wait for a single server. An example of this might be customers lining up in front of an ATM. Suppose that the time between customer arrivals is exponentially distributed with an average time of one minute and that they take an average time of one minute, exponentially distributed, to transact their business. In queuing theory, this is called an $M/M/1$ queuing system. If we calculate system performance based solely on average time, there will never be any customers waiting in the queue. Every minute that a customer arrives the previous customer finishes his or her transaction. Now if we calculate the number of customers waiting in line, taking into account the variation, we will discover that the waiting line grows to infinity (the technical term is that the system "explodes"). Who would guess that in a situation involving only one interdependent relationship that variation alone would make the difference between zero items waiting in a queue and an infinite number in the queue?

By all means, variability should be reduced and even eliminated wherever possible. System planning is much easier if you don't have to contend with it. Where it is inevitable, however, simulation can help predict the impact it will have on system performance. Likewise, simulation can help identify the degree of improvement that can be realized if variability is reduced or even eliminated. For example, it can tell you how much reduction in overall flow time and flow time variation can be achieved if operation time variation can be reduced by, say, 20 percent.

2.5 System Performance Metrics

Metrics are measures used to assess the performance of a system. At the highest level of an organization or business, metrics measure overall performance in terms of profits, revenues, costs relative to budget, return on assets, and so on. These metrics are typically financial in nature and show bottom-line performance. Unfortunately, such metrics are inherently lagging, disguise low-level operational performance, and are reported only periodically. From an operational standpoint, it is more beneficial to track such factors as time, quality, quantity, efficiency, and utilization. These operational metrics reflect immediate activity and are directly controllable. They also drive the higher financially related metrics. Key operational metrics that describe the effectiveness and efficiency of manufacturing and service systems include the following:

- *Flow time*—the average time it takes for an item or customer to be processed through the system. Synonyms include cycle time, throughput time, and manufacturing lead time. For order fulfillment systems, flow time may also be viewed as customer response time or turnaround time. A closely related term in manufacturing is *makespan,* which is the time to process a given set of jobs. Flow time can be shortened by reducing activity times that contribute to flow time such as setup, move, operation, and inspection time. It can also be reduced by decreasing work-in-process or average number of entities in the system. Since over 80 percent of cycle time is often spent waiting in storage or queues, elimination of buffers tends to produce the greatest reduction in cycle time. Another solution is to add more resources, but this can be costly.
- *Utilization*—the percentage of scheduled time that personnel, equipment, and other resources are in productive use. If a resource is not being utilized, it may be because it is idle, blocked, or down. To increase productive utilization, you can increase the demand on the resource or reduce resource count or capacity. It also helps to balance work loads. In a system with high variability in activity times, it is difficult to achieve high utilization of resources. Job shops, for example, tend to have low machine utilization. Increasing utilization for the sake of utilization is not a good objective. Increasing the utilization of nonbottleneck resources, for example, often only creates excessive inventories without creating additional throughput.
- *Value-added time*—the amount of time material, customers, and so forth spend actually receiving value, where value is defined as anything for which the customer is willing to pay. From an operational standpoint,

value-added time is considered the same as processing time or time spent actually undergoing some physical transformation or servicing. Inspection time and waiting time are considered non-value-added time.

- *Waiting time*—the amount of time that material, customers, and so on spend waiting to be processed. Waiting time is by far the greatest component of non-value-added time. Waiting time can be decreased by reducing the number of items (such as customers or inventory levels) in the system. Reducing variation and interdependencies in the system can also reduce waiting times. Additional resources can always be added, but the trade-off between the cost of adding the resources and the savings of reduced waiting time needs to be evaluated.

- *Flow rate*—the number of items produced or customers serviced per unit of time (such as parts or customers per hour). Synonyms include production rate, processing rate, or throughput rate. Flow rate can be increased by better management and utilization of resources, especially the limiting or bottleneck resource. This is done by ensuring that the bottleneck operation or resource is never starved or blocked. Once system throughput matches the bottleneck throughput, additional processing or throughput capacity can be achieved by speeding up the bottleneck operation, reducing downtimes and setup times at the bottleneck operation, adding more resources to the bottleneck operation, or off-loading work from the bottleneck operation.

- *Inventory or queue levels*—the number of items or customers in storage or waiting areas. It is desirable to keep queue levels to a minimum while still achieving target throughput and response time requirements. Where queue levels fluctuate, it is sometimes desirable to control the minimum or maximum queue level. Queuing occurs when resources are unavailable when needed. Inventory or queue levels can be controlled either by balancing flow or by restricting production at nonbottleneck operations. JIT (just-in-time) production is one way to control inventory levels.

- *Yield*—from a production standpoint, the percentage of products completed that conform to product specifications as a percentage of the total number of products that entered the system as raw materials. If 95 out of 100 items are nondefective, the yield is 95 percent. Yield can also be measured by its complement—reject or scrap rate.

- *Customer responsiveness*—the ability of the system to deliver products in a timely fashion to minimize customer waiting time. It might be measured as *fill rate,* which is the number of customer orders that can be filled immediately from inventory. In minimizing job lateness, it may be desirable to minimize the overall late time, minimize the number or percentage of jobs that are late, or minimize the maximum tardiness of jobs. In make-to-stock operations, customer responsiveness can be assured by maintaining adequate inventory levels. In make-to-order, customer responsiveness is improved by lowering inventory levels so that cycle times can be reduced.

- *Variance*—the degree of fluctuation that can and often does occur in any of the preceding metrics. Variance introduces uncertainty, and therefore risk, in achieving desired performance goals. Manufacturers and service providers are often interested in reducing variance in delivery and service times. For example, cycle times and throughput rates are going to have some variance associated with them. Variance is reduced by controlling activity times, improving resource reliability, and adhering to schedules.

These metrics can be given for the entire system, or they can be broken down by individual resource, entity type, or some other characteristic. By relating these metrics to other factors, additional meaningful metrics can be derived that are useful for benchmarking or other comparative analysis. Typical relational metrics include minimum theoretical flow time divided by actual flow time (flow time efficiency), cost per unit produced (unit cost), annual inventory sold divided by average inventory (inventory turns or turnover ratio), or units produced per cost or labor input (productivity).

2.6 System Variables

Designing a new system or improving an existing system requires more than simply identifying the elements and performance goals of the system. It requires an understanding of how system elements affect each other and overall performance objectives. To comprehend these relationships, you must understand three types of system variables:

1. Decision variables
2. Response variables
3. State variables

2.6.1 Decision Variables

Decision variables (also called *input factors* in SimRunner) are sometimes referred to as the *independent variables* in an experiment. Changing the values of a system's independent variables affects the behavior of the system. Independent variables may be either controllable or uncontrollable depending on whether the experimenter is able to manipulate them. An example of a controllable variable is the number of operators to assign to a production line or whether to work one or two shifts. Controllable variables are called decision variables because the decision maker (experimenter) controls the values of the variables. An uncontrollable variable might be the time to service a customer or the reject rate of an operation. When defining the system, controllable variables are the information about the system that is more prescriptive than descriptive (see section 2.9.3).

Obviously, all independent variables in an experiment are ultimately controllable—but at a cost. The important point here is that some variables are easier to change than others. When conducting experiments, the final solution is often based on whether the cost to implement a change produces a higher return in performance.

2.6.2 Response Variables

Response variables (sometimes called *performance* or *output variables*) measure the performance of the system in response to particular decision variable settings. A response variable might be the number of entities processed for a given period, the average utilization of a resource, or any of the other system performance metrics described in section 2.5.

In an experiment, the response variable is the dependent variable, which depends on the particular value settings of the independent variables. The experimenter doesn't manipulate dependent variables, only independent or decision variables. Obviously, the goal in systems planning is to find the right values or settings of the decision variables that give the desired response values.

2.6.3 State Variables

State variables indicate the status of the system at any specific point in time. Examples of state variables are the current number of entities waiting to be processed or the current status (busy, idle, down) of a particular resource. Response variables are often summaries of state variable changes over time. For example, the individual times that a machine is in a busy state can be summed over a particular period and divided by the total available time to report the machine utilization for that period.

State variables are dependent variables like response variables in that they depend on the setting of the independent variables. State variables are often ignored in experiments since they are not directly controlled like decision variables and are not of as much interest as the summary behavior reported by response variables.

Sometimes reference is made to the state of a system as though a system itself can be in a particular state such as busy or idle. The state of a system actually consists of "that collection of variables necessary to describe a system at a particular time, relative to the objectives of the study" (Law and Kelton 2000). If we study the flow of customers in a bank, for example, the state of the bank for a given point in time would include the current number of customers in the bank, the current status of each teller (busy, idle, or whatever), and perhaps the time that each customer has been in the system thus far.

2.7 System Optimization

Finding the right setting for decision variables that best meets performance objectives is called *optimization*. Specifically, optimization seeks the best combination of decision variable values that either minimizes or maximizes some *objective function* such as costs or profits. An objective function is simply a response variable of the system. A typical objective in an optimization problem for a manufacturing or service system might be minimizing costs or maximizing flow rate. For example, we might be interested in finding the optimum number of personnel for staffing a customer support activity that minimizes costs yet handles the call volume. In a manufacturing concern, we might be interested in maximizing the throughput that can be achieved for a given system configuration. Optimization problems often include *constraints,* limits to the values that the decision variables can take on. For example, in finding the optimum speed of a conveyor such that production cost is minimized, there would undoubtedly be physical limits to how slow or fast the conveyor can operate. Constraints can also apply to response variables. An example of this might be an objective to maximize throughput but subject to the constraint that average waiting time cannot exceed 15 minutes.

In some instances, we may find ourselves trying to achieve conflicting objectives. For example, minimizing production or service costs often conflicts with minimizing waiting costs. In system optimization, one must be careful to weigh priorities and make sure the right objective function is driving the decisions. If, for example, the goal is to minimize production or service costs, the obvious solution is to maintain only a sufficient workforce to meet processing requirements. Unfortunately, in manufacturing systems this builds up work-in-process and results in high inventory carrying costs. In service systems, long queues result in long waiting times, hence dissatisfied customers. At the other extreme, one might feel that reducing inventory or waiting costs should be the overriding goal and, therefore, decide to employ more than an adequate number of resources so that work-in-process or customer waiting time is virtually eliminated. It should be obvious that there is a point at which the cost of adding another resource can no longer be justified by the diminishing incremental savings in waiting costs that are realized. For this reason, it is generally conceded that a better strategy is to find the right trade-off or balance between the number of resources and waiting times so that the total cost is minimized (see Figure 2.3).

FIGURE 2.3

Cost curves showing optimum number of resources to minimize total cost.

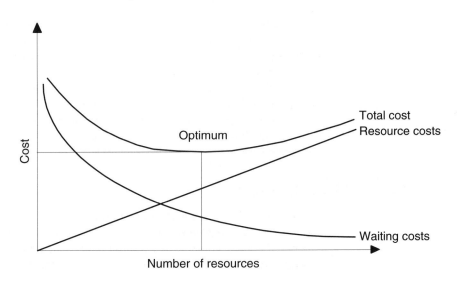

As shown in Figure 2.3, the number of resources at which the sum of the resource costs and waiting costs is at a minimum is the optimum number of resources to have. It also becomes the optimum acceptable waiting time.

In systems design, arriving at an optimum system design is not always realistic, given the almost endless configurations that are sometimes possible and limited time that is available. From a practical standpoint, the best that can be expected is a near optimum solution that gets us close enough to our objective, given the time constraints for making the decision.

2.8 The Systems Approach

Due to departmentalization and specialization, decisions in the real world are often made without regard to overall system performance. With everyone busy minding his or her own area of responsibility, often no one is paying attention to the big picture. One manager in a large manufacturing corporation noted that as many as 99 percent of the system improvement recommendations made in his company failed to look at the system holistically. He further estimated that nearly 80 percent of the suggested changes resulted in no improvement at all, and many of the suggestions actually hurt overall performance. When attempting to make system improvements, it is often discovered that localized changes fail to produce the overall improvement that is desired. Put in technical language: *Achieving a local optimum often results in a global suboptimum.* In simpler terms: *It's okay to act locally as long as one is thinking globally.* The elimination of a problem in one area may only uncover, and sometimes even exacerbate, problems in other areas.

Approaching system design with overall objectives in mind and considering how each element relates to each other and to the whole is called a systems or holistic approach to systems design. Because systems are composed of interdependent elements, it is not possible to accurately predict how a system will perform simply by examining each system element in isolation from the whole. To presume otherwise is to take a reductionist approach to systems design, which focuses on the parts rather than the whole. While structurally a system may be divisible, functionally it is indivisible and therefore requires a holistic approach to systems thinking. Kofman and Senge (1995) observe

> The defining characteristic of a system is that it cannot be understood as a function of its isolated components. First, the behavior of the system doesn't depend on what each part is doing but on how each part is interacting with the rest. . . . Second, to understand a system we need to understand how it fits into the larger system of which it is a part. . . . Third, and most important, what we call the parts need not be taken as primary. In fact, how we define the parts is fundamentally a matter of perspective and purpose, not intrinsic in the nature of the "real thing" we are looking at.

Whether designing a new system or improving an existing system, it is important to follow sound design principles that take into account all relevant variables. The activity of systems design and process improvement, also called systems

FIGURE 2.4

Four-step iterative approach to systems improvement.

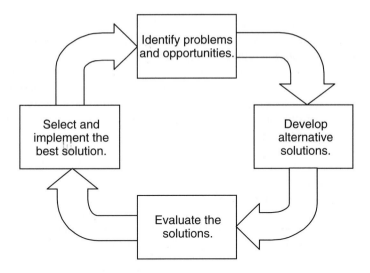

engineering, has been defined as

> The effective application of scientific and engineering efforts to transform an operational need into a defined system configuration through the top-down iterative process of requirements definition, functional analysis, synthesis, optimization, design, test and evaluation (Blanchard 1991).

To state it simply, systems engineering is the process of identifying problems or other opportunities for improvement, developing alternative solutions, evaluating the solutions, and selecting and implementing the best solutions (see Figure 2.4). All of this should be done from a systems point of view.

2.8.1 Identifying Problems and Opportunities

The importance of identifying the most significant problem areas and recognizing opportunities for improvement cannot be overstated. Performance standards should be set high in order to look for the greatest improvement opportunities. Companies making the greatest strides are setting goals of 100 to 500 percent improvement in many areas such as inventory reduction or customer lead time reduction. Setting high standards pushes people to think creatively and often results in breakthrough improvements that would otherwise never be considered. Contrast this way of thinking with one hospital whose standard for whether a patient had a quality experience was whether the patient left alive! Such lack of vision will never inspire the level of improvement needed to meet ever-increasing customer expectations.

2.8.2 Developing Alternative Solutions

We usually begin developing a solution to a problem by understanding the problem, identifying key variables, and describing important relationships. This helps identify possible areas of focus and leverage points for applying a solution. Techniques such as cause-and-effect analysis and pareto analysis are useful here.

Once a problem or opportunity has been identified and key decision variables isolated, alternative solutions can be explored. This is where most of the design and engineering expertise comes into play. Knowledge of best practices for common types of processes can also be helpful. The designer should be open to all possible alternative feasible solutions so that the best possible solutions don't get overlooked.

Generating alternative solutions requires creativity as well as organizational and engineering skills. Brainstorming sessions, in which designers exhaust every conceivably possible solution idea, are particularly useful. Designers should use every stretch of the imagination and not be stifled by conventional solutions alone. The best ideas come when system planners begin to think innovatively and break from traditional ways of doing things. Simulation is particularly helpful in this process in that it encourages thinking in radical new ways.

2.8.3 Evaluating the Solutions

Alternative solutions should be evaluated based on their ability to meet the criteria established for the evaluation. These criteria often include performance goals, cost of implementation, impact on the sociotechnical infrastructure, and consistency with organizational strategies. Many of these criteria are difficult to measure in absolute terms, although most design options can be easily assessed in terms of relative merit.

After narrowing the list to two or three of the most promising solutions using common sense and rough-cut analysis, more precise evaluation techniques may need to be used. This is where simulation and other formal analysis tools come into play.

2.8.4 Selecting and Implementing the Best Solution

Often the final selection of what solution to implement is not left to the analyst, but rather is a management decision. The analyst's role is to present his or her evaluation in the clearest way possible so that an informed decision can be made.

Even after a solution is selected, additional modeling and analysis are often needed for fine-tuning the solution. Implementers should then be careful to make sure that the system is implemented as designed, documenting reasons for any modifications.

2.9 Systems Analysis Techniques

While simulation is perhaps the most versatile and powerful systems analysis tool, other available techniques also can be useful in systems planning. These alternative techniques are usually computational methods that work well for simple systems with little interdependency and variability. For more complex systems,

Charles Harrell

Charles Harrell is an associate professor of engineering and technology at Brigham Young University and founder of PROMODEL Corporation. He received his B.S. from Brigham Young University, M.S. in industrial engineering from the University of Utah, and Ph.D. in manufacturing engineering from the Technical University of Denmark. His area of interest and expertise is in manufacturing system design and simulation. He has worked in manufacturing and systems engineering positions for Ford Motor Company and the Eaton Corporation. At BYU he teaches courses in manufacturing systems, manufacturing simulation, and manufacturing information systems. He is the author or coauthor of several simulation books and has given numerous presentations on manufacturing system design and simulation. He serves on the board of directors of PROMODEL Corporation and continues to architect their simulation and modeling tools. He is a senior member of IIE and SME. He enjoys sports and playing with his grandchildren.

Biman K. Ghosh

Biman K. Ghosh is a professor of industrial and manufacturing engineering at the California State Polytechnic University, Pomona. He has been teaching and consulting all over the United States for the past 17 years. Prior to that Dr. Ghosh worked for such multinational companies as Tata, Mercedes Benz, Sandvik, and Siemens. Dr. Ghosh has been a consultant to Northrop Grumman, McDonnell Douglas, Boeing, Rockwell, Loral, Visteon, UPS, Sikorsky, Baxter, Paccar, Johnson Controls, Powersim, and Galorath. Dr. Ghosh has published extensively. His teaching and consulting interests are in the areas of simulation, supply chain management, lean manufacturing, flexible manufacturing, cellular manufacturing, computer-aided manufacturing, design for manufacturing, design of experiments, total quality management, and work measurement. His passion is still and video photography.

Royce O. Bowden

Royce O. Bowden, Jr., is a professor of industrial engineering at Mississippi State University (MSU), where he teaches courses in simulation, statistics, production control, and artificial intelligence. Professor Bowden's degrees include a Ph.D. and M.S. in industrial engineering. Prior to joining MSU, he served in engineering positions at Texas Instruments in Dallas, Texas, and Martin Marietta Aerospace in New Orleans, Louisiana. Professor Bowden's research program centers on the use, development, and integration of simulation and artificial intelligence techniques for solving systems engineering problems and is focused on simulation output analysis for decision support. His research on combining simulation with modern optimization techniques provided the foundation for the first widely used commercial simulation optimization software. Dr. Bowden's research program has been funded by several organizations including the U.S. Department of Transportation, National Science Foundation, U.S. Department of Agriculture, National Aeronautics and Space Administration, Whirlpool Corporation, and Litton-Ingalls Shipbuilding. Professor Bowden is author (or coauthor) of numerous research papers and two simulation books and serves as an associate editor for *International Journal of Modeling and Simulation*. Dr. Bowden is a Hearin Distinguished Professor of Engineering at MSU. He is a licensed private airplane pilot and an accomplished unicyclist (rumored to have been seen riding and juggling at the same time).

these techniques still can provide rough estimates but fall short in producing the insights and accurate answers that simulation provides. Systems implemented using these techniques usually require some adjustments after implementation to compensate for inaccurate calculations. For example, if after implementing a system it is discovered that the number of resources initially calculated is insufficient to meet processing requirements, additional resources are added. This adjustment can create extensive delays and costly modifications if special personnel training or custom equipment is involved. As a precautionary measure, a safety factor is sometimes added to resource and space calculations to ensure they are adequate. Overdesigning a system, however, also can be costly and wasteful.

As system interdependency and variability increase, not only does system performance decrease, but the ability to accurately predict system performance decreases as well (Lloyd and Melton 1997). Simulation enables a planner to accurately predict the expected performance of a system design and ultimately make better design decisions.

Systems analysis tools, in addition to simulation, include simple calculations, spreadsheets, operations research techniques (such as linear programming and queuing theory), and special computerized tools for scheduling, layout, and so forth. While these tools can provide quick and approximate solutions, they tend to make oversimplifying assumptions, perform only static calculations, and are limited to narrow classes of problems. Additionally, they fail to fully account for interdependencies and variability of complex systems and therefore are not as accurate as simulation in predicting complex system performance (see Figure 2.5). They all lack the power, versatility, and visual appeal of simulation. They do provide quick solutions, however, and for certain situations produce adequate results. They are important to cover here, not only because they sometimes provide a good alternative to simulation, but also because they can complement simulation by providing initial design estimates for input to the simulation model. They also

FIGURE 2.5

Simulation improves performance predictability.

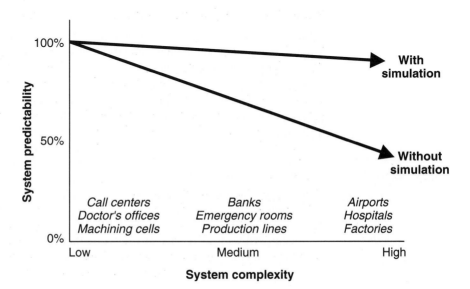

can be useful to help validate the results of a simulation by comparing them with results obtained using an analytic model.

2.9.1 Hand Calculations

Quick-and-dirty, pencil-and-paper sketches and calculations can be remarkably helpful in understanding basic requirements for a system. Many important decisions have been made as the result of sketches drawn and calculations performed on a napkin or the back of an envelope. Some decisions may be so basic that a quick mental calculation yields the needed results. Most of these calculations involve simple algebra, such as finding the number of resource units (such as machines or service agents) to process a particular workload knowing the capacity per resource unit. For example, if a requirement exists to process 200 items per hour and the processing capacity of a single resource unit is 75 work items per hour, three units of the resource, most likely, are going to be needed.

The obvious drawback to hand calculations is the inability to manually perform complex calculations or to take into account tens or potentially even hundreds of complex relationships simultaneously.

2.9.2 Spreadsheets

Spreadsheet software comes in handy when calculations, sometimes involving hundreds of values, need to be made. Manipulating rows and columns of numbers on a computer is much easier than doing it on paper, even with a calculator handy. Spreadsheets can be used to perform rough-cut analysis such as calculating average throughput or estimating machine requirements. The drawback to spreadsheet software is the inability (or, at least, limited ability) to include variability in activity times, arrival rates, and so on, and to account for the effects of interdependencies.

What-if experiments can be run on spreadsheets based on expected values (average customer arrivals, average activity times, mean time between equipment failures) and simple interactions (activity A must be performed before activity B). This type of spreadsheet simulation can be very useful for getting rough performance estimates. For some applications with little variability and component interaction, a spreadsheet simulation may be adequate. However, calculations based on only average values and oversimplified interdependencies potentially can be misleading and result in poor decisions. As one ProModel user reported, "We just completed our final presentation of a simulation project and successfully saved approximately $600,000. Our management was prepared to purchase an additional overhead crane based on spreadsheet analysis. We subsequently built a ProModel simulation that demonstrated an additional crane will not be necessary. The simulation also illustrated some potential problems that were not readily apparent with spreadsheet analysis."

Another weakness of spreadsheet modeling is the fact that all behavior is assumed to be period-driven rather than event-driven. Perhaps you have tried to figure out how your bank account balance fluctuated during a particular period when all you had to go on was your monthly statements. Using ending balances does not reflect changes as they occurred during the period. You can know the current state of the system at any point in time only by updating the state variables of the system each time an event or transaction occurs. When it comes to dynamic models, spreadsheet simulation suffers from the "curse of dimensionality" because the size of the model becomes unmanageable.

2.9.3 Operations Research Techniques

Traditional operations research (OR) techniques utilize mathematical models to solve problems involving simple to moderately complex relationships. These mathematical models include both deterministic models such as mathematical programming, routing, or network flows and probabilistic models such as queuing and decision trees. These OR techniques provide quick, quantitative answers without going through the guesswork process of trial and error. OR techniques can be divided into two general classes: prescriptive and descriptive.

Prescriptive Techniques

Prescriptive OR techniques provide an optimum solution to a problem, such as the optimum amount of resource capacity to minimize costs, or the optimum product mix that will maximize profits. Examples of prescriptive OR optimization techniques include linear programming and dynamic programming. These techniques are generally applicable when only a single goal is desired for minimizing or maximizing some objective function—such as maximizing profits or minimizing costs.

Because optimization techniques are generally limited to optimizing for a single goal, secondary goals get sacrificed that may also be important. Additionally, these techniques do not allow random variables to be defined as input data, thereby forcing the analyst to use average process times and arrival rates that can produce misleading results. They also usually assume that conditions are constant over the period of study. In contrast, simulation is capable of analyzing much more complex relationships and time-varying circumstances. With optimization capabilities now provided in simulation, simulation software has even taken on a prescriptive roll.

Descriptive Techniques

Descriptive techniques such as queuing theory are static analysis techniques that provide good estimates for basic problems such as determining the expected average number of entities in a queue or the average waiting times for entities in a queuing system. Queuing theory is of particular interest from a simulation perspective because it looks at many of the same system characteristics and issues that are addressed in simulation.

Queuing theory is essentially the science of waiting lines (in the United Kingdom, people wait in queues rather than lines). A queuing system consists of

FIGURE 2.6

*Queuing system
configuration.*

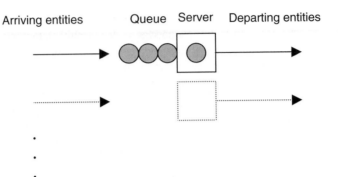

one or more queues and one or more servers (see Figure 2.6). Entities, referred to
in queuing theory as the *calling population,* enter the queuing system and either
are immediately served if a server is available or wait in a queue until a server becomes available. Entities may be serviced using one of several queuing disciplines: first-in, first-out (FIFO); last-in, first-out (LIFO); priority; and others. The
system capacity, or number of entities allowed in the system at any one time, may
be either finite or, as is often the case, infinite. Several different entity queuing behaviors can be analyzed such as balking (rejecting entry), reneging (abandoning
the queue), or jockeying (switching queues). Different interarrival time distributions (such as constant or exponential) may also be analyzed, coming from either
a finite or infinite population. Service times may also follow one of several distributions such as exponential or constant.

Kendall (1953) devised a simple system for classifying queuing systems in
the form $A/B/s$, where A is the type of interarrival distribution, B is the type of
service time distribution, and s is the number of servers. Typical distribution types
for A and B are

M	for Markovian or exponential distribution
G	for a general distribution
D	for a deterministic or constant value

An $M/D/1$ queuing system, for example, is a system in which interarrival times are
exponentially distributed, service times are constant, and there is a single server.

The arrival rate in a queuing system is usually represented by the Greek letter
lambda (λ) and the service rate by the Greek letter mu (μ). The mean interarrival
time then becomes $1/\lambda$ and the mean service time is $1/\mu$. A traffic intensity factor
λ/μ is a parameter used in many of the queuing equations and is represented by
the Greek letter rho (ρ).

Common performance measures of interest in a queuing system are based on
steady-state or long-term expected values and include

$L =$ expected number of entities in the system (number in the queue and in
service)

$L_q =$ expected number of entities in the queue (queue length)

W = expected time in the system (flow time)

W_q = expected time in the queue (waiting time)

P_n = probability of exactly n customers in the system ($n = 0, 1, \ldots$)

The $M/M/1$ system with infinite capacity and a FIFO queue discipline is perhaps the most basic queuing problem and sufficiently conveys the procedure for - analyzing queuing systems and understanding how the analysis is performed. The equations for calculating the common performance measures in an $M/M/1$ are

$$L = \frac{\rho}{(1 - \rho)} = \frac{\lambda}{(\mu - \lambda)}$$

$$L_q = L - \rho = \frac{\rho^2}{(1 - \rho)}$$

$$W = \frac{1}{\mu - \lambda}$$

$$W_q = \frac{\lambda}{\mu(\mu - \lambda)}$$

$$P_n = (1 - \rho)\rho^n \qquad n = 0, 1, \ldots$$

If either the expected number of entities in the system or the expected waiting time is known, the other can be calculated easily using Little's law (1961):

$$L = \lambda W$$

Little's law also can be applied to the queue length and waiting time:

$$L_q = \lambda W_q$$

Example: Suppose customers arrive to use an automatic teller machine (ATM) at an interarrival time of 3 minutes exponentially distributed and spend an average of 2.4 minutes exponentially distributed at the machine. What is the expected number of customers the system and in the queue? What is the expected waiting time for customers in the system and in the queue?

$$\lambda = 20 \text{ per hour}$$

$$\mu = 25 \text{ per hour}$$

$$\rho = \frac{\lambda}{\mu} = .8$$

Solving for L:

$$L = \frac{\lambda}{(\mu - \lambda)}$$

$$= \frac{20}{(25 - 20)}$$

$$= \frac{20}{5} = 4$$

Solving for L_q:

$$L_q = \frac{\rho^2}{(1 - \rho)}$$

$$= \frac{.8^2}{(1 - .8)}$$

$$= \frac{.64}{.2}$$

$$= 3.2$$

Solving for W using Little's formula:

$$W = \frac{L}{\lambda}$$

$$= \frac{4}{20}$$

$$= .20 \text{ hrs}$$

$$= 12 \text{ minutes}$$

Solving for W_q using Little's formula:

$$W_q = \frac{L_q}{\lambda}$$

$$= \frac{3.2}{20}$$

$$= .16 \text{ hrs}$$

$$= 9.6 \text{ minutes}$$

Descriptive OR techniques such as queuing theory are useful for the most basic problems, but as systems become even moderately complex, the problems get very complicated and quickly become mathematically intractable. In contrast, simulation provides close estimates for even the most complex systems (assuming the model is valid). In addition, the statistical output of simulation is not limited to only one or two metrics but instead provides information on all performance measures. Furthermore, while OR techniques give only average performance measures, simulation can generate detailed time-series data and histograms providing a complete picture of performance over time.

2.9.4 Special Computerized Tools

Many special computerized tools have been developed for forecasting, scheduling, layout, staffing, and so on. These tools are designed to be used for narrowly focused problems and are extremely effective for the kinds of problems they are intended to solve. They are usually based on constant input values and are computed using static calculations. The main benefit of special-purpose decision tools is that they are usually easy to use because they are designed to solve a specific type of problem.

2.10 Summary

An understanding of system dynamics is essential to using any tool for planning system operations. Manufacturing and service systems consist of interrelated elements (personnel, equipment, and so forth) that interactively function to produce a specified outcome (an end product, a satisfied customer, and so on). Systems are made up of entities (the objects being processed), resources (the personnel, equipment, and facilities used to process the entities), activities (the process steps), and controls (the rules specifying the who, what, where, when, and how of entity processing).

The two characteristics of systems that make them so difficult to analyze are interdependencies and variability. Interdependencies cause the behavior of one element to affect other elements in the system either directly or indirectly. Variability compounds the effect of interdependencies in the system, making system behavior nearly impossible to predict without the use of simulation.

The variables of interest in systems analysis are decision, response, and state variables. Decision variables define how a system works; response variables indicate how a system performs; and state variables indicate system conditions at specific points in time. System performance metrics or response variables are generally time, utilization, inventory, quality, or cost related. Improving system performance requires the correct manipulation of decision variables. System optimization seeks to find the best overall setting of decision variable values that maximizes or minimizes a particular response variable value.

Given the complex nature of system elements and the requirement to make good design decisions in the shortest time possible, it is evident that simulation can play a vital role in systems planning. Traditional systems analysis techniques are effective in providing quick but often rough solutions to dynamic systems problems. They generally fall short in their ability to deal with the complexity and dynamically changing conditions in manufacturing and service systems. Simulation is capable of imitating complex systems of nearly any size and to nearly any level of detail. It gives accurate estimates of multiple performance metrics and leads designers toward good design decisions.

2.11 Review Questions

1. Why is an understanding of system dynamics important to the use of simulation?
2. What is a system?
3. What are the elements of a system from a simulation perspective? Give an example of each.
4. What are two characteristics of systems that make them so complex?
5. What is the difference between a decision variable and a response variable?
6. Identify five decision variables of a manufacturing or service system that tend to be random.

7. Give two examples of state variables.

8. List three performance metrics that you feel would be important for a computer assembly line.

9. List three performance metrics you feel would be useful for a hospital emergency room.

10. Define *optimization* in terms of decision variables and response variables.

11. Is maximizing resource utilization a good overriding performance objective for a manufacturing system? Explain.

12. What is a systems approach to problem solving?

13. How does simulation fit into the overall approach of systems engineering?

14. In what situations would you use analytical techniques (like hand calculations or spreadsheet modeling) over simulation?

15. Assuming you decided to use simulation to determine how many lift trucks were needed in a distribution center, how might analytical models be used to complement the simulation study both before and after?

16. What advantages does simulation have over traditional OR techniques used in systems analysis?

17. Students come to a professor's office to receive help on a homework assignment every 10 minutes exponentially distributed. The time to help a student is exponentially distributed with a mean of 7 minutes. What are the expected number of students waiting to be helped and the average time waiting before being helped? For what percentage of time is it expected there will be more than two students waiting to be helped?

References

Blanchard, Benjamin S. *System Engineering Management*. New York: John Wiley & Sons, 1991.

Hopp, Wallace J., and M. Spearman. *Factory Physics*. New York: Irwin/McGraw-Hill, 2000, p. 282.

Kendall, D. G. "Stochastic Processes Occurring in the Theory of Queues and Their Analysis by the Method of Imbedded Markov Chains." *Annals of Mathematical Statistics* 24 (1953), pp. 338–54.

Kofman, Fred, and P. Senge. Communities of Commitment: The Heart of Learning Organizations. Sarita Chawla and John Renesch, (eds.), Portland, OR. Productivity Press, 1995.

Law, Averill M., and David W. Kelton. *Simulation Modeling and Analysis*. New York: McGraw-Hill, 2000.

Little, J. D. C. "A Proof for the Queuing Formula: $L = \lambda W$." *Operations Research* 9, no. 3 (1961), pp. 383–87.

Lloyd, S., and K. Melton. "Using Statistical Process Control to Obtain More Precise Distribution Fitting Using Distribution Fitting Software." *Simulators International XIV* 29, no. 3 (April 1997), pp. 193–98.

Senge, Peter. *The Fifth Discipline*. New York: Doubleday, 1990.

Simon, Herbert A. *Models of Man*. New York: John Wiley & Sons, 1957, p. 198.

3   SIMULATION BASICS

"Zeal without knowledge is fire without light."
—Dr. Thomas Fuller

3.1 Introduction

Simulation is much more meaningful when we understand what it is actually doing. Understanding how simulation works helps us to know whether we are applying it correctly and what the output results mean. Many books have been written that give thorough and detailed discussions of the science of simulation (see Banks et al. 2001; Hoover and Perry 1989; Law and Kelton 2000; Pooch and Wall 1993; Ross 1990; Shannon 1975; Thesen and Travis 1992; and Widman, Loparo, and Nielsen 1989). This chapter attempts to summarize the basic technical issues related to simulation that are essential to understand in order to get the greatest benefit from the tool. The chapter discusses the different types of simulation and how random behavior is simulated. A spreadsheet simulation example is given in this chapter to illustrate how various techniques are combined to simulate the behavior of a common system.

3.2 Types of Simulation

The way simulation works is based largely on the type of simulation used. There are many ways to categorize simulation. Some of the most common include

- Static or dynamic.
- Stochastic or deterministic.
- Discrete event or continuous.

The type of simulation we focus on in this book can be classified as dynamic, stochastic, discrete-event simulation. To better understand this classification, we will look at what the first two characteristics mean in this chapter and focus on what the third characteristic means in Chapter 4.

3.2.1 Static versus Dynamic Simulation

A *static* simulation is one that is not based on time. It often involves drawing random samples to generate a statistical outcome, so it is sometimes called Monte Carlo simulation. In finance, Monte Carlo simulation is used to select a portfolio of stocks and bonds. Given a portfolio, with different probabilistic payouts, it is possible to generate an expected yield. One material handling system supplier developed a static simulation model to calculate the expected time to travel from one rack location in a storage system to any other rack location. A random sample of 100 from–to relationships were used to estimate an average travel time. Had every from–to trip been calculated, a 1,000-location rack would have involved 1,000! calculations.

Dynamic simulation includes the passage of time. It looks at state changes as they occur over time. A clock mechanism moves forward in time and state variables are updated as time advances. Dynamic simulation is well suited for analyzing manufacturing and service systems since they operate over time.

3.2.2 Stochastic versus Deterministic Simulation

Simulations in which one or more input variables are random are referred to as *stochastic* or *probabilistic* simulations. A stochastic simulation produces output that is itself random and therefore gives only one data point of how the system might behave.

Simulations having no input components that are random are said to be *deterministic*. Deterministic simulation models are built the same way as stochastic models except that they contain no randomness. In a deterministic simulation, all future states are determined once the input data and initial state have been defined.

As shown in Figure 3.1, deterministic simulations have constant inputs and produce constant outputs. Stochastic simulations have random inputs and produce random outputs. Inputs might include activity times, arrival intervals, and routing sequences. Outputs include metrics such as average flow time, flow rate, and resource utilization. Any output impacted by a random input variable is going to also be a random variable. That is why the random inputs and random outputs of Figure 3.1(*b*) are shown as statistical distributions.

A deterministic simulation will always produce the exact same outcome no matter how many times it is run. In stochastic simulation, several randomized runs or replications must be made to get an accurate performance estimate because each run varies statistically. Performance estimates for stochastic simulations are obtained by calculating the average value of the performance metric across all of the replications. In contrast, deterministic simulations need to be run only once to get precise results because the results are always the same.

FIGURE 3.1

Examples of (a) a deterministic simulation and (b) a stochastic simulation.

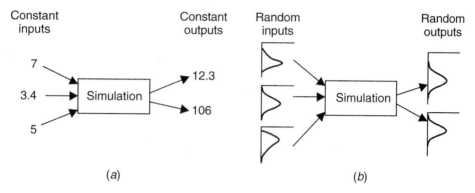

3.3 Random Behavior

Stochastic systems frequently have time or quantity values that vary within a given range and according to specified density, as defined by a probability distribution. Probability distributions are useful for predicting the next time, distance, quantity, and so forth when these values are random variables. For example, if an operation time varies between 2.2 minutes and 4.5 minutes, it would be defined in the model as a probability distribution. Probability distributions are defined by specifying the type of distribution (normal, exponential, or another type) and the parameters that describe the shape or density and range of the distribution. For example, we might describe the time for a check-in operation to be normally distributed with a mean of 5.2 minutes and a standard deviation of 0.4 minute. During the simulation, values are obtained from this distribution for successive operation times. The shape and range of time values generated for this activity will correspond to the parameters used to define the distribution. When we generate a value from a distribution, we call that value a *random variate*.

Probability distributions from which we obtain random variates may be either discrete (they describe the likelihood of specific values occurring) or continuous (they describe the likelihood of a value being within a given range). Figure 3.2 shows graphical examples of a discrete distribution and a continuous distribution.

A discrete distribution represents a finite or countable number of possible values. An example of a discrete distribution is the number of items in a lot or individuals in a group of people. A continuous distribution represents a continuum of values. An example of a continuous distribution is a machine with a cycle time that is uniformly distributed between 1.2 minutes and 1.8 minutes. An infinite number of possible values exist within this range. Discrete and continuous distributions are further defined in Chapter 6. Appendix A describes many of the distributions used in simulation.

FIGURE 3.2

Examples of (a) a discrete probability distribution and (b) a continuous probability distribution.

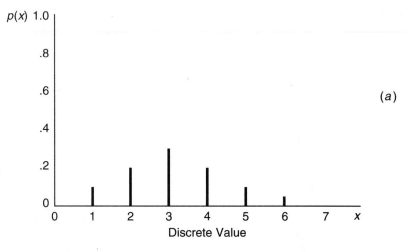

(a)

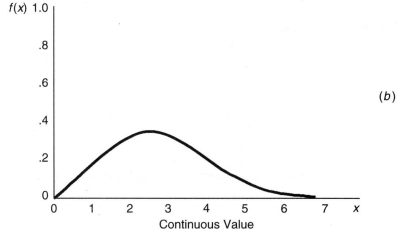

(b)

3.4 Simulating Random Behavior

One of the most powerful features of simulation is its ability to mimic random behavior or variation that is characteristic of stochastic systems. Simulating random behavior requires that a method be provided to generate random numbers as well as routines for generating random variates based on a given probability distribution. Random numbers and random variates are defined in the next sections along with the routines that are commonly used to generate them.

3.4.1 Generating Random Numbers

Random behavior is imitated in simulation by using a *random number generator*. The random number generator operates deep within the heart of a simulation model, pumping out a stream of random numbers. It provides the foundation for

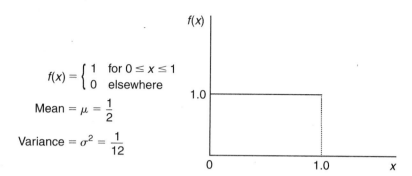

FIGURE 3.3

The uniform(0,1) distribution of a random number generator.

$$f(x) = \begin{cases} 1 & \text{for } 0 \le x \le 1 \\ 0 & \text{elsewhere} \end{cases}$$

$$\text{Mean} = \mu = \frac{1}{2}$$

$$\text{Variance} = \sigma^2 = \frac{1}{12}$$

simulating "random" events occurring in the simulated system such as the arrival time of cars to a restaurant's drive-through window; the time it takes the driver to place an order; the number of hamburgers, drinks, and fries ordered; and the time it takes the restaurant to prepare the order. The input to the procedures used to generate these types of events is a stream of numbers that are uniformly distributed between zero and one ($0 \le x \le 1$). The random number generator is responsible for producing this stream of independent and uniformly distributed numbers (Figure 3.3).

Before continuing, it should be pointed out that the numbers produced by a random number generator are not "random" in the truest sense. For example, the generator can reproduce the same sequence of numbers again and again, which is not indicative of random behavior. Therefore, they are often referred to as *pseudo-random number generators* (pseudo comes from Greek and means false or fake). Practically speaking, however, "good" pseudo-random number generators can pump out long sequences of numbers that pass statistical tests for randomness (the numbers are independent and uniformly distributed). Thus the numbers approximate real-world randomness for purposes of simulation, and the fact that they are reproducible helps us in two ways. It would be difficult to debug a simulation program if we could not "regenerate" the same sequence of random numbers to reproduce the conditions that exposed an error in our program. We will also learn in Chapter 10 how reproducing the same sequence of random numbers is useful when comparing different simulation models. For brevity, we will drop the *pseudo* prefix as we discuss how to design and keep our random number generator healthy.

Linear Congruential Generators

There are many types of established random number generators, and researchers are actively pursuing the development of new ones (L'Ecuyer 1998). However, most simulation software is based on linear congruential generators (LCG). The LCG is efficient in that it quickly produces a sequence of random numbers without requiring a great deal of computational resources. Using the LCG, a sequence of integers $Z_1, Z_2, Z_3, \ldots$ is defined by the recursive formula

$$Z_i = (a Z_{i-1} + c) \bmod(m)$$

where the constant a is called the multiplier, the constant c the increment, and the constant m the modulus (Law and Kelton 2000). The user must provide a seed or starting value, denoted Z_0, to begin generating the sequence of integer values. Z_0, a, c, and m are all nonnegative integers. The value of Z_i is computed by dividing $(aZ_{i-1} + c)$ by m and setting Z_i equal to the remainder part of the division, which is the result returned by the mod function. Therefore, the Z_i values are bounded by $0 \leq Z_i \leq m - 1$ and are uniformly distributed in the discrete case. However, we desire the continuous version of the uniform distribution with values ranging between a low of zero and a high of one, which we will denote as U_i for $i = 1, 2, 3, \ldots$. Accordingly, the value of U_i is computed by dividing Z_i by m.

In a moment, we will consider some requirements for selecting the values for a, c, and m to ensure that the random number generator produces a long sequence of numbers before it begins to repeat them. For now, however, let's assign the following values $a = 21$, $c = 3$, and $m = 16$ and generate a few pseudo-random numbers. Table 3.1 contains a sequence of 20 random numbers generated from the recursive formula

$$Z_i = (21Z_{i-1} + 3) \bmod(16)$$

An integer value of 13 was somewhat arbitrarily selected between 0 and $m - 1 = 16 - 1 = 15$ as the seed ($Z_0 = 13$) to begin generating the sequence of

TABLE 3.1   **Example LCG $Z_i = (21Z_{i-1} + 3)\bmod(16)$, with $Z_0 = 13$**

i	$21Z_{i-1} + 3$	Z_i	$U_i = Z_i/16$
0		13	
1	276	4	0.2500
2	87	7	0.4375
3	150	6	0.3750
4	129	1	0.0625
5	24	8	0.5000
6	171	11	0.6875
7	234	10	0.6250
8	213	5	0.3125
9	108	12	0.7500
10	255	15	0.9375
11	318	14	0.8750
12	297	9	0.5625
13	192	0	0.0000
14	3	3	0.1875
15	66	2	0.1250
16	45	13	0.8125
17	276	4	0.2500
18	87	7	0.4375
19	150	6	0.3750
20	129	1	0.0625

numbers in Table 3.1. The value of Z_1 is obtained as

$$Z_1 = (aZ_0 + c) \, \text{mod}(m) = (21(13) + 3) \, \text{mod}(16) = (276) \, \text{mod}(16) = 4$$

Note that 4 is the remainder term from dividing 16 into 276. The value of U_1 is computed as

$$U_1 = Z_1/16 = 4/16 = 0.2500$$

The process continues using the value of Z_1 to compute Z_2 and then U_2.

Changing the value of Z_0 produces a different sequence of uniform(0,1) numbers. However, the original sequence can always be reproduced by setting the seed value back to 13 ($Z_0 = 13$). The ability to repeat the simulation experiment under the exact same "random" conditions is very useful, as will be demonstrated in Chapter 10 with a technique called common random numbers.

Note that in ProModel the sequence of random numbers is not generated in advance and then read from a table as we have done. Instead, the only value saved is the last Z_i value that was generated. When the next random number in the sequence is needed, the saved value is fed back to the generator to produce the next random number in the sequence. In this way, the random number generator is called each time a new random event is scheduled for the simulation.

Due to computational constraints, the random number generator cannot go on indefinitely before it begins to repeat the same sequence of numbers. The LCG in Table 3.1 will generate a sequence of 16 numbers before it begins to repeat itself. You can see that it began repeating itself starting in the 17th position. The value of 16 is referred to as the cycle length of the random number generator, which is disturbingly short in this case. A long cycle length is desirable so that each replication of a simulation is based on a different segment of random numbers. This is how we collect indepentdent observations of the model's output.

Let's say, for example, that to run a certain simulation model for one replication requires that the random number generator be called 1,000 times during the simulation and we wish to execute five replications of the simulation. The first replication would use the first 1,000 random numbers in the sequence, the second replication would use the next 1,000 numbers in the sequence (the number that would appear in positions 1,001 to 2,000 if a table were generated in advance), and so on. In all, the random number generator would be called 5,000 times. Thus we would need a random number generator with a cycle length of at least 5,000.

The maximum cycle length that an LCG can achieve is m. To realize the maximum cycle length, the values of a, c, and m have to be carefully selected. A guideline for the selection is to assign (Pritsker 1995)

1. $m = 2^b$, where b is determined based on the number of bits per word on the computer being used. Many computers use 32 bits per word, making 31 a good choice for b.
2. c and m such that their greatest common factor is 1 (the only positive integer that exactly divides both m and c is 1).
3. $a = 1 + 4k$, where k is an integer.

Following this guideline, the LCG can achieve a full cycle length of over 2.1 billion (2^{31} to be exact) random numbers.

Frequently, the long sequence of random numbers is subdivided into smaller segments. These subsegments are referred to as *streams*. For example, Stream 1 could begin with the random number in the first position of the sequence and continue down to the random number in the 200,000th position of the sequence. Stream 2, then, would start with the random number in the 200,001st position of the sequence and end at the 400,000th position, and so on. Using this approach, each type of random event in the simulation model can be controlled by a unique stream of random numbers. For example, Stream 1 could be used to generate the arrival pattern of cars to a restaurant's drive-through window and Stream 2 could be used to generate the time required for the driver of the car to place an order. This assumes that no more than 200,000 random numbers are needed to simulate each type of event. The practical and statistical advantages of assigning unique streams to each type of event in the model are described in Chapter 10.

To subdivide the generator's sequence of random numbers into streams, you first need to decide how many random numbers to place in each stream. Next, you begin generating the entire sequence of random numbers (cycle length) produced by the generator and recording the Z_i values that mark the beginning of each stream. Therefore, each stream has its own starting or seed value. When using the random number generator to drive different events in a simulation model, the previously generated random number from a particular stream is used as input to the generator to generate the next random number from that stream. For convenience, you may want to think of each stream as a separate random number generator to be used in different places in the model. For example, see Figure 10.5 in Chapter 10.

There are two types of linear congruential generators: the mixed congruential generator and the multiplicative congruential generator. *Mixed congruential generators* are designed by assigning $c > 0$. *Multiplicative congruential generators* are designed by assigning $c = 0$. The multiplicative generator is more efficient than the mixed generator because it does not require the addition of c. The maximum cycle length for a multiplicative generator can be set within one unit of the maximum cycle length of the mixed generator by carefully selecting values for a and m. From a practical standpoint, the difference in cycle length is insignificant considering that both types of generators can boast cycle lengths of more than 2.1 billion.

ProModel uses the following multiplicative generator:

$$Z_i = (630,360,016Z_{i-1}) \bmod(2^{31} - 1)$$

Specifically, it is a prime modulus multiplicative linear congruential generator (PMMLCG) with $a = 630,360,016$, $c = 0$, and $m = 2^{31} - 1$. It has been extensively tested and is known to be a reliable random number generator for simulation (Law and Kelton 2000). The ProModel implementation of this generator divides the cycle length of $2^{31} - 1 = 2,147,483,647$ into 100 unique streams.

Testing Random Number Generators

When faced with using a random number generator about which you know very little, it is wise to verify that the numbers emanating from it satisfy the two important properties defined at the beginning of this section. The numbers produced by the random number generator must be (1) independent and (2) uniformly distributed between zero and one (uniform(0,1)). To verify that the generator satisfies these properties, you first generate a sequence of random numbers U_1, U_2, U_3, ... and then subject them to an appropriate test of hypothesis.

The hypotheses for testing the independence property are

H_0: U_i values from the generator are independent
H_1: U_i values from the generator are not independent

Several statistical methods have been developed for testing these hypotheses at a specified significance level α. One of the most commonly used methods is the runs test. Banks et al. (2001) review three different versions of the runs test for conducting this independence test. Additionally, two runs tests are implemented in Stat::Fit—the Runs Above and Below the Median Test and the Runs Up and Runs Down Test. Chapter 6 contains additional material on tests for independence.

The hypotheses for testing the uniformity property are

H_0: U_i values are uniform(0,1)
H_1: U_i values are not uniform(0,1)

Several statistical methods have also been developed for testing these hypotheses at a specified significance level α. The Kolmogorov-Smirnov test and the chi-square test are perhaps the most frequently used tests. (See Chapter 6 for a description of the chi-square test.) The objective is to determine if the uniform(0,1) distribution fits or describes the sequence of random numbers produced by the random number generator. These tests are included in the Stat::Fit software and are further described in many introductory textbooks on probability and statistics (see, for example, Johnson 1994).

3.4.2 Generating Random Variates

This section introduces common methods for generating observations (random variates) from probability distributions other than the uniform(0,1) distribution. For example, the time between arrivals of cars to a restaurant's drive-through window may be exponentially distributed, and the time required for drivers to place orders at the window might follow the lognormal distribution. Observations from these and other commonly used distributions are obtained by transforming the observations generated by the random number generator to the desired distribution. The transformed values are referred to as variates from the specified distribution.

There are several methods for generating random variates from a desired distribution, and the selection of a particular method is somewhat dependent on the characteristics of the desired distribution. Methods include the inverse transformation method, the acceptance/rejection method, the composition method, the convolution method, and the methods employing special properties. The inverse transformation method, which is commonly used to generate variates from both discrete and continuous distributions, is described starting first with the continuous case. For a review of the other methods, see Law and Kelton (2000).

Continuous Distributions

The application of the inverse transformation method to generate random variates from continuous distributions is straightforward and efficient for many continuous distributions. For a given probability density function $f(x)$, find the cumulative distribution function of X. That is, $F(x) = P(X \leq x)$. Next, set $U = F(x)$, where U is uniform(0,1), and solve for x. Solving for x yields $x = F^{-1}(U)$. The equation $x = F^{-1}(U)$ transforms U into a value for x that conforms to the given distribution $f(x)$.

As an example, suppose that we need to generate variates from the exponential distribution with mean β. The probability density function $f(x)$ and corresponding cumulative distribution function $F(x)$ are

$$f(x) = \begin{cases} \dfrac{1}{\beta}e^{-x/\beta} & \text{for } x > 0 \\ 0 & \text{elsewhere} \end{cases}$$

$$F(x) = \begin{cases} 1 - e^{-x/\beta} & \text{for } x > 0 \\ 0 & \text{elsewhere} \end{cases}$$

Setting $U = F(x)$ and solving for x yields

$$U = 1 - e^{-x/\beta}$$
$$e^{-x/\beta} = 1 - U$$
$$\ln(e^{-x/\beta}) = \ln(1 - U) \qquad \text{where ln is the natural logarithm}$$
$$-x/\beta = \ln(1 - U)$$
$$x = -\beta \ln(1 - U)$$

The random variate x in the above equation is exponentially distributed with mean β provided U is uniform(0,1).

Suppose three observations of an exponentially distributed random variable with mean $\beta = 2$ are desired. The next three numbers generated by the random number generator are $U_1 = 0.27$, $U_2 = 0.89$, and $U_3 = 0.13$. The three numbers are transformed into variates x_1, x_2, and x_3 from the exponential distribution with mean $\beta = 2$ as follows:

$$x_1 = -2 \ln(1 - U_1) = -2 \ln(1 - 0.27) = 0.63$$
$$x_2 = -2 \ln(1 - U_2) = -2 \ln(1 - 0.89) = 4.41$$
$$x_3 = -2 \ln(1 - U_3) = -2 \ln(1 - 0.13) = 0.28$$

Figure 3.4 provides a graphical representation of the inverse transformation method in the context of this example. The first step is to generate U, where U is uniform(0,1). Next, locate U on the y axis and draw a horizontal line from that point to the cumulative distribution function $[F(x) = 1 - e^{-x/2}]$. From this point

FIGURE 3.4

*Graphical explanation
of inverse
transformation method
for continuous
variates.*

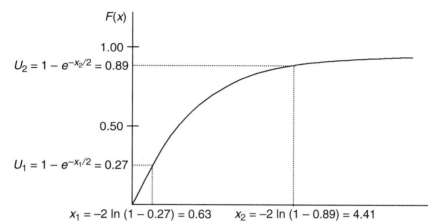

$$U_2 = 1 - e^{-x_2/2} = 0.89$$

$$U_1 = 1 - e^{-x_1/2} = 0.27$$

$$x_1 = -2 \ln (1 - 0.27) = 0.63 \qquad x_2 = -2 \ln (1 - 0.89) = 4.41$$

of intersection with $F(x)$, a vertical line is dropped down to the x axis to obtain the corresponding value of the variate. This process is illustrated in Figure 3.4 for generating variates x_1 and x_2 given $U_1 = 0.27$ and $U_2 = 0.89$.

Application of the inverse transformation method is straightforward as long as there is a closed-form formula for the cumulative distribution function, which is the case for many continuous distributions. However, the normal distribution is one exception. Thus it is not possible to solve for a simple equation to generate normally distributed variates. For these cases, there are other methods that can be used to generate the random variates. See, for example, Law and Kelton (2000) for a description of additional methods for generating random variates from continuous distributions.

Discrete Distributions

The application of the inverse transformation method to generate variates from discrete distributions is basically the same as for the continuous case. The difference is in how it is implemented. For example, consider the following probability mass function:

$$p(x) = P(X = x) = \begin{cases} 0.10 & \text{for } x = 1 \\ 0.30 & \text{for } x = 2 \\ 0.60 & \text{for } x = 3 \end{cases}$$

The random variate x has three possible values. The probability that x is equal to 1 is 0.10, $P(X = 1) = 0.10$; $P(X = 2) = 0.30$; and $P(X = 3) = 0.60$. The cumulative distribution function $F(x)$ is shown in Figure 3.5. The random variable x could be used in a simulation to represent the number of defective components on a circuit board or the number of drinks ordered from a drive-through window, for example.

Suppose that an observation from the above discrete distribution is desired. The first step is to generate U, where U is uniform(0,1). Using Figure 3.5, the

FIGURE 3.5

Graphical explanation of inverse transformation method for discrete variates.

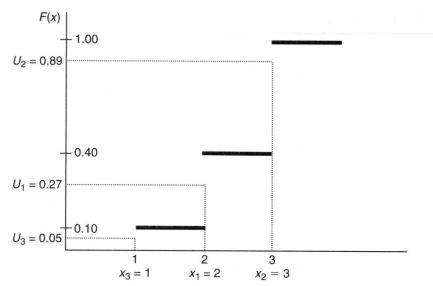

value of the random variate is determined by locating U on the y axis, drawing a horizontal line from that point until it intersects with a step in the cumulative function, and then dropping a vertical line from that point to the x axis to read the value of the variate. This process is illustrated in Figure 3.5 for x_1, x_2, and x_3 given $U_1 = 0.27$, $U_2 = 0.89$, and $U_3 = 0.05$. Equivalently, if $0 \le U_i \le 0.10$, then $x_i = 1$; if $0.10 < U_i \le 0.40$, then $x_i = 2$; if $0.40 < U_i \le 1.00$, then $x_i = 3$. Note that should we generate 100 variates using the above process, we would expect a value of 3 to be returned 60 times, a value of 2 to be returned 30 times, and a value of 1 to be returned 10 times.

The inverse transformation method can be applied to any discrete distribution by dividing the y axis into subintervals as defined by the cumulative distribution function. In our example case, the subintervals were [0, 0.10], (0.10, 0.40], and (0.40, 1.00]. For each subinterval, record the appropriate value for the random variate. Next, develop an algorithm that calls the random number generator to receive a specific value for U, searches for and locates the subinterval that contains U, and returns the appropriate value for the random variate.

Locating the subinterval that contains a specific U value is relatively straightforward when there are only a few possible values for the variate. However, the number of possible values for a variate can be quite large for some discrete distributions. For example, a random variable having 50 possible values could require that 50 subintervals be searched before locating the subinterval that contains U. In such cases, sophisticated search algorithms have been developed that quickly locate the appropriate subinterval by exploiting certain characteristics of the discrete distribution for which the search algorithm was designed. A good source of information on the subject can be found in Law and Kelton (2000).

3.5 Simple Spreadsheet Simulation

This chapter has covered some really useful information, and it will be constructive to pull it all together for our first dynamic, stochastic simulation model. The simulation will be implemented as a spreadsheet model because the example system is not so complex as to require the use of a commercial simulation software product. Furthermore, a spreadsheet will provide a nice tabulation of the random events that produce the output from the simulation, thereby making it ideal for demonstrating how simulations of dynamic, stochastic systems are often accomplished. The system to be simulated follows.

Customers arrive to use an automatic teller machine (ATM) at a mean interarrival time of 3.0 minutes exponentially distributed. When customers arrive to the system they join a queue to wait for their turn on the ATM. The queue has the capacity to hold an infinite number of customers. Customers spend an average of 2.4 minutes exponentially distributed at the ATM to complete their transactions, which is called the service time at the ATM. Simulate the system for the arrival and processing of 25 customers and estimate the expected waiting time for customers in the queue (the average time customers wait in line for the ATM) and the expected time in the system (the average time customers wait in the queue plus the average time it takes them to complete their transaction at the ATM).

Using the systems terminology introduced in Chapter 2 to describe the ATM system, the entities are the customers that arrive to the ATM for processing. The resource is the ATM that serves the customers, which has the capacity to serve one customer at a time. The system controls that dictate how, when, and where activities are performed for this ATM system consist of the queuing discipline, which is first-in, first-out (FIFO). Under the FIFO queuing discipline the entities (customers) are processed by the resource (ATM) in the order that they arrive to the queue. Figure 3.6 illustrates the relationships of the elements of the system with the customers appearing as dark circles.

FIGURE 3.6

Descriptive drawing of the automatic teller machine (ATM) system.

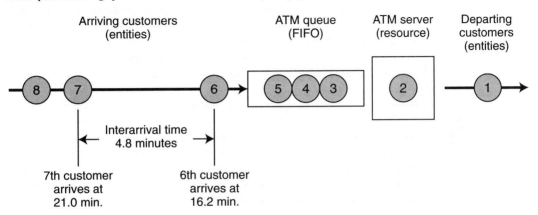

Our objective in building a spreadsheet simulation model of the ATM system is to estimate the average amount of time the 25 customers will spend waiting in the queue and the average amount of time the customers will spend in the system. To accomplish our objective, the spreadsheet will need to generate random numbers and random variates, to contain the logic that controls how customers are processed by the system, and to compute an estimate of system performance measures. The spreadsheet simulation is shown in Table 3.2 and is divided into three main sections: Arrivals to ATM, ATM Processing Time, and ATM Simulation Logic. The Arrivals to ATM and ATM Processing Time provide the foundation for the simulation, while the ATM Simulation Logic contains the spreadsheet programming that mimics customers flowing through the ATM system. The last row of the Time in Queue column and the Time in System column under the ATM Simulation Logic section contains one observation of the average time customers wait in the queue, 1.94 minutes, and one observation of their average time in the system, 4.26 minutes. The values were computed by averaging the 25 customer time values in the respective columns.

Do you think customer number 17 (see the Customer Number column) became upset over having to wait in the queue for 6.11 minutes to conduct a 0.43 minute transaction at the ATM (see the Time in Queue column and the Service Time column under ATM Simulation Logic)? Has something like this ever happened to you in real life? Simulation can realistically mimic the behavior of a system. More time will be spent interpreting the results of our spreadsheet simulation after we understand how it was put together.

3.5.1 Simulating Random Variates

The interarrival time is the elapsed time between customer arrivals to the ATM. This time changes according to the exponential distribution with a mean of 3.0 minutes. That is, the time that elapses between one customer arrival and the next is not the same from one customer to the next but averages out to be 3.0 minutes. This is illustrated in Figure 3.6 in that the interarrival time between customers 7 and 8 is much smaller than the interarrival time of 4.8 minutes between customers 6 and 7. The service time at the ATM is also a random variable following the exponential distribution and averages 2.4 minutes. To simulate this stochastic system, a random number generator is needed to produce observations (random variates) from the exponential distribution. The inverse transformation method was used in Section 3.4.2 just for this purpose.

The transformation equation is

$$X_i = -\beta \ln (1 - U_i) \qquad \text{for } i = 1, 2, 3, \ldots, 25$$

where X_i represents the ith value realized from the exponential distribution with mean β, and U_i is the ith random number drawn from a uniform$(0,1)$ distribution. The $i = 1, 2, 3, \ldots, 25$ indicates that we will compute 25 values from the transformation equation. However, we need to have two different versions of this equation to generate the two sets of 25 exponentially distributed random variates needed to simulate 25 customers because the mean interarrival time of $\beta = 3.0$ minutes is different than the mean service time of $\beta = 2.4$ minutes. Let $X1_i$ denote the interarrival

TABLE 3.2 Spreadsheet Simulation of Automatic Teller Machine (ATM)

	Arrivals to ATM			ATM Processing Time			ATM Simulation Logic						
i	Stream 1 $(Z1_i)$	Random Number $(U1_i)$	Interarrival Time $(X1_i)$	Stream 2 $(Z2_i)$	Random Number $(U2_i)$	Service Time $(X2_i)$	Customer Number (1)	Arrival Time (2)	Begin Service Time (3)	Service Time (4)	Departure Time (5) = (3) + (4)	Time in Queue (6) = (3) − (2)	Time in System (7) = (5) − (2)
0	3			122									
1	66	0.516	2.18	5	0.039	0.10	1	2.18	2.18	0.10	2.28	0.00	0.10
2	109	0.852	5.73	108	0.844	4.46	2	7.91	7.91	4.46	12.37	0.00	4.46
3	116	0.906	7.09	95	0.742	3.25	3	15.00	15.00	3.25	18.25	0.00	3.25
4	7	0.055	0.17	78	0.609	2.25	4	15.17	18.25	2.25	20.50	3.08	5.33
5	22	0.172	0.57	105	0.820	4.12	5	15.74	20.50	4.12	24.62	4.76	8.88
6	81	0.633	3.01	32	0.250	0.69	6	18.75	24.62	0.69	25.31	5.87	6.56
7	40	0.313	1.13	35	0.273	0.77	7	19.88	25.31	0.77	26.08	5.43	6.20
8	75	0.586	2.65	98	0.766	3.49	8	22.53	26.08	3.49	29.57	3.55	7.04
9	42	0.328	1.19	13	0.102	0.26	9	23.72	29.57	0.26	29.83	5.85	6.11
10	117	0.914	7.36	20	0.156	0.41	10	31.08	31.08	0.41	31.49	0.00	0.41
11	28	0.219	0.74	39	0.305	0.87	11	31.82	31.82	0.87	32.69	0.00	0.87
12	79	0.617	2.88	54	0.422	1.32	12	34.70	34.70	1.32	36.02	0.00	1.32
13	126	0.984	12.41	113	0.883	5.15	13	47.11	47.11	5.15	52.26	0.00	5.15
14	89	0.695	3.56	72	0.563	1.99	14	50.67	52.26	1.99	54.25	1.59	3.58
15	80	0.625	2.94	107	0.836	4.34	15	53.61	54.25	4.34	58.59	0.64	4.98
16	19	0.148	0.48	74	0.578	2.07	16	54.09	58.59	2.07	60.66	4.50	6.57
17	18	0.141	0.46	21	0.164	0.43	17	54.55	60.66	0.43	61.09	6.11	6.54
18	125	0.977	11.32	60	0.469	1.52	18	65.87	65.87	1.52	67.39	0.00	1.52
19	68	0.531	2.27	111	0.867	4.84	19	68.14	68.14	4.84	72.98	0.00	4.84
20	23	0.180	0.60	30	0.234	0.64	20	68.74	72.98	0.64	73.62	4.24	4.88
21	102	0.797	4.78	121	0.945	6.96	21	73.52	73.62	6.96	80.58	0.10	7.06
22	97	0.758	4.26	112	0.875	4.99	22	77.78	80.58	4.99	85.57	2.80	7.79
23	120	0.938	8.34	51	0.398	1.22	23	86.12	86.12	1.22	87.34	0.00	1.22
24	91	0.711	3.72	50	0.391	1.19	24	89.84	89.84	1.19	91.03	0.00	1.19
25	122	0.953	9.17	29	0.227	0.62	25	99.01	99.01	0.62	99.63	0.00	0.62
											Average	1.94	4.26

time and $X2_i$ denote the service time generated for the ith customer simulated in the system. The equation for transforming a random number into an interarrival time observation from the exponential distribution with mean $\beta = 3.0$ minutes becomes

$$X1_i = -3.0 \ln (1 - U1_i) \qquad \text{for } i = 1, 2, 3, \ldots, 25$$

where $U1_i$ denotes the ith value drawn from the random number generator using Stream 1. This equation is used in the Arrivals to ATM section of Table 3.2 under the Interarrival Time ($X1_i$) column.

The equation for transforming a random number into an ATM service time observation from the exponential distribution with mean $\beta = 2.4$ minutes becomes

$$X2_i = -2.4 \ln (1 - U2_i) \qquad \text{for } i = 1, 2, 3, \ldots, 25$$

where $U2_i$ denotes the ith value drawn from the random number generator using Stream 2. This equation is used in the ATM Processing Time section of Table 3.2 under the Service Time ($X2_i$) column.

Let's produce the sequence of $U1_i$ values that feeds the transformation equation ($X1_i$) for interarrival times using a linear congruential generator (LCG) similar to the one used in Table 3.1. The equations are

$$Z1_i = (21Z1_{i-1} + 3) \operatorname{mod}(128)$$
$$U1_i = Z1_i/128 \qquad \text{for } i = 1, 2, 3, \ldots, 25$$

The authors defined Stream 1's starting or seed value to be 3. So we will use $Z1_0 = 3$ to kick off this stream of 25 random numbers. These equations are used in the Arrivals to ATM section of Table 3.2 under the Stream 1 ($Z1_i$) and Random Number ($U1_i$) columns.

Likewise, we will produce the sequence of $U2_i$ values that feeds the transformation equation ($X2_i$) for service times using

$$Z2_i = (21Z2_{i-1} + 3) \operatorname{mod}(128)$$
$$U2_i = Z2_i/128 \qquad \text{for } i = 1, 2, 3, \ldots, 25$$

and will specify a starting seed value of $Z2_0 = 122$, Stream 2's seed value, to kick off the second stream of 25 random numbers. These equations are used in the ATM Processing Time section of Table 3.2 under the Stream 2 ($Z2_i$) and Random Number ($U2_i$) columns.

The spreadsheet presented in Table 3.2 illustrates 25 random variates for both the interarrival time, column ($X1_i$), and service time, column ($X2_i$). All time values are given in minutes in Table 3.2. To be sure we pull this together correctly, let's compute a couple of interarrival times with mean $\beta = 3.0$ minutes and compare them to the values given in Table 3.2.

Given $Z1_0 = 3$

$$Z1_1 = (21Z1_0 + 3) \operatorname{mod}(128) = (21(3) + 3) \operatorname{mod}(128)$$
$$= (66) \operatorname{mod}(128) = 66$$
$$U1_1 = Z1_1/128 = 66/128 = 0.516$$
$$X1_1 = -\beta \ln (1 - U1_1) = -3.0 \ln (1 - 0.516) = 2.18 \text{ minutes}$$

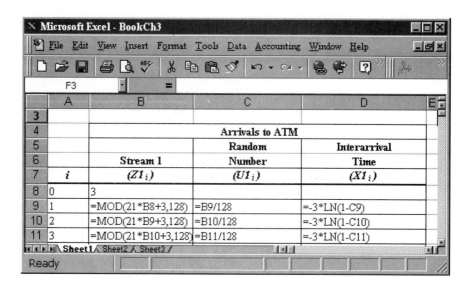

FIGURE 3.7

*Microsoft Excel
snapshot of the ATM
spreadsheet
illustrating the
equations for the
Arrivals to ATM
section.*

The value of 2.18 minutes is the first value appearing under the column, Interarrival Time $(X1_i)$. To compute the next interarrival time value $X1_2$, we start by using the value of $Z1_1$ to compute $Z1_2$.

Given $Z1_1 = 66$

$$Z1_2 = (21Z1_1 + 3)\,\mathrm{mod}(128) = (21(66) + 3)\,\mathrm{mod}(128) = 109$$
$$U1_2 = Z1_2/128 = 109/128 = 0.852$$
$$X1_2 = -3\ln(1 - U1_2) = -3.0\ln(1 - 0.852) = 5.73 \text{ minutes}$$

Figure 3.7 illustrates how the equations were programmed in Microsoft Excel for the Arrivals to ATM section of the spreadsheet. Note that the $U1_i$ and $X1_i$ values in Table 3.2 are rounded to three and two places to the right of the decimal, respectively. The same rounding rule is used for $U2_i$ and $X2_i$.

It would be useful for you to verify a few of the service time values with mean $\beta = 2.4$ minutes appearing in Table 3.2 using

$$Z2_0 = 122$$
$$Z2_i = (21Z2_{i-1} + 3)\,\mathrm{mod}(128)$$
$$U2_i = Z2_i/128$$
$$X2_i = -2.4\ln(1 - U2_i) \qquad \text{for } i = 1, 2, 3, \ldots$$

The equations started out looking a little difficult to manipulate but turned out not to be so bad when we put some numbers in them and organized them in a spreadsheet—though it was a bit tedious. The important thing to note here is that although it is transparent to the user, ProModel uses a very similar method to produce exponentially distributed random variates, and you now understand how it is done.

The LCG just given has a maximum cycle length of 128 random numbers (you may want to verify this), which is more than enough to generate 25 interarrival time values and 25 service time values for this simulation. However, it is a poor random number generator compared to the one used by ProModel. It was chosen because it is easy to program into a spreadsheet and to compute by hand to facilitate our understanding. The biggest difference between it and the random number generator in ProModel is that the ProModel random number generator manipulates much larger numbers to pump out a much longer stream of numbers that pass all statistical tests for randomness.

Before moving on, let's take a look at why we chose $Z1_0 = 3$ and $Z2_0 = 122$. Our goal was to make sure that we did not use the same uniform(0,1) random number to generate both an interarrival time and a service time. If you look carefully at Table 3.2, you will notice that the seed value $Z2_0 = 122$ is the $Z1_{25}$ value from random number Stream 1. Stream 2 was merely defined to start where Stream 1 ended. Thus our spreadsheet used a unique random number to generate each interarrival and service time. Now let's add the necessary logic to our spreadsheet to conduct the simulation of the ATM system.

3.5.2 Simulating Dynamic, Stochastic Systems

The heart of the simulation is the generation of the random variates that drive the stochastic events in the simulation. However, the random variates are simply a list of numbers at this point. The spreadsheet section labeled ATM Simulation Logic in Table 3.2 is programmed to coordinate the execution of the events to mimic the processing of customers through the ATM system. The simulation program must keep up with the passage of time to coordinate the events. The word *dynamic* appearing in the title for this section refers to the fact that the simulation tracks the passage of time.

The first column under the ATM Simulation Logic section of Table 3.2, labeled Customer Number, is simply to keep a record of the order in which the 25 customers are processed, which is FIFO. The numbers appearing in parentheses under each column heading are used to illustrate how different columns are added or subtracted to compute the values appearing in the simulation.

The Arrival Time column denotes the moment in time at which each customer arrives to the ATM system. The first customer arrives to the system at time 2.18 minutes. This is the first interarrival time value ($X1_1 = 2.18$) appearing in the Arrival to ATM section of the spreadsheet table. The second customer arrives to the system at time 7.91 minutes. This is computed by taking the arrival time of the first customer of 2.18 minutes and adding to it the next interarrival time of $X1_2 = 5.73$ minutes that was generated in the Arrival to ATM section of the spreadsheet table. The arrival time of the second customer is $2.18 + 5.73 = 7.91$ minutes. The process continues with the third customer arriving at $7.91 + 7.09 = 15.00$ minutes.

The trickiest piece to program into the spreadsheet is the part that determines the moment in time at which customers begin service at the ATM after waiting in the queue. Therefore, we will skip over the Begin Service Time column for now and come back to it after we understand how the other columns are computed.

The Service Time column simply records the simulated amount of time required for the customer to complete their transaction at the ATM. These values are copies of the service time $X2_i$ values generated in the ATM Processing Time section of the spreadsheet.

The Departure Time column records the moment in time at which a customer departs the system after completing their transaction at the ATM. To compute the time at which a customer departs the system, we take the time at which the customer gained access to the ATM to begin service, column (3), and add to that the length of time the service required, column (4). For example, the first customer gained access to the ATM to begin service at 2.18 minutes, column (3). The service time for the customer was determined to be 0.10 minutes in column (4). So, the customer departs 0.10 minutes later or at time $2.18 + 0.10 = 2.28$ minutes. This customer's short service time must be because they forgot their PIN number and could not conduct their transaction.

The Time in Queue column records how long a customer waits in the queue before gaining access to the ATM. To compute the time spent in the queue, we take the time at which the ATM began serving the customer, column (3), and subtract from that the time at which the customer arrived to the system, column (2). The fourth customer arrives to the system at time 15.17 and begins getting service from the ATM at 18.25 minutes; thus, the fourth customer's time in the queue is $18.25 - 15.17 = 3.08$ minutes.

The Time in System column records how long a customer was in the system. To compute the time spent in the system, we subtract the customer's departure time, column (5), from the customer's arrival time, column (2). The fifth customer arrives to the system at 15.74 minutes and departs the system at 24.62 minutes. Therefore, this customer spent $24.62 - 15.74 = 8.88$ minutes in the system.

Now let's go back to the Begin Service Time column, which records the time at which a customer begins to be served by the ATM. The very first customer to arrive to the system when it opens for service advances directly to the ATM. There is no waiting time in the queue; thus the value recorded for the time that the first customer begins service at the ATM is the customer's arrival time. With the exception of the first customer to arrive to the system, we have to capture the logic that a customer cannot begin service at the ATM until the previous customer using the ATM completes his or her transaction. One way to do this is with an IF statement as follows:

```
IF (Current Customer's Arrival Time < Previous Customer's
   Departure Time)
THEN (Current Customer's Begin Service Time = Previous Customer's
   Departure Time)
ELSE (Current Customer's Begin Service Time = Current Customer's
   Arrival Time)
```

Figure 3.8 illustrates how the IF statement was programmed in Microsoft Excel. The format of the Excel IF statement is

```
IF (logical test, use this value if test is true, else use this
   value if test is false)
```

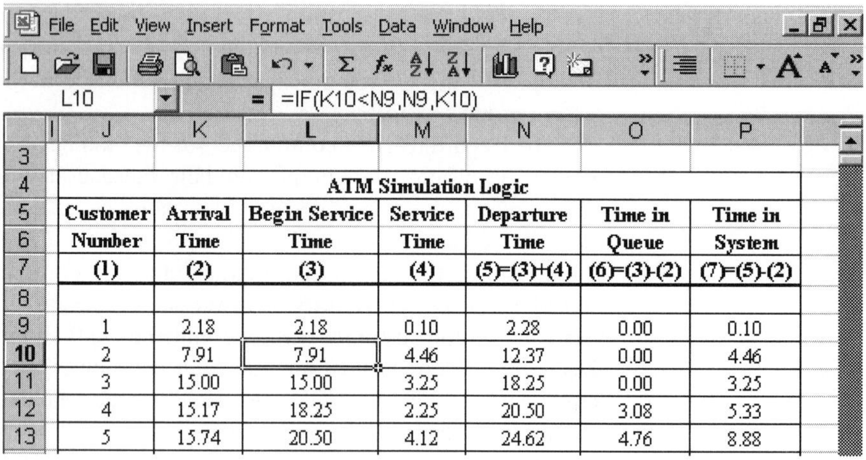

FIGURE 3.8

Microsoft Excel snapshot of the ATM spreadsheet illustrating the IF statement for the Begin Service Time column.

	File Edit View Insert Format Tools Data Window Help						
L10		=	=IF(K10<N9,N9,K10)				

	J	K	L	M	N	O	P
3							
4			**ATM Simulation Logic**				
5	**Customer**	**Arrival**	**Begin Service**	**Service**	**Departure**	**Time in**	**Time in**
6	**Number**	**Time**	**Time**	**Time**	**Time**	**Queue**	**System**
7	**(1)**	**(2)**	**(3)**	**(4)**	**(5)=(3)+(4)**	**(6)=(3)-(2)**	**(7)=(5)-(2)**
8							
9	1	2.18	2.18	0.10	2.28	0.00	0.10
10	2	7.91	7.91	4.46	12.37	0.00	4.46
11	3	15.00	15.00	3.25	18.25	0.00	3.25
12	4	15.17	18.25	2.25	20.50	3.08	5.33
13	5	15.74	20.50	4.12	24.62	4.76	8.88

The Excel spreadsheet cell L10 (column L, row 10) in Figure 3.8 is the Begin Service Time for the second customer to arrive to the system and is programmed with IF(K10<N9,N9,K10). Since the second customer's arrival time (Excel cell K10) is not less than the first customer's departure time (Excel cell N9), the logical test evaluates to "false" and the second customer's time to begin service is set to his or her arrival time (Excel cell K10). The fourth customer shown in Figure 3.8 provides an example of when the logical test evaluates to "true," which results in the fourth customer beginning service when the third customer departs the ATM.

3.5.3 Simulation Replications and Output Analysis

The spreadsheet model makes a nice simulation of the first 25 customers to arrive to the ATM system. And we have a simulation output observation of 1.94 minutes that could be used as an estimate for the average time that the 25 customers waited in the queue (see the last value under the Time in Queue column, which represents the average of the 25 individual customer queue times). We also have a simulation output observation of 4.26 minutes that could be used as an estimate for the average time the 25 customers were in the system. These results represent only one possible value of each performance measure. Why? If at least one input variable to the simulation model is random, then the output of the simulation model will also be random. The interarrival times of customers to the ATM system and their service times are random variables. Thus the output of the simulation model of the ATM system is also a random variable. Before we bet that the average time customers spend in the system each day is 4.26 minutes, we may want to run the simulation model again with a new sample of interarrival times and service times to see what happens. This would be analogous to going to the real ATM on a second day to replicate our observing the first 25 customers processed to compute a second observation of the average time that the 25 customers spend in the system. In simulation, we can accomplish this by changing the values of the seed numbers used for our random number generator.

TABLE 3.3   **Summary of ATM System Simulation Output**

Replication	Average Time in Queue	Average Time in System
1	1.94 minutes	4.26 minutes
2	0.84 minutes	2.36 minutes
Average	1.39 minutes	3.31 minutes

Changing the seed values $Z1_0$ and $Z2_0$ causes the spreadsheet program to recompute all values in the spreadsheet. When we change the seed values $Z1_0$ and $Z2_0$ appropriately, we produce another replication of the simulation. When we run replications of the simulation, we are driving the simulation with a set of random numbers never used by the simulation before. This is analogous to the fact that the arrival pattern of customers to the real ATM and their transaction times at the ATM will not likely be the same from one day to the next. If $Z1_0 = 29$ and $Z2_0 = 92$ are used to start the random number generator for the ATM simulation model, a new replication of the simulation will be computed that produces an average time in queue of 0.84 minutes and an average time in system of 2.36 minutes. Review question number eight at the end of the chapter asks you to verify this second replication.

Table 3.3 contains the results from the two replications. Obviously, the results from this second replication are very different than those produced by the first replication. Good thing we did not make bets on how much time customers spend in the system based on the output of the simulation for the first replication only. Statistically speaking, we should get a better estimate of the average time customers spend in queue and the average time they are in the system if we combine the results from the two replications into an overall average. Doing so yields an estimate of 1.39 minutes for the average time in queue and an estimate of 3.31 minutes for the average time in system (see Table 3.3). However, the large variation in the output of the two simulation replications indicates that more replications are needed to get reasonably accurate estimates. How many replications are enough? You will know how to answer the question upon completing Chapter 9.

While spreadsheet technology is effective for conducting Monte Carlo simulations and simulations of simple dynamic systems, it is ineffective and inefficient as a simulation tool for complex systems. Discrete-event simulation software technology was designed especially for mimicking the behavior of complex systems and is the subject of Chapter 4.

3.6 Summary

Modeling random behavior begins with transforming the output produced by a random number generator into observations (random variates) from an appropriate statistical distribution. The values of the random variates are combined with logical operators in a computer program to compute output that mimics the performance behavior of stochastic systems. Performance estimates for stochastic simulations are obtained by calculating the average value of the performance metric across several replications of the simulation. Models can realistically simulate a variety of systems.

3.7 Review Questions

1. What is the difference between a stochastic model and a deterministic model in terms of the input variables and the way results are interpreted?

2. Give a statistical description of the numbers produced by a random number generator.

3. What are the two statistical properties that random numbers must satisfy?

4. Given these two LCGs:

$$Z_i = (9Z_{i-1} + 3) \bmod(32)$$
$$Z_i = (12Z_{i-1} + 5) \bmod(32)$$

   a. Which LCG will achieve its maximum cycle length? Answer the question without computing any Z_i values.

   b. Compute Z_1 through Z_5 from a seed of 29 ($Z_0 = 29$) for the second LCG.

5. What is a random variate, and how are random variates generated?

6. Apply the inverse transformation method to generate three variates from the following distributions using $U_1 = 0.10$, $U_2 = 0.53$, and $U_3 = 0.15$.

   a. Probability density function:

$$f(x) = \begin{cases} \dfrac{1}{\beta - \alpha} & \text{for } \alpha \leq x \leq \beta \\ 0 & \text{elsewhere} \end{cases}$$

   where $\beta = 7$ and $\alpha = 4$.

   b. Probability mass function:

$$p(x) = P(X = x) = \begin{cases} \dfrac{x}{15} & \text{for } x = 1, 2, 3, 4, 5 \\ 0 & \text{elsewhere} \end{cases}$$

7. How would a random number generator be used to simulate a 12 percent chance of rejecting a part because it is defective?

8. Reproduce the spreadsheet simulation of the ATM system presented in Section 3.5. Set the random numbers seeds $Z1_0 = 29$ and $Z2_0 = 92$ to compute the average time customers spend in the queue and in the system.

   a. Verify that the average time customers spend in the queue and in the system match the values given for the second replication in Table 3.3.

b. Verify that the resulting random number Stream 1 and random number Stream 2 are completely different than the corresponding streams in Table 3.2. Is this a requirement for a new replication of the simulation?

References

Banks, Jerry; John S. Carson II; Barry L. Nelson; and David M. Nicol. *Discrete-Event System Simulation.* Englewood Cliffs, NJ: Prentice Hall, 2001.

Hoover, Stewart V., and Ronald F. Perry. *Simulation: A Problem-Solving Approach.* Reading, MA: Addison-Wesley, 1989.

Johnson, R. A. *Miller and Freund's Probability and Statistics for Engineers.* 5th ed. Englewood Cliffs, NJ: Prentice Hall, 1994.

Law, Averill M., and David W. Kelton. *Simulation Modeling and Analysis.* New York: McGraw-Hill, 2000.

L'Ecuyer, P. "Random Number Generation." In *Handbook of Simulation: Principles, Methodology, Advances, Applications, and Practice,* ed. J. Banks, pp. 93–137. New York: John Wiley & Sons, 1998.

Pooch, Udo W., and James A. Wall. *Discrete Event Simulation: A Practical Approach.* Boca Raton, FL: CRC Press, 1993.

Pritsker, A. A. B. *Introduction to Simulation and SLAM II.* 4th ed. New York: John Wiley & Sons, 1995.

Ross, Sheldon M. *A Course in Simulation.* New York: Macmillan, 1990.

Shannon, Robert E. *System Simulation: The Art and Science.* Englewood Cliffs, NJ: Prentice Hall, 1975.

Thesen, Arne, and Laurel E. Travis. *Simulation for Decision Making.* Minneapolis, MN: West Publishing, 1992.

Widman, Lawrence E.; Kenneth A. Loparo; and Norman R. Nielsen. *Artificial Intelligence, Simulation, and Modeling.* New York: John Wiley & Sons, 1989.

4 DISCRETE-EVENT SIMULATION

"When the only tool you have is a hammer, every problem begins to resemble a nail."

—Abraham Maslow

4.1 Introduction

Building on the foundation provided by Chapter 3 on how random numbers and random variates are used to simulate stochastic systems, the focus of this chapter is on discrete-event simulation, which is the main topic of this book. A *discrete-event simulation* is one in which changes in the state of the simulation model occur at discrete points in time as triggered by events. The events in the automatic teller machine (ATM) simulation example of Chapter 3 that occur at discrete points in time are the arrivals of customers to the ATM queue and the completion of their transactions at the ATM. However, you will learn in this chapter that the spreadsheet simulation of the ATM system in Chapter 3 was based on only a few of the methodologies used in discrete-event simulation software like ProModel.

This chapter first defines what a discrete-event simulation is compared to a continuous simulation. Next the chapter summarizes the basic technical issues related to discrete-event simulation to facilitate your understanding of how to effectively use the tool. Questions that will be answered include these:

- How does discrete-event simulation work?
- What do commercial simulation software packages provide?
- What are the differences between simulation languages and simulators?
- What is the future of simulation technology?

A manual dynamic, stochastic, discrete-event simulation of the ATM example system from Chapter 3 is given to further illustrate what goes on inside this type of simulation.

4.2 Discrete-Event versus Continuous Simulation

Commonly, simulations are classified according to the following characteristics:

- Static or dynamic
- Stochastic or deterministic
- Discrete-event or continuous

In Chapter 3 we discussed the differences between the first two sets of characteristics. Now we direct our attention to discrete-event versus continuous simulations. A *discrete-event* simulation is one in which state changes occur at discrete points in time as triggered by events. Typical simulation events might include

- The arrival of an entity to a workstation.
- The failure of a resource.
- The completion of an activity.
- The end of a shift.

State changes in a model occur when some event happens. The state of the model becomes the collective state of all the elements in the model at a particular point in time. State variables in a discrete-event simulation are referred to as *discrete-change* state variables. A restaurant simulation is an example of a discrete-event simulation because all of the state variables in the model, such as the number of customers in the restaurant, are discrete-change state variables (see Figure 4.1). Most manufacturing and service systems are typically modeled using discrete-event simulation.

In *continuous* simulation, state variables change continuously with respect to time and are therefore referred to as *continuous-change* state variables. An example of a continuous-change state variable is the level of oil in an oil tanker that is being either loaded or unloaded, or the temperature of a building that is controlled by a heating and cooling system. Figure 4.2 compares a discrete-change state variable and a continuous-change state variable as they vary over time.

FIGURE 4.1

Discrete events cause discrete state changes.

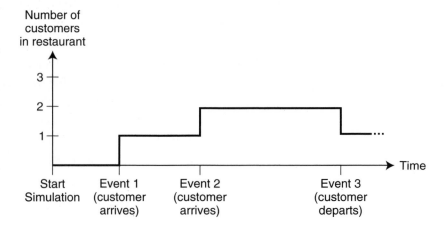

FIGURE 4.2

*Comparison of a
discrete-change state
variable and a
continuous-change
state variable.*

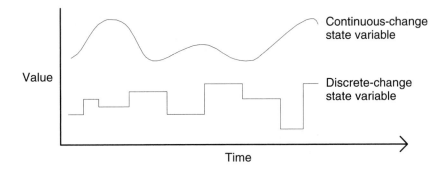

Continuous simulation products use either differential equations or difference equations to define the rates of change in state variables over time.

4.2.1 Differential Equations

The change that occurs in some continuous-change state variables is expressed in terms of the derivatives of the state variables. Equations involving derivatives of a state variable are referred to as *differential equations*. The state variable v, for example, might change as a function of both v and time t:

$$\frac{dv(t)}{dt} = v^2(t) + t^2$$

We then need a second equation to define the initial condition of v:

$$v(0) = K$$

On a computer, numerical integration is used to calculate the change in a particular response variable over time. Numerical integration is performed at the end of successive small time increments referred to as *steps*. Numerical analysis techniques, such as Runge-Kutta integration, are used to integrate the differential equations numerically for each incremental time step. One or more threshold values for each continuous-change state variable are usually defined that determine when some action is to be triggered, such as shutting off a valve or turning on a pump.

4.2.2 Difference Equations

Sometimes a continuous-change state variable can be modeled using difference equations. In such instances, the time is decomposed into periods of length t. An algebraic expression is then used to calculate the value of the state variable at the end of period $k + 1$ based on the value of the state variable at the end of period k. For example, the following difference equation might be used to express the rate of change in the state variable v as a function of the current value of v, a rate of change (r), and the length of the time period (Δt):

$$v(k + 1) = v(k) + r\,\Delta t$$

Batch processing in which fluids are pumped into and out of tanks can often be modeled using difference equations.

4.2.3 Combined Continuous and Discrete Simulation

Many simulation software products provide both discrete-event and continuous simulation capabilities. This enables systems that have both discrete-event and continuous characteristics to be modeled, resulting in a hybrid simulation. Most processing systems that have continuous-change state variables also have discrete-change state variables. For example, a truck or tanker arrives at a fill station (a discrete event) and begins filling a tank (a continuous process).

Four basic interactions occur between discrete- and continuous-change variables:

1. A continuous variable value may suddenly increase or decrease as the result of a discrete event (like the replenishment of inventory in an inventory model).
2. The initiation of a discrete event may occur as the result of reaching a threshold value in a continuous variable (like reaching a reorder point in an inventory model).
3. The change rate of a continuous variable may be altered as the result of a discrete event (a change in inventory usage rate as the result of a sudden change in demand).
4. An initiation or interruption of change in a continuous variable may occur as the result of a discrete event (the replenishment or depletion of inventory initiates or terminates a continuous change of the continuous variable).

4.3 How Discrete-Event Simulation Works

Most simulation software, including ProModel, presents a process-oriented world view to the user for defining models. This is the way most humans tend to think about systems that process entities. When describing a system, it is natural to do so in terms of the process flow. Entities begin processing at activity A then move on to activity B and so on. In discrete-event simulation, these process flow definitions are translated into a sequence of events for running the simulation: first event 1 happens (an entity begins processing at activity A), then event 2 occurs (it completes processing at activity A), and so on. Events in simulation are of two types: scheduled and conditional, both of which create the activity delays in the simulation to replicate the passage of time.

A *scheduled event* is one whose time of occurrence can be determined beforehand and can therefore be scheduled in advance. Assume, for example, that an activity has just begun that will take X amount of time, where X is a normally distributed random variable with a mean of 5 minutes and a standard deviation of

1.2 minutes. At the start of the activity, a normal random variate is generated based on these parameters, say 4.2 minutes, and an activity completion event is scheduled for that time into the future. Scheduled events are inserted chronologically into an event calendar to await the time of their occurrence. Events that occur at predefined intervals theoretically all could be determined in advance and therefore be scheduled at the beginning of the simulation. For example, entities arriving every five minutes into the model could all be scheduled easily at the start of the simulation. Rather than preschedule all events at once that occur at a set frequency, however, they are scheduled only when the next occurrence must be determined. In the case of a periodic arrival, the next arrival would not be scheduled until the current scheduled arrival is actually pulled from the event calendar for processing. This postponement until the latest possible moment minimizes the size of the event calendar and eliminates the necessity of knowing in advance how many events to schedule when the length of the simulation may be unknown.

Conditional events are triggered by a condition being met rather than by the passage of time. An example of a conditional event might be the capturing of a resource that is predicated on the resource being available. Another example would be an order waiting for all of the individual items making up the order to be assembled. In these situations, the event time cannot be known beforehand, so the pending event is simply placed into a waiting list until the conditions can be satisfied. Often multiple pending events in a list are waiting for the same condition. For example, multiple entities might be waiting to use the same resource when it becomes available. Internally, the resource would have a waiting list for all items currently waiting to use it. While in most cases events in a waiting list are processed first-in, first-out (FIFO), items could be inserted and removed using a number of different criteria. For example, items may be inserted according to item priority but be removed according to earliest due date.

Events, whether scheduled or conditional, trigger the execution of logic that is associated with that event. For example, when an entity frees a resource, the state and statistical variables for the resource are updated, the graphical animation is updated, and the input waiting list for the resource is examined to see what activity to respond to next. Any new events resulting from the processing of the current event are inserted into either the event calendar or another appropriate waiting list.

In real life, events can occur simultaneously so that multiple entities can be doing things at the same instant in time. In computer simulation, however, especially when running on a single processor, events can be processed only one at a time even though it is the same instant in simulated time. As a consequence, a method or rule must be established for processing events that occur at the exact same simulated time. For some special cases, the order in which events are processed at the current simulation time might be significant. For example, an entity that frees a resource and then tries to immediately get the same resource might have an unfair advantage over other entities that might have been waiting for that particular resource.

In ProModel, the entity, downtime, or other item that is currently being processed is allowed to continue processing as far as it can at the current simulation time. That means it continues processing until it reaches either a conditional

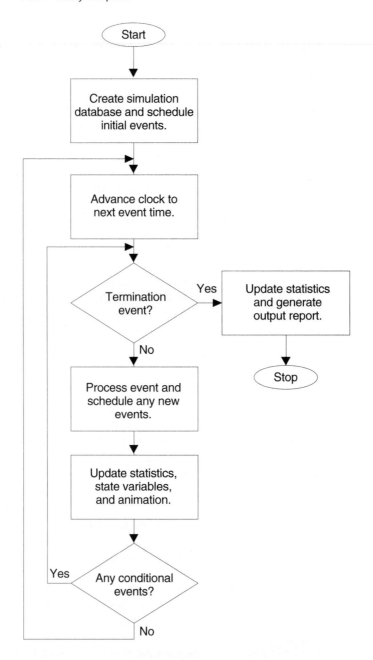

event that cannot be satisfied or a timed delay that causes a future event to be scheduled. It is also possible that the object simply finishes all of the processing defined for it and, in the case of an entity, exits the system. As an object is being processed, any resources that are freed or other entities that might have been created as byproducts are placed in an action list and are processed one at a time in a similar fashion after the current object reaches a stopping point. To deliberately suspend the current object in order to allow items in the action list to be processed, a zero delay time can be specified for the current object. This puts the current item into the future events list (event calendar) for later processing, even though it is still processed at the current simulation time.

When all scheduled and conditional events have been processed that are possible at the current simulation time, the clock advances to the next scheduled event and the process continues. When a termination event occurs, the simulation ends and statistical reports are generated. The ongoing cycle of processing scheduled and conditional events, updating state and statistical variables, and creating new events constitutes the essence of discrete-event simulation (see Figure 4.3).

4.4 A Manual Discrete-Event Simulation Example

To illustrate how discrete-event simulation works, a manual simulation is presented of the automatic teller machine (ATM) system of Chapter 3. Customers arrive to use an automatic teller machine (ATM) at a mean interarrival time of 3.0 minutes exponentially distributed. When customers arrive to the system, they join a queue to wait for their turn on the ATM. The queue has the capacity to hold an infinite number of customers. The ATM itself has a capacity of one (only one customer at a time can be processed at the ATM). Customers spend an average of 2.4 minutes exponentially distributed at the ATM to complete their transactions, which is called the service time at the ATM. Assuming that the simulation starts at time zero, simulate the ATM system for its first 22 minutes of operation and estimate the expected waiting time for customers in the queue (the average time customers wait in line for the ATM) and the expected number of customers waiting in the queue (the average number of customers waiting in the queue during the simulated time period). If you are wondering why 22 minutes was selected as the simulation run length, it is because 22 minutes of manual simulation will nicely fit on one page of the textbook. An entity flow diagram of the ATM system is shown in Figure 4.4.

4.4.1 Simulation Model Assumptions

We did not list our assumptions for simulating the ATM system back in Chapter 3 although it would have been a good idea to have done so. In any simulation,

FIGURE 4.4

Entity flow diagram for example automatic teller machine (ATM) system.

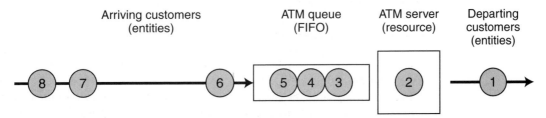

certain assumptions must be made where information is not clear or complete. Therefore, it is important that assumptions be clearly documented. The assumptions we will be making for this simulation are as follows:

- There are no customers in the system initially, so the queue is empty and the ATM is idle.
- The move time from the queue to the ATM is negligible and therefore ignored.
- Customers are processed from the queue on a first-in, first-out (FIFO) basis.
- The ATM never experiences failures.

4.4.2 *Setting Up the Simulation*

A discrete-event simulation does not generate all of the events in advance as was done in the spreadsheet simulation of Chapter 3. Instead the simulation schedules the arrival of customer entities to the ATM queue and the departure of customer entities from the ATM as they occur over time. That is, the simulation calls a function to obtain an exponentially distributed interarrival time value only when a new customer entity needs to be scheduled to arrive to the ATM queue. Likewise, the simulation calls a function to obtain an exponentially distributed service time value at the ATM only after a customer entity has gained access (entry) to the ATM and the entity's future departure time from the ATM is needed. An *event calendar* is maintained to coordinate the processing of the customer arrival and customer departure events. As arrival and departure event times are created, they are placed on the event calendar in chronological order. As these events get processed, state variables (like the contents of a queue) get changed and statistical accumulators (like the cumulative time in the queue) are updated.

With this overview, let's list the structured components needed to conduct the manual simulation.

Simulation Clock

As the simulation transitions from one discrete-event to the next, the simulation clock is fast forwarded to the time that the next event is scheduled to occur. There is no need for the clock to tick seconds away until reaching the time at which the next event in the list is scheduled to occur because nothing will happen that changes the state of the system until the next event occurs. Instead, the simulation clock advances through a series of time steps. Let t_i denote the value of the simulation clock at time step i, for $i = 0$ to the number of discrete events to process. Assuming that the simulation starts at time zero, then the initial value of the simulation clock is denoted as $t_0 = 0$. Using this nomenclature, t_1 denotes the value of the simulation clock when the first discrete event in the list is processed, t_2 denotes the value of the simulation clock when the second discrete-event in the list is processed, and so on.

Entity Attributes

To capture some statistics about the entities being processed through the system, a discrete-event simulation maintains an array of entity attribute values. *Entity attributes* are characteristics of the entity that are maintained for that entity until the entity exits the system. For example, to compute the amount of time an entity waited in a queue location, an attribute is needed to remember when the entity entered the location. For the ATM simulation, one entity attribute is used to remember the customer's time of arrival to the system. This entity attribute is called the Arrival Time attribute. The simulation program computes how long each customer entity waited in the queue by subtracting the time that the customer entity arrived to the queue from the value of the simulation clock when the customer entity gained access to the ATM.

State Variables
Two discrete-change state variables are needed to track how the status (state) of the system changes as customer entities arrive in and depart from the ATM system.

- Number of Entities in Queue at time step i, NQ_i.
- ATM Status$_i$ to denote if the ATM is busy or idle at time step i.

Statistical Accumulators
The objective of the example manual simulation is to estimate the expected amount of time customers wait in the queue and the expected number of customers waiting in the queue. The average time customers wait in queue is a *simple average*. Computing this requires that we record how many customers passed through the queue and the amount of time each customer waited in the queue. The average number of customers in the queue is a time-weighted average, which is usually called a *time average* in simulation. Computing this requires that we not only observe the queue's contents during the simulation but that we also measure the amount of time that the queue maintained each of the observed values. We record each observed value after it has been multiplied (weighted) by the amount of time it was maintained.

Here's what the simulation needs to tally at each simulation time step i to compute the two performance measures at the end of the simulation.

Simple-average time in queue.
- Record the number of customer entities processed through the queue, Total Processed. Note that the simulation may end before all customer entities in the queue get a turn at the ATM. This accumulator keeps track of how many customers actually made it through the queue.
- For a customer processed through the queue, record the time that it waited in the queue. This is computed by subtracting the value of the simulation clock time when the entity enters the queue (stored in the entity attribute array Arrival Time) from the value of the simulation clock time when the entity leaves the queue, $t_i -$ Arrival Time.

Time-average number of customers in the queue.
- For the duration of the last time step, which is $t_i - t_{i-1}$, and the number of customer entities in the queue during the last time step, which is NQ_{i-1}, record the product of $t_i - t_{i-1}$ and NQ_{i-1}. Call the product $(t_i - t_{i-1})NQ_{i-1}$ the Time-Weighted Number of Entities in the Queue.

Events

There are two types of recurring scheduled events that change the state of the system: arrival events and departure events. An arrival event occurs when a customer entity arrives to the queue. A departure event occurs when a customer entity completes its transaction at the ATM. Each processing of a customer entity's arrival to the queue includes scheduling the future arrival of the next customer entity to the ATM queue. Each time an entity gains access to the ATM, its future departure from the system is scheduled based on its expected service time at the ATM. We actually need a third event to end the simulation. This event is usually called the termination event.

To schedule the time at which the next entity arrives to the system, the simulation needs to generate an interarrival time and add it to the current simulation clock time, t_i. The interarrival time is exponentially distributed with a mean of 3.0 minutes for our example ATM system. Assume that the function $E(3.0)$ returns an exponentially distributed random variate with a mean of 3.0 minutes. The future arrival time of the next customer entity can then be scheduled by using the equation $t_i + E(3.0)$.

The customer service time at the ATM is exponentially distributed with a mean of 2.4 minutes. The future departure time of an entity gaining access to the ATM is scheduled by the equation $t_i + E(2.4)$.

Event Calendar

The event calendar maintains the list of active events (events that have been scheduled and are waiting to be processed) in chronological order. The simulation progresses by removing the first event listed on the event calendar, setting the simulation clock, t_i, equal to the time at which the event is scheduled to occur, and processing the event.

4.4.3 Running the Simulation

To run the manual simulation, we need a way to generate exponentially distributed random variates for interarrival times [the function $E(3.0)$] and service times [the function $E(2.4)$]. Rather than using our calculators to generate a random number and transform it into an exponentially distributed random variate each time one is needed for the simulation, let's take advantage of the work we did to build the spreadsheet simulation of the ATM system in Chapter 3. In Table 3.2 of Chapter 3, we generated 25 exponentially distributed interarrival times with a mean of 3.0 minutes and 25 exponentially distributed service times with a mean of 2.4 minutes in advance of running the spreadsheet simulation. So when we need a service time or an interarrival time for our manual simulation, let's use the values from the Service Time and Interarrival Time columns of Table 3.2 rather than computing them by hand. Note, however, that we do not need to generate all random variates in advance for our manual discrete-event simulation (that's one of its advantages over spreadsheet simulation). We are just using Table 3.2 to make our manual simulation a little less tedious, and it will be instructive to compare the results of the manual simulation with those produced by the spread-sheet simulation.

Notice that Table 3.2 contains a subscript i in the leftmost column. This subscript denotes the customer entity number as opposed to the simulation time step. We wanted to point this out to avoid any confusion because of the different uses of the subscript. In fact, you can ignore the subscript in Table 3.2 as you pick values from the Service Time and Interarrival Time columns.

A discrete-event simulation logic diagram for the ATM system is shown in Figure 4.5 to help us carry out the manual simulation. Table 4.1 presents the results of the manual simulation after processing 12 events using the simulation logic diagram presented in Figure 4.5. The table tracks the creation and scheduling of events on the event calendar as well as how the state of the system changes and how the values of the statistical accumulators change as events are processed from the event calendar. Although Table 4.1 is completely filled in, it was initially blank until the instructions presented in the simulation logic diagram were executed. As you work through the simulation logic diagram, you should process the information in Table 4.1 from the first row down to the last row, one row at a time (completely filling in a row before going down to the next row). A dash (—) in a cell in Table 4.1 signifies that the simulation logic diagram does not require you to update that particular cell at the current simulation time step. An arrow (↑) in a cell in the table also signifies that the simulation logic diagram does not require you to update that cell at the current time step. However, the arrows serve as a reminder to look up one or more rows above your current position in the table to determine the state of the ATM system. Arrows appear under the Number of Entities in Queue, NQ_i column, and ATM Status$_i$ column. The only exception to the use of dashes or arrows is that we keep a running total in the two Cumulative sub-columns in the table for each time step. Let's get the manual simulation started.

> $i = 0, t_0 = 0.$ As shown in Figure 4.5, the first block after the start position indicates that the model is initialized to its starting conditions. The simulation time step begins at $i = 0$. The initial value of the simulation clock is zero, $t_0 = 0$. The system state variables are set to ATM Status$_0 =$ "Idle"; Number of Entities in Queue, $NQ_0 = 0$; and the Entity Attribute Array is cleared. This reflects the initial conditions of no customer entities in the queue and an idle ATM. The statistical accumulator Total Processed is set to zero. There are two different Cumulative variables in Table 4.1: one to accumulate the time in queue values of t_i − Arrival Time, and the other to accumulate the values of the time-weighted number of entities in the queue, $(t_i - t_{i-1})NQ_{i-1}$. Recall that t_i − Arrival Time is the amount of time that entities, which gained access to the ATM, waited in queue. Both Cumulative variables $(t_i-$ Arrival Time) and $(t_i - t_{i-1})NQ_{i-1}$ are initialized to zero. Next, an initial arrival event and termination event are scheduled and placed under the Scheduled Future Events column. The listing of an event is formatted as "(Entity Number, Event, and Event Time)". Entity Number denotes the customer number that the event pertains to (such as the first, second, or third customer). Event is the type of event: a customer arrives, a

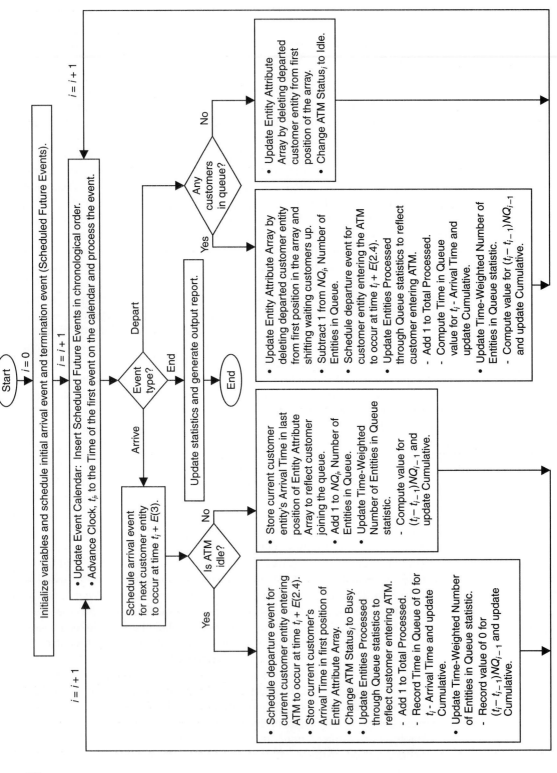

FIGURE 4.5

Discrete-event simulation logic diagram for ATM system.

TABLE 4.1 Manual Discrete-Event Simulation of ATM System

i	Event Calendar (Entity Number, Event, Time)	Processed Event: Clock t_i	Processed Event: Entity Number	Processed Event: Event	Entity Attribute Array (Entity Number, Arrival Time) *Entity Using ATM, array position 1 / Entities Waiting in Queue, array positions 2, 3, …	Number of Entities in Queue, NQ_i	ATM Status$_i$	Total Processed	Time in Queue, $t_i -$ Arrival Time	Cumulative, $\sum (t_i -$ Arrival Time$)$	$(t_i - t_{i-1})NQ_{i-1}$	Cumulative, $\sum (t_i - t_{i-1})NQ_{i-1}$	Scheduled Future Events (Entity Number, Event, Time)
0	—	0	—	—	*( ) / ( )	0	Idle	0	—	0	—	0	(1, Arrive, 2.18) (_, End, 22.00)
1	(1, Arrive, 2.18) (_, End, 22.00)	2.18	1	Arrive	*(1, 2.18) / ( )	←	Busy	1	0	0	0	0	(2, Arrival, 7.91) (1, Depart, 2.28)
2	(1, Depart, 2.28) (2, Arrive, 7.91) (_, End, 22.00)	2.28	1	Depart	*( ) / ( )	←	Idle	—	—	0	—	0	No new events
3	(2, Arrive, 7.91) (_, End, 22.00)	7.91	2	Arrive	*(2, 7.91) / ( )	←	Busy	2	0	0	0	0	(3, Arrive, 15.00) (2, Depart, 12.37)
4	(2, Depart, 12.37) (3, Arrive, 15.00) (_, End, 22.00)	12.37	2	Depart	*( ) / ( )	←	Idle	—	—	0	—	0	No new events
5	(3, Arrive, 15.00) (_, End, 22.00)	15.00	3	Arrive	*(3, 15.00) / ( )	←	Busy	3	0	0	0	0	(4, Arrive, 15.17) (3, Depart, 18.25)
6	(4, Arrive, 15.17) (3, Depart, 18.25) (_, End, 22.00)	15.17	4	Arrive	*(3, 15.00) / (4, 15.17)	1	←	—	—	0	0	0	(5, Arrive, 15.74)
7	(5, Arrive, 15.74) (3, Depart, 18.25) (_, End, 22.00)	15.74	5	Arrive	*(3, 15.00) / (4, 15.17) (5, 15.74)	2	←	—	—	0	0.57	0.57	(6, Arrive, 18.75)
8	(3, Depart, 18.25) (6, Arrive, 18.75) (_, End, 22.00)	18.25	3	Depart	*(4, 15.17) / (5, 15.74)	1	←	4	3.08	3.08	5.02	5.59	(4, Depart, 20.50)
9	(6, Arrive, 18.75) (4, Depart, 20.50) (_, End, 22.00)	18.75	6	Arrive	*(4, 15.17) / (5, 15.74) (6, 18.75)	2	←	—	—	3.08	0.50	6.09	(7, Arrive, 19.88)
10	(7, Arrive, 19.88) (4, Depart, 20.50) (_, End, 22.00)	19.88	7	Arrive	*(4, 15.17) / (5, 15.74) (6, 18.75) (7, 19.88)	3	←	—	—	3.08	2.26	8.35	(8, Arrive, 22.53)
11	(4, Depart, 20.50) (_, End, 22.00) (8, Arrive, 22.53)	20.50	4	Depart	*(5, 15.74) / (6, 18.75) (7, 19.88)	2	←	5	4.76	7.84	1.86	10.21	(5, Depart, 24.62)
12	(_, End, 22.00) (8, Arrive, 22.53) (5, Depart, 24.62)	22.00		End	—	←	←	5	—	7.84	3.00	13.21	—

customer departs, or the simulation ends. Time is the future time that the event is to occur. The event "(1, Arrive, 2.18)" under the Scheduled Future Events column prescribes that the first customer entity is scheduled to arrive at time 2.18 minutes. The arrival time was generated using the equation $t_0 + E(3.0)$. To obtain the value returned from the function $E(3)$, we went to Table 3.2, read the first random variate from the Interarrival Time column (a value of 2.18 minutes), and added it to the current value of the simulation clock, $t_0 = 0$. The simulation is to be terminated after 22 minutes. Note the "(__, End, 22.00)" under the Scheduled Future Events column. For the termination event, no value is assigned to Entity Number because it is not relevant.

$i = 1, t_1 = 2.18$. After the initialization step, the list of scheduled future events is added to the event calendar in chronological order in preparation for the next simulation time step $i = 1$. The simulation clock is fast forwarded to the time that the next event is scheduled to occur, which is $t_1 = 2.18$ (the arrival time of the first customer to the ATM queue), and then the event is processed. Following the simulation logic diagram, arrival events are processed by first scheduling the future arrival event for the next customer entity using the equation $t_1 + E(3.0) = 2.18 + 5.73 = 7.91$ minutes. Note the value of 5.73 returned by the function $E(3.0)$ is the second random variate listed under the Interarrival Time column of Table 3.2. This future event is placed under the Scheduled Future Events column in Table 4.1 as "(2, Arrive, 7.91)". Checking the status of the ATM from the previous simulation time step reveals that the ATM is idle (ATM Status$_0$ = "Idle"). Therefore, the arriving customer entity immediately flows through the queue to the ATM to conduct its transaction. The future departure event of this entity from the ATM is scheduled using the equation $t_1 + E(2.4) = 2.18 + 0.10 = 2.28$ minutes. See "(1, Depart, 2.28)" under the Scheduled Future Events column, denoting that the first customer entity is scheduled to depart the ATM at time 2.28 minutes. Note that the value of 0.10 returned by the function $E(2.4)$ is the first random variate listed under the Service Time column of Table 3.2. The arriving customer entity's arrival time is then stored in the first position of the Entity Attribute Array to signify that it is being served by the ATM. The ATM Status$_1$ is set to "Busy," and the statistical accumulators for Entities Processed through Queue are updated. Add 1 to Total Processed and since this entity entered the queue and immediately advanced to the idle ATM for processing, record zero minutes in the Time in Queue, $t_1 -$ Arrival Time, subcolumn and update this statistic's cumulative value. The statistical accumulators for Time-Weighted Number of Entities in the Queue are updated next. Record zero for $(t_1 - t_0)NQ_0$ since there were no entities in queue during the previous time step, $NQ_0 = 0$, and update this statistic's cumulative value. Note the arrow "↑" entered under the Number of Entities in Queue, NQ_1 column. Recall that the arrow is placed there to signify that the number of entities waiting in the queue has not changed from its previous value.

$i = 2, t_2 = 2.28.$ Following the loop back around to the top of the simulation logic diagram, we place the two new future events onto the event calendar in chronological order in preparation for the next simulation time step $i = 2$. The simulation clock is fast forwarded to $t_2 = 2.28$, and the departure event for the first customer entity arriving to the system is processed. Given that there are no customers in the queue from the previous time step, $NQ_1 = 0$ (follow the arrows up to get this value), we simply remove the departed customer from the first position of the Entity Attribute Array and change the status of the ATM to idle, ATM Status$_2$ = "Idle". The statistical accumulators do not require updating because there are no customer entities waiting in the queue or leaving the queue. The dashes (—) entered under the statistical accumulator columns indicate that updates are not required. No new future events are scheduled.

As before, we follow the loop back to the top of the simulation logic diagram, and place any new events, of which there are none at the end of time step $i = 2$, onto the event calendar in chronological order in preparation for the next simulation time step $i = 3$. The simulation clock is fast forwarded to $t_3 = 7.91$, and the arrival of the second customer entity to the ATM queue is processed. Complete the processing of this event and continue the manual simulation until the termination event "(__, End, 22.00)" reaches the top of the event calendar.

As you work through the simulation logic diagram to complete the manual simulation, you will see that the fourth customer arriving to the system requires that you use logic from a different path in the diagram. When the fourth customer entity arrives to the ATM queue, simulation time step $i = 6$, the ATM is busy (see ATM Status$_5$) processing customer entity 3's transaction. Therefore, the fourth customer entity joins the queue and waits to use the ATM. (Don't forget that it invoked the creation of the fifth customer's arrival event.) The fourth entity's arrival time of 15.17 minutes is stored in the last position of the Entity Attribute Array in keeping with the first-in, first-out (FIFO) rule. The Number of Entities in Queue, NQ_6, is incremented to 1. Further, the Time-Weighted Number of Entities in the Queue statistical accumulators are updated by first computing $(t_6 - t_5)NQ_5 = (15.17 - 15.00)0 = 0$ and then recording the result. Next, this statistic's cumulative value is updated. Customers 5, 6, and 7 also arrive to the system finding the ATM busy and therefore take their place in the queue to wait for the ATM.

The fourth customer waited a total of 3.08 minutes in the queue (see the t_i − Arrival Time subcolumn) before it gained access to the ATM in time step $i = 8$ as the third customer departed. The value of 3.08 minutes in the queue for the fourth customer computed in time step $i = 8$ by t_8 − Arrival Time = $18.25 - 15.17 = 3.08$ minutes. Note that Arrival Time is the time that the fourth customer arrived to the queue, and that the value is stored in the Entity Attribute Array.

At time $t_{12} = 22.00$ minutes the simulation terminates and tallies the final values for the statistical accumulators, indicating that a total of five customer entities were processed through the queue. The total amount of time that these five customers waited in the queue is 7.84 minutes. The final cumulative value for Time-Weighted Number of Entities in the Queue is 13.21 minutes. Note that at the end of the simulation, two customers are in the queue (customers 6 and 7) and one is at the ATM (customer 5). A few quick observations are worth considering before we discuss how the accumulated values are used to calculate summary statistics for a simulation output report.

This simple and brief (while tedious) manual simulation is relatively easy to follow. But imagine a system with dozens of processes and dozens of factors influencing behavior such as downtimes, mixed routings, resource contention, and others. You can see how essential computers are for performing a simulation of any magnitude. Computers have no difficulty tracking the many relationships and updating the numerous statistics that are present in most simulations. Equally as important, computers are not error prone and can perform millions of instructions per second with absolute accuracy. We also want to point out that the simulation logic diagram (Figure 4.5) and Table 4.1 were designed to convey the essence of what happens inside a discrete-event simulation program. When you view a trace report of a ProModel simulation in Lab Chapter 8 you will see the similarities between the trace report and Table 4.1. Although the basic process presented is sound, its efficiency could be improved. For example, there is no need to keep both a "scheduled future events" list and an "event calendar." Instead, future events are inserted directly onto the event calendar as they are created. We separated them to facilitate our describing the flow of information in the discrete-event framework.

4.4.4 *Calculating Results*

When a simulation terminates, statistics that summarize the model's behavior are calculated and displayed in an output report. Many of the statistics reported in a simulation output report consist of average values, although most software reports a multitude of statistics including the maximum and minimum values observed during the simulation. Average values may be either simple averages or time averages.

Simple-Average Statistic
A simple-average statistic is calculated by dividing the sum of all observation values of a response variable by the number of observations:

$$\text{Simple average} = \frac{\sum_{i=1}^{n} x_i}{n}$$

where n is the number of observations and x_i is the value of ith observation. Example simple-average statistics include the average time entities spent in the system (from system entry to system exit), the average time entities spent at a specific location, or the average time per use for a resource. An average of an observation-based response variable in ProModel is computed as a simple average.

The average time that customer entities waited in the queue for their turn on the ATM during the manual simulation reported in Table 4.1 is a simple-average statistic. Recall that the simulation processed five customers through the queue. Let x_i denote the amount of time that the ith customer processed spent in the queue. The average waiting time in queue based on the $n = 5$ observations is

$$\text{Average time in queue} = \frac{\sum_{i=1}^{5} x_i}{5} = \frac{0 + 0 + 0 + 3.08 + 4.76}{5}$$

$$= \frac{7.84 \text{ minutes}}{5} = 1.57 \text{ minutes}$$

The values necessary for computing this average are accumulated under the Entities Processed through Queue columns of the manual simulation table (see the last row of Table 4.1 for the cumulative value $\sum(t_i - \text{Arrival Time}) = 7.84$ and Total Processed $= 5$).

Time-Average Statistic

A time-average statistic, sometimes called a time-weighted average, reports the average value of a response variable weighted by the time duration for each observed value of the variable:

$$\text{Time average} = \frac{\sum_{i=1}^{n} (T_i x_i)}{T}$$

where x_i denotes the value of the ith observation, T_i denotes the time duration of the ith observation (the weighting factor), and T denotes the total duration over which the observations were collected. Example time-average statistics include the average number of entities in a system, the average number of entities at a location, and the average utilization of a resource. An average of a time-weighted response variable in ProModel is computed as a time average.

The average number of customer entities waiting in the queue location for their turn on the ATM during the manual simulation is a time-average statistic. Figure 4.6 is a plot of the number of customer entities in the queue during the manual simulation recorded in Table 4.1. The 12 discrete-events manually simulated in Table 4.1 are labeled $t_1, t_2, t_3, \ldots, t_{11}, t_{12}$ on the plot. Recall that t_i denotes the value of the simulation clock at time step i in Table 4.1, and that its initial value is zero, $t_0 = 0$.

Using the notation from the time-average equation just given, the total simulation time illustrated in Figure 4.6 is $T = 22$ minutes. The T_i denotes the duration of time step i (distance between adjacent discrete-events in Figure 4.6). That is, $T_i = t_i - t_{i-1}$ for $i = 1, 2, 3, \ldots, 12$. The x_i denotes the queue's contents (number of customer entities in the queue) during each T_i time interval. Therefore, $x_i = NQ_{i-1}$ for $i = 1, 2, 3, \ldots, 12$ (recall that in Table 4.1, NQ_{i-1} denotes the number of customer entities in the queue from t_{i-1} to t_i). The time-average

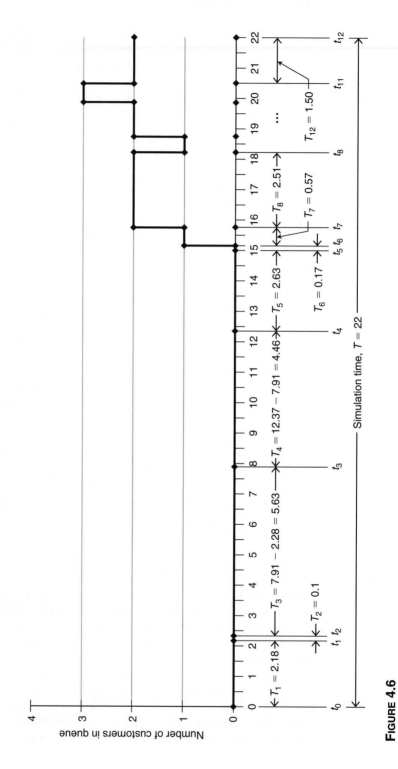

FIGURE 4.6
Number of customers in the queue during the manual simulation.

number of customer entities in the queue (let's call it Average *NQ*) is

$$\text{Average } NQ = \frac{\sum_{i=1}^{12}(T_i x_i)}{T} = \frac{\sum_{i=1}^{12}(t_i - t_{i-1})NQ_{i-1}}{T}$$

$$\text{Average } NQ = \frac{(2.18)(0) + (0.1)(0) + (5.63)(0) + (4.46)(0) + (2.63)(0) + (0.17)(0) + (0.57)(1) + (2.51)(2) + \cdots + (1.5)(2)}{22}$$

$$\text{Average } NQ = \frac{13.21}{22} = 0.60 \text{ customers}$$

You may recognize that the numerator of this equation $(\sum_{i=1}^{12}(t_i - t_{i-1})NQ_{i-1})$ calculates the area under the plot of the queue's contents during the simulation (Figure 4.6). The values necessary for computing this area are accumulated under the Time-Weighted Number of Entities in Queue column of Table 4.1 (see the Cumulative value of 13.21 in the table's last row).

4.4.5 Issues

Even though this example is a simple and somewhat crude simulation, it provides a good illustration of basic simulation issues that need to be addressed when conducting a simulation study. *First,* note that the simulation start-up conditions can bias the output statistics. Since the system started out empty, queue content statistics are slightly less than what they might be if we began the simulation with customers already in the system. *Second,* note that we ran the simulation for only 22 minutes before calculating the results. Had we ran longer, it is very likely that the long-run average time in the queue would have been somewhat different (most likely greater) than the time from the short run because the simulation did not have a chance to reach a steady state.

These are the kinds of issues that should be addressed whenever running a simulation. The modeler must carefully analyze the output and understand the significance of the results that are given. This example also points to the need for considering beforehand just how long a simulation should be run. These issues are addressed in Chapters 9 and 10.

4.5 Commercial Simulation Software

While a simulation model may be programmed using any development language such as C++ or Java, most models are built using commercial simulation software such as ProModel. Commercial simulation software consists of several modules for performing different functions during simulation modeling. Typical modules are shown in Figure 4.7.

4.5.1 Modeling Interface Module

A modeler defines a model using an input or modeling interface module. This module provides graphical tools, dialogs, and other text editing capabilities for

FIGURE 4.7

Typical components of simulation software.

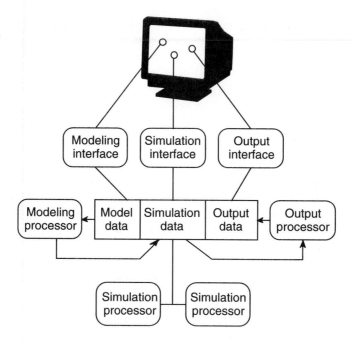

entering and editing model information. External files used in the simulation are specified here as well as run-time options (number of replications and so on).

4.5.2 Model Processor

When a completed model is run, the model processor takes the model database, and any other external data files that are used as input data, and creates a simulation database. This involves data translation and possibly language compilation. This data conversion is performed because the model database and external files are generally in a format that cannot be used efficiently by the simulation processor during the simulation. Some simulation languages are *interpretive,* meaning that the simulation works with the input data the way they were entered. This allows simulations to start running immediately without going through a translator, but it slows down the speed at which the simulation executes.

In addition to translating the model input data, other data elements are created for use during simulation, including statistical counters and state variables. Statistical counters are used to log statistical information during the simulation such as the cumulative number of entries into the system or the cumulative activity time at a workstation.

4.5.3 Simulation Interface Module

The simulation interface module displays the animation that occurs during the simulation run. It also permits the user to interact with the simulation to control the current animation speed, trace events, debug logic, query state variables, request snapshot reports, pan or zoom the layout, and so forth. If visual interactive capability is provided, the user is even permitted to make changes dynamically to model variables with immediate visual feedback of the effects of such changes.

The animation speed can be adjusted and animation can even be disabled by the user during the simulation. When unconstrained, a simulation is capable of running as fast as the computer can process all of the events that occur within the simulated time. The simulation clock advances instantly to each scheduled event; the only central processing unit (CPU) time of the computer that is used is what is necessary for processing the event logic. This is how simulation is able to run in compressed time. It is also the reason why large models with millions of events take so long to simulate. Ironically, in real life activities take time while events take no time. In simulation, events take time while activities take no time. To slow down a simulation, delay loops or system timers are used to create pauses between events. These techniques give the appearance of elapsing time in an animation. In some applications, it may even be desirable to run a simulation at the same rate as a real clock. These real-time simulations are achieved by synchronizing the simulation clock with the computer's internal system clock. Human-in-the-loop (such as operator training simulators) and hardware-in-the-loop (testing of new equipment and control systems) are examples of real-time simulations.

4.5.4 Simulation Processor

The simulation processor processes the simulated events and updates the statistical accumulators and state variables. A typical simulation processor consists of the following basic components:

- *Clock variable*—a variable used to keep track of the elapsed time during the simulation.
- *Event calendar*—a list of scheduled events arranged chronologically according to the time at which they are to occur.
- *Event dispatch manager*—the internal logic used to update the clock and manage the execution of events as they occur.
- *Event logic*—algorithms that describe the logic to be executed and statistics to be gathered when an event occurs.
- *Waiting lists*—one or more lists (arrays) containing events waiting for a resource or other condition to be satisfied before continuing processing.
- *Random number generator*—an algorithm for generating one or more streams of pseudo-random numbers between 0 and 1.
- *Random variate generators*—routines for generating random variates from specified probability distributions.

4.5.5 Animation Processor

The animation processor interacts with the simulation database to update the graphical data to correspond to the changing state data. Animation is usually displayed during the simulation itself, although some simulation products create an animation file that can be played back at the end of the simulation. In addition to animated figures, dynamic graphs and history plots can be displayed during the simulation.

Animation and dynamically updated displays and graphs provide a visual representation of what is happening in the model while the simulation is running. Animation comes in varying degrees of realism from three-dimensional animation to simple animated flowcharts. Often, the only output from the simulation that is of interest is what is displayed in the animation. This is particularly true when simulation is used for facilitating conceptualization or for communication purposes.

A lot can be learned about model behavior by watching the animation (a picture is worth a thousand words, and an animation is worth a thousand pictures). Animation can be as simple as circles moving from box to box, to detailed, realistic graphical representations. The strategic use of graphics should be planned in advance to make the best use of them. While insufficient animation can weaken the message, excessive use of graphics can distract from the central point to be made. It is always good to dress up the simulation graphics for the final presentation; however, such embellishments should be deferred at least until after the model has been debugged.

For most simulations where statistical analysis is required, animation is no substitute for the postsimulation summary, which gives a quantitative overview of the entire system performance. Basing decisions on the animation alone reflects shallow thinking and can even result in unwarranted conclusions.

4.5.6 Output Processor

The output processor summarizes the statistical data collected during the simulation run and creates an output database. Statistics are reported on such performance measures as

- Resource utilization.
- Queue sizes.
- Waiting times.
- Processing rates.

In addition to standard output reporting, most simulation software provides the ability to write user-specified data out to external files.

4.5.7 Output Interface Module

The output interface module provides a user interface for displaying the output results from the simulation. Output results may be displayed in the form of reports or charts (bar charts, pie charts, or the like). Output data analysis capabilities such as correlation analysis and confidence interval calculations also are often provided. Some simulation products even point out potential problem areas such as bottlenecks.

4.6 Simulation Using ProModel

ProModel is a powerful, yet easy-to-use, commercial simulation package that is designed to effectively model any discrete-event processing system. It also has continuous modeling capabilities for modeling flow in and out of tanks and other vessels. The online tutorial in ProModel describes the building, running, and output analysis of simulation models. A brief overview of how ProModel works is presented here. The labs in this book provide a more detailed, hands-on approach for actually building and running simulations using ProModel.

4.6.1 Building a Model

A model is defined in ProModel using simple graphical tools, data entry tables, and fill-in-the-blank dialog boxes. In ProModel, a model consists of entities (the items being processed), locations (the places where processing occurs), resources (agents used to process and move entities), and paths (aisles and pathways along which entities and resources traverse). Dialogs are associated with each of these modeling elements for defining operational behavior such as entity arrivals and processing logic. Schedules, downtimes, and other attributes can also be defined for entities, resources, and locations.

Most of the system elements are defined graphically in ProModel (see Figure 4.8). A graphic representing a location, for example, is placed on the layout to create a new location in the model (see Figure 4.9). Information about this location can then be entered such as its name, capacity, and so on. Default values are provided to help simplify this process. Defining objects graphically provides a highly intuitive and visual approach to model building. The use of graphics is optional, and a model can even be defined without using any graphics. In addition to graphic objects provided by the modeling software, import capability is available to bring in graphics from other packages. This includes complete facility layouts created using CAD software such as AutoCAD.

4.6.2 Running the Simulation

When running a model created in ProModel, the model database is translated or compiled to create the simulation database. The animation in ProModel is displayed concurrently with the simulation. Animation graphics are classified as either static or dynamic. Static graphics include walls, aisles, machines, screen text, and others. Static graphics provide the background against which the animation

FIGURE 4.8

Sample of ProModel graphic objects.

FIGURE 4.9

ProModel animation provides useful feedback.

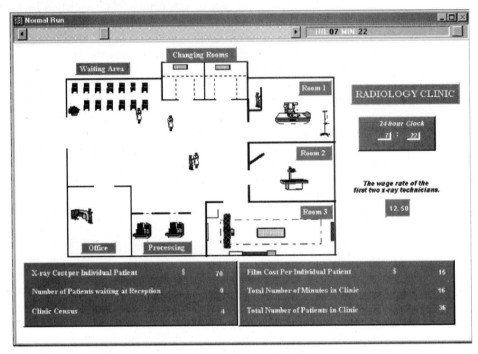

takes place. This background might be a CAD layout imported into the model. The dynamic animation objects that move around on the background during the simulation include entities (parts, customers, and so on) and resources (people, fork trucks, and so forth). Animation also includes dynamically updated counters, indicators, gauges, and graphs that display count, status, and statistical information (see Figure 4.9).

4.6.3 Output Analysis

The output processor in ProModel provides both summary and detailed statistics on key performance measures. Simulation results are presented in the form of reports, plots, histograms, pie charts, and others. Output data analysis capabilities such as confidence interval estimation are provided for more precise analysis. Outputs from multiple replications and multiple scenarios can also be summarized and compared. Averaging performance across replications and showing multiple scenario output side-by-side make the results much easier to interpret.

Summary Reports

Summary reports show totals, averages, and other overall values of interest. Figure 4.10 shows an output report summary generated from a ProModel simulation run.

FIGURE 4.10

Summary report of simulation activity.

```
--------------------------------------------------------------------------------
General Report
Output from C:\ProMod4\models\demos\Mfg_cost.mod [Manufacturing Costing Optimization]
Date: Feb/27/2003   Time: 06:50:05 PM
--------------------------------------------------------------------------------
Scenario        : Model Parameters
Replication     : 1 of 1
Warmup Time     : 5 hr
Simulation Time : 15 hr
--------------------------------------------------------------------------------
```

LOCATIONS

Location Name	Scheduled Hours	Capacity	Total Entries	Average Minutes Per Entry	Average Contents	Maximum Contents	Current Contents
Receive	10	2	21	57.1428	2	2	2
NC Lathe 1	10	1	57	10.1164	0.961065	1	1
NC Lathe 2	10	1	57	9.8918	0.939725	1	1
Degrease	10	2	114	10.1889	1.9359	2	2
Inspect	10	1	113	4.6900	0.883293	1	1
Bearing Que	10	100	90	34.5174	5.17762	13	11
Loc1	10	5	117	25.6410	5	5	5

RESOURCES

Resource Name	Units	Scheduled Hours	Number Of Times Used	Average Minutes Per Usage	Average Minutes Travel To Use	Average Minutes Travel To Park	% Blocked In Travel	% Util
CellOp.1	1	10	122	2.7376	0.1038	0.0000	0.00	57.76
CellOp.2	1	10	118	2.7265	0.1062	0.0000	0.00	55.71
CellOp.3	1	10	115	2.5416	0.1020	0.0000	0.00	50.67
CellOp	3	30	355	2.6704	0.1040	0.0000	0.00	54.71

ENTITY ACTIVITY

Entity Name	Total Exits	Current Quantity In System	Average Minutes In System	Average Minutes Moving	Average Minutes Wait for Res, etc.	Average Minutes In Operation	Average Minutes Blocked
Pallet	19	2	63.1657	0.0000	31.6055	1.0000	30.5602
Blank	0	7	-	-	-	-	-
Cog	79	3	52.5925	0.8492	3.2269	33.5332	14.9831
Reject	33	0	49.5600	0.8536	2.4885	33.0656	13.1522
Bearing	78	12	42.1855	0.0500	35.5899	0.0000	6.5455

FIGURE 4.11

Time-series graph showing changes in queue size over time.

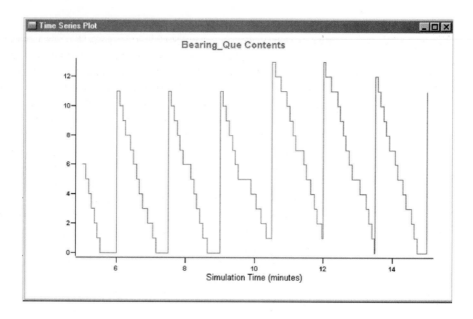

FIGURE 4.12

Histogram of queue contents.

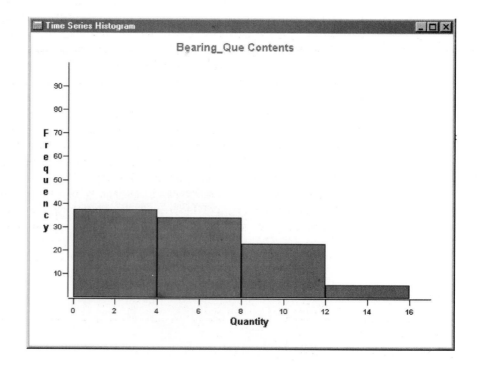

Time Series Plots and Histograms

Sometimes a summary report is too general to capture the information being sought. Fluctuations in model behavior over time can be viewed using a time series report. In ProModel, time series output data are displayed graphically so that patterns can be identified easily. Time series plots can show how inventory levels fluctuate throughout the simulation or how changes in workload affect resource utilization. Figure 4.11 is a time series graph showing how the length of a queue fluctuates over time.

Once collected, time series data can be grouped in different ways to show patterns or trends. One common way of showing patterns is using a histogram. The histogram in Figure 4.12 breaks down contents in a queue to show the percentage of time different quantities were in that queue.

As depicted in Figure 4.12, over 95 percent of the time, there were fewer than 12 items in the queue. This is more meaningful than simply knowing what the average length of the queue was.

4.7 Languages versus Simulators

A distinction is frequently made between simulation languages and simulators. Simulation languages are often considered more general purpose and have fewer predefined constructs for modeling specific types of systems. Simulators, on the other hand, are designed to handle specific problems—for example, a job shop simulator or a clinic simulator. The first simulators that appeared provided little, if any, programming capability, just as the first simulation languages provided few, if any, special modeling constructs to facilitate modeling. Consequently, simulators acquired the reputation of being easy to use but inflexible, while simulation languages were branded as being very flexible but difficult to use (see Figure 4.13).

Over time, the distinction between simulation languages and simulators has become blurred as specialized modeling constructs have been added to general-purpose simulation languages, making them easier to use. During the same period, general programming extensions have been added to simulators, making them more flexible. The most popular simulation tools today combine powerful industry-specific constructs with flexible programming capabilities, all accessible from an intuitive graphical user interface (Bowden 1998). Some tools are even configurable, allowing the software to be adapted to specific applications yet still retaining programming capability. Rather than put languages and simulators on opposite ends of the same spectrum as though flexibility and ease of use were mutually exclusive, it is more appropriate to measure the flexibility and ease of use for all simulation software along two separate axes (Figure 4.14).

FIGURE 4.13

Old paradigm that polarized ease of use and flexibility.

Simulators   Languages

Ease of use   Flexibility

FIGURE 4.14

New paradigm that views ease of use and flexibility as independent characteristics.

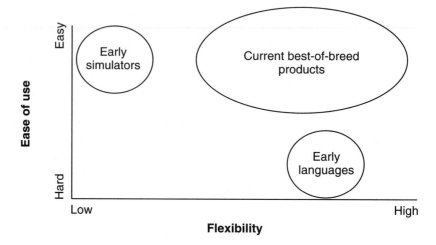

4.8 Future of Simulation

Simulation products have evolved to provide more than simply stand-alone simulation capability. Modern simulation products have open architectures based on component technology and standard data access methods (like SQL) to provide interfacing capability with other applications such as CAD programs and enterprise planning tools. Surveys reported annually in *Industrial Engineering Solutions* show that most simulation products have the following features:

- Input data analysis for distribution fitting.
- Point-and-click graphical user interface.
- Reusable components and templates.
- Two- (2-D) and three-dimensional (3-D) animation.
- Online help and tutorials.
- Interactive debugging.
- Automatic model generation.
- Output analysis tools.
- Optimization.
- Open architecture and database connectivity.

Simulation is a technology that will continue to evolve as related technologies improve and more time is devoted to the development of the software. Products will become easier to use with more intelligence being incorporated into the software itself. Evidence of this trend can already be seen by optimization and other time-saving utilities that are appearing in simulation products. Animation and other graphical visualization techniques will continue to play an important role in simulation. As 3-D and other graphic technologies advance, these features will also be incorporated into simulation products.

Simulation products targeted at vertical markets are on the rise. This trend is driven by efforts to make simulation easier to use and more solution oriented. Specific areas where dedicated simulators have been developed include call center management, supply chain management, and high-speed processing. At the same time many simulation applications are becoming more narrowly focused, others are becoming more global and look at the entire enterprise or value chain in a hierarchical fashion from top to bottom.

Perhaps the most dramatic change in simulation will be in the area of software interoperability and technology integration. Historically, simulation has been viewed as a stand-alone, project-based technology. Simulation models were built to support an analysis project, to predict the performance of complex systems, and to select the best alternative from a few well-defined alternatives. Typically these projects were time-consuming and expensive, and relied heavily on the expertise of a simulation analyst or consultant. The models produced were generally "single use" models that were discarded after the project.

In recent years, the simulation industry has seen increasing interest in extending the useful life of simulation models by using them on an ongoing basis (Harrell and Hicks 1998). Front-end spreadsheets and push-button user interfaces are making such models more accessible to decision makers. In these flexible simulation models, controlled changes can be made to models throughout the system life cycle. This trend is growing to include dynamic links to databases and other data sources, enabling entire models actually to be built and run in the background using data already available from other enterprise applications.

The trend to integrate simulation as an embedded component in enterprise applications is part of a larger development of software components that can be distributed over the Internet. This movement is being fueled by three emerging information technologies: (1) component technology that delivers true object orientation; (2) the Internet or World Wide Web, which connects business communities and industries; and (3) Web service technologies such as J2EE and Microsoft's .NET ("DOTNET"). These technologies promise to enable parallel and distributed model execution and provide a mechanism for maintaining distributed model repositories that can be shared by many modelers (Fishwick 1997). The interest in Web-based simulation, like all other Web-based applications, continues to grow.

4.9 Summary

Most manufacturing and service systems are modeled using dynamic, stochastic, discrete-event simulation. Discrete-event simulation works by converting all activities to events and consequent reactions. Events are either time-triggered or condition-triggered, and are therefore processed either chronologically or when a satisfying condition has been met.

Simulation models are generally defined using commercial simulation software that provides convenient modeling constructs and analysis tools. Simulation software consists of several modules with which the user interacts. Internally, model data are converted to simulation data, which are processed during the simulation. At the end of the simulation, statistics are summarized in an output database that can be tabulated or graphed in various forms. The future of simulation is promising and will continue to incorporate exciting new technologies.

4.10 Review Questions

1. Give an example of a discrete-change state variable and a continuous-change state variable.

2. In simulation, the completion of an activity time that is random must be known at the start of the activity. Why is this necessary?

3. Give an example of an activity whose completion is a scheduled event and one whose completion is a conditional event.

4. For the first 10 customers processed completely through the ATM spreadsheet simulation presented in Table 3.2 of Chapter 3, construct a table similar to Table 4.1 as you carry out a manual discrete-event simulation of the ATM system to
   a. Compute the average amount of time the first 10 customers spent in the system. Hint: Add time in system and corresponding cumulative columns to the table.
   b. Compute the average amount of time the first 10 customers spent in the queue.
   c. Plot the number of customers in the queue over the course of the simulation and compute the average number of customers in the queue for the simulation.
   d. Compute the utilization of the ATM. Hint: Define a utilization variable that is equal to zero when the ATM is idle and is equal to 1 when the ATM is busy. At the end of the simulation, compute a time-weighted average of the utilization variable.

5. Identify whether each of the following output statistics would be computed as a simple or as a time-weighted average value.
   a. Average utilization of a resource.
   b. Average time entities spend in a queue.
   c. Average time entities spend waiting for a resource.
   d. Average number of entities waiting for a particular resource.
   e. Average repair time for a machine.

6. Give an example of a situation where a time series graph would be more useful than just seeing an average value.

7. In real life, activities take time and events take no time. During a simulation, activities take no time and events take all of the time. Explain this paradox.

8. What are the main components of simulation software?

9. Look up a simulation product on the Internet or in a trade journal and describe two promotional features of the product being advertised.

10. For each of the following simulation applications identify one discrete- and one continuous-change state variable.

    a. Inventory control of an oil storage and pumping facility.

    b. Study of beverage production in a soft drink production facility.

    c. Study of congestion in a busy traffic intersection.

References

Bowden, Royce. "The Spectrum of Simulation Software." *IIE Solutions,* May 1998, pp. 44–46.

Fishwick, Paul A. "Web-Based Simulation." In *Proceedings of the 1997 Winter Simulation Conference,* ed. S. Andradottir, K. J. Healy, D. H. Withers, and B. L. Nelson. Institute of Electric and Electronics Engineers, Piscataway, NJ, 1997. pp. 100–109.

Gottfried, Byron S. *Elements of Stochastic Process Simulation.* Englewood Cliffs, NJ: Prentice Hall, 1984, p. 8.

Haider, S. W., and J. Banks. "Simulation Software Products for Analyzing Manufacturing Systems." *Industrial Engineering,* July 1986, p. 98.

Harrell, Charles R., and Don Hicks. "Simulation Software Component Architecture for Simulation-Based Enterprise Applications." *Proceedings of the 1998 Winter Simulation Conference,* ed. D. J. Medeiros, E. F. Watson, J. S. Carson, and M. S. Manivannan. Institute of Electrical and Electronics Engineers, Piscataway, New Jersey, 1998, pp. 1717–21.

5 GETTING STARTED

"For which of you, intending to build a tower, sitteth not down first, and counteth the cost, whether he have sufficient to finish it? Lest haply, after he hath laid the foundation, and is not able to finish it, all that behold it begin to mock him, Saying, This man began to build, and was not able to finish."
—Luke 14:28–30

5.1 Introduction

In this chapter we look at how to begin a simulation project. Specifically, we discuss how to select a project and set up a plan for successfully completing it. Simulation is not something you do simply because you have a tool and a process to which it can be applied. Nor should you begin a simulation without forethought and preparation. A simulation project should be carefully planned following basic project management principles and practices. Questions to be answered in this chapter are

- How do you prepare to do a simulation study?
- What are the steps for doing a simulation study?
- What are typical objectives for a simulation study?
- What is required to successfully complete a simulation project?
- What are some pitfalls to avoid when doing simulation?

While specific tasks may vary from project to project, the basic procedure for doing simulation is essentially the same. Much as in building a house, you are better off following a time-proven methodology than approaching it haphazardly. In this chapter, we present the preliminary activities for preparing to conduct a simulation study. We then cover the steps for successfully completing a simulation project. Subsequent chapters elaborate on these steps. Here we focus primarily on the first step: defining the objective, scope, and requirements of the study. Poor planning, ill-defined objectives, unrealistic expectations, and unanticipated costs can turn a simulation project sour. For a simulation project to succeed, the objectives and scope should be clearly defined and requirements identified and quantified for conducting the project.

5.2 Preliminary Activities

Simulation is not a tool to be applied indiscriminately with little or no forethought. The decision to use simulation itself requires some consideration. Is the application appropriate for simulation? Are other approaches equally as effective yet less expensive? These and other questions should be raised when exploring the potential use of simulation. Once it has been determined that simulation is the right approach, other preparations should be made to ensure that the necessary personnel and resources are in place to conduct the study. Personnel must be identified and trained. The right software should also be carefully selected. The nature of the study and the policies of the organization will largely dictate how to prepare for a simulation project.

5.2.1 Selecting an Application

The decision to use simulation usually begins with a systemic problem for which a solution is being sought. The problem might be as simple as getting better utilization from a key resource or as complex as how to increase throughput while at the same time reducing cycle time in a large factory. Such problems become opportunities for system improvement through the use of simulation.

While many opportunities may exist for using simulation, beginners in simulation should select an application that is not too complicated yet can have an impact on an organization's bottom line. On the one hand, you want to focus on problems of significance, but you don't want to get in over your head. The first simulation project should be one that is well defined and can be completed within a few weeks. This pilot project might consist of a small process such as a manufacturing cell or a service operation with only a few process steps. The goal at this point should be focused more on gaining experience in simulation than making dramatic, enterprisewide improvements. Conducting a pilot project is much like a warm-up exercise. It should be treated as a learning experience as well as a confidence and momentum builder.

As one gains experience in the use of simulation, more challenging projects can be undertaken. As each project is considered, it should be evaluated first for its suitability for simulation. The following questions can help you determine whether a process is a good candidate for simulation:

- Is the process well defined?
- Is process information readily available?

- Does the process have interdependencies?
- Does the process exhibit variability?
- Are the potential cost savings greater than the cost of doing the study?
- If it is a new process, is there time to perform a simulation analysis?
- If it is an existing process, would it be less costly to experiment on the actual system?
- Is management willing to support the project?

The wisdom of asking these questions should be clear. You don't want to waste time on efforts or problems that are too unstructured or for which simulation is not well suited. The last question, which relates to management support, is a reminder that no simulation project should be undertaken if management is unwilling to support the study or take the recommendations of the study seriously. Obviously, multiple projects may fit the criteria just listed, in which case the list of candidates for simulation should be prioritized using a cost–benefit analysis (a.k.a. biggest-bang-for-the-buck analysis). Management preferences and timing may also come into play in prioritization.

5.2.2 Personnel Identification

One of the first preparations for a simulation project is to have management assign a project leader who works alone or with a project team. When assessing personnel requirements for the simulation, consideration should be given to factors such as level of expertise, length of the assignment, and organizational structure. Large companies, for example, often have a central simulation group that provides simulation services to entities within the organization. In some organizations, simulation is used no more than once or twice a year, such as when a major change in operation is being considered. In such situations, it may be more cost-effective to hire a simulation consultant than to do the simulation in-house. While the decision to use a consultant is usually based on the level and availability of in-house simulation expertise as well as the difficulty of the project, the opportunity to train additional people or acquire increased simulation expertise to apply to future projects should not be overlooked.

Assuming the simulation will be done internally, either a simulation specialist should be assigned to lead the project or an engineer or manager who works closely with the specialist should be assigned. The person or group doing the modeling should be well trained and proficient in the use of one or more simulation packages. If no such experienced individual exists, management must be willing to invest in employee training in simulation.

The process owner should take an active role in the project and, if possible, even lead the effort. If the process owner refuses or doesn't have time to get involved in the simulation project, the effort will be largely wasted. It is imperative to have someone involved who not only understands the process and what possible design options are feasible, but who also has responsibility for the system's performance.

5.2.3 Software Selection

Once the responsibility for doing simulation has been assigned, software tools, if not already in possession, need to be acquired to develop the simulation model. Having the right tool for the job can mean the difference between success and failure in a simulation project. Modeling efforts sometimes go on for months without much progress because of limited capabilities in the software. Simulation software can't make a simulation succeed, but it can make it fail.

Selecting the right simulation software requires that an assessment first be made of the simulation requirements. Then alternative products should be evaluated against those requirements. If simulation is to be used in multiple applications that are quite diverse, it may be desirable to have more than one simulation product so that the right tool is used for the right job.

In selecting simulation software, attention should be given to the following criteria:

- *Quality*—Reliable algorithms, absence of bugs, and graceful handling of user input errors.
- *Features and capability*—Ability to model systems of any size and complexity and meet simulation needs in terms of external interfacing, animation, reporting, and so forth.
- *Ease of use*—Intuitive editing and built-in constructs that minimize the time and effort to learn and use the product.
- *Services*—Quality and level of training and technical support, including the long-term sustainability of this service.
- *Cost*—The total cost of ownership (TCO), which includes cost of procurement, training, support, and labor to use the software.

Although fewer than a dozen major products are on the market, there are over a hundred products from which to choose. It is best to conduct an initial screening to develop a short list. Then a more in-depth evaluation can be conducted on the three or four finalists. In conducting the initial screening, different products should be screened in terms of their ability to meet minimum requirements in the areas just identified. Often vendor literature or buying guides are helpful at this stage of the evaluation. Magazines such as *IIE Solutions* and *OR/MS Today* run regular reviews of simulation products.

In the final evaluation phase, the products should be examined with greater scrutiny in each of the areas just listed. Since all major software products have bugs, it is best to look for a product that is mature and has a large user base. This will help ensure that even rare bugs have been identified and fixed. Simulation features are pretty easy to assess because most vendors have feature lists available. One of the most difficult criteria to assess in a simulation product is ease of use. Most vendors will likely claim that their product is the easiest to use, so it is always a good idea to have the vendor being considered build a simple model in your presence that is representative of the target application. This is the best way to test the suitability and ease of use of the software. At a minimum, it is good to experiment building simple models with the software to see how user friendly it is.

Many vendors offer guarantees on their products so they may be returned after some trial period. This allows you to try out the software to see how well it fits your needs.

The services provided by the software provider can be a lifesaver. If working late on a project, it may be urgent to get immediate help with a modeling or software problem. Basic and advanced training classes, good documentation, and lots of example models can provide invaluable resources for becoming proficient in the use of the software.

When selecting simulation software, it is important to assess the total cost of ownership. There often tends to be an overemphasis on the purchase price of the software with little regard for the cost associated with learning and using the software. It has been recommended that simulation software be purchased on the basis of productivity rather than price (Banks and Gibson 1997). The purchase price of the software can sometimes be only a small fraction of the cost in time and labor that results from having a tool that is difficult to use or inadequate for the application.

Other considerations that may come into play when selecting a product include quality of the documentation, hardware requirements (for example, is a graphics accelerator card required?), and available consulting services.

5.3 Simulation Procedure

Once a suitable application has been selected and appropriate tools and personnel are in place, the simulation study can begin. Simulation is much more than building and running a model of the process. Successful simulation projects are well planned and coordinated. While there are no strict rules on how to conduct a simulation project, the following steps are generally recommended:

Step 1: Define objective, scope, and requirements. Define the purpose of the simulation project and what the scope of the project will be. Requirements need to be defined in terms of resources, time, and budget for carrying out the project.

Step 2: Collect and analyze system data. Identify, gather, and analyze the data defining the system to be modeled. This step results in a conceptual model and a data document on which all can agree.

Step 3: Build the model. Develop a simulation model of the system.

Step 4: Validate the model. Debug the model and make sure it is a credible representation of the real system.

Step 5: Conduct experiments. Run the simulation for each of the scenarios to be evaluated and analyze the results.

Step 6: Present the results. Present the findings and make recommendations so that an informed decision can be made.

Each step need not be completed in its entirety before moving to the next step. The procedure for doing a simulation is an iterative one in which activities are refined and sometimes redefined with each iteration. The decision to push toward further refinement should be dictated by the objectives and constraints of the study as well as by sensitivity analysis, which determines whether additional refinement will yield meaningful results. Even after the results are presented, there are often requests to conduct additional experiments. Describing this iterative process, Pritsker and Pegden (1979) observe,

> The stages of simulation are rarely performed in a structured sequence beginning with problem definition and ending with documentation. A simulation project may involve false starts, erroneous assumptions which must later be abandoned, reformulation of the problem objectives, and repeated evaluation and redesign of the model. If properly done, however, this iterative process should result in a simulation model which properly assesses alternatives and enhances the decision making process.

Figure 5.1 illustrates this iterative process.

FIGURE 5.1

Iterative nature of simulation.

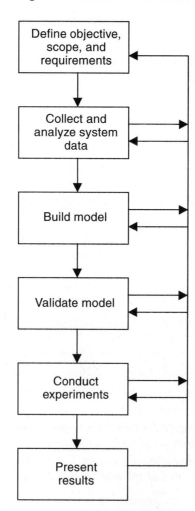

The remainder of this chapter focuses on the first step of defining the objective, scope, and requirements of the study. The remaining steps will be discussed in later chapters.

5.4 Defining the Objective

The objective of a simulation defines the purpose or reason for conducting the simulation study. It should be realistic and achievable, given the time and resource constraints of the study. Simulation objectives can be grouped into the following general categories:

- *Performance analysis*—What is the all-around performance of the system in terms of resource utilization, flow time, output rate, and so on?
- *Capacity/constraint analysis*—When pushed to the maximum, what is the processing or production capacity of the system and where are the bottlenecks?
- *Configuration comparison*—How well does one system configuration meet performance objectives compared to another?
- *Optimization*—What settings for particular decision variables best achieve desired performance goals?
- *Sensitivity analysis*—Which decision variables are the most influential on performance measures, and how influential are they?
- *Visualization*—How can system dynamics be most effectively visualized?

Naturally, a simulation project may have multiple objectives and therefore fall under two or more of the categories listed here. For example, a sensitivity analysis may be conducted as part of an optimization study to determine the highest-leveraging variables on which to optimize. After the model is optimized, it may be necessary to focus attention on enhancing the visualization of the model to more effectively sell the solution to management.

Sometimes the objectives change or expand as the project progresses. The insights provided by simulation give rise to previously unforeseen opportunities for improvement. Furthermore, the very process of doing a simulation helps refine and clarify objectives. Staying as close as possible to the initial scope of the project is important, and changes to the scope should be documented and assessed in terms of their impact on budget and schedule. Otherwise, "scope creep" sets in and the project never ends.

Defining objectives for a simulation study should take into account the ultimate intended use of the model. Some models are built as "throwaway" models that are used only once and then discarded. Other models will be reused for ongoing what-if analyses. Some simulations are run to obtain a single summary statistical measure. Others require realistic animation to convince a skeptical customer. Some models are only for the use of the analyst and need not have a user-friendly interface. Other models are intended for use by managers with little simulation background and therefore require a customized, easy-to-use interface. Some models are used to make decisions of minor consequence. Other models serve as the basis for major financial decisions.

Following is a list of sample design and operational questions that simulation can help answer. They are intended as examples of specific objectives that might be defined for a simulation study.

Design Decisions

1. What division and sequence of processing activities provide the best flow?
2. What is the best layout of offices, machines, equipment, and other work areas for minimizing travel time and congestion?
3. How many operating personnel are needed to meet required production or service levels?
4. What level of automation is the most cost-effective?
5. How many machines, tools, fixtures, or containers are needed to meet throughput requirements?
6. What is the least-cost method of material handling or transportation that meets processing requirements?
7. What are the appropriate number and location of pickup and drop-off points in a material handling or transportation system that minimizes waiting times?
8. What are the optimum number and size of waiting areas, storage areas, queues, and buffers?
9. What is the effect of localizing rather than centralizing material storage, resource pools, and so forth?
10. What automation control logic provides the best utilization of resources?
11. What is the optimum unit load or batch size for processing?
12. Where are the bottlenecks in the system, and how can they be eliminated?
13. How many shifts are needed to meet a specific production or service level?
14. What is the best speed to operate conveyors and other handling or transportation equipment to meet move demands?

Operational Decisions

1. What is the best way to route material, customers, or calls through the system?
2. What is the best way to allocate personnel for a particular set of tasks?
3. What is the best schedule for preventive maintenance?
4. How much preventive maintenance should be performed?
5. What is the best priority rule for selecting jobs and tasks?

6. What is the best schedule for producing a mix of products or customers?
7. What is the best inventory replenishment policy?
8. How much raw material and work-in-process inventory should be maintained?
9. What is the best production control method (kanban, for example)?
10. What is the optimum sequence for producing a particular set of jobs or processing a given set of customers?
11. How many personnel should be scheduled for a particular shift?

When the goal is to analyze some aspect of system performance, the tendency is to think in terms of the mean or expected value of the performance metric. For example, we are frequently interested in the average contents of a queue or the average utilization of a resource. There are other metrics that may have equal or even greater meaning that can be obtained from a simulation study. For example, we might be interested in variation as a metric, such as the standard deviation in waiting times. Extreme values can also be informative, such as the minimum and maximum number of contents in a storage area. We might also be interested in a percentile such as the percentage of time that the utilization of a machine is less than a particular value, say, 80 percent. While frequently we speak of designing systems to be able to handle peak periods, it often makes more sense to design for a value above which values only occur less than 5 or 10 percent of the time. It is more economical, for example, to design a staging area on a shipping dock based on 90 percent of peak time usage rather than based on the highest usage during peak time. Sometimes a single measure is not as descriptive as a trend or pattern of performance. Perhaps a measure has increasing and decreasing periods such as the activity in a restaurant. In these situations, a detailed time series report would be the most meaningful.

It is always best if objectives can be clearly, completely, and precisely stated. You need to be able to set your sight clearly on what you are trying to accomplish in the simulation study. You also need to be able to know whether you hit your target objective. To be effective, an objective should be one that

- Has high potential impact (*reduce throughput rate costs,* not *reduce backtraveling*).
- Is achievable (*20 percent inventory reduction,* not *zero inventory*).
- Is specific (*reduce waiting time in a particular queue,* not *eliminate waste*).
- Is quantifiable (*reduce flow time by 40 percent,* not *reduce flow time*).
- Is measurable (*increase yield by 10 percent,* not *improve morale by 10 percent*).
- Identifies any relevant constraints (*reduce turnaround time by 20 percent without adding resources,* not just *reduce turnaround time by 20 percent*).

Examples of well-stated objectives might be the following:

- Find the lowest-cost solution to reduce patient waiting time in an emergency room so that no more than 10 percent of patients wait longer than 15 minutes.
- Increase the throughput rate of an assembly line by an average of 20 percent without additional capital expenditure.

While well-defined and clearly stated objectives are important to guide the simulation effort, they should not restrict the simulation or inhibit creativity. Michael Schrage (1999) observes that "the real value of a model or simulation may stem less from its ability to test a hypothesis than from its power to generate useful surprise. Louis Pasteur once remarked that 'chance favors the prepared mind.' It holds equally true that chance favors the prepared prototype: models and simulations can and should be media to create and capture surprise and serendipity. . . . The challenge is to devise transparent models that also make people shake their heads and say 'Wow!'" The right "experts" can be "hypervulnerable to surprise but well situated to turn surprise to their advantage. That's why Alexander Fleming recognized the importance of a mold on an agar plate and discovered penicillin." Finally, he says, "A prototype should be an invitation to play. You know you have a successful prototype when people who see it make useful suggestions about how it can be improved."

5.5 Defining the Scope of Work

With a realistic, meaningful, and well-defined objective established, a scope of work can be defined for achieving the stated objective. The scope of work is important for guiding the study as well as providing a specification of the work to be done upon which all can agree. The scope is essentially a project specification that helps set expectations by clarifying to others exactly what the simulation will include and exclude. Such a specification is especially important if an outside consultant is performing the simulation so that there is mutual understanding of the deliverables required.

An important part of the scope is a specification of the models that will be built. When evaluating improvements to an existing system, it is often desirable to model the current system first. This is called an "as-is" model. Results from the as-is model are statistically compared with outputs of the real-world system to validate the simulation model. This as-is model can then be used as a benchmark or baseline to compare the results of "to-be" models. For reengineering or process improvement studies, this two-phase modeling approach is recommended. For entirely new facilities or processes, there will be no as-is model. There may, however, be several to-be models to compare.

To ensure that the budget and schedule are realistic, a detailed specification should be drafted to include

- Model scope.
- Level of detail.

- Data-gathering responsibilities.
- Experimentation.
- Form of results.

5.5.1 Determining Model Scope

Model scope refers to what system elements should be represented in the model. Determining the scope of a model should be based on how much impact a particular activity has on achieving the objectives of the simulation.

One decision regarding model scope relates to the model boundaries. A common tendency for beginners is to model the entire system, even when the problem area and all relevant variables are actually isolated within a smaller subsystem. If, for example, the objective is to find the number of operators needed to meet a required production level for a machining cell, it probably isn't necessary to model what happens to parts after leaving the cell. Figure 5.2 illustrates how the extent of the model should be confined to only those activities whose interactions have a direct bearing on the problem being studied. If the objective of the simulation is to find how many operators are needed to perform activity C, upstream and downstream activities that do not impact this decision should be omitted from the model. In Figure 5.2, since the input rate to activity B is reasonably predictable, based on knowledge of how activity A behaves, activity A is excluded from the model and only modeled as an arrival rate to location 2. Since activity E never causes entities to back up at D, activity E can also be ignored.

In addition to limiting the model boundaries, other system elements may be excluded from a model due to their irrelevance. Factors such as off-shift maintenance activities or machine setups that are brief and occur only infrequently can usually be safely ignored. The determination of what elements can be eliminated should be based on their relevance to the objectives of the study.

5.5.2 Deciding on Level of Detail

Where model scope is concerned more with model breadth, the *level of detail* determines the depth of a model. More specifically, it defines the granularity or resolution of the model. At one extreme, an entire factory can be modeled as a single "black box" operation with a random activity time. At the other extreme is a

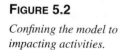

FIGURE 5.2

Confining the model to impacting activities.

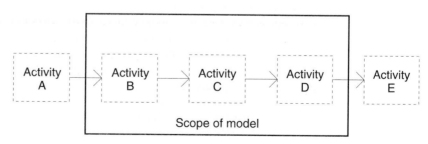

Figure 5.3

Effect of level of detail on model development time.

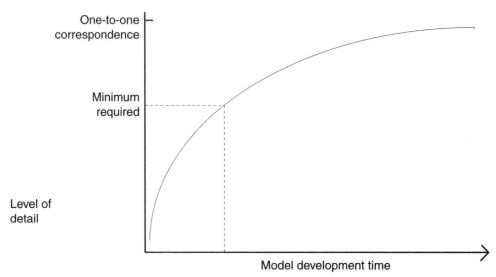

"white box" model that is very detailed and produces a one-to-one correspondence between the model and the system.

Determining the appropriate level of detail is an important decision. Too much detail makes it difficult and time-consuming to develop and debug the model. Too little detail may make the model unrealistic by oversimplifying the process. Figure 5.3 illustrates how the time to develop a model is affected by the level of detail. It also highlights the importance of including only enough detail to meet the objectives of the study.

The level of detail is determined largely by the degree of accuracy required in the results. If only a rough estimate is being sought, it may be sufficient to model just the flow sequence and processing times. If, on the other hand, a close answer is needed, all of the elements that drive system behavior should be precisely modeled.

5.5.3 *Assigning Data-Gathering Responsibilities*

In any simulation project, data gathering is almost always the most difficult and time-consuming task. Identifying data requirements and who will be responsible for gathering the data is essential if this activity is to succeed. If the modeler is responsible for gathering data, it is important to have the full cooperation of those who possess the data, including equipment vendors, engineers, technicians, direct labor personnel, and supervisors. It is also a good idea to get those individuals involved who have a stake in the results, such as managers and possibly even customers. The more fingerprints you can get on the knife, the stronger the commitment you'll have to the success of the project.

Dates should be set for completing the data-gathering phase because, left uncontrolled, it could go on indefinitely. One lesson you learn quickly is that good data are elusive and you can always spend more time trying to refine the data. Nearly all models are based partially on assumptions, simply because complete and accurate information is usually lacking. The project team, in cooperation with stakeholders in the system, will need to agree on the assumptions to be made in the model.

5.5.4 Planning the Experimentation

The number and nature of the scenarios or alternative configurations to be evaluated should be planned from the outset to ensure that adequate time is allotted. Often this decision itself is influenced by the time constraints of the study. Where only slightly different scenarios are being evaluated, a single model can be developed and modified appropriately to run each scenario. If alternative configurations are significantly different, separate models may need to be built. It may require nearly as much effort modeling each configuration as it does developing the initial model.

For studies considering improvements to an existing system, it is often helpful and effective to model the current system as well as the proposed system. The basic premise is that you are not ready to improve a system until you understand how the current system operates. Information on the current system operation is easier to obtain than information on the proposed change of operation. Once a model of the current system is built, it is easier to visualize what changes need to be made for the modified system. Both systems may even be modeled together in the same simulation and made to run side by side. During the final presentation of the results, being able to show both the as-is and to-be versions of the system effectively demonstrates the differences in performance.

5.5.5 Determining the Form of Results

The form in which the results are to be presented can significantly affect the time and effort involved in the simulation study. If a detailed animation or an extensive written report is expected, the project can easily stretch on for several weeks after the experimentation phase has been completed. Many times the only result required is a simple verification of whether a system is capable of meeting a production requirement or service level. In such cases, a single number may suffice. In other instances, a complete document may be required that details all of the goals, data sources, assumptions, modeling procedures, experiments, analysis of results, and recommendations. The key to determining the kind and quantity of information to present to the decision maker or stakeholder is to ask what decision is being made with the simulation and what the background is of the person(s) who will be making the decision. Attention should then be focused on providing adequate information and effective visualization to enable the right decision to be made.

Sometimes the simulation animation can effectively show bottleneck areas or compare alternative scenarios. When showing animation, use graphics and displays that focus attention on the area of interest. If lots of color and detail are added to the animation, it may detract from the key issues. Usually the best approach is to keep stationary or background graphics simple, perhaps displaying only a schematic of the layout using neutral colors. Entities or other dynamic graphics can then be displayed more colorfully to make them stand out. Sometimes the most effective presentation is a realistic 3-D animation. Other times the flow of entities along a flowchart consisting of simple boxes and arrows may be more effective.

Another effective use of animation in presenting the results is to run two or more scenarios side by side, displaying a scoreboard that shows how they compare on one or two key performance measures. The scoreboard may even include a bar graph or other chart that is dynamically updated to compare results.

Most decision makers such as managers need to have only a few key items of information for making the decision. It should be remembered that people, not the model, make the final decision. With this in mind, every effort should be made to help the decision maker clearly understand the options and their associated consequences. The use of charts helps managers visualize and focus their attention on key decision factors. Charts are attention grabbers and are much more effective in making a point than plowing through a written document or sheets of computer printout.

5.6 Defining Project Requirements

With the scope of work defined, resource, budget, and time requirements can be determined for the project. Most of the resources (personnel, hardware, and so on) should have already been put in place when the decision was made to do a simulation. The primary task at this point is to develop a budget and schedule for the project. A budget or schedule overrun can sour an otherwise great modeling effort. It does little good if simulation solves a problem but the time to do the simulation extends beyond the deadline for applying the solution, or the cost to find the solution exceeds the benefit that is derived. Realistic budgeting and scheduling are based on a complete knowledge of the requirements of the study. It is not uncommon to begin a simulation project with aspirations of developing an impressively detailed model or of creating a stunningly realistic animation only to scramble at the last minute to throw together a crude model that barely meets the deadline.

Obviously, the time to perform a simulation project will vary depending on the size and difficulty of the project. If input data cannot be readily obtained, it may be necessary to add several additional weeks to the project. Assuming input data are readily available, a small project typically takes two to four weeks, while large projects can take two to four months. A simulation schedule should be based on realistic projections of the time requirements, keeping in mind that

- Data gathering and analysis can take over 50 percent of the overall project time.
- Model building usually takes the least amount of time (10 to 20 percent).

- Once a base model is built, several additional weeks may be needed to conduct all of the desired experiments, especially if the purpose of the project is to make system comparisons.
- At least several days must be allowed to clean up the models and develop the final presentation.
- Simulation projects follow the 90–10 rule, where the first 90 percent takes 10 percent of the time, and the last 10 percent takes the other 90 percent.

5.7 Reasons Why Simulation Projects Fail

No matter how good the intentions, simulation projects can fail if not carefully planned and executed. Many specific reasons contribute to simulation failure. Some of the most common causes of *un*successful projects include

- Unclear objectives.
- Unskilled modelers.
- Unavailable data.
- Unmanaged expectations.
- Unsupportive management.
- Underestimated requirements.
- Uninvolved process owner(s).

If the right procedure is followed, and the necessary time and resources are committed to the project, simulation is always going to provide some benefit to the decision-making process. The best way to ensure success is to make sure that everyone involved is educated in the process and understands the benefits and limitations of simulation.

5.8 Summary

Simulation projects are almost certain to fail if there is little or no planning. Doing simulation requires some preliminary work so that the appropriate resources and personnel are in place. Beginning a simulation project requires selecting the right application, defining objectives, acquiring the necessary tools and resources, and planning the work to be performed. Applications should be selected that hold the greatest promise for achieving company goals.

Simulation is most effective when it follows a logical procedure. Objectives should be clearly stated and a plan developed for completing the project. Data gathering should focus on defining the system and formulating a conceptual model. A simulation should then be built that accurately yet minimally captures the system definition. The model should be verified and validated to ensure that the results can be relied upon. Experiments should be run that are oriented toward meeting the original objectives. Finally, the results should be presented in a way that clearly represents the findings of the study. Simulation is an iterative process that requires redefinition and fine-tuning at all stages in the process.

Objectives should be clearly defined and agreed upon to avoid wasted efforts. They should be documented and followed to avoid "scope creep." A concisely stated objective and a written scope of work can help keep a simulation study on track.

Specific items to address when planning a study include defining the model scope, describing the level of detail required, assigning data-gathering responsibilities, specifying the types of experiments, and deciding on the form of the results. Simulation objectives together with time, resources, and budget constraints drive the rest of the decisions that are made in completing a simulation project. Finally, it is people, not the model, who ultimately make the decision.

The importance of involvement of process owners and stakeholders throughout the project cannot be overemphasized. Management support throughout the project is vital. Ultimately, it is only when expectations are met or exceeded that a simulation can be deemed a success.

5.9 Review Questions

1. Why are good project management skills important to the success of a simulation project?

2. If you are new to simulation and have the option of simulating either a small system with a small payback or a large system with a large payback, which project would you choose and why?

3. How can simulation software make or break a simulation study?

4. When evaluating ease of use in simulation software, how can you make an informed decision if every vendor advertises ease of use?

5. Why is it important to have clearly defined objectives for a simulation that everyone understands and can agree on?

6. For each of the following systems, define one possible simulation objective:
   a. A manufacturing cell.
   b. A fast-food restaurant.
   c. An emergency room of a hospital.
   d. A tire distribution center.
   e. A public bus transportation system.
   f. A post office.
   g. A bank of elevators in a high-rise office building.
   h. A time-shared, multitasking computer system consisting of a server, many slave terminals, a high-speed disk, and a processor.
   i. A rental car agency at a major airport.

7. How can you effectively manage expectations during a simulation project?

8. Suppose you are determining the size of a waiting area for a service facility (such as a clinic, a car repair shop, or a copy center) where the

arrival rate fluctuates throughout the day. On which of the following metrics would you base the waiting area capacity and why?

a. The average number of customers waiting.

b. The peak number of customers waiting.

c. The 95th percentile of customers waiting.

9. What step in doing a simulation study generally takes the longest?

10. What stage in model building usually gets cut short?

11. What factors should determine the scope of a model?

12. What factors should determine the level of detail to include in a model?

13. What is the advantage of building as-is and to-be models?

14. What should determine the form of results of a simulation study?

5.10 Case Studies

CASE STUDY A
AST COMPUTES BIG BENEFITS USING SIMULATION

John Perry
Manufacturing Engineer

Source: AST Research Inc. Used with permission by ProModel, 1999.

In the world of computer manufacturing and assembly, success is measured in terms of seconds. The PC business poses some unique challenges: it is intensely competitive with thin profit margins and is characterized by constantly changing technologies and products. The race is to excel in rapid delivery to the marketplace through high-velocity manufacturing. A manufacturer that can minimize the time needed to assemble a computer, move nimbly to respond to change, and get maximum productivity out of its capital investment has an edge. At AST Research, we use computer simulation to realize these goals and, so far, have achieved significant benefits. The application area, which is product assembly, is an area where almost anyone can benefit from the use of simulation.

The Problem

AST Research Inc., founded in 1980, has become a multibillion-dollar PC manufacturer. We assemble personal computers and servers in Fort Worth, Texas, and offshore. For a long time, we "optimized" (planned) our assembly procedures using traditional methods—gathering time-and-motion data via stopwatches and videotape, performing simple arithmetic calculations to obtain information about the operation and performance of the assembly line, and using seat-of-the-pants guesstimates to "optimize" assembly line output and labor utilization.

The Model

In December 1994 a new vice president joined AST. Management had long been committed to increasing the plant's efficiency and output, and our new vice president had experience in using simulation to improve production. We began using ProModel simulation as a tool for optimizing our assembly lines, and to improve our confidence that changes proposed in the assembly process would work out in practice. The results have been significant. They include

- Reduced labor input to each unit.
- Shortened production cycle times.
- Reduced reject rate.
- Increased ability to change assembly instructions quickly when changes become necessary.

Now when we implement changes, we are confident that those changes in fact will improve things.

The first thing we did was learn how to use the simulation software. Then we attempted to construct a model of our current operations. This would serve two important functions: it would tell us whether we really understood how to use the software, and it would validate our understanding of our own assembly lines. If we could not construct a model of our assembly line that agreed with the real one, that would mean that there was a major flaw either in our ability to model the system or in our understanding of our own operations.

Building a model of our own operations sounded simple enough. After all, we had an exhaustive catalog of all the steps needed to assemble a computer, and (from previous data collection efforts) information on how long each operation took. But building a model that agreed with our real-world situation turned out to be a challenging yet tremendously educational activity.

For example, one of our early models showed that we were producing several thousand units in a few hours. Since we were not quite that good—we were off by at least a factor of 10—we concluded that we had a major problem in the model we had built, so we went back to study things in more detail. In every case, our early models failed because we had overlooked or misunderstood how things actually worked on our assembly lines.

Eventually, we built a model that worked and agreed reasonably well with our real-world system out in the factory. To make use of the model, we generated ideas that we thought would cut down our assembly time and then simulated them in our model.

We examined a number of changes proposed by our engineers and others, and then simulated the ones that looked most promising. Some proposed changes were counterproductive according to our simulation results. We also did a detailed investigation of our testing stations to determine whether it was more efficient to move computers to be tested into the testing station via FIFO (first-in, first-out) or LIFO (last-in, first-out). Modeling showed us that FIFO was more efficient. When we implemented that change, we realized the gain we had predicted.

Simulation helped us avoid buying more expensive equipment. Some of our material-handling specialists predicted, based on their experience, that if we increased throughput by 30 percent, we would have to add some additional, special equipment to the assembly floor or risk some serious blockages. Simulation showed us that was not true and in practice the simulation turned out to be correct.

We determined that we could move material faster if we gave material movers a specific pattern to follow instead of just doing things sequentially. For example, in moving certain items from our testing area, we determined that the most time-efficient way would be to move shelf 1 first, followed by shelf 4, then shelf 3, and so on.

After our first round of making "serious" changes to our operation and simulating them, our actual production was within a few percentage points of our predicted production. Also, by combining some tasks, we were able to reduce our head count on each assembly line significantly.

We have completed several rounds of changes, and today, encouraged by the experience of our new investor, Samsung, we have made a significant advance that we call Vision 5. The idea of Vision 5 is to have only five people in each cell assembling computers. Although there was initially some skepticism about whether this concept would work, our simulations showed that it would, so today we have converted one of our "focused factories" to this concept and have experienced additional benefits. Seeing the benefits from that effort has caused our management to increase its commitment to simulation.

The Results

Simulation has proven its effectiveness at AST Research. We have achieved a number of useful, measurable goals. For competitive reasons, specific numbers cannot be provided; however, in order of importance, the benefits we have achieved are

- Reduced the reject rate.
- Reduced blockage by 25 percent.
- Increased operator efficiency by 20 percent.
- Increased overall output by 15 percent.
- Reduced the labor cost of each computer.

Other benefits included increased ability to explain and justify proposed changes to management through the use of the graphic animation. Simulation helped us make fewer missteps in terms of implementing changes that could have impaired our output. We were able to try multiple scenarios in our efforts to improve productivity and efficiency at comparatively low cost and risk. We also learned that the best simulation efforts invite participation by more disciplines in the factory, which helps in terms of team-building. All of these benefits were accomplished at minimal cost. These gains have also caused a cultural shift at AST, and because we have a tool that facilitates production changes, the company is now committed to continuous improvement of our assembly practices.

Our use of simulation has convinced us that it produces real, measurable results—and equally important, it has helped us avoid making changes that we thought made common sense, but when simulated turned out to be ineffective. Because of that demonstrable payoff, simulation has become a key element of our toolkit in optimizing production.

Questions

1. What were the objectives for using simulation at AST?
2. Why was simulation better than the traditional methods they were using to achieve these objectives?
3. What common-sense solution was disproved by using simulation?
4. What were some of the unexpected side benefits from using simulation?
5. What insights on the use of simulation did you gain from this case study?

CASE STUDY B
DURHAM REGIONAL HOSPITAL SAVES $150,000 ANNUALLY USING SIMULATION TOOLS

Bonnie Lowder
Management Engineer, Premier

Used with permission by ProModel, 1999.

Durham Regional Hospital, a 450-bed facility located in Durham, North Carolina, has been serving Durham County for 25 years. This public, full-service, acute-care facility is facing the same competition that is now a part of the entire health care industry. With that in mind, Durham Regional Hospital is making a conscious effort to provide the highest quality of care while also controlling costs.

To assist with cost control efforts, Durham Regional Hospital uses Premier's Customer-Based Management Engineering program. Premier's management engineers are very involved with the hospital's reengineering and work redesign projects. Simulation is one of the tools the management engineers use to assist in the redesign of hospital services and processes. Since the hospital was preparing to add an outpatient services area that was to open in May 1997, a MedModel simulation project was requested by Durham Regional Hospital to see how this Express Services area would impact their other outpatient areas.

This project involved the addition of an outpatient Express Services addition. The Express Services area is made up of two radiology rooms, four phlebotomy lab stations, a patient interview room, and an EKG room. The model was set up to examine which kind of patients would best be serviced in that area, what hours the clinic would operate, and what staffing levels would be necessary to provide optimum care.

The Model

Data were collected from each department with potential Express Services patients. The new Express Services area would eliminate the current reception desk in the main radiology department; all radiology outpatients would have their order entry at Express Services. In fiscal year 1996, the radiology department registered 21,159 outpatients. Of those, one-third could have had their procedure performed in Express Services. An average of 18 outpatient surgery patients are seen each week for their preadmission testing. All these patients could have their preadmission tests performed in the Express Services area. The laboratory sees approximately 14 walk-in phlebotomy patients per week. Of those, 10 patients are simple collections and 4 are considered complex. The simple collections can be performed by anyone trained in phlebotomy. The complex collections should be done by skilled lab personnel. The collections for all of these patients could be performed in Express Services. Based on the data, 25 patients a day from the Convenient Care Clinic will need simple X rays and could also use the Express Services area. Procedure times for each patient were determined from previous data collection and observation.

The model was built in two months. Durham Regional Hospital had used simulation in the past for both Emergency Department and Ambulatory Care Unit redesign projects and thus the management team was convinced of its efficacy. After the model was completed, it was presented to department managers from all affected areas. The model was presented to the assembled group in order to validate the data and assumptions. To test for validity, the model was run for 30 replications, with each replication lasting a period of two weeks. The results were measured against known values.

The Results

The model showed that routing all Convenient Care Clinic patients through Express Services would create a bottleneck in the imaging rooms. This would create unacceptable wait times for the radiology patients and the Convenient Care patients. Creating a model scenario where Convenient Care patients were accepted only after 5:00 P.M. showed that the anticipated problem could be eliminated. The model also showed that the weekend volume would be very low. Even at minimum staffing levels, the radiology technicians and clerks would be underutilized. The recommendation was made to close the Express Services area on the weekends. Finally, the model showed that the staffing levels could be lower than had been planned. For example, the workload for the outpatient lab tech drops off after 6:00 P.M. The recommendation was to eliminate outpatient lab techs after 6:00 P.M. Further savings could be achieved by cross-training the radiology technicians and possibly the clerks to perform simple phlebotomy. This would also provide for backup during busy times. The savings for the simulation efforts were projected to be $148,762 annually. These savings were identified from the difference in staffing levels initially requested for Express Services and the levels that were validated after the simulation model results were analyzed, as well as the closing of the clinic on weekends. This model would also be used in the future to test possible changes to Express Services. Durham Regional Hospital would be able to make minor adjustments to the area and visualize the outcome before implementation. Since this was a new area, they would also be able to test minor changes before the area was opened.

The results of the model allowed the hospital to avoid potential bottlenecks in the radiology department, reduce the proposed staffing levels in the Express Services area, and validate that the clinic should be closed on weekends. As stated by Dottie Hughes, director of Radiology Services: "The simulation model allowed us to see what changes needed to be made before the area is opened. By making these changes now, we anticipate a shorter wait time for our patients than if the simulation had not been used." The simulation results were able to show that an annual savings of $148,762 could be expected by altering some of the preconceived Express Services plan.

Future Applications

Larry Suitt, senior vice president, explains, "Simulation has proved to be a valuable tool for our hospital. It has allowed us to evaluate changes in processes before money is spent on construction. It has also helped us to redesign existing services to better meet the needs of our patients." Durham Regional Hospital will continue to use simulation in new projects to improve its health care processes. The hospital is responsible for the ambulance service for the entire county. After the 911 call is received, the hospital's ambulance service picks up the patient and takes him or her to the nearest hospital. Durham Regional Hospital is planning to use simulation to evaluate how relocating some of the ambulances to other stations will affect the response time to the 911 calls.

Questions

1. Why was this a good application for simulation?
2. What key elements of the study made the project successful?
3. What specific decisions were made as a result of the simulation study?
4. What economic benefit was able to be shown from the project?
5. What insights did you gain from this case study about the way simulation is used?

References

Banks, Jerry, and Randall R. Gibson. "Selecting Simulation Software." *IIE Solutions,* May 1997, pp. 29–32.

Kelton, W. D. "Statistical Issues in Simulation." In *Proceedings of the 1996 Winter Simulation Conference,* ed. J. Charnes, D. Morrice, D. Brunner, and J. Swain, 1996, pp. 47–54.

Pritsker, Alan B., and Claude Dennis Pegden. *Introduction to Simulation and SLAM.* New York: John Wiley & Sons, 1979.

Schrage, Michael. *Serious Play: How the World's Best Companies Simulate to Innovate.* Cambridge, MA: Harvard Business School Press, 1999.

6   DATA COLLECTION AND ANALYSIS

"You can observe a lot just by watching."

—Yogi Berra

6.1 Introduction

In the previous chapter, we discussed the importance of having clearly defined objectives and a well-organized plan for conducting a simulation study. In this chapter, we look at the data-gathering phase of a simulation project that defines the system being modeled. The result of the data-gathering effort is a conceptual or mental model of how the system is configured and how it operates. This conceptual model may take the form of a written description, a flow diagram, or even a simple sketch on the back of an envelope. It becomes the basis for the simulation model that will be created.

Data collection is the most challenging and time-consuming task in simulation. For new systems, information is usually very sketchy and only roughly estimated. For existing systems, there may be years of raw, unorganized data to sort through. Information is seldom available in a form that is directly usable in building a simulation model. It nearly always needs to be filtered and massaged to get it into the right format and to reflect the projected conditions under which the system is to be analyzed. Many data-gathering efforts end up with lots of data but little useful information. Data should be gathered purposefully to avoid wasting not only the modeler's time but also the time of individuals who are supplying the data.

This chapter presents guidelines and procedures for gathering data. Statistical techniques for analyzing data and fitting probability distributions to data are also discussed. The following questions are answered:

- What is the best procedure to follow when gathering data?
- What types of data should be gathered?

- What sources should be used when gathering data?
- What types of analyses should be performed on the data?
- How do you select the right probability distribution that represents the data?
- How should data be documented?

6.2 Guidelines for Data Gathering

Data gathering should not be performed haphazardly with the idea that you will sort through all the information once it is collected. Data should be collected systematically, seeking specific items of information that are needed for building the model. If approached with the end in mind, data gathering will be much more productive. Following are a few guidelines to keep in mind when gathering data that will help maintain a purposeful focus.

1. *Identify triggering events.* When defining the activities that occur in the system to be modeled, it is important to identify the causes or conditions that trigger the activities. In gathering downtime data, for example, it is useful to distinguish between downtimes due to failure, planned or scheduled downtimes, and downtimes that are actually idle periods due to unavailability of stock. In other words, you need to know the cause of the downtime. Questions should be asked such as What triggers the movement of entities? and What triggers the use of a particular resource? To be valid, a model needs to capture the right triggering events that initiate the activities within the system.

2. *Focus only on key impact factors.* Discrimination should be used when gathering data to avoid wasting time tracking down information that has little or no impact on system performance. For example, off-shift activities such as preventive maintenance that do not delay the process flow should be ignored because they are performed during off-hours. Similarly, extremely rare downtimes, negligible move times, on-the-fly inspections, external setups, and other activities that have little or no impact on the overall flow may be safely ignored.

3. *Isolate actual activity times.* In determining activity times, it is important to isolate only the time it takes to do the activity itself, excluding any extraneous time waiting for material and resources so the activity can be performed. For example, filling a customer order or performing an assembly operation requires that items or component parts first be available. The time to wait for items or parts should not be included in the time to do the actual assembly. This waiting time is going to vary based on conditions in the system and therefore can't be known in advance. It will be known only as the simulation is run. In some instances, isolating actual activity times may be difficult, especially when dealing with historical data reflecting aggregate times. Historical

records of repair times, for example, often lump together the time spent waiting for repair personnel to become available and the actual time spent performing the repair. What you would like to do is separate the waiting time from the actual repair time because the waiting time is a function of the availability of the repair person, which may vary depending on the system operation.

4. *Look for common groupings.* When dealing with lots of variety in a simulation such as hundreds of part types or customer profiles, it helps to look for common groupings or patterns. If, for example, you are modeling a process that has 300 entity types, it may be difficult to get information on the exact mix and all of the varied routings that can occur. Having such detailed information is usually too cumbersome to work with even if you did have it. The solution is to reduce the data to common behaviors and patterns. One way to group common data is to first identify general categories into which all data can be assigned. Then the percentage of cases that fall within each category is calculated or estimated. It is not uncommon for beginning modelers to attempt to use actual logged input streams such as customer arrivals or material shipments when building a model. After struggling to define hundreds of individual arrivals or routings, it begins to dawn on them that they can group this information into a few categories and assign probabilities that any given instance will fall within a particular category. This allows dozens and sometimes hundreds of unique instances to be described in a few brief commands. The secret to identifying common groupings is to "think probabilistically."

5. *Focus on essence rather than substance.* A system definition for modeling purposes should capture the cause-and-effect relationships and ignore the meaningless (to simulation) details. This is called *system abstraction* and seeks to define the essence of system behavior rather than the substance. A system should be abstracted to the highest level possible while still preserving the essence of the system operation. Using this "black box" approach to system definition, we are not concerned about the nature of the activity being performed, such as milling, grinding, or inspection. We are interested only in the impact that the activity has on the use of resources and the delay of entity flow. A proficient modeler constantly is thinking abstractly about the system operation and avoids getting caught up in the mechanics of the process.

6. *Separate input variables from response variables.* First-time modelers often confuse input variables that define the operation of the system with response variables that report system performance. Input variables define how the system works (activity times, routing sequences, and the like) and should be the focus of the data gathering. Response variables describe how the system responds to a given set of input variables (amount of work in process, resource utilization, throughput times, and so on).

Response variables do not "drive" model behavior. Performance or response data should be of interest only if you are building an as-is model and want to validate the simulation by comparing the simulation output with actual performance.

Following these guidelines, the data-gathering effort will be much more productive. In addition to keeping these guidelines in mind, it is also helpful to follow a logical sequence of steps when gathering data. This ensures that time and efforts are well spent. The steps to gathering data should follow roughly this sequence:

Step 1: Determine data requirements.
Step 2: Identify data sources.
Step 3: Collect the data.
Step 4: Make assumptions where necessary.
Step 5: Analyze the data.
Step 6: Document and approve the data.

Of course, most of these steps overlap, and some iteration will occur as objectives change and assumptions are refined. The balance of this chapter is devoted to providing recommendations and examples of how to perform each of these steps.

6.3 Determining Data Requirements

The first step in gathering data is to determine the data required for building the model. This should be dictated primarily by the model scope and level of detail required for achieving the objectives of the simulation. System data can be categorized as structural data, operational data, or numerical data.

6.3.1 Structural Data

Structural data involve all of the objects in the system to be modeled. This includes such elements as entities (products, customers, and so on), resources (operators, machines) and locations (waiting areas, workstations). Structural information basically describes the layout or configuration of the system as well as identifies the items that are processed. It is important that all relevant components that affect the behavior of the system be included.

6.3.2 Operational Data

Operational data explain how the system operates—that is, when, where, and how events and activities take place. Operational data consist of all the logical or behavioral information about the system such as routings, schedules, downtime behavior, and resource allocation. If the process is structured and well controlled, operational information is easy to define. If, on the other hand, the process has evolved into an informal operation with no set rules, it can be very difficult to define. For a system to be simulated, operating policies that are undefined and ambiguous must be codified into defined procedures and rules. If decisions and outcomes vary, it is important to at least define this variability statistically using probability expressions or distributions.

6.3.3 Numerical Data

Numerical data provide quantitative information about the system. Examples of numerical data include capacities, arrival rates, activity times, and time between failures. Some numerical values are easily determined, such as resource capacities and working hours. Other values are more difficult to assess, such as time between failures or routing probabilities. This is especially true if the system being modeled is new and data are unavailable. It is important when gathering numerical data to ask exactly what a particular data value represents. For example, in collecting machining times, one should ask whether the time reflects load and unload time or tool adjustments, and whether it is for processing a batch of parts rather than individual parts.

6.3.4 Use of a Questionnaire

To help focus data-gathering efforts on the right information and to ensure productive meetings with others on whom you depend for model information, it may be useful to prepare a questionnaire for use in interviews. It is much easier for people to respond to specific questions than to puzzle over and guess at what information is needed. When conducting an interview, sometimes giving the contact person a specific list of questions in advance reduces the length of the interview and the number of follow-up visits that would otherwise be needed.

Below is a list of questions that might be included in a questionnaire. For any questions requiring numeric information that could take the form of a random variable, it is best if either a good sample of data or the actual defining probability distribution can be obtained. If sample data are unavailable, it is useful to get at least estimates of the minimum, most likely, and maximum values until more precise data can be obtained. Average values should be used only as preliminary estimates.

System Description Questionnaire

1. What types of entities are processed in the system?
2. What is the routing sequence (each stopping or decision point) for each entity type?
3. Where, when, and in what quantities do entities enter the system?
4. What are the time and resource requirements for each operation and move?
5. In what quantities are entities processed and moved? (Define for each location.)

6. What triggers entity movement from location to location (completion of the operation, accumulation of a batch, a signal from a downstream location, etc.)?

7. How do locations and resources determine which job to do next (oldest waiting, highest priority, etc.)?

8. How are alternative routing and operation decisions made (percentage, condition, etc.)?

9. How often do interruptions occur (setups, downtimes, etc.) and what resources and time are required when they happen?

10. What is the schedule of availability for locations and resources (define in terms of shifts, break times, scheduled maintenance intervals, etc.)?

Answers to these questions should provide most if not all of the information needed to build a model. Depending on the nature of the system and level of detail required, additional questions may be added.

6.4 Identifying Data Sources

Rarely is all the information needed to build a model available from a single source. It is usually the result of reviewing reports, conducting personal interviews, making personal observations, and making lots of assumptions. "It has been my experience," notes Carson (1986), "that for large-scale real systems, there is seldom any one individual who understands how the system works in sufficient detail to build an accurate simulation model. The modeler must be willing to be a bit of a detective to ferret out the necessary knowledge." This is where persistence, careful observation, and good communication skills pay off.

The types of sources to use when gathering data will depend largely on whether you are simulating an existing system or a new system. For existing systems, there will be a much wider range of sources from which to draw since there are records and individuals with firsthand knowledge of the system. The fact that the system exists also permits the modeler to personally observe the system and therefore be less dependent on the input of others—although collaboration in data gathering is always recommended. For new systems, the sources of information are usually limited to those individuals directly involved in the design of the system. Most of the information is obtained from system specifications and personal interviews with the design team. Depending on the circumstances, good sources of data include

- *Historical records*—production, sales, scrap rates, equipment reliability.
- *System documentation*—process plans, facility layouts, work procedures.
- *Personal observation*—facility walk-through, time studies, work sampling.
- *Personal interviews*—operators (work methods), maintenance personnel (repair procedures), engineers (routings), managers (schedules and forecasts).
- *Comparison with similar systems*—within the company, within the same industry, within any industry.

- *Vendor claims*—process times, reliability of new machines.
- *Design estimates*—process times, move times, and so on for a new system.
- *Research literature*—published research on learning curves, predetermined time studies.

In deciding whether to use a particular data source, it is important to consider the reliability and accessibility of the source. Reliability of the source directly impacts the validity of the model. A manager's recollection, for example, may not be as reliable as actual production logs. Opinions are frequently biased and tend to provide distorted information. Objectivity should be sought when using any data source.

Accessibility is also an important factor in source selection. If the source is difficult to access, such as historical data filed away in some corporate archive or buried somewhere among hundreds of reports, it may have to be overlooked. If the system spans several businesses, such as a supply chain, there even may be an unwillingness to disclose information. The principle to follow is to find out what sources are accessible and then use the sources that provide the most reliable and complete information. Secondary sources can always validate the primary source.

6.5 Collecting the Data

When gathering data, it is best to go from general to specific. In practice, data gathering continues through to the end of the simulation project as objectives change and information that was unavailable at the beginning of the project begins to materialize. Consistent with this progressive refinement approach, data should be gathered in the following sequence:

1. Define the overall entity flow.
2. Develop a description of operation.
3. Define incidental details and firm up data values.

This doesn't necessarily mean that data will become available in this sequence. It only means that the focus of attention should follow this order. For example, some assumptions may need to be made initially to come up with a complete definition of entity flow. Other incidental information obtained early on, such as downtimes, may need to be set aside until it is needed at a later stage of model building.

6.5.1 Defining the Entity Flow

The first area of focus in data collection should be in defining the basic entity flow through the system. This establishes a skeletal framework to which additional data can be attached. The entity flow is defined by following the entity movement through the system, taking the perspective of the entity itself. The overall entity flow is best described using an entity flow diagram (see Figure 6.1) or perhaps by superimposing the entity flow on an actual layout of the system.

An entity flow diagram is slightly different from a process flowchart. A *process flowchart* shows the logical sequence of activities through which entities

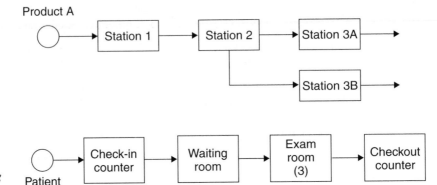

FIGURE 6.1

Entity flow diagram.

FIGURE 6.2

Entity flow diagram for patient processing at Dr. Brown's office.

go and defines *what* happens to the entity, not *where* it happens. An *entity flow diagram,* on the other hand, is more of a routing chart that shows the physical movement of entities through the system from location to location. An entity flow diagram should depict any branching that may occur in the flow such as routings to alternative work centers or rework loops. The purpose of the entity flow diagram is to document the overall flow of entities in the system and to provide a visual aid for communicating the entity flow to others. A flow diagram is easy to understand and gets everyone thinking in the same way about the system. It can easily be expanded as additional information is gathered to show activity times, where resources are used, and so forth.

6.5.2 Developing a Description of Operation

Once an entity flow diagram has been created, a *description of operation* should be developed that explains how entities are processed through the system. A description of operation may be written in a step-by-step, brief narrative form, or it may be represented in tabular form. For simple systems, the entity flow diagram itself can be annotated with operational information. Whatever the form, it should identify for each type of entity at each location in the system

- The time and resource requirements of the activity or operation.
- Where, when, and in what quantities entities get routed next.
- The time and resource requirements for moving to the next location.

Example: Patients enter Dr. Brown's office and sign in at the check-in counter before taking a seat to wait for the nurse's call. Patients are summoned to one of three examination rooms when one becomes available. The nurse escorts patients to the examination room, where they wait for Dr. Brown to see them. After treatment, patients return on their own to the checkout counter, where they make payment and possibly reschedule another visit. This process is shown in the entity flow diagram in Figure 6.2.

Using the entity flow diagram and information provided, a summary description of operation is created using a table format, as shown in Table 6.1.

TABLE 6.1   Process Description for Dr. Brown's Office

Location	Activity Time	Activity Resource	Next Location	Move Trigger	Move Time	Move Resource
Check-in counter	$N(1,.2)$ min.	Secretary	Waiting room	None	0.2 min.	None
Waiting room	None	None	Exam room	When room is available	0.8 min.*	Nurse
Exam room	$N(15,4)$ min.	Doctor	Checkout counter	None	0.2 min.	None
Checkout counter	$N(3,.5)$ min.	Secretary	Exit	None	None	None

*Stops to get weighed on the way.

Notice that the description of operation really provides the details of the entity flow diagram. This detail is needed for defining the simulation model. The times associated with activities and movements can be just estimates at this stage. The important thing to accomplish at this point is simply to describe how entities are processed through the system.

The entity flow diagram, together with the description of operation, provides a good data document that can be expanded as the project progresses. At this point, it is a good idea to conduct a structured walk-through of the operation using the entity flow diagram as the focal point. Individuals should be involved in this review who are familiar with the operation to ensure that the description of operation is accurate and complete.

Based on this description of operation, a first cut at building the model can begin. Using ProModel, a model for simulating Dr. Brown's practice can be built in a matter of just a few minutes. Little translation is needed as both the diagram (Figure 6.2) and the data table (Table 6.1) can be entered pretty much in the same way they are shown. The only additional modeling information required to build a running model is the interarrival time of patients.

Getting a basic model up and running early in a simulation project helps hold the interest of stakeholders. It also helps identify missing information and motivates team members to try to fill in these information gaps. Additional questions about the operation of the system are usually raised once a basic model is running. Some of the questions that begin to be asked are Have all of the routings been accounted for? and Have any entities been overlooked? In essence, modeling the system actually helps define and validate system data.

6.5.3 Defining Incidental Details and Refining Data Values

Once a basic model has been constructed and tested, additional details of the process such as downtimes, setups, and work priorities can be added. This information is not essential to getting a running model, but is necessary for a complete and accurate model. Sometimes the decision of whether to include a particular element of the system such as a downtime is easier to make once the basic model is running. The potential impact of the element and the time available to implement it in the model are much clearer than at the start of the project.

At this point, any numerical values such as activity times, arrival rates, and others should also be firmed up. Having a running model enables estimates and other assumptions to be tested to see if it is necessary to spend additional time getting more accurate information. For existing systems, obtaining more accurate data is usually accomplished by conducting time studies of the activity or event under investigation. A sample is gathered to represent all conditions under which the activity or event occurs. Any biases that do not represent normal operating conditions are eliminated. The sample size should be large enough to provide an accurate picture yet not so large that it becomes costly without really adding additional information.

6.6 Making Assumptions

A simulation model can run with incorrect data, but it can't run with incomplete data. Using an architectural metaphor, a computer model is less like an artist's rendition of a building and much more like a full set of blueprints—complete with dimensions and material lists. This is one of the reasons why simulation is so beneficial. It gets down to the operational details that actually define how the system runs.

It doesn't take long after data gathering has commenced to realize that certain information is unavailable or perhaps unreliable. Complete, accurate, and up-to-date information is rarely obtainable, especially when modeling a new system about which little is known. Even for existing systems, it may be impractical to obtain certain types of data, such as long time intervals between machine failures.

Because the reliability of information on new systems is often questionable, the validity of the model itself is going to be suspect. Accurate information is going to be lacking in at least one or more areas of almost any model—even in models of existing systems. Because a simulation nearly always is run for some future anticipated situation, there are always going to be assumptions made regarding any unknown future conditions. In simulation, we must be content to begin with assumptions and be grateful whenever we can make assertions.

Since a model that is based on assumptions is automatically invalid, the question might be raised, What good is a simulation that is based on assumptions? We forget that we routinely make decisions in life based on assumptions and therefore can never be quite sure of the outcome. Any prediction about how a system will operate in the future is based on assumptions about the future. There is nothing wrong with making assumptions when doing simulation as long as our confidence in the results never exceeds our confidence in the underlying assumptions. The reason for doing simulation is to be able to predict system performance based on given assumptions. It is better to make decisions knowing the implications of our assumptions than to compound the error of our decision by drawing erroneous inferences from these assumptions. Incorporating assumptions into a model can actually help validate our assumptions by seeing whether they make sense in the overall operation of the model. Seeing absurd behavior may tell us that certain assumptions just don't make sense.

For comparative studies in which two design alternatives are evaluated, the fact that assumptions are made is less significant because we are evaluating *relative* performance, not *absolute* performance. For example, if we are trying to determine whether on-time deliveries can be improved by assigning tasks to a resource by due date rather than by first-come, first-served, a simulation can provide useful information to make this decision without necessarily having completely accurate data. Because both models use the same assumptions, it may be possible to compare relative performance. We may not know what the absolute performance of the best option is, but we should be able to assess fairly accurately how much better one option is than another.

Some assumptions will naturally have a greater influence on the validity of a model than others. For example, in a system with large processing times compared to move times, a move time that is off by 20 percent may make little or no difference in system throughput. On the other hand, an activity time that is off by 20 percent could make a 20 percent difference in throughput. One way to assess the influence of an assumption on the validity of a model is through sensitivity analysis. *Sensitivity analysis,* in which a range of values is tested for potential impact on model performance, can indicate just how accurate an assumption needs to be. A decision can then be made to firm up the assumption or to leave it as is. If, for example, the degree of variation in a particular activity time has little or no impact on system performance, then a constant activity time may be used. At the other extreme, it may be found that even the type of distribution has a noticeable impact on model behavior and therefore needs to be selected carefully.

A simple approach to sensitivity analysis for a particular assumption is to run three different scenarios showing (1) a "best" or most optimistic case, (2) a "worst" or most pessimistic case, and (3) a "most likely" or best-guess case. These runs will help determine the extent to which the assumption influences model behavior. It will also help assess the risk of relying on the particular assumption.

6.7 Statistical Analysis of Numerical Data

To be of use in a simulation model, raw data must be analyzed and interpreted so that the system operation is correctly represented in the model. Triggering events and other cause-and-effect relationships need to be identified. Irrelevant or insignificant data need to be discarded. Actual activity times need to be isolated from system-caused delays. The numerous instances of a particular activity need to be generalized into a few defining patterns. Complex operation descriptions need to be translated into simplified abstractions. Input or defining system variables need to be separated from output or response variables. Finally, summarized historical data should be projected forward to the time period being studied by the simulation. Information to help adjust data for the future can be obtained from sales forecasts, market trends, or business plans.

Prior to developing a representation of the data, the data should be analyzed to ascertain their suitability for use in the simulation model. Data characteristics such as independence (randomness), homogeneity (the data come from the same

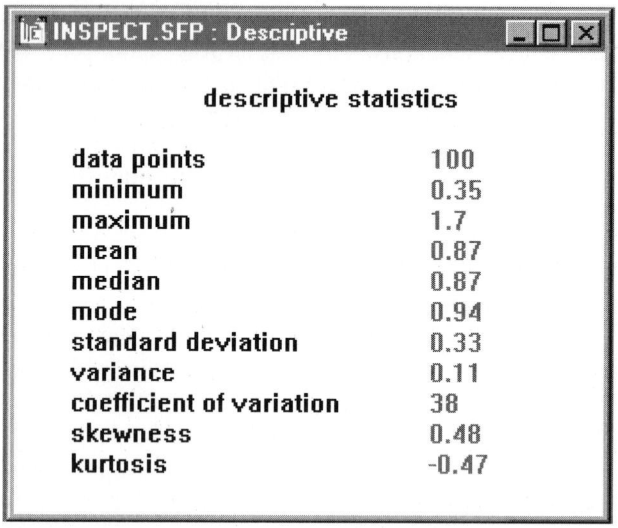

FIGURE 6.3

Descriptive statistics
for a sample data set
of 100 observations.

```
┌─────────────────────────────────────────────────┐
│ INSPECT.SFP : Descriptive              _ □ ✕    │
├─────────────────────────────────────────────────┤
│                                                  │
│              descriptive statistics              │
│                                                  │
│        data points              100              │
│        minimum                  0.35             │
│        maximum                  1.7              │
│        mean                     0.87             │
│        median                   0.87             │
│        mode                     0.94             │
│        standard deviation       0.33             │
│        variance                 0.11             │
│        coefficient of variation 38               │
│        skewness                 0.48             │
│        kurtosis                 -0.47            │
│                                                  │
└─────────────────────────────────────────────────┘
```

TABLE 6.2   100 Observed Inspection Times (Minutes)

0.99	0.41	0.89	0.59	0.98	0.47	0.70	0.94	0.39	0.92
1.30	0.67	0.64	0.88	0.57	0.87	0.43	0.97	1.20	1.50
1.20	0.98	0.89	0.62	0.97	1.30	1.20	1.10	1.00	0.44
0.67	1.70	1.40	1.00	1.00	0.88	0.52	1.30	0.59	0.35
0.67	0.51	0.72	0.76	0.61	0.37	0.66	0.75	1.10	0.76
0.79	0.78	0.49	1.10	0.74	0.97	0.93	0.76	0.66	0.57
1.20	0.49	0.92	1.50	1.10	0.64	0.96	0.87	1.10	0.50
0.60	1.30	1.30	1.40	1.30	0.96	0.95	1.60	0.58	1.10
0.43	1.60	1.20	0.49	0.35	0.41	0.54	0.83	1.20	0.99
1.00	0.65	0.82	0.52	0.52	0.80	0.72	1.20	0.59	1.60

distribution), and stationarity (the distribution of the data doesn't change with time) should be determined. Using data analysis software such as Stat::Fit, data sets can be automatically analyzed, tested for usefulness in a simulation, and matched to the best-fitting underlying distribution. Stat::Fit is bundled with ProModel and can be accessed from the Tools menu after opening ProModel. To illustrate how data are analyzed and converted to a form for use in simulation, let's take a look at a data set containing 100 observations of an inspection operation time, shown in Table 6.2.

By entering these data or importing them from a file into Stat::Fit, a descriptive analysis of the data set can be performed. The summary of this analysis is displayed in Figure 6.3. These parameters describe the entire sample collection. Because reference will be made later to some of these parameters, a brief definition of each is given below.

Mean—the average value of the data.

Median—the value of the middle observation when the data are sorted in ascending order.

Mode—the value that occurs with the greatest frequency.

Standard deviation—a measure of the average deviation from the mean.

Variance—the square of the standard deviation.

Coefficient of variation—the standard deviation divided by the mean. This value provides a relative measure of the standard deviation to the mean.

Skewness—a measure of symmetry. If the maximum data value is farther from the mean than the minimum data value, the data will be skewed to the right, or positively skewed. Otherwise they are skewed to the left, or negatively skewed.

Kurtosis—a measure of the flatness or peakedness of the distribution.

Another value of interest is the *range,* which is simply the difference between the maximum and minimum values.

A descriptive analysis tells us key characteristics about the data set, but it does not tell us how suitable the data are for use in a simulation model. What we would like to be able to do is fit a theoretical distribution, such as a normal or beta distribution, to the sample data set. This requires that the data be independent (truly random) and identically distributed (they all come from the same distribution). When gathering data from a dynamic and possibly time-varying system, one must be sensitive to trends, patterns, and cycles that may occur with time. The sample data points may be somewhat codependent, or they may come from multiple sources and therefore may not be identically distributed. Specific tests can be performed to determine whether data are independent and identically distributed.

6.7.1 Tests for Independence

Data are independent if the value of one observation is not influenced by the value of another observation. Dependence is common in sampling from a finite population when sampling is performed without replacement. For example, in sampling from a deck of cards to find the probability of drawing an ace, the probability of the next card being an ace increases with each draw that is not an ace. The probability of drawing an ace in the first draw is $\frac{4}{52}$. If no ace is drawn, the probability of drawing an ace in the second draw is $\frac{4}{51}$. If, however, the drawn card is replaced each time, the probability remains the same ($\frac{4}{52}$) for each draw.

For very large or infinite populations such as ongoing activities or repeating events, data dependence usually manifests itself when the value of an observation is influenced by the value of the previous observation. In sampling the probability that an item will be defective, for example, sometimes there is a greater probability of a defective item occurring if the previous item was defective (defectives frequently occur in waves). Perhaps the time to perform a manual cutting operation increases each time due to the blade becoming dull. Finally, when the blade is sharpened, the activity time is reduced back to the original starting time. When the value of an observation is dependent on the value of a previous observation, the data are said to be *correlated.* Table 6.3 shows outdoor temperature recordings taken randomly throughout the day from 8:00 A.M. to 8:00 P.M. (read left to right). As would be expected, sampled temperatures are not independent because they depend largely on the time of day.

TABLE 6.3   **100 Outdoor Temperature Readings from 8:00 A.M. to 8:00 P.M.**

57	57	58	59	59	60	60	62	62	62
63	63	64	64	65	66	66	68	68	69
70	71	72	72	73	74	73	74	75	75
75	75	76	77	78	78	79	80	80	81
80	81	81	82	83	83	84	84	83	84
83	83	83	82	82	81	81	80	81	80
79	79	78	77	77	76	76	75	74	75
75	74	73	73	72	72	72	71	71	71
71	70	70	70	70	69	69	68	68	68
67	67	66	66	66	65	66	65	65	64

Several techniques are used to determine data dependence or correlation. Three techniques discussed here include the scatter plot, autocorrelation plot, and runs test. Any of these procedures can be performed on data sets using Stat::Fit. They are each specific to different scales of randomness, so all must be passed to conclude data independence. The scatter plot and autocorrelation plot are parametric statistical tests in which the nature of the plot depends on the underlying distribution of the data. The runs test is nonparametric and makes no assumption about the distribution of the data. We explain how each is used.

Scatter Plot

This is a plot of adjacent points in the sequence of observed values plotted against each other. Thus each plotted point represents a pair of consecutive observations (X_i, X_{i+1}) for $i = 1, 2, \ldots, n - 1$. This procedure is repeated for all adjacent data points so that 100 observations would result in 99 plotted points. If the X_i's are independent, the points will be scattered randomly. If, however, the data are dependent on each other, a trend line will be apparent. If the X_i's are positively correlated, a positively sloped trend line will appear. Negatively correlated X_i's will produce a negatively sloped line. A scatter plot is a simple way to detect strongly dependent behavior.

Figure 6.4 shows a plot of paired observations using the 100 inspection times from Table 6.2. Notice that the points are scattered randomly with no defined pattern, indicating that the observations are independent.

Figure 6.5 is a plot of the 100 temperature observations shown in Table 6.3. Notice the strong positive correlation.

Autocorrelation Plot

If observations in a sample are independent, they are also uncorrelated. Correlated data are dependent on each other and are said to be autocorrelated. A measure of the autocorrelation, rho (ρ), can be calculated using the equation

$$\rho = \sum_{i=1}^{n-j} \frac{(x_i - \bar{x})(x_{i+j} - \bar{x})}{\sigma^2 (n - j)}$$

FIGURE 6.4

Scatter plot showing uncorrelated data.

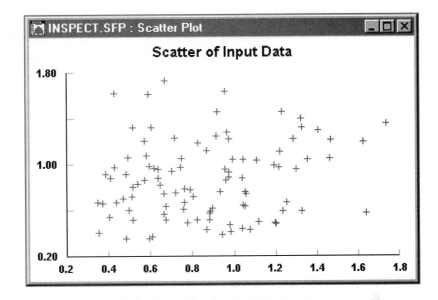

FIGURE 6.5

Scatter plot showing correlated temperature data.

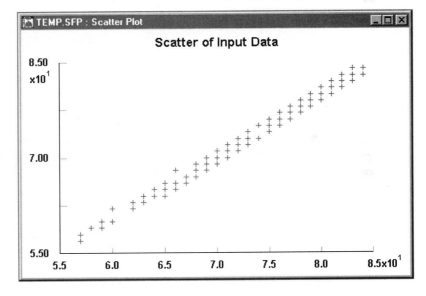

where j is the lag or distance between data points; σ is the standard deviation of the population, approximated by the standard deviation of the sample; and $\bar{X}$ is the sample mean. The calculation is carried out to $\frac{1}{5}$ of the length of the data set, where diminishing pairs start to make the calculation unreliable.

This calculation of autocorrelation assumes that the data are taken from a stationary process; that is, the data would appear to come from the same distribution regardless of when the data were sampled (that is, the data are time invariant). In the case of a time series, this implies that the time origin may be shifted without

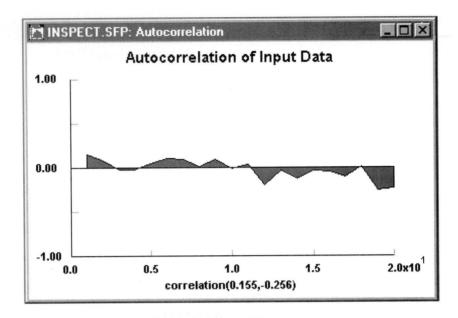

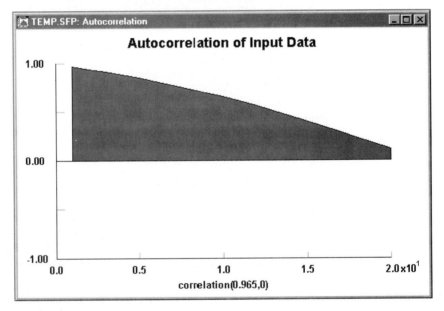

affecting the statistical characteristics of the series. Thus the variance for the whole sample can be used to represent the variance of any subset. If the process being studied is not stationary, the calculation of autocorrelation is more complex.

The autocorrelation value varies between 1 and -1 (that is, between positive and negative correlation). If the autocorrelation is near either extreme, the data are autocorrelated. Figure 6.6 shows an autocorrelation plot for the 100 inspection time observations from Table 6.2. Notice that the values are near zero, indicating

FIGURE 6.8

Runs test based on points above and below the median and number of turning points.

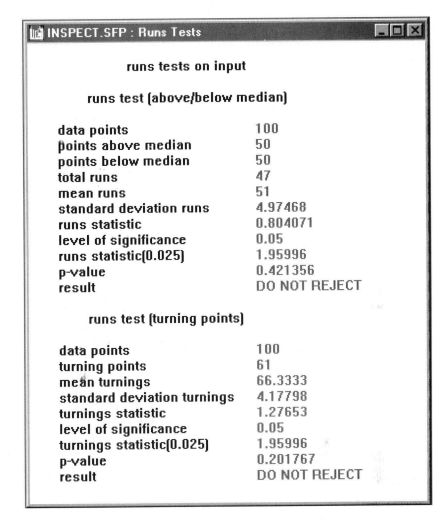

little or no correlation. The numbers in parentheses below the *x* axis are the maximum autocorrelation in both the positive and negative directions.

Figure 6.7 is an autocorrelation plot for the sampled temperatures in Table 6.3. The graph shows a broad autocorrelation.

Runs Tests

The runs test looks for runs in the data that might indicate data correlation. A *run* in a series of observations is the occurrence of an uninterrupted sequence of numbers showing the same trend. For instance, a consecutive set of increasing or decreasing numbers is said to provide runs "up" or "down" respectively. Two types of runs tests that can be made are the *median test* and the *turning point test*. Both of these tests can be conducted automatically on data using Stat::Fit. The runs test for the 100 sample inspection times in Table 6.2 is summarized in Figure 6.8. The result of each test is either *do not reject* the hypothesis that the series is random or *reject* that hypothesis with the level of significance given. The level of significance is the probability that a rejected hypothesis is actually true—that is, that the test rejects the randomness of the series when the series is actually random.

The median test measures the number of runs—that is, sequences of numbers, above and below the median. The run can be a single number above or below the median if the numbers adjacent to it are in the opposite direction. If there are too many or too few runs, the randomness of the series is rejected. This median runs test uses a normal approximation for acceptance or rejection that requires that the number of data points above or below the median be greater than 10.

The turning point test measures the number of times the series changes direction (see Johnson, Kotz, and Kemp 1992). Again, if there are too many turning points or too few, the randomness of the series is rejected. This turning point runs test uses a normal approximation for acceptance or rejection that requires more than 12 data points.

While there are many other runs tests for randomness, some of the most sensitive require larger data sets, in excess of 4,000 numbers (see Knuth 1981).

The number of runs in a series of observations indicates the randomness of those observations. A few runs indicate strong correlation, point to point. Several runs may indicate cyclic behavior.

6.7.2 Tests for Identically Distributed Data

When collecting sample data, often it is necessary to determine whether the data in a single data set come from the same population (they are identically distributed) or whether they represent multiple populations. In other instances, it is desirable to determine whether two or more data sets can be treated as having come from the same population or whether they need to be kept as separate populations. There are several ways to determine whether data come from the same population. These are sometimes called *tests for homogeneity*.

Let's look at the first case of determining whether data from a single data set are identically distributed. The first inclination when gathering data defining a particular activity time or event interval is to assume that they are identically distributed. Upon closer examination, however, it is sometimes discovered that they are actually nonhomogeneous data. Examples of data sets that tend to be nonhomogeneous include

- Activity times that are longer or shorter depending on the type of entity being processed.
- Interarrival times that fluctuate in length depending on the time of day or day of the week.
- Time between failures and time to repair where the failure may result from a number of different causes.

In each of these instances, the data collected may actually represent multiple populations.

Testing to see if data are identically distributed can be done in several ways. One approach is to visually inspect the distribution to see if it has more than one mode. You may find, for example, that a plot of observed downtime values is bimodal, as shown in Figure 6.9. The fact that there are two clusters of data indicates that there are at least two distinct causes of downtimes, each producing different distributions for repair times. Perhaps after examining the cause of the downtimes, it is discovered that some were due to part jams that were quickly fixed while others were due to mechanical failures that took longer to repair.

One type of nonhomogeneous data occurs when the distribution changes over time. This is different from two or more distributions manifesting themselves over the same time period such as that caused by mixed types of downtimes. An example of a time-changing distribution might result from an operator who works 20 percent faster during the second hour of a shift than during the first hour. Over long periods of time, a *learning curve* phenomenon occurs where workers perform at a faster rate as they gain experience on the job. Such distributions are called *nonstationary* or *time variant* because of their time-changing nature. A common example of a distribution that changes with time is the arrival rate of customers to a service facility. Customer arrivals to a bank or store, for example, tend to occur at a rate that fluctuates throughout the day. Nonstationary distributions can be detected by plotting subgroups of data that occur within successive time intervals. For example, sampled arrivals between 8 A.M. and 9 A.M. can be plotted separately from arrivals between 9 A.M. and 10 A.M., and so on. If the distribution is of a different type or is the same distribution but shifted up or down such that the mean changes value over time, the distribution is nonstationary. This fact will need to be taken into account when defining the model behavior. Figure 6.10 is a plot of customer arrival rates for a department store occurring by half-hour interval between 10 A.M. and 6 P.M.

FIGURE 6.9

Bimodal distribution of downtimes indicating multiple causes.

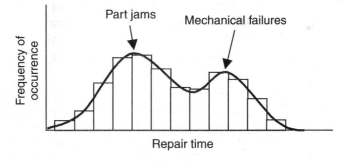

FIGURE 6.10

Change in rate of customer arrivals between 10 A.M. and 6 P.M.

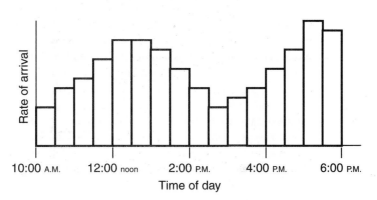

Note that while the type of distribution (Poisson) is the same for each period, the rate (and hence the mean interarrival time) changes every half hour.

The second case we look at is where two sets of data have been gathered and we desire to know whether they come from the same population or are identically distributed. Situations where this type of testing is useful include the following:

- Interarrival times have been gathered for different days and you want to know if the data collected for each day come from the same distribution.
- Activity times for two different operators were collected and you want to know if the same distribution can be used for both operators.
- Time to failure has been gathered on four similar machines and you are interested in knowing if they are all identically distributed.

One easy way to tell whether two sets of data have the same distribution is to run Stat::Fit and see what distribution best fits each data set. If the same distribution fits both sets of data, you can assume that they come from the same population. If in doubt, they can simply be modeled as separate distributions.

Several formal tests exist for determining whether two or more data sets can be assumed to come from identical populations. Some of them apply to specific families of distributions such as analysis of variance (ANOVA) tests for normally distributed data. Other tests are distribution independent and can be applied to compare data sets having any distribution, such as the Kolmogorov-Smirnov two-sample test and the chi-square multisample test (see Hoover and Perry 1990). The Kruskal-Wallis test is another nonparametric test because no assumption is made about the distribution of the data (see Law and Kelton 2000).

6.8 Distribution Fitting

Once numerical data have been tested for independence and correlation, they can be converted to a form suitable for use in the simulation model. Sample numerical data that have been gathered on activity times, arrival intervals, batch quantities, and so forth can be represented in a simulation model in one of three ways. First, the data can be used exactly the way they were recorded. Second, an empirical distribution can be used that characterizes the data. The third and preferred method is to select a theoretical distribution that best fits the data.

Using the data exactly as they were recorded is relatively straightforward but problematic. During the simulation, data values are read directly from the sample data file in the order in which they were collected. The problem with this technique is that the data set is usually not large enough to represent the population. It also presents a problem for running multiple replications of a simulation experiment because a separate data set, which usually isn't available, is needed for each replication. Sample data sets tend to be used directly in a simulation when they are sampled data from another related simulation model from which it receives its input. This is a model partitioning technique sometimes used in simulating large systems. The exit times for entities in one simulation become the arrival stream used for the model into which it feeds.

Using an empirical distribution in ProModel requires that the data be converted to either a continuous or discrete frequency distribution. A continuous frequency distribution summarizes the percentage of values that fall within given intervals. In the case of a discrete frequency distribution, it is the percentage of times a particular value occurs. For continuous frequency distributions the intervals need not be of equal width. During the simulation, random variates are generated using a continuous, piecewise-linear empirical distribution function based on the grouped data (see Law and Kelton, 2000). The drawbacks to using an empirical distribution as input to a simulation are twofold. First, an insufficient sample size may create an artificial bias or "choppiness" in the distribution that does not represent the true underlying distribution. Second, empirical distributions based on a limited sample size often fail to capture rare extreme values that may exist in the population from which they were sampled. As a general rule, empirical distributions should be used only for rough-cut modeling or when the shape is very irregular and doesn't permit a good distribution fit.

Representing the data using a theoretical distribution involves fitting a theoretical distribution to the data. During the simulation, random variates are generated from the probability distribution to provide the simulated random values. Fitting a theoretical distribution to sample data smooths artificial irregularities in the data and ensures that extreme values are included. For these reasons, it is best to use theoretical distributions if a reasonably good fit exists. Most popular simulation software provide utilities for fitting distributions to numerical data, thus relieving the modeler from performing this complicated procedure. A modeler should be careful when using a theoretical distribution to ensure that if an unbounded distribution is used, the extreme values that can be generated are realistic. Techniques for controlling the range of unbounded distributions in a simulation model are presented in Chapter 7.

6.8.1 Frequency Distributions

Regardless of whether an empirical or theoretical distribution is used, it is a good idea to construct a frequency distribution to get a summary view of the data. Then a fitness test can be conducted to evaluate whether a standard theoretical distribution, such as a normal distribution, fits the data. A frequency distribution groups or bins the data into intervals or cells according to frequency of occurrence. Showing the frequency with which values occur within each cell gives a clear idea of how observations are distributed throughout the range of the data. Frequency distributions may be either discrete or continuous.

Discrete Frequency Distributions

A discrete frequency distribution is limited to specific values. That is, only frequencies of a finite set of specific values are shown. To illustrate a discrete frequency distribution, suppose a sample is collected on the number of customer entries to a fast-food restaurant per five-minute interval. Table 6.4 summarizes these observations, which were made during lunchtime (between 11 A.M. and 1 P.M.) for five workdays over a 20-week period.

TABLE 6.4   **Data for Customer Arrival at a Restaurant**

Arrivals per 5-Minute Interval	Frequency	Arrivals per 5-Minute Interval	Frequency
0	15	6	7
1	11	7	4
2	19	8	5
3	16	9	3
4	8	10	3
5	8	11	1

Frequency distributions can be graphically shown using a histogram. A histogram depicting the discrete frequency distribution in Table 6.4 is shown in Figure 6.11.

Continuous Frequency Distributions

A continuous frequency distribution defines ranges of values within which sample values fall. Going back to our inspection time sample consisting of 100 observations, we can construct a continuous distribution for these data because values can take on any value within the interval specified. A frequency distribution for the data has been constructed using Stat::Fit (Figure 6.12).

Note that a relative frequency or density is shown (third column) as well as cumulative (ascending and descending) frequencies. All of the relative densities add up to 1, which is verified by the last value in the ascending cumulative frequency column.

A histogram based on this frequency distribution was also created in Stat::Fit and is shown in Figure 6.13.

Note that the frequency distribution and histogram for our sample inspection times are based on dividing the data into six even intervals or cells. While there are guidelines for determining the best interval or cell size, the most important thing is to make sure that enough cells are defined to show a gradual transition in values, yet not so many cells that groupings become obscured. The number of intervals should be based on the total number of observations and the variance in the data. One rule of thumb is to set the number of intervals to the cube root of twice the number of samples—that is, $(2N)^{\frac{1}{3}}$. This is the default method used in Stat::Fit. The goal is to use the minimum number of intervals possible without losing information about the spread of the data.

6.8.2 Theoretical Distributions

As mentioned earlier, it may be appropriate to use the frequency distribution in the simulation model as it is, as an empirical distribution. Whenever possible, however, the underlying distribution from which the sample data came should be determined. This way the actual distribution of the population can be used to

FIGURE 6.11

Histogram showing arrival count per five-minute interval.

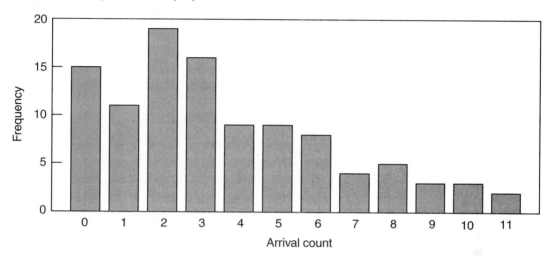

FIGURE 6.12

Frequency distribution for 100 observed inspection times.

INSPECT.SFP : Binned Data

binned data

data points   100
precision   2

continuous relative frequency

intervals   6

end points	mid points	density	ascending cumulative	descending cumulative
0.35				1
0.58	0.47	0.22	0.22	0.78
0.81	0.7	0.25	0.47	0.53
1	0.93	0.24	0.71	0.29
1.3	1.2	0.16	0.87	0.13
1.5	1.4	0.09	0.96	0.04
1.7	1.6	0.04	1	

FIGURE 6.13

*Histogram distribution
for 100 observed
inspection times.*

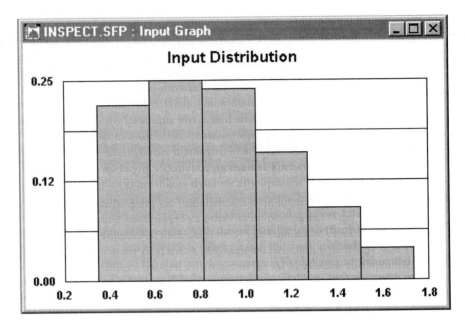

generate random variates in the simulation rather than relying only on a sampling of observations from the population. Defining the theoretical distribution that best fits sample data is called *distribution fitting*. Before discussing how theoretical distributions are fit to data, it is helpful to have a basic understanding of at least the most common theoretical distributions.

There are about 12 statistical distributions that are commonly used in simulation (Banks and Gibson 1997). Theoretical distributions can be defined by a simple set of parameters usually defining dispersion and density. A normal distribution, for example, is defined by a mean value and a standard deviation value. Theoretical distributions are either discrete or continuous, depending on whether a finite set of values within the range or an infinite continuum of possible values within a range can occur. Discrete distributions are seldom used in manufacturing and service system simulations because they can usually be defined by simple probability expressions. Below is a description of a few theoretical distributions sometimes used in simulation. These particular ones are presented here purely because of their familiarity and ease of understanding. Beginners to simulation usually feel most comfortable using these distributions, although the precautions given for their use should be noted. An extensive list of theoretical distributions and their applications is given in Appendix A.

Binomial Distribution

The binomial distribution is a discrete distribution that expresses the probability (p) that a particular condition or outcome can occur in n trials. We call an occurrence of the outcome of interest a *success* and its nonoccurrence a *failure*. For a binomial distribution to apply, each trial must be a *Bernoulli trial:* it must be independent and have only two possible outcomes (success or failure), and the probability of a success must remain constant from trial to trial. The mean of a binomial distribution is given by np, where n is the number of trials and p is the probability of success on any given trial. The variance is given by $np(1 - p)$.

A common application of the binomial distribution in simulation is to test for the number of defective items in a lot or the number of customers of a particular type in a group of customers. Suppose, for example, it is known that the probability of a part being defective coming out of an operation is .1 and we inspect the parts in batch sizes of 10. The number of defectives for any given sample can be determined by generating a binomial random variate. The probability mass function for the binomial distribution in this example is shown in Figure 6.14.

Uniform Distribution
A uniform or rectangular distribution is used to describe a process in which the outcome is equally likely to fall between the values of a and b. In a uniform distribution, the mean is $(a + b)/2$. The variance is expressed by $(b - a)^2/12$. The probability density function for the uniform distribution is shown in Figure 6.15.

FIGURE 6.14

The probability mass function of a binomial distribution ($n = 10, p = .1$).

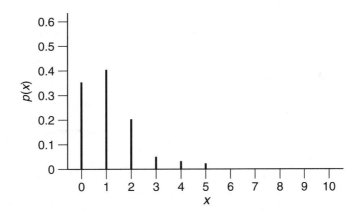

FIGURE 6.15

The probability density function of a uniform distribution.

FIGURE 6.16

*The probability density
function of a
triangular distribution.*

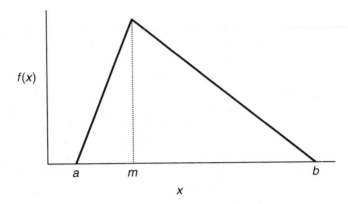

The uniform distribution is often used in the early stages of simulation projects because it is a convenient and well-understood source of random variation. In the real world, it is extremely rare to find an activity time that is uniformly distributed because nearly all activity times have a central tendency or mode. Sometimes a uniform distribution is used to represent a worst-case test for variation when doing sensitivity analysis.

Triangular Distribution

A triangular distribution is a good approximation to use in the absence of data, especially if a minimum, maximum, and most likely value (mode) can be estimated. These are the three parameters of the triangular distribution. If a, m, and b represent the minimum, mode, and maximum values respectively of a triangular distribution, then the mean of a triangular distribution is $(a + m + b)/3$. The variance is defined by $(a^2 + m^2 + b^2 - am - ab - mb)/18$. The probability density function for the triangular distribution is shown in Figure 6.16.

The weakness of the triangular distribution is that values in real activity times rarely taper linearly, which means that the triangular distribution will probably create more variation than the true distribution. Also, extreme values that may be rare are not captured by a triangular distribution. This means that the full range of values of the true distribution of the population may not be represented by the triangular distribution.

Normal Distribution

The normal distribution (sometimes called the *Gaussian distribution*) describes phenomena that vary symmetrically above and below the mean (hence the bell-shaped curve). While the normal distribution is often selected for defining activity times, in practice manual activity times are rarely ever normally distributed. They are nearly always skewed to the right (the ending tail of the distribution is longer than the beginning tail). This is because humans can sometimes take significantly longer than the mean time, but usually not much less than the mean

FIGURE 6.17

The probability density function for a normal distribution.

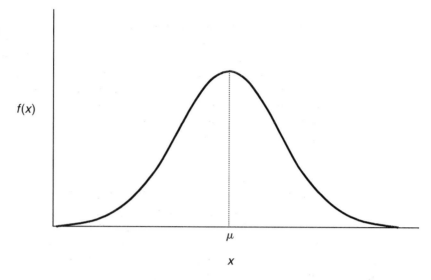

$f(x)$

μ

x

time. Examples of normal distributions might be

- Physical measurements—height, length, diameter, weight.
- Activities involving multiple tasks (like loading a truck or filling a customer order).

The mean of the normal distribution is designated by the Greek letter mu (μ). The variance is σ^2 where σ (sigma) is the standard deviation. The probability density function for the normal distribution is shown in Figure 6.17.

Exponential Distribution

Sometimes referred to as the *negative exponential,* this distribution is used frequently in simulations to represent event intervals. The exponential distribution is defined by a single parameter, the mean (μ). This distribution is related to the Poisson distribution in that if an occurrence happens at a rate that is Poisson distributed, the time between occurrences is exponentially distributed. In other words, the mean of the exponential distribution is the inverse of the Poisson rate. For example, if the rate at which customers arrive at a bank is Poisson distributed with a rate of 12 per hour, the time between arrivals is exponentially distributed with a mean of 5 minutes. The exponential distribution has a memoryless or forgetfulness property that makes it well suited for modeling certain phenomena that occur independently of one another. For example, if arrival times are exponentially distributed with a mean of 5 minutes, then the expected time before the next arrival is 5 minutes regardless of how much time has elapsed since the previous arrival. It is as though there is no memory of how much time has already elapsed when predicting the next event—hence the term *memoryless.* Examples

FIGURE 6.18

The probability density function for an exponential distribution.

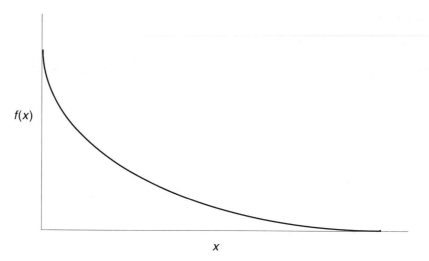

of this distribution are

- Time between customer arrivals at a bank.
- Duration of telephone conversations.
- Time between arrivals of planes at a major airport.
- Time between failures of certain types of electronic devices.
- Time between interrupts of the CPU in a computer system.

For an exponential distribution, the variance is the same as the mean. The probability density function of the exponential distribution is shown in Figure 6.18.

6.8.3 Fitting Theoretical Distributions to Data

Fitting a theoretical distribution to data is essentially an attempt to identify the underlying distribution from which the data were generated. Finding the best distribution that fits the sample data can be quite complicated and is not an exact science. Fortunately, software such as Stat::Fit (accessible under the ProModel Tools menu) is available for automatically fitting distributions. Lab 6 at the end of the book provides exercises on how to use Stat::Fit. Distribution fitting software follows basically the same procedures as we discuss next for manual fitting. Many software tools provide the added benefit of being able to iterate through all applicable distributions searching for the best fit. Figure 6.19 shows a relative ranking in Stat::Fit of how well different distributions fit the observed inspection times in Table 6.2.

Distribution fitting is largely a trial-and-error process. The basic procedure consists of three steps: (1) one or more distributions are selected as candidates for being good fits to the sample data; (2) estimates of the parameters for each distribution must be calculated; and (3) goodness-of-fit tests are performed to ascertain how well each distribution fits the data.

FIGURE 6.19

Ranking distributions by goodness of fit for inspection time data set.

INSPECT.SFP : Auto::Fit		
Auto::Fit Distributions		
distribution	**rank**	**acceptance**
Beta(0.3, 1.73, 1.55, 2.36)	98.9	accept
Weibull(0.3, 1.79, 0.641)	84.9	accept
Triangular(0.3, 1.81, 0.496)	75.7	accept
Gamma(0.3, 2.47, 0.231)	31.3	accept
Pearson 6(0.3, 4.2e+05, 2.47, 1.82e+06)	31.3	accept
Log-Logistic(0.3, 2.45, 0.494)	15.2	reject
Erlang(0.3, 2, 0.285)	14.9	accept
Lognormal(0.3, -0.777, 0.732)	2.68	reject
Inverse Gaussian(0.3, 0.79, 0.571)	0.123	reject
Pearson 5(0.3, 1.67, 0.554)	0.0534	reject
Pareto(0.3, 1.01)	0	reject
Exponential(0.3, 0.571)	0	reject
Uniform(0.3, 1.73)	0	reject

Choosing a distribution that appears to be a good fit to the sample data requires a basic knowledge of the types of distributions available and their properties. It is also helpful to have some intuition about the variable whose data are being fit. Sometimes creating a histogram of the data can reveal important characteristics about the distribution of the data. If, for example, a histogram of sampled assembly times appears symmetric on either side of the mean, a normal distribution may be inferred. Looking at the histogram in Figure 6.13, and knowing the basic shapes of different theoretical distributions, we might hypothesize that the data come from a beta, lognormal, or perhaps even a triangular distribution.

After a particular type of distribution has been selected, we must estimate the defining parameter values of the distribution based on the sample data. In the case of a normal distribution, the parameters to be estimated are the mean and standard deviation, which can be estimated by simply calculating the average and standard deviation of the sample data. Parameter estimates are generally calculated using moment equations or the maximum likelihood equation (see Law and Kelton 2000).

Once a candidate distribution with its associated parameters has been defined, a goodness-of-fit test can be performed to evaluate how closely the distribution fits the data. A *goodness-of-fit test* measures the deviation of the sample distribution from the inferred theoretical distribution. Three commonly used goodness-of-fit tests are the chi-square (χ^2), the Kolmogorov-Smirnov, and the Anderson-Darling tests (see Breiman 1973; Banks et al. 2001; Law and Kelton 2000; Stuart and Ord 1991).

Each of these goodness-of-fit tests starts with the null hypothesis that the fit is good and calculates a test statistic for comparison to a standard. To test this null hypothesis, a *level of significance* value is selected, which is the likelihood of making a type I error—that is, rejecting the null hypothesis when it is true. Stated in a different manner, it is the probability that you will make a mistake and reject a distribution that is actually a good fit. Therefore, the smaller this value, the less likely you are to reject when you should accept.

The oldest and one of the most common statistical tests for goodness of fit is the chi-square test. The chi-square, goodness-of-fit test is a versatile test that can be used to perform hypothesis tests for either discrete or continuous distributions. It is also a useful test to determine data independence (see Chapter 3). The chi-square test basically compares the frequency distribution of the sample data to the way the same number of data points would be distributed if the data came from the inferred distribution. The chi-square, goodness-of-fit test can be broken down into nine steps:

1. Analyze the data and infer an underlying distribution.
2. Create a frequency distribution of the data with equiprobable cells based on the inferred distribution.
3. Calculate the expected frequency (e_i) for each cell.
4. Adjust cells if necessary so that all expected frequencies are at least 5.
5. Calculate the chi-square test statistic.
6. Determine the number of degrees of freedom ($k - 1$).
7. Choose a desired level of significance (α).
8. Find the critical chi-square value from the chi-square table (Appendix D).
9. Reject the distribution if the chi-square statistic exceeds the critical value.

To illustrate how the chi-square, goodness-of-fit test is performed, let's look at an example problem in which 40 observations were made of activity times for a bagging operation. These observations are shown in Table 6.5. What we would like to know is what distribution best fits the data.

Step 1: Analyze the data and infer an underlying distribution. A distribution should be chosen as a candidate for the true, underlying distribution that has similar characteristics to those exhibited by the data. A descriptive analysis, like the one described for the data in Table 6.2, can help reveal some of the defining characteristics of the data. This includes calculating parameters such as the mean, range, standard deviation, and so on. If, for example, the mean and median are roughly the same, and the standard deviation is roughly one-sixth of the range, we might infer that the data are normally distributed. Constructing a histogram of the data like that shown in Figure 6.13 can also be a good way to match a distribution to the data.

TABLE 6.5   40 Observed Bagging Times (seconds)

11.3	8.2	13.8	10.3
7.2	8.6	15.2	9.6
12.5	7.4	13.5	11.1
14.3	11.1	9.2	11.8
12.8	12.3	16.3	9.5
10.2	16.8	14.9	16.3
7.7	15.2	12.9	12.4
11.0	8.3	14.3	16.9
13.2	14.5	7.5	13.2
14.4	10.7	15.1	10.7

TABLE 6.6   Frequency Distribution for Bagging Times

Cell (i)	Interval	Observed Frequency (o_i)	H_0 Probability (p_i)	H_0 Expected Frequency (e_i)	$\dfrac{(o_i - e_i)^2}{e_i}$
1	7–9	7	.20	8	0.125
2	9–11	7	.20	8	0.125
3	11–13	10	.20	8	0.50
4	13–15	8	.20	8	0
5	15–17	8	.20	8	0

For our example, the data appear to be evenly distributed throughout the range, so we will hypothesize that the observations are uniformly distributed between 7 and 17. This is our null hypothesis (H_0). The alternate hypothesis (H_1) is that the observations are *not* uniformly distributed between 7 and 17.

Step 2: Create a frequency distribution of the data with equiprobable cells based on the inferred distribution. To create a frequency distribution, recall that a good rule of thumb for determining the appropriate number of cells (k) is $k = (2n)^{\frac{1}{3}}$. Normally a frequency distribution is divided into cells of equal intervals. When conducting a chi-square test, however, it often works better if cell intervals have equal probabilities rather than equal interval ranges. If, for example, you want to create a frequency distribution with six cells for a particular data set, and you suspect that the data come from a normal distribution, you would create equiprobable intervals that each take in one-sixth of the population of the corresponding normal distribution.

For this example, $k = (2n)^{\frac{1}{3}} = (2 \times 40)^{\frac{1}{3}} = 4.3 \approx 5$. Because we are assuming a uniform distribution in which equal interval widths also have equal probabilities, we can divide the range into five equal intervals each with a cell width of $10/5 = 2$ and a probability of $1/5 = .20$. For each cell (i), the resultant observed frequency (o_i) is shown in Table 6.6 together with the hypothesized probability (p_i). That each cell has a probability of .2 based on a cell width of 2 can be verified by calculating the probability of the first interval as

$$p(7 \le x < 9) = \int_7^9 f(x)\,dx = \int_7^9 \frac{1}{10}\,dx = \left[\frac{x}{10}\right]_7^9 = \frac{9}{10} - \frac{7}{10} = \frac{2}{10} = .20.$$

For a uniform distribution the probabilities for all intervals are equal, so the remaining intervals also have a hypothesized probability of .20.

Step 3: Calculate the expected frequency for each cell (e_i). The expected frequency (e) for each cell (i) is the expected number of observations that would fall into each interval if the null hypothesis were true. It is calculated by multiplying the total number of observations (n) by the probability (p) that an observation would fall within each cell. So for each cell, the expected frequency (e_i) equals np_i.

In our example, since the hypothesized probability (p) of each cell is the same, the expected frequency for every cell is $e_i = np_i = 40 \times .2 = 8$.

Step 4: Adjust cells if necessary so that all expected frequencies are at least 5. If the expected frequency of any cell is less than 5, the cells must be adjusted. This "rule of five" is a conservative rule that provides satisfactory validity of the chi-square test. When adjusting cells, the easiest approach is to simply consolidate adjacent cells. After any consolidation, the total number of cells should be at least 3; otherwise you no longer have a meaningful differentiation of the data and, therefore, will need to gather additional data. If you merge any cells as the result of this step, you will need to adjust the observed frequency, hypothesized probability, and expected frequency of those cells accordingly.

In our example, the expected frequency of each cell is 8, which meets the minimum requirement of 5, so no adjustment is necessary.

Step 5: Calculate the chi-square statistic. The equation for calculating the chi-square statistic is $\chi^2_{calc} = \sum_{i=1}^{k} (o_i - e_i)^2 / e_i$. If the fit is good the chi-square statistic will be small.

For our example,

$$\chi^2_{calc} = \sum_{i=1}^{5} \frac{(o_i - e_i)^2}{e_i} = .125 + .125 + .50 + 0 + 0 = 0.75.$$

Step 6: Determine the number of degrees of freedom ($k - 1$). A simple way of determining a conservative number of degrees of freedom is to take the number of cells minus 1, or $k - 1$.

For our example, the number of degrees of freedom is

$$k - 1 = 5 - 1 = 4.$$

Note: The number of degrees of freedom is often computed to be $k - s - 1$, where k is the number of cells and s is the number of parameters estimated from the data for defining the distribution. So for a normal distribution with two parameters (mean and standard deviation) that are estimated from the data, $s = 2$. In Stat::Fit the number of degrees of freedom is assumed to be $k - 1$.

Step 7: Choose a desired level of significance (α). The significance level is the probability of rejecting the null hypothesis when it is true—that is, when the hypothesized distribution is correct. A typical level of significance is .05; however, it could also be .10 or any other desired value.

For our example we will use a level of significance (α) of .05.

Step 8: Find the critical chi-square value from the chi-square table. A chi-square table can be found in Appendix D. The critical values in the chi-square table are based on the number of degrees of freedom ($k - 1$) and the desired level of significance (α).

For our example,

$$\chi^2_{k-1,\alpha} = \chi^2_{4,0.05} = 9.488$$

Step 9: Reject the distribution if the chi-square statistic exceeds the critical value. If the calculated test statistic is larger than the critical value from the chi-square table, then the observed and expected values are not close and the model is a poor fit to the data. Therefore, we reject H_0 if $\chi^2_{calc} > \chi^2_{k-1,\alpha}$.

For our example, 0.75 is not greater than 9.488; therefore, we fail to reject H_0, which means we can assume that a uniform distribution is a good fit. If we are satisfied with our find we can terminate our testing; otherwise, we can return to step 2 and try another distribution.

Note: While the test statistic for the chi-square test can be useful, the p-value is actually more useful in determining goodness of fit. The p-value is defined as the probability that another sample will compare the same as the current sample given that the fit is appropriate. A small p-value indicates that another sample would not likely compare the same, and therefore, the fit should be rejected. Conversely, a high p-value indicates that another sample would likely compare the same, and, therefore, the fit should not be rejected. Thus the higher the p-value, the more likely that the fit is appropriate. When comparing two different distributions for goodness of fit, the distribution with the higher p-value is likely to be the better fit regardless of the level of significance. Most distribution-fitting software uses the p-value for ranking goodness of fit.

If very few data are available, then a goodness-of-fit test is unlikely to reject any candidate distribution. However, if a lot of data are available, then a goodness-of-fit test may reject all candidate distributions. Therefore, failing to reject a candidate distribution should be taken as one piece of evidence in favor of the choice, while rejecting a particular distribution is only one piece of evidence against the choice. Before making a decision, it is a good idea to look at a graphical display that compares the theoretical distribution to a histogram of the data. Figure 6.20, for example, overlays a beta distribution (the distribution with the highest fit ranking for this particular data set) on top of a histogram of the data.

FIGURE 6.20

Visual comparison between beta distribution and a histogram of the 100 sample inspection time values.

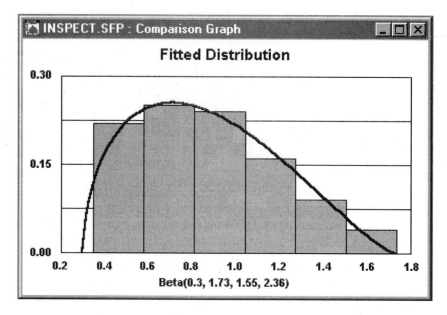

6.9 Selecting a Distribution in the Absence of Data

There are a couple of ways to obtain reasonable estimates of system behavior when no historical records exist or when direct measurement isn't possible. The best source is the people who have expertise in some aspect of the system being modeled. Operators, maintenance personnel, vendors, and others who are experienced in similar processes can often give good approximations of what to expect in new but similar situations. In addition to getting expert opinions, sometimes the nature of the process itself can indicate what to expect. For example, in attempting to select an appropriate distribution to represent a manual task for which a most likely time has been estimated, experience has shown that such tasks tend to have a triangular or beta distribution. As another example, customer interarrival times tend to be exponentially distributed.

For numerical data such as activity times, estimates are usually given in one of the following forms:

- A single most likely or mean value.
- Minimum and maximum values defining a range.
- Minimum, most likely, and maximum values.

6.9.1 Most Likely or Mean Value

Sometimes the only estimate that can be obtained for a system variable is a mean or most likely value. This is a single "best guess" estimate of what the value is.

Examples of single-value estimates include

- About 10 customer arrivals per hour.
- Approximately 20 minutes to assemble parts in a kit.
- Around five machine failures each day.

A simple sensitivity test for these types of estimates would be to increase and decrease the value by 20 percent to see how much of a difference it makes in the simulation results. One of the problems with single-value estimates is that no consideration is given to possible variation. Fluctuations in activity times can have a significant impact on model performance. Sensitivity analysis should be performed looking at the potential impact of variation. For example, if the estimated time for a manual activity is 1.2 minutes, a test run might be made using a normal distribution with a mean of 1.2 minutes and a somewhat large standard deviation value of, say, .3 minute (the coefficient of variation = .25). Comparing this run with a run using a constant value of 1.2 minutes will help determine how sensitive the model is to variation in the activity time.

6.9.2 Minimum and Maximum Values

Sometimes only the range of values can be given as an estimate for a particular variable. Examples of minimum and maximum values include

- 1.5 to 3 minutes to inspect items.
- 5 to 10 customer arrivals per hour.
- Four to six minutes to set up a machine.

Often a uniform distribution is used to represent a range of values because no central tendency is specified. The problem with using a uniform distribution is that it tends to give unrealistic weight to extreme values and therefore exaggerates the random behavior beyond what it really is. A normal distribution may be more appropriate in this situation because it recognizes that a higher density of occurrences exists near the mean. If a normal distribution is used, the standard deviation could be reasonably set at $\frac{1}{6}$ of the range. This ensures that 99.73 percent of all values will fall within the specified range. So for a time range of .25 to 1.75 minutes, a normal distribution could be used with a mean of 1 minute and a standard deviation of .25 minute (see Figure 6.21).

6.9.3 Minimum, Most Likely, and Maximum Values

The best estimate to obtain for a random variable when historical data can't be obtained is the minimum, most likely, and maximum values. Because most activity times are not symmetrical about a mean, this type of estimate is probably going to be the most realistic. Minimum, most likely, and maximum values closely resemble a triangular or a beta distribution. Examples of minimum, most likely, and

FIGURE 6.21

Normal distribution with mean = 1 and standard deviation = .25.

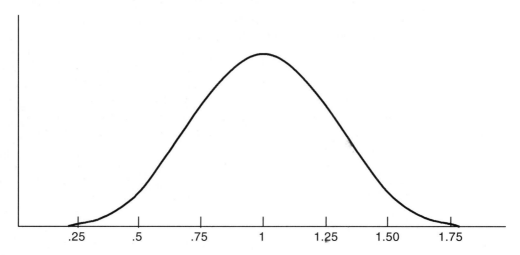

FIGURE 6.22

A triangular distribution with minimum = 2, mode = 5, and maximum = 15.

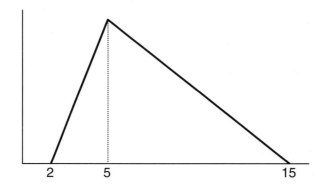

maximum values might be

- 2 to 15 with most likely 5 minutes to repair a machine.
- 3 to 5 with most likely 2.5 minutes between customer arrivals.
- 1 to 3 with most likely 1.5 minutes to complete a purchase order.

Minimum, most likely, and maximum values can be easily set up as a triangular distribution. Looking at the first repair time example, if we fit these times to a triangular distribution, it would produce a minimum of 2 minutes, a mode of 5 minutes and a maximum of 15 minutes (see Figure 6.22).

Because these values are still based only on estimates, it is a good idea to perform a sensitivity analysis, especially on the mode value (assuming the minimum and maximum values represent best- and worst-case possibilities, respectively).

6.10 Bounded versus Boundless Distributions

Continuous theoretical distributions may be bounded (doubly or singly) or unbounded. An unbounded distribution can have values that go to either negative or positive infinity. The normal distribution, for example, ranges from $-\infty$ to $+\infty$ and is therefore unbounded. A bounded distribution may be bounded at both ends or only on one end. The exponential distribution, for example, is a nonnegative distribution (it can't take on negative values) and is therefore bounded only on the lower end of the distribution. The uniform and triangular distributions are doubly bounded, having both lower and upper boundaries. Empirical continuous distributions are always doubly bounded by the mere fact that they are defined by a finite number of cell intervals.

Sampling during a simulation from a boundless distribution may cause values to be generated that are unrealistic in real-world situations. An operation time that is normally distributed with a mean of three minutes and a standard deviation of one minute would have a possibility, for example, of being negative on the low side as well as a remote chance of being absurdly large on the high side. It is often desirable, therefore, to draw random samples during a simulation from a *truncated* continuous distribution where either the lower or upper tail is cut short. ProModel automatically converts negative values to 0 when an unbounded continuous distribution is specified for a time value. It has no way of knowing, however, what a reasonably acceptable upper bound should be. This must be specified by the modeler. Lower and upper bounds can be imposed on a distribution by simply screening for values outside of the acceptable range. Either values generated outside the acceptable range can be rejected and another value generated or the generated value can be brought within the acceptable value range.

6.11 Modeling Discrete Probabilities
Using Continuous Distributions

Discrete probability distributions are used much less in simulation than continuous probability distributions. Most discrete probabilities—such as the probability of routing to a particular location or the probability of an item being defective—are modeled using simple probability expressions like .34. In simulation, most occasions for generating values from a discrete probability distribution involve generating integer values rather than real values. For example, the discrete random variable might represent the number of items in a batch. One useful way to generate discretely distributed integer values is to use a continuous distribution where fractions are simply truncated. In a manufacturing system, for example, suppose that when filling containers or pallets there is sometimes a remaining amount that fills only a partial container or pallet. If a full pallet holds 20 boxes, the number of boxes on a partial pallet would likely be uniformly distributed between 1 and 19. To model the partial pallet quantity, you would simply define a continuous uniform distribution with a range between 1 and 20. A uniform random variable X will have values generated for X where min $\leq X <$ max. Thus values will never be equal to the maximum value (in this case 20). Because generated values are automatically truncated when used in a context requiring an integer, only integer values that are evenly distributed from 1 to 19 will occur (this is effectively a discrete uniform distribution).

6.12 Data Documentation and Approval

When it is felt that all relevant data have been gathered, analyzed, and converted into a usable form, it is advisable to document the data using tables, relational diagrams, and assumption lists. Sources of data should also be noted. This document should then be reviewed by those in a position to evaluate the validity of the data and approve the assumptions made. Where more formal documentation is required, a separate document will need to be created. This document will be helpful later if modifications need to be made to the model or to analyze why the actual system ends up working differently than the simulation.

In addition to identifying data used to build the model, the document should also specify factors that were deliberately excluded from the model because they were deemed insignificant or irrelevant. For example, if break times are not included in the system description because of their perceived insignificance, the document should state this. Justification for omissions should also be included if necessary. Stating why certain factors are being excluded from the system description will help resolve later questions that may arise regarding the model premises.

Reviewing and approving input data can be a time-consuming and difficult task, especially when many assumptions are made. In practice, data validation ends up being more of a consensus-building process where agreement is reached that the information is good enough for the purposes of the simulation. The data document is not a static document but rather a dynamic one that often changes as model building and experimentation get under way. Much if not all of the data documentation can be scripted right into the model. Most software provides the capability to write comments in the model. Where more formal documentation is required, a separate data document will need to be created.

6.12.1 Data Documentation Example

To illustrate how system data might be documented, imagine you have just collected information for an assembly operation for three different monitor sizes. Here is an example of how the data collected for this system might be diagrammed and tabulated. The diagram and data should be sufficiently clear and complete for those familiar with the system to validate the data and for a model to be constructed. Review the data and see if they are sufficiently clear to formulate a mental image of the system being modeled.

Objective

The objective of the study is to determine station utilization and throughput of the system.

Entity Flow Diagram

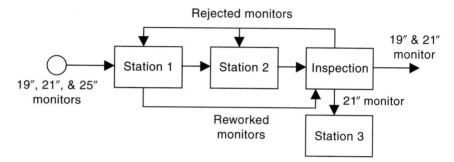

Entities

> 19" monitor
>
> 21" monitor
>
> 25" monitor

Workstation Information

Workstation	Buffer Capacity	Defective Rate
Station 1	5	5%
Station 2	8	8%
Station 3	5	0%
Inspection	5	0%

Processing Sequence

Entity	Station	Operating Time in Minutes (min, mode, max)
19" monitor	Station 1	0.8, 1, 1.5
	Station 2	0.9, 1.2, 1.8
	Inspection	1.8, 2.2, 3
21" monitor	Station 1	0.8, 1, 1.5
	Station 2	1.1, 1.3, 1.9
	Inspection	1.8, 2.2, 3
25" monitor	Station 1	0.9, 1.1, 1.6
	Station 2	1.2, 1.4, 2
	Inspection	1.8, 2.3, 3.2
	Station 3	0.5, 0.7, 1

Handling Defective Monitors

- Defective monitors are detected at inspection and routed to whichever station created the problem.
- Monitors waiting at a station for rework have a higher priority than first-time monitors.
- Corrected monitors are routed back to inspection.
- A reworked monitor has only a 2 percent chance of failing again, in which case it is removed from the system.

Arrivals

A cartload of four monitor assemblies arrives every four hours normally distributed with a standard deviation of 0.2 hour. The probability of an arriving monitor being of a particular size is

Monitor Size	Probability
19"	.6
21"	.3
25"	.1

Move Times

All movement is on an accumulation conveyor with the following times:

From	To	Time (seconds)
Station 1	Station 2	12
Station 2	Inspection	15
Inspection	Station 3	12
Inspection	Station 1	20
Inspection	Station 2	14
Station 1	Inspection	18

Move Triggers

Entities move from one location to the next based on available capacity of the input buffer at the next location.

Work Schedule

Stations are scheduled to operate eight hours a day.

Assumption List

- No downtimes (downtimes occur too infrequently).
- Dedicated operators at each workstation are always available during the scheduled work time.
- Rework times are half of the normal operation times.

Simulation Time and Replications

The simulation is run for 40 hours (10 hours of warm-up). There are five replications.

6.13 Summary

Data for building a model should be collected systematically with a view of how the data are going to be used in the model. Data are of three types: structural, operational, and numerical. Structural data consist of the physical objects that make up the system. Operational data define how the elements behave. Numerical data quantify attributes and behavioral parameters.

When gathering data, primary sources should be used first, such as historical records or specifications. Developing a questionnaire is a good way to request information when conducting personal interviews. Data gathering should start with structural data, then operational data, and finally numerical data. The first piece of the puzzle to be put together is the routing sequence because everything else hinges on the entity flow.

Numerical data for random variables should be analyzed to test for independence and homogeneity. Also, a theoretical distribution should be fit to the data if there is an acceptable fit. Some data are best represented using an empirical distribution. Theoretical distributions should be used wherever possible.

Data should be documented, reviewed, and approved by concerned individuals. This data document becomes the basis for building the simulation model and provides a baseline for later modification or for future studies.

6.14 Review Questions

1. Give two examples of structural data, operational data, and numerical data to be gathered when building a model.
2. Why is it best to begin gathering data by defining entity routings?
3. Of the distributions shown in the chapter, which theoretical distribution most likely would be representative of time to failure for a machine?
4. Why would a normal distribution likely be a poor representation of an activity time?
5. Assume a new system is being simulated and the only estimate available for a manual operation is a most likely value. How would you handle this situation?
6. Under what circumstances would you use an empirical distribution instead of a standard theoretical distribution for an activity time?
7. Why is the distribution for interarrival times often nonstationary?
8. Assuming you had historical data on truck arrivals for the past year, how would you arrive at an appropriate arrival distribution to model the system for the next six months?

9. A new machine is being considered for which the company has no reliability history. How would you obtain the best possible estimate of reliability for the machine?

10. Suppose you are interested in looking at the impact of having workers inspect their own work instead of having a dedicated inspection station. If this is a new system requiring lots of assumptions to be made, how can simulation be useful in making the comparison?

11. State whether the following are examples of a discrete probability distribution or a continuous probability distribution.
    a. Activity times.
    b. Batch sizes.
    c. Time between arrivals.
    d. Probability of routing to one of six possible destinations.

12. Conceptually, how would you model a random variable X that represents an activity time that is normally distributed with a mean of 10 minutes and a standard deviation of 3 minutes but is never less than 8 minutes?

13. Using Stat::Fit, generate a list of 50 random values between 10 and 100. Choose the Scatter Plot option and plot the data. Now put the data in ascending order using the Input/Operate commands and plot the data. Explain the correlation, if any, that you see in each scatter plot.

14. How can you check to see if a distribution in ProModel is giving the right values?

15. Since many theoretical distributions are unbounded on the bottom, what happens in ProModel if a negative value is sampled for an activity time?

16. Go to a small convenience store or the university bookstore and collect data on the interarrival and service times of customers. Make histograms of the number of arrivals per time period and the number of service completions per period. Note if these distributions vary by the time of the day and by the day of the week. Record the number of service channels available at all times. Make sure you secure permission to perform the study.

17. The following are throughput time values for 30 simulation runs. Calculate an estimate of the mean, variance, standard deviation, and coefficient of variation for the throughput time. Construct a histogram that has six cells of equal width.

    10.7, 5.4, 7.8, 12.2, 6.4, 9.5, 6.2, 11.9, 13.1, 5.9, 9.6, 8.1, 6.3, 10.3, 11.5, 12.7, 15.4, 7.1, 10.2, 7.4, 6.5, 11.2, 12.9, 10.1, 9.9, 8.6, 7.9, 10.3, 8.3, 11.1

18. Customers calling into a service center are categorized according to the nature of their problems. Five types of problem categories (A through E) have been identified. One hundred observations were made

of customers calling in during a day, with a summary of the data shown here. By inspection, you conclude that the data are most likely uniformly distributed. Perform a chi-square goodness-of-fit test of this hypothesis.

Type	A	B	C	D	E
Observations	10	14	12	9	5

19. While doing your homework one afternoon, you notice that you are frequently interrupted by friends. You decide to record the times between interruptions to see if they might be exponentially distributed. Here are 30 observed times (in minutes) that you have recorded; conduct a goodness-of-fit test to see if the data are exponentially distributed. (Hint: Use the data average as an estimate of the mean. For the range, assume a range between 0 and infinity. Divide the cells based on equal probabilities (p_i) for each cell rather than equal cell intervals.)

2.08	6.86	4.86	2.55	5.94
2.96	0.91	2.13	2.20	1.40
16.17	2.11	2.38	0.83	2.81
14.57	0.29	2.73	0.73	1.76
2.79	11.69	18.29	5.25	7.42
2.15	0.96	6.28	0.94	13.76

6.15 Case Study

COLLECTING AND DOCUMENTING DATA FOR HARRY'S DRIVE-THROUGH RESTAURANT

This exercise is intended to help you practice sorting through and organizing data on a service facility. The facility is a fast-food restaurant called Harry's Drive-Through and is shown on the next page.

Harry's Drive-Through caters to two types of customers, walk-in and drive-through. During peak times, walk-in customers arrive exponentially every 4 minutes and place their order, which is sent to the kitchen. Nonpeak arrivals occur every 8–10 minutes. Customers wait in a pickup queue while their orders are filled in the kitchen. Two workers are assigned to take and deliver orders. Once customers pick up their orders, 60 percent of the customers stay and eat at a table, while the other 40 percent leave with their orders. There is seating for 30 people, and the number of people in each customer party is one

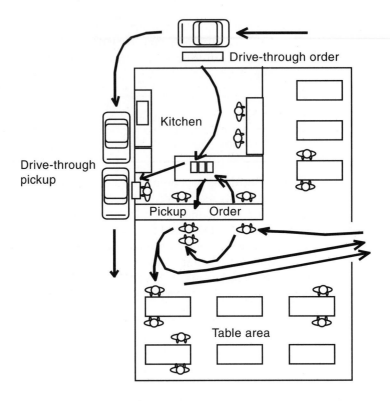

40 percent of the time, two 30 percent of the time, three 18 percent of the time, four 10 percent of the time, and five 2 percent of the time. Eating time is normally distributed with a mean of 15 minutes and a standard deviation of 2 minutes. If a walk-in customer enters and sees that more than 15 customers are waiting to place their orders, the customer will balk (that is, leave).

Harry's is especially popular as a drive-through restaurant. Cars enter at a rate of 10 per hour during peak times, place their orders, and then pull forward to pick up their orders. No more than five cars can be in the pickup queue at a time. One person is dedicated to taking orders. If over seven cars are at the order station, arriving cars will drive on.

The time to take orders is uniformly distributed between 0.5 minute and 1.2 minutes including payment. Orders take an average of 8.2 minutes to fill with a standard deviation of 1.2 minutes (normal distribution). These times are the same for both walk-in and drive-through customers.

The objective of the simulation is to analyze performance during peak periods to see how long customers spend waiting in line, how long lines become, how often customers balk (pass by), and what the utilization of the table area is.

Problem

Summarize these data in table form and list any assumptions that need to be made in order to conduct a meaningful simulation study based on the data given and objectives specified.

References

Banks, Jerry; John S. Carson II; Barry L. Nelson; and David M. Nicol. *Discrete-Event System Simulation.* Englewood Cliffs, NJ: Prentice Hall, 2001.

Banks, Jerry, and Randall R. Gibson. "Selecting Simulation Software." *IIE Solutions,* May 1997, pp. 29–32.

————. Stat::Fit. South Kent, CT: Geer Mountain Software Corporation, 1996.

Breiman, Leo. *Statistics: With a View toward Applications.* New York: Houghton Mifflin, 1973.

Brunk, H. D. *An Introduction to Mathematical Statistics.* 2nd ed. N.Y.: Blaisdell Publishing Co. 1965.

Carson, John S. "Convincing Users of Model's Validity Is Challenging Aspect of Modeler's Job." *Industrial Engineering,* June 1986, p. 77.

Hoover, S. V., and R. F. Perry. *Simulation: A Problem Solving Approach.* Reading, MA, Addison-Wesley, 1990.

Johnson, Norman L.; Samuel Kotz; and Adrienne W. Kemp. *Univariate Discrete Distributions.* New York: John Wiley & Sons, 1992, p. 425.

Knuth, Donald E. *Seminumerical Algorithms.* Reading, MA: Addison-Wesley, 1981.

Law, Averill M., and W. David Kelton. *Simulation Modeling & Analysis.* New York: McGraw-Hill, 2000.

Stuart, Alan, and J. Keith Ord. *Kendall's Advanced Theory of Statistics.* vol. 2. Cambridge: Oxford University Press, 1991.

7 MODEL BUILDING

"Every theory [model] should be stated [built] as simply as possible, but not simpler."

—Albert Einstein

7.1 Introduction

In this chapter we look at how to translate a conceptual model of a system into a simulation model. The focus is on elements common to both manufacturing and service systems such as entity flow and resource allocation. Modeling issues more specific to either manufacturing or service systems will be covered in later chapters.

Modeling is more than knowing how to use a simulation software tool. Learning to use modern, easy-to-use software is one of the least difficult aspects of modeling. Indeed, current simulation software makes poor and inaccurate models easier to create than ever before. Unfortunately, software cannot make decisions about how the elements of a particular system operate and how they should interact with each other. This is the role of the modeler.

Modeling is considered an art or craft as much as a science. Knowing the theory behind simulation and understanding the statistical issues are the science part. But knowing how to effectively and efficiently represent a system using a simulation tool is the artistic part of simulation. It takes a special knack to be able to look at a system in the abstract and then creatively construct a representative logical model using a simulation tool. If three different people were to model the same system, chances are three different modeling approaches would be taken. Modelers tend to use the techniques with which they are most familiar. So the best way to develop good modeling skills is to look at lots of good examples and, most of all, practice, practice, practice! Skilled simulation analysts are able to quickly translate a process into a simulation model and begin conducting experiments.

Consider the following statement by Robert Shannon (1998):

> If you want to become proficient in an art you must take up the tools (palette, canvas, paint, and brushes) and begin to paint. As you do so, you will begin to see what works and what doesn't. The same thing is true in simulation. You learn the art of simulation by simulating. Having a mentor to help show you the way can shorten the time and effort.

The purpose of this chapter is to provide a little mentoring by showing how typical modeling issues are handled during simulation. Some of the questions answered in this chapter include

- How do I convert a conceptual model to a simulation model?
- What is the relationship between model simplicity and model usefulness?
- How do I determine which system elements to include in a model?
- How do I determine the best way to represent system elements in a model?
- How are common situations modeled using simulation?

Because a simulation model relies on a simulation language for its expression, (and since this textbook is titled *Simulation Using ProModel*), ProModel will be used to illustrate how system elements are modeled.

For the most part, the discussion of modeling issues in this chapter is kept at a conceptual level. Specific modeling examples in ProModel can be found in Lab 7 of Part II.

7.2 Converting a Conceptual Model to a Simulation Model

The conceptual model is the result of the data-gathering effort and is a formulation in one's mind (supplemented with notes and diagrams) of how a particular system operates. Building a simulation model requires that this conceptual model be converted to a simulation model. Making this translation requires two important transitions in one's thinking. First, the modeler must be able to think of the system in terms of the modeling paradigm supported by the particular modeling software that is being used. Second, the different possible ways to model the system should be evaluated to determine the most efficient yet effective way to represent the system.

7.2.1 Modeling Paradigms

A simulation model is a computer representation of how elements in a particular system behave and interact. The internal representation of a system may not be radically different between simulation products. However, the way that model information is defined using a particular product may be quite different. How a user defines a model using a particular simulation product is based on the modeling paradigm of the product. A *modeling paradigm* consists of the constructs and associated language that dictate how the modeler should think about the system being modeled. In this regard, a modeling paradigm determines the particular "world view" that one should have when modeling a system. Learning a simulation product requires that one first learn the particular modeling paradigm used by the product.

Modeling paradigms differ among simulation products—although many differences are minor. Historically, simulation products required models to be defined line by line, specifying a list of instructions like a script for processing entities. Most modern products have a process orientation and support object-based modeling. Some products view a process as an activity sequence, while others view it from the standpoint of an entity routing.

Conceptually, object-based modeling is quite simple. An *object* represents a real-world object such as a machine, an operator, a process part, or a customer. An object is defined in terms of attributes and behaviors. *Attributes* are essentially variables that are associated with an object such as its size, condition, time in the system, and so on. Attributes are used to carry information about the object, either for decision making or for output reporting. Attributes may be modified during the simulation to reflect changing values. *Behaviors* define the operational logic associated with the object. This logic gets executed either whenever certain events occur or when explicitly called in the model. Examples of behavioral logic include operations, machine setup, routing decisions, and material movement. Examples of events or conditions that may trigger a behavior might be the completion of an operation, the elapse of a specified period of time, or the drop of a tank or inventory to a specified level.

Objects are organized into classes according to similarity of attributes and behavior. In this way, all objects of the same class inherit the attributes and behaviors defined for the class. This simplifies object definitions and allows changes to be quickly made to a set of similar objects. It also allows objects that are defined once to be reused in other models. In ProModel, predefined object classes include entities, resources, locations, and paths.

While object-based modeling is consistent with modern object-oriented programming (OOP) practices, OOP itself is quite complicated and requires a significant amount of training to become proficient. The easiest simulation languages are object-based, but they are not pure OOP languages. Otherwise, a modeler might as well pick up an OOP language like Java and begin programming his or her models.

ProModel is object-based but also provides an intuitive entity-flow modeling paradigm. The natural way to think about most systems being modeled is to think of them from the perspective of the entity as it flows through each workstation, storage area, or some other location. In fact, it actually helps when building the model to put yourself in the place of the entity flowing through the system and describe what you do and what resources you require as you move from place to place in the system. For modeling purposes, it is useful to think of a system in terms of the same structural and operational elements that were described in Chapter 6. This is essentially how models are defined using ProModel.

7.2.2 Model Definition

A model is a simplified representation of reality, with emphasis on the word *simplified*. This means that the exact way in which an operation is performed is not so important as the way in which the operation impacts the rest of the system. An activity should always be viewed in terms of its effect on other system elements rather than in terms of the detailed way in which it is performed. Such detailed mechanics are inconsequential to the overall flow of entities and utilization of resources.

Most models, especially those built by beginners, tend to err on the side of being overly detailed rather than being too general. The tendency is to reproduce the precise way that the system operates (sometimes referred to as *emulation*). Not only is it difficult to create extremely detailed models, but it is also difficult to debug and maintain them. Furthermore, all of the detail may obscure the key issues being analyzed so that it actually weakens the model. The power of a model is more a function of its simplicity rather than its complexity. A point can be made more effectively when it is reduced to its simplest form rather than disguised in a morass of detail. Lou Keller of PROMODEL Corporation has suggested that the Laffer curve, borrowed from economics, effectively illustrates the relationship between model complexity and model utility (see Figure 7.1). Notice that a certain degree of complexity is essential to capture the key cause-and-effect relationships in the system. However, a system has many more cause-and-effect relationships than should be included in a model. There is an optimum amount of complexity for any given model beyond which additional utility begins to diminish.

The remaining sections in this chapter give recommendations on how to effectively yet minimally represent system elements in a simulation model using a simulation language like ProModel.

FIGURE 7.1

Relationship between model complexity and model utility (also known as the Laffer curve).

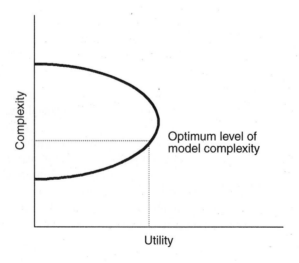

7.3 Structural Elements

For the most part, model objects represent the structural elements in a system such as machines, people, work items, and work areas. For purposes of our discussion, we will be using the simple object classification employed by ProModel:

- *Entities*—the items processed in the system.
- *Locations*—places where entities are processed or held.
- *Resources*—agents used in the processing of entities.
- *Paths*—the course of travel for entities and resources in the system.

Not all simulation products provide the same set or classification of modeling elements. Even these elements could be further subdivided to provide greater differentiation. Locations, for example, could be subdivided into workstations, buffers, queues, and storage areas. From a system dynamics perspective, however, these are all still simply places to which entities are routed and where operations or activities may be performed. For this reason, it is easier just to think of them all in the generic sense as locations. The object classification used by ProModel is simple yet broad enough to encompass virtually any object encountered in most manufacturing and service systems.

These model elements or objects have behavior associated with them (discussed in Section 7.4, "Operational Elements") and attributes. Most common behaviors and attributes are selectable from menus, which reduces the time required to build models. The user may also predefine behaviors and attributes that are imported into a model.

7.3.1 Entities

Entities are the objects processed in the model that represent the inputs and outputs of the system. Entities in a system may have special characteristics such as speed, size, condition, and so on. Entities follow one or more different routings in a system and have processes performed on them. They may arrive from outside the system or be created within the system. Usually, entities exit the system after visiting a defined sequence of locations.

Simulation models often make extensive use of entity attributes. For example, an entity may have an attribute called Condition that may have a value of 1 for defective or 0 for nondefective. The value of this attribute may determine where the entity gets routed in the system. Attributes are also frequently used to gather information during the course of the simulation. For example, a modeler may define an attribute called ValueAddedTime to track the amount of value-added time an entity spends in the system.

The statistics of interest that are generally collected for entities include time in the system (flow time), quantity processed (output), value-added time, time spent waiting to be serviced, and the average number of entities in the system.

Entities to Include

When deciding what entities to include in a model, it is best to look at every kind of entity that has a bearing on the problem being addressed. For example, if a component part is assembled to a base item at an assembly station, and the station is always stocked with the component part, it is probably unnecessary to model the component part. In this case, what is essential to simulate is just the time delay to perform the assembly. If, however, the component part may not always be available due to delays, then it might be necessary to simulate the flow of component parts as well as the base items. The rule is that if you can adequately capture the dynamics of the system without including the entity, don't include it.

Entity Aggregating

It is not uncommon for some manufacturing systems to have hundreds of part types or for a service system to have hundreds of different customer types. Modeling each one of these entity types individually would be a painstaking task that would yield little, if any, benefit. A better approach is to treat entity types in the aggregate whenever possible (see Figure 7.2). This works especially well when all entities have the same processing sequence. Even if a slight difference in processing exists, it often can be handled through use of attributes or by using probabilities. If statistics by entity type are not required and differences in treatment can be defined using attributes or probabilities, it makes sense to aggregate entity types into a single generic entity and perhaps call it *part* or *customer*.

Entity Resolution

Each individual item or person in the system need not always be represented by a corresponding model entity. Sometimes a group of items or people can be represented by a single entity (see Figure 7.3). For example, a single entity might be used to represent a batch of parts processed as a single unit or a party of people eating together in a restaurant. If a group of entities is processed as a group and moved as a group, there is no need to model them individually. Activity times or statistics that are a function of the size of the group can be handled using an attribute that keeps track of the items represented by the single entity.

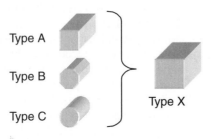

FIGURE 7.2

Treating different entity types as a single type.

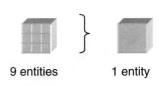

FIGURE 7.3

Treating multiple entities as a single entity.

High-Rate Entity Processing

In some systems, especially in the food and beverage industries, it is not uncommon to process entities at a rate of hundreds per minute. In such situations, modeling each individual entity can significantly slow down the simulation. For high-rate processing in which individual entity tracking isn't critical, it may be preferable to merely track entity production at various stages of a process using variables or attributes instead of individual entities. Simulation languages having a tank or accumulator construct can make this easier. Having thousands of entities in the system at one time (especially those with multiple attributes) can consume lots of computer memory and may slow the simulation significantly.

An alternative approach to modeling high-rate entity processing is to adjust the resolution of the entity, as discussed earlier. Using this approach, you might model every carton of 48 bottles or containers in a beverage-processing facility as a single entity. This approach also works for continuously flowing substances such as liquids or granules, which can often be converted, for simulation, into discrete units of measure such as gallons, pounds, or barrels.

7.3.2 Locations

Locations are places in the system that entities visit for processing, waiting, or decision making. A location might be a treatment room, workstation, check-in point, queue, or storage area. Locations have a holding capacity and may have certain times that they are available. They may also have special input and output such as input based on highest priority or output based on first-in, first out (FIFO).

In simulation, we are often interested in the average contents of a location such as the average number of customers in a queue or the average number of parts in a storage rack. We might also be interested in how much time entities spend at a particular location for processing. There are also location state statistics that are of interest such as utilization, downtime, or idle time.

Locations to Include

Deciding what to model as a route location depends largely on what happens at the location. If an entity merely passes through a location en route to another without spending any time, it probably isn't necessary to include the location. For example, a water spray station through which parts pass without pausing probably doesn't need to be included in a model. In considering what to define as a location, any point in the flow of an entity where one or more of the following actions take place may be a candidate:

- Place where an entity is detained for a specified period of time while undergoing an activity (such as fabrication, inspection, or cleaning).
- Place where an entity waits until some condition is satisfied (like the availability of a resource or the accumulation of multiple entities).
- Place or point where some action takes place or logic gets executed, even though no time is required (splitting or destroying an entity, sending a signal, incrementing an attribute or variable).

• Place or point where a decision is made about further routing (a branch in a conveyor or path, an area in a store where customers decide to which checkout counter they will go).

Location Resolution

Depending on the level of resolution needed for the model, a location may be an entire factory or service facility at one extreme, or individual positions on a desk or workbench at the other. The combination of locations into a single location is done differently depending on whether the locations are parallel or serial locations.

When combining parallel locations having identical processing times, the resulting location should have a capacity equal to the combined capacities of the individual locations; however, the activity time should equal that of only one of the locations (see Figure 7.4). A situation where multiple locations might be combined in this way is a parallel workstation. An example of a parallel workstation is a work area where three machines perform the same operation. All three machines could be modeled as a single location with a capacity equal to three. Combining locations can significantly reduce model size, especially when the number of parallel units gets very large, such as a 20-station checkout area in a large shopping center or a 100-seat dining area in a restaurant.

When combining serial locations, the resultant location should have a capacity equal to the sum of the individual capacities and an activity time equal to the sum of the activity times. An example of a combined serial sequence of locations

FIGURE 7.4

Example of combining three parallel stations into a single station.

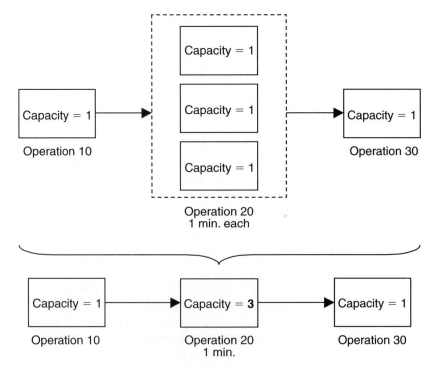

FIGURE 7.5

Example of combining three serial stations into a single station.

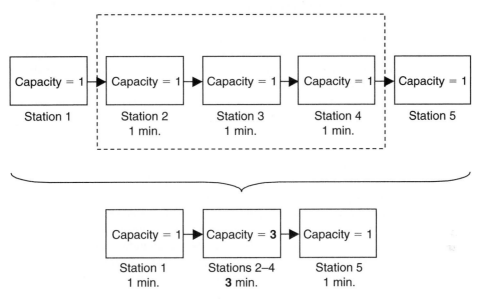

might be a synchronous transfer line that has multiple serial stations. All of them could be represented as a single location with a capacity equal to the number of stations (see Figure 7.5). Parts enter the location, spend an amount of time equal to the sum of all the station times, and then exit the location. The behavior may not be exactly the same as having individual stations, such as when the location becomes blocked and up to three parts may be finished and waiting to move to station 5. The modeler must decide if the representation is a good enough approximation for the intended purpose of the simulation.

7.3.3 Resources

Resources are the agents used to process entities in the system. Resources may be either static or dynamic depending on whether they are stationary (like a copy machine) or move about in the system (like an operator). Dynamic resources behave much like entities in that they both move about in the system. Like entities, resources may be either animate (living beings) or inanimate (a tool or machine). The primary difference between entities and resources is that entities enter the system, have a defined processing sequence, and, in most cases, finally leave the system. Resources, however, usually don't have a defined flow sequence and remain in the system (except for off-duty times). Resources often respond to requests for their use, whereas entities are usually the objects requiring the use of resources.

In simulation, we are interested in how resources are utilized, how many resources are needed, and how entity processing is affected by resource availability. The response time for acquiring a resource may also be of interest.

Resources to Include

The decision as to whether a resource should be included in a model depends largely on what impact it has on the behavior of the system. If the resource is dedicated to a particular workstation, for example, there may be little benefit in including it in the model since entities never have to wait for the resource to become available before using it. You simply assign the processing time to the workstation. If, on the other hand, the resource may not always be available (it experiences downtime) or is a shared resource (multiple activities compete for the same resource), it should probably be included. Once again, the consideration is how much the resource is likely to affect system behavior.

Resource Travel Time

One consideration when modeling the use of resources is the travel time associated with mobile resources. A modeler must ask whether a resource is immediately accessible when available, or if there is some travel time involved. For example, a special piece of test equipment may be transported around to several locations in a facility as needed. If the test equipment is available when needed at some location, but it takes 10 minutes to get the test equipment to the requesting location, that time should be accounted for in the model. The time for the resource to move to a location may also be a function of the distance it must travel.

Consumable Resources

Depending on the purpose of the simulation and degree of influence on system behavior, it may be desirable to model consumable resources. Consumable resources are used up during the simulation and may include

- *Services* such as electricity or compressed air.
- *Supplies* such as staples or tooling.

Consumable resources are usually modeled either as a function of time or as a step function associated with some event such as the completion of an operation. This can be done by defining a variable or attribute that changes value with time or by event. A variable representing the consumption of packaging materials, for example, might be based on the number of entities processed at a packaging station.

Transport Resources

Transport resources are resources used to move entities within the system. Examples of transport resources are lift trucks, elevators, cranes, buses, and airplanes. These resources are dynamic and often are capable of carrying multiple entities. Sometimes there are multiple pickup and drop-off points to deal with. The transporter may even have a prescribed route it follows, similar to an entity routing. A common example of this is a bus route.

In advanced manufacturing systems, the most complex element to model is often the transport or material handling system. This is because of the complex operation that is associated with these computer-controlled systems such as conveyor systems and automated guided vehicle systems (AGVS). Modeling advanced material handling systems can be simplified if the modeling language provides special constructs for defining them; otherwise the logic describing their behavior must be defined by the modeler. ProModel provides facilities for modeling material handling systems that are described in Chapter 13.

7.3.4 Paths

Paths define the course of travel for entities and resources. Paths may be isolated, or they may be connected to other paths to create a path network. In ProModel simple paths are automatically created when a routing path is defined. A routing path connecting two locations becomes the default path of travel if no explicitly defined path or path network connects the locations.

Paths linked together to form path networks are common in manufacturing and service systems. In manufacturing, aisles are connected to create travel ways for lift trucks and other material handlers. An AGVS sometimes has complex path networks that allow controlled traffic flow of the vehicles in the system. In service systems, office complexes have hallways connecting other hallways that connect to offices. Transportation systems use roadways, tracks, and so on that are often interconnected.

When using path networks, there can sometimes be hundreds of routes to take to get from one location to another. ProModel is able to automatically navigate entities and resources along the shortest path sequence between two locations. Optionally, you can explicitly define the path sequence to take to get from one point to any other point in the network.

7.4 Operational Elements

Operational elements define the behavior of the different physical elements in the system and how they interact. These include routings, operations, arrivals, entity and resource movement, task selection rules, resource schedules, and downtimes and repairs.

Most of the operational elements of a model can be defined using constructs that are specifically provided for modeling such elements. The operational rules for each can usually be selected from menus. There may be situations, however, that require the modeler to use special logic such as "if–then" statements to achieve the special operating behavior that is desired.

7.4.1 Routings

Routings define the sequence of flow for entities from location to location. When entities complete their activity at a location, the routing defines where the entity goes next and specifies the criterion for selecting from among multiple possible locations.

Frequently entities may be routed to more than one possible location. When choosing from among multiple alternative locations, a rule or criterion must be defined for making the selection. A few typical rules that might be used for selecting the next location in a routing decision include

- *Probabilistic*—entities are routed to one of several locations according to a frequency distribution.
- *First available*—entities go to the first available location in the order they are listed.
- *By turn*—the selection rotates through the locations in the list.
- *Most available capacity*—entities select the location that has the most available capacity.
- *Until full*—entities continue to go to a single location until it is full and then switch to another location, where they continue to go until it is full, and so on.
- *Random*—entities choose randomly from among a list of locations.
- *User condition*—entities choose from among a list of locations based on a condition defined by the user.

Recirculation

Sometimes entities revisit or pass through the same location multiple times. The best approach to modeling this situation is to use an entity attribute to keep track of the number of passes through the location and determine the operation or routing accordingly. When using an entity attribute, the attribute is incremented either on entry to or on exit from a location and tested before making the particular operation or routing decision to see which pass the entity is currently on. Based on the value of the attribute, a different operation or routing may be executed.

Unordered Routings

Certain systems may not require a specific sequence for visiting a set of locations but allow activities to be performed in any order as long as they all eventually get performed. An example is a document requiring signatures from several departments. The sequence in which the signatures are obtained may be unimportant as long as all signatures are obtained.

In unordered routing situations, it is important to keep track of which locations have or haven't been visited. Entity attributes are usually the most practical way of tracking this information. An attribute may be defined for each possible location and then set to 1 whenever that location is visited. The routing is then based on which of the defined attributes are still set to zero.

7.4.2 Entity Operations

An entity operation defines what happens to an entity when it enters a location. For modeling purposes, the exact nature of the operation (machining, patient check-in, or whatever) is unimportant. What is essential is to know what happens in terms of the time required, the resources used, and any other logic that impacts system performance. For operations requiring more than a time and resource designation, detailed logic may need to be defined using if–then statements, variable assignment statements, or some other type of statement (see Section 7.4.8, "Use of Programming Logic").

An entity operation is one of several different types of activities that take place in a system. As with any other activity in the system, the decision to include an entity operation in a model should be based on whether the operation impacts entity flow in some way. For example, if a labeling activity is performed on entities in motion on a conveyor, the activity need not be modeled unless there are situations where the labeler experiences frequent interruptions.

Consolidation of Entities

Entities often undergo operations where they are consolidated or become either physically or logically connected with other entities. Examples of entity consolidation include batching and stacking. In such situations, entities are allowed to simply accumulate until a specified quantity has been gathered, and then they are grouped together into a single unit. Entity consolidation may be temporary, allowing them to later be separated, or permanent, in which case the consolidated entities no longer retain their individual identities. Figure 7.6 illustrates these two types of consolidation.

Examples of consolidating multiple entities to a single entity include

- Accumulating multiple items to fill a container.
- Gathering people together into groups of five for a ride at an amusement park.
- Grouping items to load them into an oven for heating.

In ProModel, entities are consolidated permanently using the COMBINE command. Entities may be consolidated temporarily using the GROUP command.

Attachment of Entities

In addition to consolidating accumulated entities at a location, entities can also be attached to a specific entity at a location. Examples of attaching entities might be

FIGURE 7.6

Consolidation of entities into a single entity. In (a) permanent consolidation, batched entities get destroyed. In (b) temporary consolidation, batched entities are preserved for later unbatching.

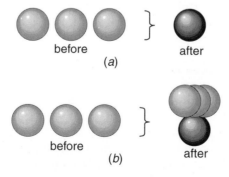

FIGURE 7.7

Attachment of one or more entities to another entity. In (a) permanent attachment, the attached entities get destroyed. In (b) temporary attachment, the attached entities are preserved for later detachment.

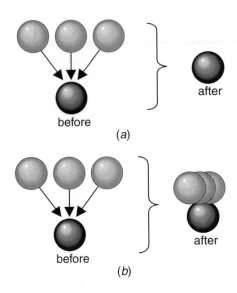

attaching a shipping document to an order ready for delivery, assembling wheels to a chassis, or placing items into a container. The difference between attaching entities and consolidating entities is that entities become attached to an existing base or main entity that must be present at the location. It is the presence of the main entity at the location that triggers the routing of the entities to be attached. In consolidation, entities are combined in whichever order they happen to enter the location. Like consolidation, attachment may be temporary, where the attached entities are later detached, or permanent, where the attached entities are destroyed and only the initial entity to which they were attached continues. Figure 7.7 illustrates these two types of attachment.

Examples of attaching entities to another entity include

- Attaching component parts to a base assembly.
- Delivering a completed order to a waiting customer.
- Loading material into a container.

In ProModel, entities are attached to another entity using the LOAD command for temporary attachments or the JOIN command for permanent attachments. Corresponding LOAD and JOIN routings must also be defined for the entities to be loaded or joined.

Dividing Entities

In some entity processes, a single entity is converted into two or more new entities. An example of entity splitting might be an item that is cut into smaller pieces or a purchase order that has carbon copies removed for filing or sending to accounting. Entities are divided in one of two ways: either the entity is split up into two or more new entities and the original entity no longer exists; or additional entities are merely created (cloned) from the original entity, which continues to exist. These two methods are shown in Figure 7.8.

FIGURE 7.8

*Multiple entities
created from a single
entity. Either (a) the
entity splits into
multiple entities (the
original entity is
destroyed) or (b) the
entity creates one or
more entities (the
original entity
continues).*

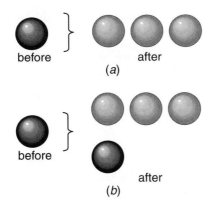

Examples of entities being split or creating new entities from a single entity
include

- A container or pallet load being broken down into the individual items
  comprising the load.
- Driving in and leaving a car at an automotive service center.
- Separating a form from a multiform document.
- A customer placing an order that is processed while the customer waits.
- A length of bar stock being cut into smaller pieces.

In ProModel, entities are split using a SPLIT statement. New entities are created
from an existing entity using a CREATE statement. Alternatively, entities can be conveniently split or created using the routing options provided in ProModel.

7.4.3 Entity Arrivals

Entity arrivals define the time, quantity, frequency, and location of entities entering the system. Examples of arrivals are customers arriving at a post office or cars
arriving at an intersection. Entities may arrive to a manufacturing or service system in one of several different ways:

- *Periodic*—they arrive at a periodic interval.
- *Scheduled*—they arrive at specified times.
- *Fluctuating*—the rate of arrival fluctuates with time.
- *Event triggered*—they arrive when triggered by some event.

In any case, entities can arrive individually or in batches.

Periodic Arrivals

Periodic arrivals occur more or less at the same interval each time. They may
occur in varying quantities, and the interval is often defined as a random variable.
Periodic arrivals are often used to model the output of an upstream process that
feeds into the system being simulated. For example, computer monitors might
arrive from an assembly line to be packaged at an interval that is normally

distributed with a mean of 1.6 minutes and a standard deviation of 0.2 minute. Examples of periodic arrivals include

- Parts arriving from an upstream operation that is not included in the model.
- Customers arriving to use a copy machine.
- Phone calls for customer service during a particular part of the day.

Periodic arrivals are defined in ProModel by using the arrivals table.

Scheduled Arrivals

Scheduled arrivals occur when entities arrive at specified times with possibly some defined variation (that is, a percentage will arrive early or late). Scheduled arrivals may occur in quantities greater than one such as a shuttle bus transporting guests at a scheduled time. It is often desirable to be able to read in a schedule from an external file, especially when the number of scheduled arrivals is large and the schedule may change from run to run. Examples of scheduled arrivals include

- Customer appointments to receive a professional service such as counseling.
- Patients scheduled for lab work.
- Production release times created through an MRP (material requirements planning) system.

Scheduled arrivals sometime occur at intervals, such as appointments that occur at 15-minute intervals with some variation. This may sound like a periodic arrival; however, periodic arrivals are autocorrelated in that the absolute time of each arrival is dependent on the time of the previous arrival. In scheduled arrival intervals, each arrival occurs independently of the previous arrival. If one appointment arrives early or late, it will not affect when the next appointment arrives.

ProModel provides a straightforward way for defining scheduled arrivals using the arrivals table. A variation may be assigned to a scheduled arrival to simulate early or late arrivals for appointments.

Fluctuating Arrivals

Sometimes entities arrive at a rate that fluctuates with time. For example, the rate at which customers arrive at a bank usually varies throughout the day with peak and lull times. This pattern may be repeated each day (see Figure 7.9). Examples of fluctuating arrivals include

- Customers arriving at a restaurant.
- Arriving flights at an international airport.
- Arriving phone calls for customer service.

In ProModel, fluctuating arrivals are specified by defining an arrival cycle pattern for a time period that may be repeated as often as desired.

Event-Triggered Arrivals

In many situations, entities are introduced to the system by some internal trigger such as the completion of an operation or the lowering of an inventory level to a

FIGURE 7.9

*A daily cycle pattern
of arrivals.*

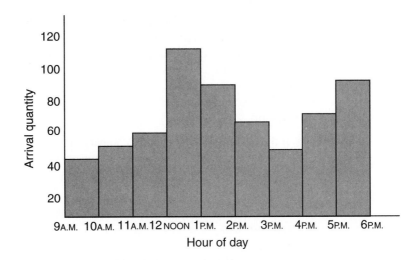

reorder point. Triggered arrivals might occur when

- A kanban or delivery signal is received.
- Inventory drops to the reorder-point level.
- Conditions have been met to start processing a new entity.

Sometimes the signal to initiate an arrival or to begin production of an item is triggered by some other arrival. An example of this is the arrival of a customer order for a product. In such instances, it isn't the order that flows through the system but the product the order triggers. To model this situation, you define the characteristic of the order arrival (frequency, type, or the like) and then have the arriving order trigger production of the product.

In ProModel, entity arrivals can be triggered using an ORDER statement. This statement can be used anywhere in the model so that any situation can cause a new entity to be ordered into the model.

7.4.4 Entity and Resource Movement

Entities and resources seldom remain stationary in a system. Entities move through the system from location to location for processing. Resources also move to different locations where they are requested for use. Additionally, resources frequently move or escort entities in the system.

Movement can be handled in three basic ways in a simulation:

1. Ignore the movement.
2. Model the move using a simple move time (which may also be defined by speed and distance).
3. Model the move using a path network that requires the moving entity or resource to contend with traffic.

A path network essentially imitates the network of aisles or hallways found on plant floors and in office facilities. A path network reduces the number of paths that need to be defined if there are a lot of different routings yet all movement shares common path segments.

The decision as to which method to use depends once again on the level of detail needed for a valid model. In making this determination, the following rules may be useful:

- If the move time is negligible compared to activity times, or if it makes sense to simply include it as part of the operation time, it may be ignored.
- If move times are significant but traffic congestion is light, a simple move time (or speed and distance) should be defined.
- If move times are significant and traffic congestion is heavy, a path network should be defined.

7.4.5 Accessing Locations and Resources

Much of the activity in a simulation is governed by how entities are able to access locations and resources for processing. Entities may be given priorities when contending with other entities for a particular location or resource. The location or resource might also be given decision-making capability for selecting from among multiple items awaiting input.

Use of Priorities

Locations and resources may be requested with a particular priority in ProModel. Priorities range from 0 to 999, with higher values having higher priority. If no priority is specified, it is assumed to be 0. For simple prioritizing, you should use priorities from 0 to 99. Priorities greater than 99 are for preempting entities and downtimes currently in control of a location or resource. The command Get Operator, 10 will attempt to get the resource called Operator with a priority of 10. If the Operator is available when requested, the priority has no significance. If, however, the Operator is currently busy, this request having a priority of 10 will get the Operator before any other waiting request with a lower priority.

Preemption

Sometimes it is desirable to have a resource or location respond immediately to a task, interrupting the current activity it is doing. This ability to bump another activity or entity that is using a location or resource is referred to as *preemption*. For example, an emergency patient requesting a particular doctor may preempt a patient receiving routine treatment by the doctor. Manufacturing and service organizations frequently have "hot" jobs that take priority over any routine job being done. Preemption is achieved by specifying a preemptive priority (100 to 999) for entering the location or acquiring the resource.

Priority values in ProModel are divided into 10 levels (0 to 99, 100 to 199, . . . , 900 to 999). Levels higher than 99 are used to preempt entities or downtimes of a lower level. Multiple preemptive levels make it possible to preempt entities or downtimes that are themselves preemptive.

The level of preemption to specify is determined by whether an entity or a downtime is being preempted. To preempt an entity that is currently using a location or resource, the preempting entity or downtime must have a priority that is at least one level higher than the current entity. For example, a priority of 100 would preempt any entity that had acquired a location or resource with a priority of 0 to 99. To preempt a downtime currently in effect at a location or resource, the preempting entity or downtime must have a priority that is at least two levels higher than the downtime (the exception to this is a setup downtime, which is preempted using the same rules as those for preempting an entity).

The primary issue associated with preemption is what to do with the preempted entity. The common way of handling a preempted entity is to simply put it on hold until the resource or location becomes available again for use. Then any remaining time is completed. This is the default way that ProModel handles preemption. Alternatively, it may be desirable to get a different resource instead of waiting for the one being used by the preempting entity. Sometimes it may be necessary to delay the preemption or possibly even ignore it, depending on certain circumstances in the model. For example, it may be unallowable to interrupt a machining cycle or a surgical operation. In such instances, a preemptive attempt may just have to wait. ProModel allows you to write special preemption logic for achieving these kinds of effects.

Task Selection Rules

Locations and resources are often discriminating with respect to which waiting entity or activity they next select to service. While priorities determine the order in which entities line up to use a location or resource, task selection rules determine which entity waiting for a location or resource is actually granted permission to use the location or resource when it becomes available. For example, parts waiting to be processed at a machine may also be competing with parts waiting to be reworked on the same machine. In such a situation, it may be desirable to first select the part requiring rework.

Task selection plays a crucial role in simulation-based scheduling (Thompson 1994). Rules range from simple priority rules such as *shortest processing time, critical ratio,* or *earliest due date* to combinatorial rules and user-defined decision logic. User-defined decision logic provides the most flexible means for defining task selection. It allows the user to define the logic for selecting the next task by examining the nature of all of the tasks waiting to be performed, looking upstream or looking ahead to see what entities or tasks will be coming along, and looking downstream to see what tasks will best serve the needs of subsequent waiting resources. By default locations and resources in ProModel respond to the highest-priority request that has been waiting the longest. Other criteria may be defined, however, to force a different response.

7.4.6 Resource Scheduling

Resources, as well as locations, frequently have scheduled times during which they are unavailable. These include off-shift periods, breaks, and preventive maintenance. If it is necessary to model periods of availability and unavailability to have a valid model, some of the issues that need to be addressed when modeling scheduled availability include

- Deciding what to do with a task that is only half completed when the end of a shift occurs.
- Making sure that resource statistics are based only on the scheduled available time and not on the entire simulation time.
- Deciding what to do with arrivals that occur at a location that is off shift.

Going off Schedule in the Middle of a Task

It is not uncommon when running a simulation with work schedules to have a resource go off schedule in the middle of a task (the current task is preempted). For short tasks, this is generally not a problem; you simply allow the resource to complete the task and then allow the schedule to take control. For longer tasks that may take hours, it usually isn't desirable to keep the resource until the task is completed. There are at least three options to take:

1. Don't start the task in the first place.
2. Interrupt the task and go off schedule.
3. If the task is nearly complete, go ahead and finish the task.

To determine whether to start the task in the first place, the modeler needs to test for the time before the next schedule change and compare it with the anticipated time to complete the task at hand. Interrupting a task usually requires only that a preemptive priority be given to the schedule. To determine whether the task is almost complete, a variable should be assigned to the completion time, which can be checked at the time of the schedule change. A related situation is when several tasks are waiting that must be processed before going off schedule (for example, a checkout stand in a grocery store closes but must service current customers in line). This situation is facilitated if some logic can be executed at the time a schedule is about to be executed. This logic needs to be able to delay the schedule execution. ProModel allows these types of delays to be defined.

Basing Resource Statistics on Scheduled Time

When including work schedules in a model, it is generally desirable to gather statistics on the resource, such as utilization statistics, based on the scheduled time available—not on total simulation time. ProModel automatically excludes off-scheduled time when calculating statistics. This way, statistics are reported for resources and locations only for the time during which they were scheduled to be available.

Handling Arrivals during Off-Shift Times

It is usually desirable to prevent arrivals (especially human entities) from occurring during off-schedule times because you don't want them to have to wait until a location or resource is back on schedule. To turn off arrivals during off-shift times, one solution is to try to synchronize the arrivals with the work schedule. This usually complicates the way arrivals are defined. Another solution, and usually an easier one, is to have the arrivals enter a preliminary location where they test whether the facility is closed and, if so, exit the system. In ProModel, if a location where entities are scheduled to arrive is unavailable at the time of an arrival, the arriving entities are simply discarded.

7.4.7 Downtimes and Repairs

It is not uncommon for resources and even locations to unexpectedly go down or become unavailable for one reason or another, such as a mechanical failure or a personal interruption. Downtimes usually occur periodically as a function of total elapsed time, time in use, or number of times used.

Downtimes Based on Total Elapsed Time
An example of a periodic downtime based on elapsed clock time might be a worker who takes a break every two hours. Scheduled maintenance is also a type of downtime performed at periodic intervals based on elapsed clock time. Figure 7.10 illustrates how a downtime based on elapsed time would be simulated. Notice that the calculation of the interval between downtimes takes into account not only busy time, but also idle time and downtime. In other words, it is the total elapsed time from the start of one downtime to the start of the next. Downtimes based on total elapsed time are often scheduled downtimes during which operational statistics on the location or resource are suspended. ProModel allows you to designate whether a particular downtime is to be counted as scheduled downtime or unscheduled downtime.

Sometimes it may be desirable to use elapsed time to define random equipment failures. This is especially true if this is how historical data were gathered on the downtime. When using historical data, it is important to determine if the time between failure was based on (1) the total elapsed time from one failure to the next, (2) the time between the repair of one failure to the time of the next failure (operational time between failures), or (3) the time that the machine was actually in operation (operating time between failures). ProModel accepts downtime definitions based on cases 1 and 3, but requires that case 2 be converted to case 1. This is done by adding the time until the next failure to the repair time of the last failure. For example, if the operational time between failures is exponentially distributed with a mean of 10 minutes and the repair time is exponentially distributed with a

FIGURE 7.10

Resource downtime occurring every 20 minutes based on total elapsed time.

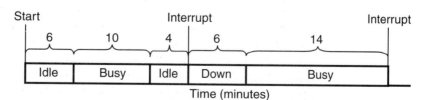

FIGURE 7.11

*Resource downtime
occurring every 20
minutes, based on
operating time.*

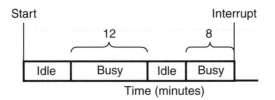

mean of 2 minutes, the time between failures should be defined as $x_{last} + E(10)$ where x_{last} is the last repair time generated using $E(2)$ minutes.

Downtimes Based on Time in Use

Most equipment and machine failures occur only when the resource is in use. A mechanical or tool failure, for example, generally happens only when a machine is running, not while a machine is idle. In this situation, the interval between downtimes would be defined relative to actual machine operation time. A machine that goes down every 20 minutes of operating time for a three-minute repair is illustrated in Figure 7.11. Note that any idle times and downtimes are not included in determining when the next downtime occurs. The only time counted is the actual operating time.

Because downtimes usually occur randomly, the time to failure is most accurately defined as a probability distribution. Studies have shown, for example, that the operating time to failure is often exponentially distributed.

Downtimes Based on the Number of Times Used

The last type of downtime occurs based on the number of times a location was used. For example, a tool on a machine may need to be replaced every 50 cycles due to tool wear, or a copy machine may need paper added after a mean of 200 copies with a standard deviation of 25 copies. ProModel permits downtimes to be defined in this manner by selecting ENTRY as the type of downtime and then specifying the number of entity entries between downtimes.

Downtime Resolution

Unfortunately, data are rarely available on equipment downtime. When they are available, they are often recorded as overall downtime and seldom broken down into number of times down and time between failures. Depending on the nature of the downtime information and degree of resolution required for the simulation, downtimes can be treated in the following ways:

- Ignore the downtime.
- Simply increase processing times to adjust for downtime.
- Use average values for mean time between failures (MTBF) and mean time to repair (MTTR).
- Use statistical distributions for time between failures and time to repair.

Ignoring Downtime.   There are several situations where it might make sense to ignore downtimes in building a simulation model. Obviously, one situation is where absolutely no data are unavailable on downtimes. If there is no knowledge of resource downtimes, it is appropriate to model the resource with no downtimes and document it as such in the final write-up. When there are downtimes, but they are extremely infrequent and not likely to affect model performance for the period of the study, it is safe to ignore them. For example, if a machine fails only two or three times a year and you are trying to predict processing capacity for the next workweek, it doesn't make sense to include the downtime. It is also safe to ignore occasional downtimes that are very small compared to activity times. If, for example, a downtime takes only seconds to correct (like clearing a part in a machine or an occasional paper jam in a copy machine), it could be ignored.

Increasing Processing Times.   A common way of treating downtime, due in part to the lack of good downtime data, is to simply reduce the production capacity of the machine by the downtime percentage. In other words, if a machine has an effective capacity of 100 parts per hour and experiences a 10 percent downtime, the effective capacity is reduced to 90 parts per hour. This spreads the downtime across each machine cycle so that both the mean time between failures and the mean time to repair are very small and both are constant. Thus no consideration is given for the variability in both time between failures and time to repair that typifies most production systems. Law (1986) has shown that this deterministic adjustment for downtime can produce results that differ greatly from the results based on actual machine downtimes.

MTBF/MTTR.   Two parts to any downtime should be defined when modeling downtime. One, *time between failures,* defines the interval between failures. The other, *time to repair,* defines the time required to bring a resource back online whenever it goes down. Often downtimes are defined in terms of mean time between failures (MTBF) and mean time to repair (MTTR). Using average times for these intervals presents the same problems as using average times for any activity in a simulation: it fails to account for variability, which can have a significant impact on system performance.

Using Statistical Distributions.   Whenever possible, time between failures and time to repair should be represented by statistical distributions that reflect the variation that is characteristic of these elements. Studies have shown that the time until failure, particularly due to items (like bearings or tooling) that wear, tends to follow a Weibull distribution. Repair times often follow a lognormal distribution.

Elapsed Time or Usage Time?

When determining the distribution for time to failure, a distinction should be made between downtime events that can occur anytime whether the resource is operating or idle and downtime events that occur only when a resource is in use. As explained earlier, downtimes that can occur anytime should be defined as a function of clock time. If the resource goes down only while in operation, it should be defined as a function of time in use.

Erroneously basing downtime on elapsed time when it should be based on operating time artificially inflates time between failures by the inclusion of idle time. It also implies that during periods of high equipment utilization, the same amount of downtime occurs as during low utilization periods. Equipment failures should generally be based on operating time and not on elapsed time because elapsed time includes operating time, idle time, and downtime. It should be left to the simulation to determine how idle time and downtime affect the overall elapsed time between failures.

To illustrate the difference this can make, let's assume that the following times were logged for a given operation:

Status	Time (Hours)
In use	20
Down	5
Idle	15
Total time	40

If it is assumed that downtime is a function of total time, then the percentage of downtime would be calculated to be 5 hours/40 hours, or 12.5 percent. If, however, it is assumed that the downtime is a function of usage time, then the downtime percentage would be 5 hours/20 hours, or 25 percent. Now let's suppose that the system is modeled with increased input to the operation so that it is never starved (the idle time $= 0$). If downtime is assumed to be 12.5 percent, the total time down will be $.125 \times 40 = 5$ hours. If, on the other hand, we use the assumption that it is 25 percent, then the time down will end up being $.25 \times 40$ hours $= 10$ hours. This is a difference of five hours, which means that if downtime is falsely assumed to be a function of total time, the simulation would realize five extra hours of production in a 40-hour period that shouldn't have happened.

Handling Interrupted Entities

When a resource goes down, there might be entities that were in the middle of being processed that are now left dangling (that is, they have been preempted). For example, a machine might break down while running the seventh part of a batch of 20 parts. The modeler must decide what to do with these entities. Several alternatives may be chosen, and the modeler must select which alternative is the most appropriate:

- Resume processing the entity after the downtime is over.
- Find another available resource to continue the process.
- Scrap the entity.
- Delay start of the downtime until the entity is processed.

The last option is the easiest way to handle downtimes and, in fact, may be adequate in situations where either the processing times or the downtimes are relatively short. In such circumstances, the delay in entity flow is still going to closely approximate what would happen in the actual system.

If the entity resumes processing later using either the same or another resource, a decision must be made as to whether only the remaining process time is used or if additional time must be added. By default, ProModel suspends processing until the location or resource returns to operation. Alternatively, other logic may be defined.

7.4.8 Use of Programming Logic

Sometimes the desired model behavior can't be achieved using a "canned" construct provided by the software. In these instances, it is necessary to use programming logic that tests probabilities, variables, and attributes to make behavioral decisions. A brief explanation of how these elements can be used in developing a simulation model is given here.

Using Probabilities to Model Probabilistic Behavior

Sometimes operational elements (routings, operations, or others) exhibit random behavior. To model a routing that occurs randomly, it is easy just to select a probabilistic routing rule and specify the probability value. The routing is then randomly chosen according to the probability specified. For operations that occur randomly at a particular location, however, a probability function must be used in connection with if–then logic. At an inspection and rework station, for example, suppose that a product is defective 10 percent of the time and requires 3 ± 1 minutes (uniformly distributed) to correct the defect. Assuming rand() is a function that returns a random value between 0 and 1, the code for defining this behavior might be as follows:

```
if rand() <= .10
then wait U(3,1) min
```

A similar situation occurs in activity times that have more than one distribution. For example, when a machine goes down, 30 percent of the time it takes Triangular(0.2, 1.5, 3) minutes to repair and 70 percent of the time it takes Triangular(3, 7.5, 15) minutes to repair. The logic for the downtime definition might be

```
if rand() <= .30
then wait T(.2, 1.5, 3) min
else wait T(3, 7.5, 15) min
```

Another example of multidistribution activity is a call center that handles three different types of calls that we will designate as A, B, and C. The time to handle calls is exponentially distributed, but the mean time is different depending on the type of call (see Table 7.1). To model this situation, the following code

TABLE 7.1 **Table of Service Times for Different Call Types**

Call Type	Probability of Occurrence	Service Time (Minutes)
A	.20	$E(5)$
B	.50	$E(8)$
C	.30	$E(12)$

might be entered ("//" are used at the beginning of a comment line):

```
// set a real variable (rValue) to a random number
real rValue = rand()
// test for call type A
if rValue <=.20
then wait E(5) min
else if rValue <=.70
then wait E(8) min
else wait E(12) min
```

Using Attributes to Model Special Decision Logic

Sometimes an item that has passed through the same location a second time requires less time to be reworked than it took for the initial operation. In this situation, an attribute of the entity should be the basis of the decision because we need to test how many times that particular entity has visited the location. An attribute we will call Pass will need to be defined in the attribute module. If a normal operation takes five minutes but a rework operation takes only two minutes, the following code would be entered in the operation logic for the location:

```
// increment the value of Pass by one
INC Pass
if Pass = 1
then wait 5 min
else wait 2 min
```

Using Global Variables to Gather Statistics

Global variables are placeholders for values that may change during the simulation. What makes them global is that they are accessible from any place in the model by any object in the model, and they exist during the entire simulation.

Global variables are defined by the user for user-defined purposes. For example, a variable may be defined to keep track of the total number of entities in a certain area of the system (work in process). Each time an entity enters the area, it increments the variable. Also, each time an entity leaves the area, it decrements the variable. Because the user has the option of collecting statistics on global variables, either the simple or time-weighted average value of the variable can be reported at the end of the simulation. A time series report showing all of the changes in the variable over time can also be reported.

Using Local Variables for Looping

Local variables are variables that are declared within the logic itself, either right before the point of use or at the beginning of the logic block. A local variable exists only for the current object executing the block of logic. When the object finishes executing the logic block in which the local variable was defined, the variable disappears. A local variable may be thought of as a temporary attribute of the object executing the logic because multiple objects executing the same logic at the same time would each have their own local variables assigned to them while executing the block of logic. Local variables are useful primarily for executing a logic loop. Suppose, for example, that you wanted to assign the value of 4 to the first 10 elements of an array called NumOfBins that was defined in the Array module. To do this within operation logic (or in any other logic), you would enter something like the following, where Count is defined as a local variable:

```
int Count = 1
while Count < 11 do
        {
        NumOfBins[Count] = 4
        Inc Count
        }
```

The braces "{" and "}" are the ProModel notation (also used in C++ and Java) for starting and ending a block of logic. In this case it is the block of statements to be executed repeatedly by an object as long as the local variable Count is less than 11.

7.5 Miscellaneous Modeling Issues

This section addresses special issues that may be encountered in simulation. They don't fit well under previous headings, so they are put in a collection here even though they are not necessarily related.

7.5.1 Modeling Rare Occurrences

Often there exist situations in the real world that occur only rarely. For example, a machine may break down once every two months, or only one in a thousand parts is rejected at an inspection station. In simulation analysis, we are generally interested in the normal behavior of the system, not extremely rare behavior. It is advisable to ignore rare situations and exclude them from the simulation model. This approach not only reduces modeling time but also simplifies the model and helps maintain the focus of everyone involved in the model on key input variables.

But some rare situations have a significant impact on the operation of the system. For example, a plant shutdown may bring the entire operation to a stop. If the focus of interest is to evaluate the effects of the rare occurrence, such as how long it takes for inventories to be depleted if a shutdown occurs, then it makes sense to include the rare occurrence in the model. The easiest way to model a rare event is not to let the model run for six months or a year before the event occurs, but to go ahead and force the event to happen. The point of modeling the rare event is to see what impact it has on system behavior, not to predict when it actually might occur. So it really doesn't matter when it happens as long as the state of the model is typical of what the system might be when it does occur.

7.5.2 Large-Scale Modeling

Occasionally it may be desirable to model a large system such as an entire factory or the entire activity in an international airport. The tendency (especially for novice simulators) is to painstakingly piece together a huge, complex model only to find that it runs nowhere near the way that was expected and may not even run at all. The disappointment of not getting the model to run correctly the first time is soon overshadowed by the utter despair in trying to debug such an enormous model. When faced with building a supermodel, it is always a good idea to partition the model into several submodels and tackle the problem on a smaller scale first. Once each of the submodels has been built and validated, they can be merged into a larger composite model. This composite model can be structured either as a single monolithic model or as a hierarchical model in which the details of each submodel are hidden unless explicitly opened for viewing. Several ways have been described for merging individual submodels into a composite model (Jayaraman and Agarwal 1996). Three of the most common ways that might be considered for integrating submodels are

- *Option 1: Integrate all of the submodels just as they have been built.* This approach preserves all of the detail and therefore accuracy of the individual submodels. However, the resulting composite model may be enormous and cause lengthy execution times. The composite model may be structured as a flat model or, to reduce complexity, as a hierarchical model.

- *Option 2: Use only the recorded output from one or more of the submodels.* By simulating and recording the time at which each entity exits the model for a single submodel, these exit times can be used in place of the submodel for determining the arrival times for the larger model. This eliminates the need to include the overhead of the individual submodel in the composite model. This technique, while drastically reducing the complexity of the composite model, may not be possible if the interaction with the submodel is two-way. For submodels representing subsystems that simply feed into a larger system (in which case the subsystem operates fairly independently of downstream activities), this technique is valid. An example is an assembly facility in which fabricated components or even subassemblies feed into a final assembly line. Basically, each feeder line is viewed as a "black box" whose output is read from a file.

- *Option 3: Represent the output of one or more of the submodels as statistical distributions.* This approach is the same as option 2, but instead of using the recorded output times from the submodel in the composite model, a statistical distribution is fit to the output times and used to generate the input to the composite model. This technique eliminates the need for using data files that, depending on the submodel, may be quite large. Theoretically, it should also be more accurate because the true underlying distribution is used instead of just a sample unless there are discontinuities in the output. Multiple sample streams can also be generated for running multiple replications.

7.5.3 Cost Modeling

Often it is desirable to include cost in a model to determine the most cost-effective solution to a design problem. If, for example, two operators on an assembly line result in the same performance as using three robots, the decision may end up being based on cost rather than performance. There are two approaches to modeling cost.

One is to include cost factors in the model itself and dynamically update cost collection variables during the simulation. ProModel includes a cost module for assigning costs to different factors in the simulation such as entity cost, waiting cost, and operation cost. The alternative approach is to run a cost analysis after the simulation, applying cost factors to collected cost drivers such as resource utilization or time spent in storage. The first method is best when it is difficult to summarize cost drivers. For example, the cost per unit of production may be based on the types of resources used and the time for using each type. This may be a lot of information for each entity to carry using attributes. It is much easier to simply update the entity's cost attribute dynamically whenever a particular resource has been used. Dynamic cost tracking suffers, however, from requiring cost factors to be considered during the modeling stage rather than the analysis stage. For some models, it may be difficult to dynamically track costs during a simulation, especially when relationships become very complex.

The preferred way to analyze costs, whenever possible, is to do a postsimulation analysis and to treat cost modeling as a follow-on activity to system modeling rather than as a concurrent activity (see Lenz and Neitzel 1995). There are several advantages to separating the logic model from the cost model. First, the model is not encumbered with tracking information that does not directly affect how the model operates. Second, and perhaps more importantly, post analysis of costs gives more flexibility for doing "what-if" scenarios with the cost model. For example, different cost scenarios can be run based on varying labor rates in a matter of seconds when applied to simulation output data that are immediately available. If modeled during the simulation, a separate simulation would have to be run applying each labor rate.

7.6 Summary

Model building is a process that takes a conceptual model and converts it to a simulation model. This requires a knowledge of the modeling paradigm of the particular simulation software being used and a familiarity with the different modeling constructs that are provided in the software. Building a model involves knowing what elements to include in the model and how to best express those elements in the model. The principle of parsimony should always be followed, which results in the most minimal model possible that achieves the simulation objectives. Finally, the keys to successful modeling are seeing lots of examples and practice, practice, practice!

7.7 Review Questions

1. How is modeling an art as well as a science?
2. What is a modeling paradigm?
3. Describe the modeling paradigm used in a specific simulation product.

4. Suppose you were modeling an inspection station as an object. Identify two attributes and one behavior you might want to define for it.

5. Identify five different things an entity might represent in a manufacturing or service simulation.

6. For a manufacturing operation that produces 300 different part types, how would you represent the different part types?

7. A stand and power cord are assembled to a monitor at an assembly station. Stands are produced in-house and are not always available when needed. Power cords are purchased finished and are always available at the station. What entity types (monitor, cord, or stand) would you include in the model and, conceptually, how would you model this assembly operation?

8. Every hour, a stack of purchase orders is delivered to the purchasing department for processing. The stack is routed sequentially to three different people who review, approve, and send out the orders. The orders move together as a stack, although they are processed individually at each station. The processing time at each activity is a function of the number of orders in the stack. The number of orders in a stack can be determined by a probability distribution. Conceptually, how would you model this process and still capture the essence of what is happening?

9. What criteria would you use to identify the route stops or locations in a flow sequence to include in a model?

10. Suppose you are modeling a system with bins in which parts are transported. Would you model the bins as resources or entities? Justify your answer.

11. How would you model a consumable resource such as energy consumption that occurs based on the length of time a particular machine is in operation?

12. What is the danger of using simple time values to model material movement if the material handling system often encounters traffic delays?

13. How would you use an entity attribute to model multiple routing passes through the same location in which a different operation time is required depending on the pass?

14. An operator must inspect every fifth part at a manual workstation. Conceptually, how would you model this activity if the operation takes five minutes and inspection takes two minutes?

15. When attaching entities to another entity, under what circumstances would you want to preserve the identities of the entities being attached?

16. Define four types of arrivals and give an example of each.

17. A factory operates with two eight-hour shifts (changeover occurs without interruption) five days a week. If you wanted to simulate the system for four weeks, how would you set the run length in hours?

18. What is the problem with modeling downtimes in terms of mean time between failures (MTBF) and mean time to repair (MTTR)?

19. Why should unplanned downtimes or failures be defined as a function of usage time rather than total elapsed time on the clock?

20. In modeling repair times, how should the time spent waiting for a repairperson be modeled?

21. What is preemption? What activities or events might preempt other activities in a simulation?

22. A boring machine experiences downtimes every five hours (exponentially distributed). It also requires routine preventive maintenance (PM) after every eight hours (fixed) of operation. If a downtime occurs within two hours of the next scheduled PM, the PM is performed as part of the repair time (no added time is needed) and, after completing the repair coupled with the PM, the next PM is set for eight hours away. Conceptually, how would you model this situation?

23. A real estate agent schedules six customers (potential buyers) each day, one every 1.5 hours, starting at 8 A.M. Customers are expected to arrive for their appointments at the scheduled times. However, past experience shows that customer arrival times are normally distributed with a mean equal to the scheduled time and a standard deviation of five minutes. The time the agent spends with each customer is normally distributed with a mean of 1.4 hours and a standard deviation of .2 hours. Develop a simulation model to calculate the expected waiting time for customers.

References

Jayaraman, Arun, and Arun Agarwal. "Simulating an Engine Plant." *Manufacturing Engineering,* November 1996, pp. 60–68.

Law, A. M. "Introduction to Simulation: A Powerful Tool for Analyzing Complex Manufacturing Systems." *Industrial Engineering,* 1986, 18(5):57–58.

Lenz, John, and Ray Neitzel. "Cost Modeling: An Effective Means to Compare Alternatives." *Industrial Engineering,* January 1995, pp. 18–20.

Shannon, Robert E. "Introduction to the Art and Science of Simulation." In *Proceedings of the 1998 Winter Simulation Conference,* ed. D. J. Medeiros, E. F. Watson, J. S. Carson, and M. S. Manivannan. Piscataway, NJ: Institute of Electrical and Electronics Engineers, 1998.

Thompson, Michael B. "Expanding Simulation beyond Planning and Design." *Industrial Engineering,* October 1994, pp. 64–66.

8   MODEL VERIFICATION AND VALIDATION

"Truth comes out of error more easily than out of confusion."
—Francis Bacon

8.1 Introduction

Building a simulation model is much like developing an architectural plan for a house. A good architect will review the plan and specifications with the client or owner of the house to ensure that the design meets the client's expectations. The architect will also carefully check all dimensions and other specifications shown on the plan for accuracy. Once the architect is reasonably satisfied that the right information has been accurately represented, a contractor can be given the plan to begin building the house. In a similar way the simulation analyst should examine the validity and correctness of the model before using it to make implementation decisions.

In this chapter we cover the importance and challenges associated with model verification and validation. We also present techniques for verifying and validating models. Balci (1997) provides a taxonomy of more than 77 techniques for model verification and validation. In this chapter we give only a few of the more common and practical methods used. The greatest problem with verification and validation is not one of failing to use the right techniques but failing to use *any* technique. Questions addressed in this chapter include the following:

- What are model verification and validation?
- What are obstacles to model verification and validation?
- What techniques are used to verify and validate a model?
- How are model verification and validation maintained?

Two case studies are presented at the end of the chapter showing how verification and validation techniques have been used in actual simulation projects.

8.2 Importance of Model Verification and Validation

Model building is, by nature, very error prone. The modeler must translate the real-world system into a conceptual model that, in turn, must be translated into a simulation model. This translation process is iterative as the modeler's conception of the real system changes and the modeling approach continues to be refined (see Figure 8.1). In this translation process, there is plenty of room for making errors. Verification and validation processes have been developed to reduce and ideally eliminate these errors.

Model *verification* is the process of determining whether the simulation model correctly reflects the conceptual model. Model *validation* is the process of determining whether the conceptual model correctly reflects the real system. Model verification and validation are critical to the success of a simulation project. Important decisions may be based on the outcome of the simulation experiment, and, therefore, demonstrable evidence should exist for the validity of the model. For simulations that are used on an ongoing basis for operational decisions such as production scheduling, model validation is even more important. The Department of Defense (DoD) has recognized the importance of model validation and has established a defined set of criteria for determining model accreditation. *Accreditation,* as defined by DoD (http://www.dmso.mil/docslib/mspolicy/glossary/9801glss.doc), is "The official certification that a model or simulation is acceptable for use for a specific purpose."

8.2.1 Reasons for Neglect

Despite their importance, seldom is adequate attention given to verification and validation when doing simulation. The primary reasons for neglecting this

FIGURE 8.1

Two-step translation process to convert a real-world system to a simulation model.

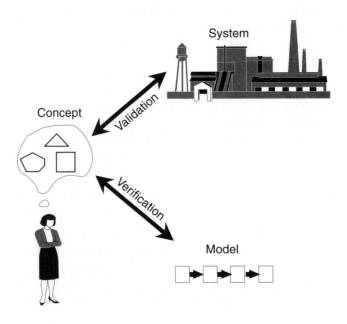

important activity are

- Time and budget pressures.
- Laziness.
- Overconfidence.
- Ignorance.

Time and budget pressures can be overcome through better planning and increased proficiency in the validation process. Because validation doesn't *have* to be performed to complete a simulation study, it is the activity that is most likely to get shortchanged when pressure is being felt to complete the project on time or within budget. Not only does the modeler become pressed for time and resources, but often others on whom the modeler relies for feedback on model validity also become too busy to get involved the way they should.

Laziness is a bit more difficult to deal with because it is characteristic of human nature and is not easily overcome. Discipline and patience must be developed before one is willing to painstakingly go back over a model once it has been built and is running. Model building is a creative activity that can actually be quite fun. Model verification and validation, on the other hand, are a laborious effort that is too often dreaded.

The problem of overconfidence can be dealt with only by developing a more critical and even skeptical attitude toward one's own work. Too often it is assumed that if the simulation runs, it must be okay. A surprising number of decisions are based on invalid simulations simply because the model runs and produces results. Computer output can give an aura of validity to results that may be completely in error. This false sense of security in computer output is a bit like saying, "It must be true, it came out of the computer." Having such a naive mentality can be dangerously misleading.

The final problem is simply a lack of knowledge of verification and validation procedures. This is a common problem particularly among newcomers to simulation. This seems to be one of the last areas in which formal training is received because it is presumed to be nonessential to model building and output analysis. Another misconception of validation is that it is just another phase in a simulation project rather than a continuous activity that should be performed throughout the entire project. The intent of this chapter is to dispel common misconceptions, help shake attitudes of indifference, and provide insight into this important activity.

8.2.2 Practices That Facilitate Verification and Validation

A major problem in verifying and validating models occurs because of poor modeling practices. Beginners often build models with little or no thought about being able to verify or validate the model. Often these models contain spaghetti code that is difficult for anyone, including the modeler, to follow. This problem becomes even more acute as models grow in complexity. When the original model builder moves on and others inherit these models, the desperate efforts to unravel the tangled mess and figure out what the model creator had in mind become almost futile. It is especially discouraging when attempting to use a poorly constructed model for future experimentation. Trying to figure out what changes need to be made to model new scenarios becomes difficult if not impossible.

The solution to creating models that ease the difficulty of verification and validation is to first reduce the amount of complexity of the model. Frequently the most complex models are built by amateurs who do not have sense enough to know how to abstract system information. They code way too much detail into the model. Once a model has been simplified as much as possible, it needs to be coded so it is easily readable and understandable. Using object-oriented techniques such as encapsulation can help organize model data. The right simulation software can also help keep model data organized and readable by providing table entries and intelligent, parameterized constructs rather than requiring lots of low-level programming. Finally, model data and logic code should be thoroughly and clearly documented. This means that every subroutine used should have an explanation of what it does, where it is invoked, what the parameters represent, and how to change the subroutine for modifications that may be anticipated.

8.3 Model Verification

Verification is the process of determining whether the model operates as intended. It doesn't necessarily mean that the model is valid, only that it runs correctly. According to Banks, et al. (2001), "Verification is concerned with building the model right. It is utilized in the comparison of the conceptual model to the computer representation that implements that conception." Model verification involves the modeler more than the customer. During the verification process, the modeler tries to detect unintended errors in the model data and logic and remove them. In essence, verification is the process of debugging the model. A verified model is a bug-free model.

Errors or bugs in a simulation model are of two types: syntax errors and semantic errors. Syntax errors are like grammatical errors and include the unintentional addition, omission, or misplacement of notation that either prevents the model from running or causes it to run incorrectly. The omission or misplacement of a decimal point in a number or parentheses in an expression can dramatically impact the outcome of a simulation. Most simulation software has built-in error detection that can identify inconsistencies and omissions. This feature is much like spelling and grammar checking in a word processor. As with a word processor, however, it still remains the responsibility of the user to ensure that the intended meaning has been conveyed.

Semantic errors are associated with the meaning or intention of the modeler and are therefore difficult to detect. Often they are logical errors that cause behavior that is different than what was intended. For example, an if–then test on a variable may branch to the wrong logic simply because the condition was incorrectly specified or the branching was not set up correctly. Or numerical values may be misentered and therefore result in the wrong intended behavior. Verification techniques discussed here, such as event tracing and code debugging, can help identify and correct semantic errors.

8.3.1 Preventive Measures

Obviously, the preferred method of dealing with model bugs is to avoid building models with bugs in the first place. In practice, this isn't always possible as bugs are often sneaky and difficult to prevent. However, a few basic practices can minimize the number of bugs that find their way into models. The most important practice is to simply be careful. This means focusing on what you are doing and eliminating distractions that may cause you to make silly typing mistakes. You should check and double-check the data and your entry of those data into the model to make sure there are no transcription errors. Another practice that has been used successfully for years in software development for ensuring good code is called *structured programming*. Hoover and Perry (1990) describe five basic principles of structured programming:

1. *Top-down design.* The simulation model and the code begin at a high level, and then lower-level areas are added that are congruent with, yet subordinate to, the top level. It is a hierarchical approach to modeling. A routing to a six-machine cell may initially be defined to the cell as a whole and then later decomposed to route to each individual workstation.

2. *Modularity.* The model is built in modules or logical divisions to simplify model development and debugging and to facilitate reuse. Where possible, data modules should be separate from the logic modules so that data can be easily accessed.

3. *Compact modules.* Modules should be kept as short and simple as possible. Subroutines consisting of over 25 statements, for example, should probably be broken into multiple subroutines.

4. *Stepwise refinement.* The model is built with complexity being progressively added. In stepwise refinement, one goes from general to specific so that generalities and assumptions are refined into precise and accurate details. The idea is to create the model in multiple passes, filling in additional details with each pass. At each pass, a test for reasonableness can be made. It is easier to verify a model when the model is built incrementally than when it is built all at once.

5. *Structured control.* GOTO statements or other unstructured branching of control should be avoided wherever possible as they may lead to unexpected results. Logic control should be based on structured control statements such as

```
IF-THEN-ELSE
WHILE...DO
DO...WHILE
(etc.)
```

8.3.2 Establishing a Standard for Comparison

Verifying a model presupposes that some standard exists for comparison, providing a sense of whether the model is running correctly. Obviously, if you knew exactly how the simulation should run and what results should be expected, you wouldn't need to be doing the simulation in the first place. Several reasonable standards can be used for model comparison. One simple standard is common sense (although someone has said that common sense isn't all that common). If your simulation is giving totally bizarre results, you know you have a problem somewhere in your model. Say, for example, that you define a machine downtime to occur every two hours for two minutes and yet the simulation results show that the machine was down 80 percent of the time. Common sense tells you that the results don't make sense and indicates that you should check how you defined the downtime behavior.

Another way to establish a sense of correctness is to construct an analytic model of the problem if at all possible. Because of the simplified assumptions made in most analytic techniques, it may be necessary to scale back the model to verify only its basic operation. It may be necessary, for example, to run the simulation without downtimes because failures may complicate an analytic formulation. As additional examples, we may operate a queuing system without balking, or we may use instantaneous replenishment rather than random lead times in an inventory model. In these cases, the simplified model may be one that can be analyzed without simulation, and the mathematical results can then be compared to those from the simulation. In this way we can get an indication of whether the simulation model, or at least a simplified version of it, is correct.

8.3.3 Verification Techniques

Once everything has been done at the outset to ensure that the model has been built correctly and a standard for comparison has been established, several techniques can be used to verify the model. Some of the more common ways include

- Conduct model code reviews.
- Check the output for reasonableness.
- Watch the animation for correct behavior.
- Use the trace and debug facilities provided with the software.

Reviewing Model Code

A model code review can be conducted by either the modeler or someone else familiar with both modeling and the system being modeled. The purpose of the model code review is to check for errors and inconsistencies. The simulation model can be tested in either a bottom-up or top-down fashion. In the bottom-up testing method, also called *unit testing,* the lowest modules are tested and verified first. Next, interfaces between two or more modules are tested. This is continued until the model can be tested as a single system.

In the top-down approach, the verification testing begins with the main module and moves down gradually to lower modules. At the top level, you are more interested that the outputs of modules are as expected given the inputs. If discrepancies arise, lower-level code analysis is conducted.

In both approaches sample test data are used to test the program code. The bottom-up approach typically requires a smaller set of data to begin with. After exercising the model using test data, the model is stress tested using extreme input values. With careful selection, the results of the simulation under extreme conditions can be predicted fairly well and compared with the test results.

Checking for Reasonable Output

In any simulation model, there are operational relationships and quantitative values that are predictable during the simulation. An example of an operational relationship is a routing that occurs whenever a particular operation is completed. A predictable quantitative value might be the number of boxes on a conveyor always being between zero and the conveyor capacity. Often the software itself will flag values that are out of an acceptable range, such as an attempt to free more resources than were captured.

For simple models, one way to help determine reasonableness of the output is to replace random times and probabilistic outcomes with constant times and deterministic outcomes in the model. This allows you to predict precisely what the results will be because the analytically determined results should match the results of the simulation.

Watching the Animation

Animation can be used to visually verify whether the simulation operates the way you think it should. Errors are detected visually that can otherwise go unnoticed. The animation can be made to run slowly enough for the analyst to follow along visually. However, the amount of time required to observe an entire simulation run can be extremely long. If the animation is sped up, the runtime will be smaller, but inconsistent behavior will be more difficult to detect.

To provide the most meaningful information, it helps to have interactive simulation capability, which allows the modeler to examine data associated with simulation objects during runtime. For example, during the simulation, the user can view the current status of a workstation or resource to see if state variables are being set correctly.

Animation is usually more helpful in identifying a problem than in discovering the cause of a problem. The following are some common symptoms apparent from the animation that reveal underlying problems in the model:

- The simulation runs fine for hours and even days and then suddenly freezes.
- A terminating simulation that should have emptied the system at the end of the simulation leaves some entities stranded at one or more locations.
- A resource sits idle when there is plenty of work waiting to be done.
- A particular routing never gets executed.

Using Trace and Debugging Facilities

Trace and debugging information provides detailed textual feedback of what is happening during a simulation. This allows the modeler to look under the hood and see exactly what is going on inside the simulation. Most simulation software comes with some sort of trace and debugging facility. At a minimum, manual trace or debugging information can be coded into the simulation model that is either deleted or disabled once the model is debugged. The following logic, for example, might be inserted somewhere in the model logic to test whether a resource is available at a specific time.

```
If OperatorState = busy
Then display "Operator is busy"
Else display "Operator is NOT busy"
```

Trace messages describe chronologically what happens during the simulation, event by event. A typical trace message might be the time that an entity enters a particular location or the time that a specific resource is freed. Trace messaging can be turned on or off, and trace messages can usually be either displayed directly on the screen or written to a file for later analysis. In ProModel, tracing can also be turned on and off programmatically by entering a Trace command at appropriate places within the model logic. A segment of an example of a trace message listing is shown in Figure 8.2. Notice that the simulation time is shown in the left column followed by a description of what is happening at that time in the right column.

Another way in which behavior can be tracked during a simulation run is through the use of a debugger. Anyone who has used a modern commercial compiler to do programming is familiar with debuggers. A simulation debugger is a utility that displays and steps through the actual logic entered by the user to define the model. As logic gets executed, windows can be opened to display variable and

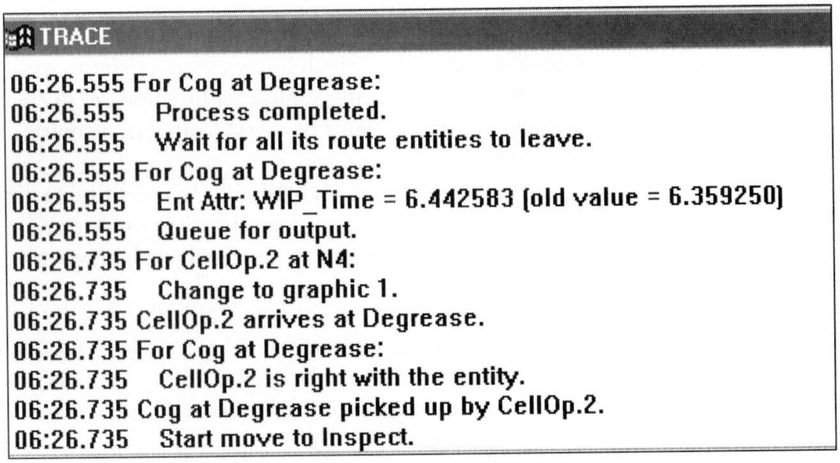

FIGURE 8.2

Fragment of a trace listing.

```
TRACE
06:26.555 For Cog at Degrease:
06:26.555    Process completed.
06:26.555    Wait for all its route entities to leave.
06:26.555 For Cog at Degrease:
06:26.555    Ent Attr: WIP_Time = 6.442583 (old value = 6.359250)
06:26.555    Queue for output.
06:26.735 For CellOp.2 at N4:
06:26.735    Change to graphic 1.
06:26.735 CellOp.2 arrives at Degrease.
06:26.735 For Cog at Degrease:
06:26.735    CellOp.2 is right with the entity.
06:26.735 Cog at Degrease picked up by CellOp.2.
06:26.735    Start move to Inspect.
```

FIGURE 8.3
*ProModel debugger
window.*

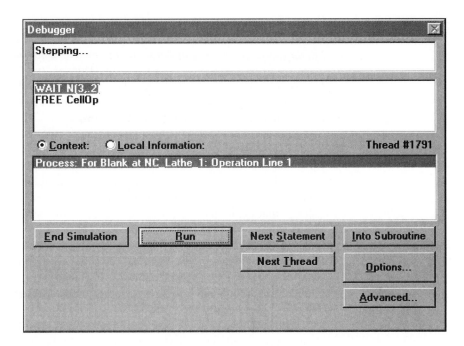

state values as they dynamically change. Like trace messaging, debugging can be
turned on either interactively or programmatically. An example of a debugger
window is shown in Figure 8.3.

Experienced modelers make extensive use of trace and debugging capabilities. Animation and output reports are good for detecting problems in a simulation, but trace and debug messages help uncover why problems occur.

Using trace and debugging features, event occurrences and state variables
can be examined and compared with hand calculations to see if the program is operating as it should. For example, a trace list might show when a particular operation began and ended. This can be compared against the input operation time to
see if they match.

One type of error that a trace or debugger is useful in diagnosing is *gridlock*.
This situation is caused when there is a circular dependency where one action depends on another action, which, in turn, is dependent on the first action. You may
have experienced this situation at a busy traffic intersection or when trying to
leave a crowded parking lot after a football game. It leaves you with a sense of
utter helplessness and frustration. An example of gridlock in simulation sometimes occurs when a resource attempts to move a part from one station to another,
but the second station is unavailable because an entity there is waiting for the
same resource. The usual symptom of this situation is that the model appears to
freeze up and no more activity takes place. Meanwhile the simulation clock races
rapidly forward. A trace of events can help detect why the entity at the second station is stuck waiting for the resource.

8.4 Model Validation

Validation is the process of determining whether the model is a meaningful and accurate representation of the real system (Hoover and Perry 1990). Where verification is concerned with building the model right, validation is concerned with building the right model. For this reason, stakeholders and customers should become heavily involved in the validation process.

Because the process of validation can be very time-consuming and ultimately somewhat elusive, we are not interested in achieving absolute validity but only functional validity. As Neelamkavil (1987) explains, "True validation is a philosophical impossibility and all we can do is either invalidate or 'fail to invalidate.'" From a functional standpoint, model validation is viewed as the process of establishing "that the model's output behavior has sufficient accuracy for the model's intended purpose over the domain of the model's intended applicability" (Sargent 1998). Through validation we convey to the user or the customer that the simulation results can be trusted and used to make real-world decisions. A model can be valid even when many simplifications may have been made and details omitted while building the model. What we are seeking is not certainty but rather a comfort level in the results.

For existing systems or when building an "as-is" model, the model behavior should correspond to that of the actual system. In the case of a new system that is being designed, and therefore has no demonstrated performance, the input data should accurately reflect the design specification of the system and the model should be carefully verified.

A model may run correctly (it has been verified) but still not be an accurate representation of the system (it is invalid). Validation can be thought of as a reality check (not to be confused with a sanity check, which people who do simulation may frequently need). If a model is based on accurate information, and the model has been verified to work as intended, then the model can be said to be a valid model. A model that has bugs cannot be a valid model (unless purely by accident), thus the importance of verification in the validation process.

Validation actually begins with the data-gathering stage of a simulation project before a model is even built. It is the garbage-in, garbage-out syndrome that should be avoided. If you use erroneous data in the model, the model may work perfectly fine (it has been verified) yet still give misleading results. Data validation issues were addressed in Chapter 6, "Data Collection and Analysis." By way of review, data validation is a matter of verifying that the sources of information are reliable and that source data have been correctly interpreted. For example, the process used to fit distributions to sample data may be checked. Any assumptions should also be reviewed and approved by those in a position to give an informed opinion about the validity of the assumptions. Having valid information on the system being modeled means that the conceptual model of the system is valid. At this point, creating a valid simulation model essentially consists of creating a model that is verified or, in other words, that corresponds with the conceptual model.

8.4.1 *Determining Model Validity*

There is no simple test to determine the validity of a model. Validation is an inductive process in which the modeler draws conclusions about the accuracy of the model based on the evidence available. Several techniques are given here that are described by Sargent (1998). Some of them tend to be rather subjective, while others lend themselves to statistical tests such as hypothesis tests and confidence intervals. Many of the techniques are the same as those used to verify a model, only now the customer and others who are knowledgeable about the system need to be involved. As with model verification, it is common to use a combination of techniques when validating a model.

- *Watching the animation.* The visual animation of the operational behavior of the model is compared with one's knowledge about how the actual system behaves. This can include dynamic plots and counters that provide dynamic visual feedback.

- *Comparing with the actual system.* Both the model and the system are run under the same conditions and using the same inputs to see if the results match.

- *Comparing with other models.* If other valid models have been built of the process such as analytic models, spreadsheet models, and even other valid simulation models, the output of the simulation can be compared to these known results. For example, a simple simulation model might be compared to a validated queuing model.

- *Conducting degeneracy and extreme condition tests.* There are known situations for which model behavior degenerates, causing a particular response variable to grow infinitely large. An example is the number of items in a queue prior to a server where the service rate is less than the arrival rate. These situations can be forced in the model (by increasing the number of arrivals) to see if the model degenerates as expected. The model can also be run under extreme conditions to see if it behaves as would be expected. For example, if arrivals are cut off, the model should eventually run dry of entities.

- *Checking for face validity.* Face validity is checked by asking people who are knowledgeable about the system whether the model and its behavior appear reasonable. This technique can be used in determining if the logic in the conceptual model is correct and if a model's input–output relationships are reasonable.

- *Testing against historical data.* If historical information exists for both operating and performance data, the model can be tested using the same operating data and comparing the results with the historical performance data. This is sort of an "as-once-was" model.

- *Performing sensitivity analysis.* This technique consists of changing model input values to determine the effect on the model's behavior and its output. The same relationships should occur in the model as in the real system. The modeler should have at least an intuitive idea of how the model will react to a given change. It should be obvious, for example, that doubling the number of resources for a bottleneck operation should increase, though not necessarily double, throughput.

- *Running traces.* An entity or sequence of events can be traced through the model processing logic to see if it follows the behavior that would occur in the actual system.
- *Conducting Turing tests.* People who are knowledgeable about the operations of a system are asked if they can discriminate between system and model outputs. If they are unable to detect which outputs are the model outputs and which are the actual system outputs, this is another piece of evidence to use in favor of the model being valid.

A common method of validating a model of an existing system is to compare the model performance with that of the actual system. This approach requires that an as-is simulation be built that corresponds to the current system. This helps "calibrate" the model so that it can be used for simulating variations of the same model. After running the as-is simulation, the performance data are compared to those of the real-world system. If sufficient data are available on a performance measure of the real system, a statistical test can be applied called the Student's *t* test to determine whether the sampled data sets from both the model and the actual system come from the same distribution. An *F* test can be performed to test the equality of variances of the real system and the simulation model. Some of the problems associated with statistical comparisons include these:

- Simulation model performance is based on very long periods of time. On the other hand, real system performances are often based on much shorter periods and therefore may not be representative of the long-term statistical average.
- The initial conditions of the real system are usually unknown and are therefore difficult to replicate in the model.
- The performance of the real system is affected by many factors that may be excluded from the simulation model, such as abnormal downtimes or defective materials.

The problem of validation becomes more challenging when the real system doesn't exist yet. In this case, the simulation analyst works with one or more experts who have a good understanding of how the real system *should* operate. They team up to see whether the simulation model behaves reasonably in a variety of situations. Animation is frequently used to watch the behavior of the simulation and spot any nonsensical or unrealistic behavior.

The purpose of validation is to mitigate the risk associated with making decisions based on the model. As in any testing effort, there is an optimum amount of validation beyond which the return may be less valuable than the investment (see Figure 8.4). The optimum effort to expend on validation should be based on minimizing the total cost of the validation effort plus the risk cost associated

FIGURE 8.4

Optimum level of validation looks at the trade-off between validation cost and risk cost.

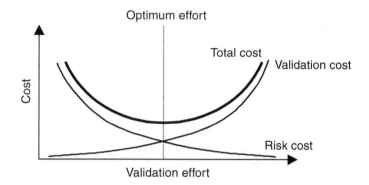

with making a decision based on an invalid model. As mentioned previously, the problem is seldom one of devoting too much time to validation, but in devoting too little.

8.4.2 Maintaining Validation

Once a model is validated for a given system specification, it is sometimes necessary to maintain validation as system specifications tend to evolve right up to, and often even after, system implementation. When simulation is used for system design, rarely is the system implemented the way it was originally simulated, especially when traditional, over-the-wall handoff procedures are used. These continuous changes make it nearly impossible to validate the simulation against the final implemented system. Unfortunately, results of the implemented system sometimes get compared with the early simulation model, and the simulation is unfairly branded as being invalid. This is another good reason for continuing to update the model to continually reflect current system design specifications.

One method that works effectively for maintaining validation of a model designed for ongoing use is to deny users access to the model data so that modifications to the model are controlled. ProModel, for example, allows a modeler to protect the model data using the packaging option under the File menu. This prevents anyone from changing the actual model data directly. Modifications to the model can be made only through a custom interface that controls precisely what parameters of the model can be changed as well as the extent to which they can be changed.

8.4.3 Validation Examples

To illustrate how the validation process might be performed for a typical simulation, we look at two simulation case studies. The first one was conducted at St. John Hospital and Medical Center for an obstetrical renovation project (O'Conner 1994). The second project was conducted at HP for a surface mount assembly line (Baxter and Johnson 1993). The St. John case study uses a more informal and intuitive approach to validation, while the HP case study relies on more formal validation techniques.

St. John Hospital and Medical Center Obstetrical Unit

At St. John Hospital and Medical Center, a simulation study was conducted by the Management Engineering Department to plan a renovation of the obstetrical unit to accommodate modern family demands to provide a more private, homelike atmosphere with a family-oriented approach to patient care during a mother's hospital stay. The current layout did not support this approach and had encountered inefficiencies when trying to accommodate such an approach (running newborns back and forth and the like).

The renovation project included provision for labor, delivery, recovery, and postpartum (LDRP) rooms with babies remaining with their mothers during the postpartum stay. The new LDRPs would be used for both low- and high-risk mothers. The hospital administration wanted to ensure that there would be enough room for future growth while still maintaining high utilization of every bed.

The purpose of the simulation was to determine the appropriate number of beds needed in the new LDRPs as well as in other related areas of the hospital. A secondary objective was to determine what rules should be used to process patients through the new system to make the best utilization of resources. Because this was a radically new configuration, an as-is model was not built. This, of course, made model validation more challenging. Data were gathered based on actual operating records, and several assumptions were made such as the length of stay for different patient classifications and the treatment sequence of patients.

The validation process was actually a responsibility given to a team that was assembled to work with the simulation group. The team consisted of the nursing managers responsible for each OB area and the Women's Health program director. The first phase of the review phase was to receive team approval for the assumptions document and flowcharts of the processes. This review was completed prior to developing the model. This assumptions document was continually revised and reviewed throughout the duration of the project.

The next phase of the review came after all data had been collected and analyzed. These reviews not only helped ensure the data were valid but established trust and cooperation between the modeling group and the review team. This began instilling confidence in the simulation even before the results were obtained.

Model verification was performed during model building and again at model completion. To help verify the model, patients, modeled as entities, were traced through the system to ensure that they followed the same processing sequence shown on the flowcharts with the correct percentages of patients following each probabilistic branch. In addition, time displays were placed in the model to verify the length of stay at each step in the process.

Model validation was performed with both the nursing staff and physician representatives. This included comparing actual data collected with model results where comparisons made sense (remember, this was a new configuration being modeled). The animation that showed the actual movement of patients and use of resources was a valuable tool to use during the validation process and increased the credibility of the model.

The use of a team approach throughout the data-gathering and model development stages proved invaluable. It both gave the modeler immediate feedback regarding incorrect assumptions and boosted the confidence of the team members in the simulation results. The frequent reviews also allowed the group to bounce ideas off one another. As the result of conducting a sound and convincing simulation study, the hospital's administration was persuaded by the simulation results and implemented the recommendations in the new construction.

HP Surface Mount Assembly Line

At HP a simulation study was undertaken to evaluate alternative production layouts and batching methods for a surface mount printed circuit assembly line based on a projected growth in product volume/mix. As is typical in a simulation project, only about 20 percent of the time was spent building the model. The bulk of the time was spent gathering data and validating the model. It was recognized that the level of confidence placed in the simulation results was dependent on the degree to which model validity could be established.

Surface mount printed circuit board assembly involves placing electronic components onto raw printed circuit boards. The process begins with solder paste being stenciled onto the raw circuit board panel. Then the panel is transported to sequential pick-and-place machines, where the components are placed on the panel. The component feeders on the pick-and-place machines must be changed depending on the components required for a particular board. This adds setup time to the process. A third of the boards require manual placement of certain components. Once all components are placed on the board and solder pasted, the board proceeds into an oven where the solder paste is cured, bonding the components to the panel. Finally, the panel is cleaned and sent to testing. Inspection steps occur at several places in the line.

The HP surface mount line was designed for a high-mix, low-volume production. More than 100 different board types are produced on the line with about 10 different types produced per day. The batch size for a product is typically less than five panels, with the typical panel containing five boards.

A simulation expert built the model working closely with the process owners—that is, the managers, engineers, and production workers. Like many simulations, the model was first built as an as-is model to facilitate model validation. The model was then expanded to reflect projected requirements and proposed configurations.

The validation process began early in the data-gathering phase with meetings being held with the process owners to agree on objectives, review the assumptions, and determine data requirements. Early involvement of the process owners gave them motivation to see that the project was successful. The first step was to clarify the objectives of the simulation that would guide the data-gathering and modeling efforts. Next a rough model framework was developed, and the initial assumptions were defined based on process information gathered from the process owners. As the model development progressed, further assumptions were made, and decisions were made and agreed upon regarding the level of detail required to capture the essence of the process.

The flow logic and move times were built into the model, but processing and setup times, since they were so varied and numerous, were put into a common lookup table so they could be verified and modified easily. The particular board sequence and batch size were also defined for the model.

The next step was to verify the model to ensure that it operated as intended. The simulation output was reviewed for reasonableness and a trace was used to check the model logic where it was complex. In reality, much of this verification was performed as the model was built because it was necessary for getting the model just to run. Note that the modeler was the one primarily responsible for verifying or debugging the model. This makes sense inasmuch as the modeler had a clear specification and was the one responsible for fashioning the model to fit the specification.

After the modeler was satisfied that the model was running correctly, all of the assumptions were reviewed again by the production operators, supervisors, and engineers. These assumptions were more meaningful as they were able to watch their impact on the simulation performance. Animation was used to communicate model behavior and demonstrate the scope of the model. Those who were familiar with the operation of the real line were able to confirm whether the model appeared to be working reasonably correctly. When explaining how the model was built, it was important to describe the modeling constructs using familiar terminology because the process owners were not versed in modeling terminology.

At this point it would be easy to conclude that the model was valid and ready for use in conducting experiments. The challenging question with the model became "Is it the right model?" A simulation model is viewed much more critically once it is running and it is realized that important decisions will soon be based on the results of the simulation experiments. With the model running correctly and providing output that was consistent with the assumptions that were made, it was felt that the assumptions should be challenged more critically. Because the initial data that were gathered—that is, processing times and setup times—were based on average times with no reflection of variability, decisions had to be made about how much more data should be gathered or what modifications in assumptions needed to be made before there was sufficient confidence in the model. Component placement times were studied using classical time-study techniques to achieve more realistic time estimates. Next the average placement times were used to calculate the average run time for a particular panel, and these were compared to the recorded run times.

Historical data and further time studies were used to validate the output. A particular sequence of panels was studied as it was processed, recording the batch size, the start time, the finish time, and some intermediate times. The model was then run using this exact same scenario, and the output was compared with the output of the real process. The initial model predicted 25 shifts to complete a specific sequence of panels, while the process actually took 35 shifts. This led to further process investigations of interruptions or other processing delays that weren't being accounted for in the model. It was discovered that feeder replacement times were underestimated in the model. A discrepancy between the model and the actual system was also discovered in the utilization of the pick-and-place machines. After tracking down the causes of these discrepancies and making appropriate adjustments to the model, the simulation results were closer to the real process. The challenge at this stage was not to yield to the temptation of making arbitrary changes to the model just to get the desired results. Then the model would lose its integrity and become nothing more than a sham for the real system.

The final step was to conduct sensitivity analysis to determine how model performance was affected in response to changes in model assumptions. By changing the input in such a way that the impact was somewhat predictable, the change in simulation results could be compared with intuitive guesses. Any bizarre results such as a decrease in work in process when increasing the arrival rate would raise an immediate flag. Input parameters were systematically changed based on a knowledge of both the process behavior and the model operation. Knowing just how to stress the model in the right places to test its robustness was an art in itself. After everyone was satisfied that the model accurately reflected the actual operation and that it seemed to respond as would be expected to specific changes in input, the model was ready for experimental use.

8.5 Summary

For a simulation model to be of greatest value, it must be an accurate representation of the system being modeled. Verification and validation are two activities that should be performed with simulation models to establish their credibility.

Model verification is basically the process of debugging the model to ensure that it accurately represents the conceptual model and that the simulation runs correctly. Verification involves mainly the modeler without the need for much input from the customer. Verification is not an exact science, although several proven techniques can be used.

Validating a model begins at the data-gathering stage and may not end until the system is finally implemented and the actual system can be compared to the model. Validation involves the customer and other stakeholders who are in a position to provide informed feedback on whether the model accurately reflects the real system. Absolute validation is philosophically impossible, although a high degree of face or functional validity can be established.

While formal methods may not always be used to validate a model, at least some time devoted to careful review should be given. The secret to validation is not so much the technique as it is the attitude. One should be as skeptical as can be tolerated—challenging every input and questioning every output. In the end, the decision maker should be confident in the simulation results, not because he or she trusts the software or the modeler, but because he or she trusts the input data and knows how the model was built.

8.6 Review Questions

1. What is the difference between model verification and model validation?
2. Why are verification and validation so important in a simulation study?
3. Why are these steps often ignored or given only little attention?
4. How can one prevent syntax and semantic errors from creeping into a simulation model?
5. How does one establish a standard for verifying a model?
6. What techniques are used to verify a model?
7. Why is it philosophically impossible to prove model validity?
8. What is functional validity?
9. What techniques are used to validate a model?
10. Why is it useful to build an as-is model if a change to an existing system is being evaluated?
11. What specific things were done in the St. John Hospital simulation study to validate the model?
12. What specific steps were taken in the HP simulation study to validate the model?

References

Balci, Osman. "Verification, Validation and Accreditation of Simulation Models." In *Proceedings of the 1997 Winter Simulation Conference,* ed. S. Andradottir, K. J. Healy, D. H. Withers, and B. L. Nelson, 1997, pp. 135–41.

Banks, Jerry. "Simulation Evolution." *IIE Solutions,* November 1998, pp. 26–29.

Banks, Jerry, John Carson, Barry Nelson, and David Nicol. *Discrete-Event Simulation,* 3rd ed. Englewood Cliffs, NJ: Prentice-Hall, 1995.

Baxter, Lori K., and Johnson, Eric. "Don't Implement before You Validate." *Industrial Engineering,* February 1993, pp. 60–62.

Hoover, Stewart, and Ronald Perry. *Simulation: A Problem Solving Approach.* Reading, MA: Addison-Wesley, 1990.

Neelamkavil, F. *Computer Simulation and Modeling.* New York: John Wiley & Sons, 1987.

O'Conner, Kathleen. "The Use of a Computer Simulation Model to Plan an Obstetrical Renovation Project." The Fifth Annual PROMODEL Users Conference, Park City, UT, 1994.

Sargent, Robert G. "Verifying and Validating Simulation Models." In *Proceedings of the 1998 Winter Simulation Conference,* ed. D. J. Medeiros, E. F. Watson, J. S. Carson, and M. S. Manivannan, 1998, pp. 121–30.

9　SIMULATION OUTPUT ANALYSIS

"Nothing is more terrible than activity without insight."
—Thomas Carlyle

9.1 Introduction

In analyzing the output from a simulation model, there is room for both rough analysis using judgmental procedures as well as statistically based procedures for more detailed analysis. The appropriateness of using judgmental procedures or statistically based procedures to analyze a simulation's output depends largely on the nature of the problem, the importance of the decision, and the validity of the input data. If you are doing a go/no-go type of analysis in which you are trying to find out whether a system is capable of meeting a minimum performance level, then a simple judgmental approach may be adequate. Finding out whether a single machine or a single service agent is adequate to handle a given workload may be easily determined by a few runs unless it looks like a close call. Even if it is a close call, if the decision is not that important (perhaps there is a backup worker who can easily fill in during peak periods), then more detailed analysis may not be needed. In cases where the model relies heavily on assumptions, it is of little value to be extremely precise in estimating model performance. It does little good to get six decimal places of precision for an output response if the input data warrant only precision to the nearest tens. Suppose, for example, that the arrival rate of customers to a bank is roughly estimated to be 30 plus or minus 10 per hour. In this situation, it is probably meaningless to try to obtain a precise estimate of teller utilization in the facility. The precision of the input data simply doesn't warrant any more than a rough estimate for the output.

These examples of rough estimates are in no way intended to minimize the importance of conducting statistically responsible experiments, but rather to emphasize the fact that the average analyst or manager can gainfully use simulation without having to be a professional statistician. Anyone who understands the cause-and-effect relationships of a system will be able to draw meaningful conclusions about it that will help in design and management decisions. Where precision is important and warranted, however, a statistical background or at least consulting with a statistical advisor will help avoid drawing invalid inferences from simulation.

In this chapter, we present basic statistical issues related to the analysis and interpretation of simulation output. The intent is not to make you an "expert" statistician, but to give you sound guidelines to follow when dealing with the output generated by simulation models. For more comprehensive coverage of statistical methods, see Hines and Montgomery (1990).

9.2 Statistical Analysis of Simulation Output

Because random input variables (such as activity times) are used to drive the model, the output measures of the simulation (throughput rate, average waiting times) are also going to be random. This means that performance metrics based on a model's output are only estimates and not precise measures. Suppose, for example, we are interested in determining the throughput for a model of a manufacturing system (like the number of widgets produced per hour). Suppose further that the model has a number of random input variables such as machining times, time between machine failures, and repair times. Running the simulation once provides a single random sample of the model's fluctuating throughput that may or may not be representative of the expected or average model throughput. This would be like rolling a die, getting a 1, and concluding that the expected average value of the die is 1 instead of the true average of 3.5.

Statistical analysis of simulation output is based on inferential or descriptive statistics. In descriptive statistics, we deal with a population, samples from the population, and a sample size. The population is the entire set of items or outcomes (such as all possible values that can be obtained when rolling a die). A sample would be one unbiased observation or instance of the population (a single roll of a die). The sample size is the number of samples we are working with (a particular number of rolls of a die). The collected samples are independent (each roll of the die is unaffected by previous rolls of the die). The idea behind inferential statistics is to gather a large enough sample size to draw valid inferences about the population (such as the average value of the die) while keeping the sample-gathering time and cost at a minimum. In some instances, collecting samples can be very time-consuming and costly.

The same statistical concepts apply to conducting an experiment with a simulation model of a system that apply to sampling from a population such as rolls of a die or a group of people:

- Samples are generated by running the experiment.
- The population size is often very large or infinite (there are an infinite number of possible outcomes from the experiment).

- A random variable is used to express each possible outcome of an experiment as a continuous or discrete number.
- Experimental samples, called *replications,* are independent.

Multiple independent runs or replications of a simulation provide multiple observations that can be used to estimate the expected value of the model's output response (like rolling the die multiple times before estimating an expected or mean value). With random simulation output results, we can't determine for certain what the true expected output will be from the model unless in some cases we run an infinite number of replications, which is not possible or even desirable.

9.2.1 Simulation Replications

In conducting a simulation experiment, one run of the simulation constitutes one replication of the experiment. The outcome of a replication represents a single sample. To obtain a sample size *n*, we need to run *n* independent replications of the experiment. A representative sample size will be a good indicator of what can be expected to happen for any subsequent replication. Not every replication will produce exactly the same behavior due to random variability. To understand how independent replications are produced in simulation, it is essential to understand the concepts of random number generation and initial seed values. These subjects were elaborated on in Chapter 3 but will be reviewed here.

Random Number Generation

At the heart of a stochastic simulation is the random number generator. The random number generator produces a cycling sequence of pseudo-random numbers that are used to generate observations from various distributions (normal, exponential, or others) that drive the stochastic processes represented in the simulation model such as activity times, machine failure times, and travel times. As pseudo-random numbers, the random number sequence, or stream, can be repeated if desired. After the random number generator sequences through all random numbers in its cycle, the cycle starts over again through the same sequence (see Figure 9.1). The length of the cycle before it repeats is called the *cycle period,* which is usually extremely long (the cycle length for the random number generator used in ProModel is over 2.1 billion).

Initial Seed Values

The random number generator requires an initial seed value before it can begin producing random numbers. Each random number produced by the generator is fed back to the generator to produce the next random number, and so on. Where the sequence begins in the cycle depends on the initial seed value given to the generator. In Figure 9.1, the seed value of $Z_0 = 17$ marks the beginning of this sequence of numbers.

You might wonder how replications are produced if the exact same sequence of random numbers drives the simulation. Obviously, identical random number sequences produce identical results. This may prove useful for repeating an exact

FIGURE 9.1

Example of a cycling pseudo-random number stream produced by a random number generator with a very short cycle length.

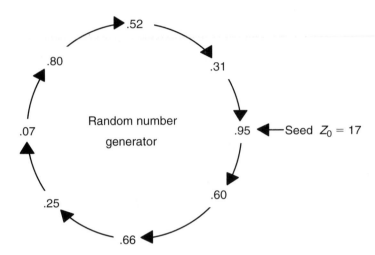

experiment but does not give us an independent replication. On the other hand, if a different seed value is appropriately selected to initialize the random number generator, the simulation will produce different results because it will be driven by a different segment of numbers from the random number stream. This is how the simulation experiment is replicated to collect statistically independent observations of the simulation model's output response. Recall that the random number generator in ProModel can produce over 2.1 billion different values before it cycles.

To replicate a simulation experiment, then, the simulation model is initialized to its starting conditions, all statistical variables for the output measures are reset, and a new seed value is appropriately selected to start the random number generator. Each time an appropriate seed value is used to start the random number generator, the simulation produces a unique output response. Repeating the process with several appropriate seed values produces a set of statistically independent observations of a model's output response. With most simulation software, users need only specify how many times they wish to replicate the experiment; the software handles the details of initializing variables and ensuring that each simulation run is driven by nonoverlapping segments of the random number cycle.

9.2.2 Performance Estimation

Assuming we have an adequate collection of independent observations, we can apply statistical methods to estimate the expected or mean value of the model's output response. There are two types of estimates: (1) point estimates (mean and standard deviation, for example) and (2) interval estimates (also called confidence intervals). This is not to say that mean and standard deviation are the only parameter estimates in which we may be interested. For example, we may compute point estimates for parameters such as the proportion of time a queue had more than 10 entities waiting in it or the maximum number of entities to simultaneously wait in a queue location. Most simulation software automatically calculates these and other estimates for you; however, it is useful to know the basic procedures used to better understand their meaning.

Point Estimates

A *point estimate* is a single value estimate of a parameter of interest. Point estimates are calculated for the mean and standard deviation of the population. To estimate the mean of the population (denoted as μ), we simply calculate the average of the sample values (denoted as $\bar{x}$):

$$\bar{x} = \frac{\sum_{i=1}^{n} x_i}{n}$$

where n is the sample size (number of observations) and x_i is the value of ith observation. The sample mean $\bar{x}$ estimates the population mean μ.

The standard deviation for the population (denoted as σ), which is a measure of the spread of data values in the population, is similarly estimated by calculating a standard deviation of the sample of values (denoted as s):

$$s = \sqrt{\frac{\sum_{i=1}^{n} [x_i - \bar{x}]^2}{n - 1}}$$

The sample variance s^2, used to estimate the variance of the population σ^2, is obtained by squaring the sample standard deviation.

Let's suppose, for example, that we are interested in determining the mean or average number of customers getting a haircut at Buddy's Style Shop on Saturday morning. Buddy opens his barbershop at 8:00 A.M. and closes at noon on Saturday. In order to determine the exact value for the true average number of customers getting a haircut on Saturday morning (μ), we would have to compute the average based on the number of haircuts given on all Saturday mornings that Buddy's Style Shop has been and will be open (that is, the complete population of observations). Not wanting to work that hard, we decide to get an estimate of the true mean μ by spending the next 12 Saturday mornings watching TV at Buddy's and recording the number of customers that get a haircut between 8:00 A.M. and 12:00 noon (Table 9.1).

TABLE 9.1   Number of Haircuts Given on 12 Saturdays

Replication (i)	Number of Haircuts (x_i)
1	21
2	16
3	8
4	11
5	17
6	16
7	6
8	14
9	15
10	16
11	14
12	10
Sample mean $\bar{x}$	13.67
Sample standard deviation s	4.21

We have replicated the experiment 12 times and have 12 sample observations of the number of customers getting haircuts on Saturday morning $(x_1, x_2, \ldots, x_{12})$. Note that there are many different values for the random variable (number of haircuts) in the sample of 12 observations. So if only one replication had been conducted and the single observation was used to estimate the true but unknown mean number of haircuts given, the estimate could potentially be way off. Using the observations in Table 9.1, we calculate the sample mean as follows:

$$\bar{x} = \frac{\sum_{i=1}^{12} x_i}{12} = \frac{21 + 16 + 8 + \cdots + 10}{12} = 13.67 \text{ haircuts}$$

The sample mean of 13.67 haircuts estimates the unknown true mean value μ for the number of haircuts given on Saturday at the barbershop. Taking additional samples generally gives a more accurate estimate of the unknown μ. However, the experiment would need to be replicated on each Saturday that Buddy's has been and will be open to determine the true mean or expected number of haircuts given.

In addition to estimating the population mean, we can also calculate an estimate for the population standard deviation based on the sample size of 12 observations and a sample mean of 13.67 haircuts. This is done by calculating the standard deviation s of the observations in Table 9.1 as follows:

$$s = \sqrt{\frac{\sum_{i=1}^{12} [x_i - 13.67]^2}{12 - 1}} = 4.21 \text{ haircuts}$$

The sample standard deviation s provides only an estimate of the true but unknown population standard deviation σ. $\bar{x}$ and s are single values; thus they are referred to as *point estimators* of the population parameters μ and σ. Also, note that $\bar{x}$ and s are random variables. As such, they will have different values if based on another set of 12 independent observations from the barbershop.

Interval Estimates

A point estimate, by itself, gives little information about how accurately it estimates the true value of the unknown parameter. *Interval estimates* constructed using $\bar{x}$ and s, on the other hand, provide information about how far off the point estimate $\bar{x}$ might be from the true mean μ. The method used to determine this is referred to as *confidence interval estimation*.

A *confidence interval* is a range within which we can have a certain level of confidence that the true mean falls. The interval is symmetric about $\bar{x}$, and the distance that each endpoint is from $\bar{x}$ is called the half-width (hw). A confidence interval, then, is expressed as the probability P that the unknown true mean μ lies within the interval $\bar{x} \pm hw$. The probability P is called the *confidence level*.

If the sample observations used to compute $\bar{x}$ and s are independent and normally distributed, the following equation can be used to calculate the half-width of a confidence interval for a given level of confidence:

$$hw = \frac{(t_{n-1,\alpha/2})s}{\sqrt{n}}$$

where $t_{n-1,\alpha/2}$ is a factor that can be obtained from the Student's t table in Appendix B. The values are identified in the Student's t table according to the value of $\alpha/2$ and the degrees of freedom $(n-1)$. The term α is the complement of P. That is, $\alpha = 1 - P$ and is referred to as the *significance level*. The significance level may be thought of as the "risk level" or probability that μ will fall outside the confidence interval. Therefore, the probability that μ will fall within the confidence interval is $1 - \alpha$. Thus confidence intervals are often stated as $P(\bar{x} - hw \leq \mu \leq \bar{x} + hw) = 1 - \alpha$ and are read as the probability that the true but unknown mean μ falls between the interval $(\bar{x} - hw)$ to $(\bar{x} + hw)$ is equal to $1 - \alpha$. The confidence interval is traditionally referred to as a $100(1 - \alpha)$ percent confidence interval.

Assuming the data from the barbershop example are independent and normally distributed (this assumption is discussed in Section 9.3), a 95 percent confidence interval is constructed as follows:

Given: P = confidence level = 0.95
α = significance level = $1 - P = 1 - 0.95 = 0.05$
n = sample size = 12
$\bar{x} = 13.67$ haircuts
$s = 4.21$ haircuts

From the Student's t table in Appendix B, we find $t_{n-1,\alpha/2} = t_{11,0.025} = 2.201$. The half-width is computed as follows:

$$hw = \frac{(t_{11,0.025})s}{\sqrt{n}} = \frac{(2.201)4.21}{\sqrt{12}} = 2.67 \text{ haircuts}$$

The lower and upper limits of the 95 percent confidence interval are calculated as follows:

Lower limit = $\bar{x} - hw = 13.67 - 2.67 = 11.00$ haircuts

Upper limit = $\bar{x} + hw = 13.67 + 2.67 = 16.34$ haircuts

It can now be asserted with 95 percent confidence that the true but unknown mean falls between 11.00 haircuts and 16.34 haircuts ($11.00 \leq \mu_{\text{haircuts}} \leq 16.34$). A risk level of 0.05 ($\alpha = 0.05$) means that if we went through this process for estimating the number of haircuts given on a Saturday morning 100 times and computed 100 confidence intervals, we would expect 5 of the confidence intervals (5 percent of 100) to exclude the true but unknown mean number of haircuts given on a Saturday morning.

The width of the interval indicates the accuracy the point estimate. It is desirable to have a small interval with high confidence (usually 90 percent or greater). The width of the confidence interval is affected by the variability in the output of the system and the number of observations collected (sample size). It can be seen from the half-width equation that, for a given confidence level, the half-width will shrink if (1) the sample size n is increased or (2) the variability in the output of the system (standard deviation s) is reduced. Given that we have little control over the variability of the system, we resign ourselves to increasing the sample size (run more replications) to increase the accuracy of our estimates.

Actually, when dealing with the output from a simulation model, techniques can sometimes reduce the variability of the output from the model without changing the expected value of the output. These are called *variance reduction techniques* and are covered in Chapter 10.

9.2.3 Number of Replications (Sample Size)

Often we are interested in knowing the sample size or number of replications needed to establish a particular confidence interval for a specified amount of absolute error (denoted by e) between the point estimate of the mean $\bar{x}$ and the unknown true mean μ. In our barbershop example, suppose we want to estimate the number of replications n' needed to be able to assert with 95 percent ($1 - \alpha = .95$) confidence that the sample mean $\bar{x}$ we compute is off by at most 2.00 haircuts ($e = 2.00$) from the true but unknown mean μ. In other words, we would like to reduce the half-width of 2.67 haircuts for the confidence interval we previously computed to 2.00 haircuts.

Note that α represents the probability that the difference between $\bar{x}$ and μ will exceed a specified error amount e. In fact, e is equal to the hw of a confidence interval ($\bar{x} \pm e$). Setting $e = hw$ and solving for n would give an equation that could be used to estimate the number of replications necessary to meet a specified absolute error amount e and significance level α:

$$hw = e$$

$$\frac{(t_{n-1,\alpha/2})s}{\sqrt{n}} = e$$

$$n = \left[\frac{(t_{n-1,\alpha/2})s}{e}\right]^2$$

However, this cannot be completed because n appears on each side of the equation. So, to estimate the number of replications needed, we replace $t_{n-1,\alpha/2}$ with $Z_{\alpha/2}$, which depends only on α. The $Z_{\alpha/2}$ is from the standard normal distribution, and its value can be found in the last row of the Student's t table in Appendix B. Note that $Z_{\alpha/2} = t_{\infty,\alpha/2}$, where ∞ denotes infinity. The revised equation is

$$n' = \left[\frac{(Z_{\alpha/2})s}{e}\right]^2$$

where n' is a rough approximation for the number of replications that will provide an adequate sample size for meeting the desired absolute error amount e and significance level α.

Before this equation can be used, an initial sample of observations must be collected to compute the sample standard deviation s of the output from the system. Actually, when using the $Z_{\alpha/2}$ value, an assumption is made that the true standard deviation for the complete population of observations σ is known. Of course, σ is not known. However, it is reasonable to approximate σ with the sample standard deviation s to get a rough approximation for the required number of replications.

Going back to the barbershop example, we want to estimate the number of replications n' needed to be able to assert with 95 percent $(1 - \alpha = .95)$ confidence that the sample mean $\bar{x}$ we compute is off by at most 2.00 haircuts $(e = 2.00)$ from the true but unknown mean μ. Based on the initial sample of 12 observations, we know that $s = 4.21$ haircuts. We compute n' as follows:

Given: $P = $ confidence level $ = 0.95$

$\quad\quad\quad \alpha = $ significance level $ = 1 - P = 1 - 0.95 = 0.05$

$\quad\quad\quad e = 2.00$

$\quad\quad\quad s = 4.21$ haircuts

From the last row of the Student's t table in Appendix B, we find $Z_{\alpha/2} = Z_{0.025} = t_{\infty,0.025} = 1.96$. Using the previous equation for n' we obtain

$$n' = \left[\frac{(Z_{0.025})s}{e} \right]^2 = \left[\frac{(1.96)4.21}{2.00} \right]^2 = 17.02 \text{ observations}$$

A fractional number of replications cannot be conducted, so the result is typically rounded up. Thus $n' \approx 18$ observations. Given that we already have 12 observations, the experiment needs to be replicated an additional 6 ($n' - n = 18 - 12 = 6$) times to collect the necessary observations. After the additional 6 replications are conducted, the 6 observations are combined with the original 12 observations (Table 9.2), and a new $\bar{x}$, s, and confidence interval are computed using all 18 observations.

TABLE 9.2   Number of Haircuts Given on 18 Saturdays

Replication (i)	Number of Haircuts x_i
1	21
2	16
3	8
4	11
5	17
6	16
7	6
8	14
9	15
10	16
11	14
12	10
13	7
14	9
15	18
16	13
17	16
18	8
Sample mean $\bar{x}$	13.06
Sample standard deviation s	4.28

Given: P = confidence level = 0.95

α = significance level = $1 - P = 1 - 0.95 = 0.05$

n = sample size = 18

$\bar{x}$ = 13.06 haircuts

s = 4.28 haircuts

From the Student's t table in Appendix B, we find $t_{n-1,\alpha/2} = t_{17,0.025} = 2.11$. The half-width is computed as follows:

$$hw = \frac{(t_{17,0.025})s}{\sqrt{n}} = \frac{(2.11)4.28}{\sqrt{18}} = 2.13 \text{ haircuts}$$

The lower and upper limits for the new 95 percent confidence interval are calculated as follows:

Lower limit = $\bar{x} - hw = 13.06 - 2.13 = 10.93$ haircuts

Upper limit = $\bar{x} - hw = 13.06 + 2.13 = 15.19$ haircuts

It can now be asserted with 95 percent confidence that the true but unknown mean falls between 10.93 haircuts and 15.19 haircuts ($10.93 \le \mu_{\text{haircuts}} \le 15.19$).

Note that with the additional observations, the half-width of the confidence interval has indeed decreased. However, the half-width is larger than the absolute error value planned for ($e = 2.0$). This is just luck, or bad luck, because the new half-width could just as easily have been smaller. Why? First, 18 replications was only a rough estimate of the number of observations needed. Second, the number of haircuts given on a Saturday morning at the barbershop is a random variable. Each collection of observations of the random variable will likely differ from previous collections. Therefore, the values computed for $\bar{x}$, s, and hw will also differ. This is the nature of statistics and why we deal only in estimates.

We have been expressing our target amount of error e in our point estimate $\bar{x}$ as an absolute value ($hw = e$). In the barbershop example, we selected an absolute value of $e = 2.00$ haircuts as our target value. However, it is sometimes more convenient to work in terms of a relative error (re) value ($hw = re|\mu|$). This allows us to talk about the percentage error in our point estimate in place of the absolute error. Percentage error is the relative error multiplied by 100 (that is, $100re$ percent). To approximate the number of replications needed to obtain a point estimate $\bar{x}$ with a certain percentage error, we need only change the denominator of the n' equation used earlier. The relative error version of the equation becomes

$$n' = \left[\frac{(z_{\alpha/2})s}{\left(\frac{re}{(1+re)}\right)\bar{x}} \right]^2$$

where re denotes the relative error. The $re/(1 + re)$ part of the denominator is an adjustment needed to realize the desired re value because we use $\bar{x}$ to estimate μ (see Chapter 9 of Law and Kelton 2000 for details). The appeal of this approach is that we can select a desired percentage error without prior knowledge of the magnitude of the value of μ.

As an example, say that after recording the number of haircuts given at the barbershop on 12 Saturdays ($n = 12$ replications of the experiment), we wish to determine the approximate number of replications needed to estimate the mean number of haircuts given per day with an error percentage of 17.14 percent and a confidence level of 95 percent. We apply our equation using the sample mean and sample standard deviation from Table 9.1.

Given: $P = $ confidence level $= 0.95$
$\alpha = $ significance level $= 1 - P = 1 - 0.95 = 0.05$
$Z_{\alpha/2} = Z_{0.025} = 1.96$ from Appendix B
$re = 0.1714$
$\bar{x} = 13.67$
$s = 4.21$

$$n' = \left[\frac{(z_{0.025})s}{\left(\frac{re}{(1+re)}\right)\bar{x}} \right]^2 = \left[\frac{(1.96)4.21}{\left(\frac{0.1714}{(1+0.1714)}\right)13.67} \right]^2 = 17.02 \text{ observations}$$

Thus $n' \approx 18$ observations. This is the same result computed earlier on page 229 for an absolute error of $e = 2.00$ haircuts. This occurred because we purposely selected $re = 0.1714$ to produce a value of 2.00 for the equation's denominator in order to demonstrate the equivalency of the different methods for approximating the number of replications needed to achieve a desired level of precision in the point estimate $\bar{x}$. The SimRunner software uses the relative error methodology to provide estimates for the number of replications needed to satisfy the specified level of precision for a given significance level α.

In general, the accuracy of the estimates improves as the number of replications increases. However, after a point, only modest improvements in accuracy (reductions in the half-width of the confidence interval) are made through conducting additional replications of the experiment. Therefore, it is sometimes necessary to compromise on a desired level of accuracy because of the time required to run a large number of replications of the model.

The ProModel simulation software automatically computes confidence intervals. Therefore, there is really no need to estimate the sample size required for a desired half-width using the method given in this section. Instead, the experiment could be replicated, say, 10 times, and the half-width of the resulting confidence interval checked. If the desired half-width is not achieved, additional replications of the experiment are made until it is. The only real advantage of estimating the number of replications in advance is that it may save time over the trial-and-error approach of repeatedly checking the half-width and running additional replications until the desired confidence interval is achieved.

9.2.4 Real-World Experiments versus Simulation Experiments

You may have noticed that we have been working with the observations as though they came from the real barbershop and not from a simulation model of the shop.

Actually, the observations were generated from a simulation of the shop by making independent runs (replications) of the model using the method described in Section 9.2.1. Note, however, that had the observations been collected from the real barbershop, we would have processed them in the same manner. Experimental results are generally handled in the same manner, regardless of their origin, if the observations are unbiased, independent, and normally distributed.

Note that very different results could be obtained from the real barbershop if the simulation model does not accurately mimic the behavior of the real shop. Statistical statements, no matter how precise, derived from the output of a simulation model provide estimates for the true but unknown population parameters of the model and not of the real system. Whether or not the model behavior actually corresponds to real-world behavior is a matter of model validity, which was discussed in Chapter 8.

9.3 Statistical Issues with Simulation Output

A single simulation run (replication) represents only a single sample of possible outcomes (values of a random variable) from the simulation. Averaging the observations from a large number of runs comes closer to the true expected performance of the model, but it is still only a statistical estimate. Constructing a confidence interval is a good idea because it provides a measure of how far off the estimate is from the model's true expected performance. However, the following three statistical assumptions must be met regarding the sample of observations used to construct the confidence interval:

- Observations are *independent* so that no correlation exists between consecutive observations.
- Observations are *identically distributed* throughout the entire duration of the process (that is, they are time invariant).
- Observations are *normally distributed.*

Table 9.3 presents a framework for thinking about the output produced from a series of simulation runs (independent replications) that will aid in the discussion of these three important statistical assumptions. Suppose a system has been

TABLE 9.3   Simulation Output from a Series of Replications

Replication (i)	Within Run Observations y_{ij}	Average x_i
1	$y_{11}, y_{12}, y_{13}, y_{14}, \ldots, y_{1m}$	x_1
2	$y_{21}, y_{22}, y_{23}, y_{24}, \ldots, y_{2m}$	x_2
3	$y_{31}, y_{32}, y_{33}, y_{34}, \ldots, y_{3m}$	x_3
4	$y_{41}, y_{42}, y_{43}, y_{44}, \ldots, y_{4m}$	x_4
$\vdots$	$\vdots$	$\vdots$
n	$y_{n1}, y_{n2}, y_{n3}, y_{n4}, \ldots, y_{nm}$	x_n

simulated to estimate the mean time that entities wait in queues in that system. The experiment is replicated several times to collect n independent sample observations, as was done in the barbershop example. As an entity exits the system, the time that the entity spent waiting in queues is recorded. The waiting time is denoted by y_{ij} in Table 9.3, where the subscript i denotes the replication from which the observation came and the subscript j denotes the value of the counter used to count entities as they exit the system. For example, y_{32} is the waiting time for the second entity that was processed through the system in the third replication. These values are recorded during a particular run (replication) of the model and are listed under the column labeled "Within Run Observations" in Table 9.3.

Note that the within run observations for a particular replication, say the ith replication, are not usually independent because of the correlation between consecutive observations. For example, when the simulation starts and the first entity begins processing, there is no waiting time in the queue. Obviously, the more congested the system becomes at various times throughout the simulation, the longer entities will wait in queues. If the waiting time observed for one entity is long, it is highly likely that the waiting time for the next entity observed is going to be long and vice versa. Observations exhibiting this correlation between consecutive observations are said to be *autocorrelated*. Furthermore, the within run observations for a particular replication are often *nonstationary* in that they do not follow an identical distribution throughout the simulation run. Therefore, they cannot be directly used as observations for statistical methods that require independent and identically distributed observations such as those used in this chapter.

At this point, it seems that we are a long way from getting a usable set of observations. However, let's focus our attention on the last column in Table 9.3, labeled "Average," which contains the x_1 through x_n values. x_i denotes the average waiting time of the entities processed during the ith simulation run (replication) and is computed as follows:

$$x_i = \frac{\sum_{j=1}^{m} y_{ij}}{m}$$

where m is the number of entities processed through the system and y_{ij} is the time that the jth entity processed through the system waited in queues during the ith replication. Although not as formally stated, you used this equation in the last row of the ATM spreadsheet simulation of Chapter 3 to compute the average waiting time of the $m = 25$ customers (Table 3.2). An x_i value for a particular replication represents only one possible value for the mean time an entity waits in queues, and it would be risky to make a statement about the true waiting time from a single observation. However, Table 9.3 contains an x_i value for each of the n independent replications of the simulation. These x_i values are statistically independent if different seed values are used to initialize the random number generator for each replication (as discussed in Section 9.2.1). The x_i values are often identically distributed as well. Therefore, the sample of x_i values can be used for statistical methods requiring independent and identically distributed observations. Thus we can use the x_i values to estimate the true but unknown average time entities wait in queues by computing the sample mean as follows:

$$\bar{x} = \frac{\sum_{i=1}^{n} x_i}{n}$$

You used this equation in Table 3.3 to average the results from the two replications of the ATM spreadsheet simulation of Chapter 3.

The standard deviation of the average time entities wait in queues is estimated by the sample standard deviation and is computed by

$$s = \sqrt{\frac{\sum_{i=1}^{n} [x_i - \bar{x}]^2}{n-1}}$$

The third assumption for many standard statistical methods is that the sample observations (x_1 through x_n) are normally distributed. For example, the normality assumption is implied at the point for which a $t_{n-1,\alpha/2}$ value from the Student's t distribution is used in computing a confidence interval's half-width. The *central limit theorem* of statistics provides a basis for making this assumption. If a variable is defined as the sum of several independent and identically distributed random values, the central limit theorem states that the variable representing the sum tends to be normally distributed. This is helpful because in computing x_i for a particular replication, the (y_{ij}) waiting times for the m entities processed in the ith replication are summed together. However, the y_{ij} observations are autocorrelated and, therefore, not independent. But there are also central limit theorems for certain types of correlated data that suggest that x_i will be *approximately* normally distributed if the sample size used to compute it (the number of entities processed through the system, m from Table 9.3) is not too small (Law and Kelton 2000). Therefore, we can compute a confidence interval using

$$\bar{x} \pm \frac{(t_{n-1,\alpha/2})s}{\sqrt{n}}$$

where $t_{n-1,\alpha/2}$ is a value from the Student's t table in Appendix B for an α level of significance.

ProModel automatically computes point estimates and confidence intervals using this method. Figure 9.2 presents the output produced from running the ProModel version of the ATM simulation of Lab Chapter 3 for five replications. The column under the Locations section of the output report labeled "Average Minutes per Entry" displays the average amount of time that customers waited in the queue during each of the five replications. These values correspond to the x_i values in Table 9.3. Note that in addition to the sample mean and standard deviation, ProModel also provides a 95 percent confidence interval.

Sometimes the output measure being evaluated is not based on the mean, or sum, of a collection of random values. For example, the output measure may be the maximum number of entities that simultaneously wait in a queue. (See the "Maximum Contents" column in Figure 9.2.) In such cases, the output measure may not be normally distributed, and because it is not a sum, the central limit

FIGURE 9.2

Replication technique used on ATM simulation of Lab Chapter 3.

```
------------------------------------------------------------------
General Report
Output from C:\Bowden\2nd Edition\ATM System Ch9.MOD [ATM System]
Date: Nov/26/2002   Time: 04:24:27 PM
------------------------------------------------------------------
Scenario       : Normal Run
Replication    : All
Period         : Final Report (1000 hr to 1500 hr Elapsed: 500 hr)
Warmup Time    : 1000
Simulation Time : 1500 hr
------------------------------------------------------------------

LOCATIONS
```

Location Name	Scheduled Hours	Capacity	Total Entries	Average Minutes Per Entry	Average Contents	Maximum Contents	Current Contents	% Util	
ATM Queue	500	999999	9903	**9.457**	3.122	26	18	0.0	**(Rep 1)**
ATM Queue	500	999999	9866	**8.912**	2.930	24	0	0.0	**(Rep 2)**
ATM Queue	500	999999	9977	**11.195**	3.723	39	4	0.0	**(Rep 3)**
ATM Queue	500	999999	10006	**8.697**	2.900	26	3	0.0	**(Rep 4)**
ATM Queue	500	999999	10187	**10.841**	3.681	32	0	0.0	**(Rep 5)**
ATM Queue	500	999999	9987.8	**9.820**	3.271	29.4	5	0.0	**(Average)**
ATM Queue	0	0	124.654	**1.134**	0.402	6.148	7.483	0.0	**(Std. Dev.)**
ATM Queue	500	999999	9833.05	**8.412**	2.772	21.767	-4.290	0.0	**(95% C.I. Low)**
ATM Queue	500	999999	10142.6	**11.229**	3.771	37.032	14.290	0.0	**(95% C.I. High)**

theorem cannot be used to imply that it is normally distributed. Some refuge may be taken in knowing that the method given in the chapter for constructing a confidence interval does not completely break down when the normality assumption is violated. So if the normality assumption is questionable, use the confidence interval with the knowledge that it is only approximate. However, you must not violate the independence assumption. This is not an obstacle because it is easy to get a set of observations that are statistically independent by replicating the simulation.

9.4 Terminating and Nonterminating Simulations

Determining experimental procedures, such as how many replications and how to gather statistics, depends on whether we should conduct a terminating or nonterminating simulation. A terminating simulation is conducted when we are interested in the behavior of the system over a particular period—for example, the movement of cars on a roadway system during rush hour. A nonterminating simulation is conducted when we are interested in the steady-state behavior of a system—for example, the long-term average behavior of a production system that steadily produces spark plugs for automobile engines. The purpose of this section is to give guidelines for deciding which is most appropriate for a given analysis, terminating simulations or nonterminating simulations.

9.4.1 Terminating Simulations

A *terminating simulation* is one in which the simulation starts at a defined state or time and ends when it reaches some other defined state or time. An initial state might be the number of parts in the system at the beginning of a workday. A terminating state or event might occur when a particular number of parts have been completed. Consider, for example, an aerospace manufacturer that receives an order to manufacture 200 airplanes of a particular model. The company might be interested in knowing how long it will take to produce the 200 airplanes. The simulation run starts with the system empty and is terminated when the 200th plane is completed. A point in time that would bring a terminating simulation to an end might be the closing of shop at the end of a business day or the completion of a weekly or monthly production period. It may be known, for example, that a production schedule for a particular item changes weekly. At the end of each 40-hour cycle, the system is "emptied" and a new production cycle begins. In this situation, a terminating simulation would be run in which the simulation run length would be 40 hours.

A common type of terminating system is one that starts empty, runs for a while, and then empties out again before shutting down. Examples of this type of system are job shops, department stores, restaurants, banks, hair salons, and post offices. All these systems open up for business in the morning with no customers in the system and with all resources initially idle. At the end of the day the system gradually closes down. Usually, at a predetermined time, the customer arrival process is terminated. (The front door is locked to prevent new arrivals.) Service for all existing customers is completed, and then the system finally shuts down for the day. Most service systems tend to operate in this fashion and are therefore modeled as terminating systems.

Terminating simulations are not intended to measure the steady-state behavior of a system, although there may be periods during a terminating simulation during which the system is in a steady state. In a terminating simulation, average measures of performance based on the entire duration of the simulation are of little meaning. Because a terminating simulation always contains transient periods that are part of the analysis, utilization figures, for example, have the most meaning if reported for successive time intervals during the simulation.

9.4.2 Nonterminating Simulations

A *nonterminating* or *steady-state simulation* is one in which the steady-state (long-term average) behavior of the system is being analyzed. A nonterminating simulation does not mean that the simulation never ends, nor does it mean that the system being simulated has no eventual termination. It only means that the simulation could theoretically go on indefinitely with no statistical change in behavior. For such systems, a suitable length of time to run the model in order to collect statistics on the steady-state behavior of the system must be determined.

An example of a nonterminating simulation is a model of a manufacturing operation in which automotive oil filters are produced continually. The operation runs two shifts with an hour break during each shift in which everything momentarily stops. Break and third-shift times are excluded from the model because work always continues exactly as it left off before the break or end of shift. The length of the simulation is determined by how long it takes to get a representative steady-state reading of the model behavior.

Nonterminating simulations can, and often do, change operating characteristics after a period of time, but usually only after enough time has elapsed to establish a steady-state condition. Take, for example, a production system that runs 10,000 units per week for 5 weeks and then increases to 15,000 units per week for the next 10 weeks. The system would have two different steady-state periods. Oil and gas refineries and distribution centers are additional examples of nonterminating systems.

Contrary to what one might think, a steady-state condition is *not* one in which the observations are all the same, or even one for which the variation in observations is any less than during a transient condition. It means only that all observations throughout the steady-state period will have *approximately the same distribution*. Once in a steady state, if the operating rules change or the rate at which entities arrive changes, the system reverts again to a transient state until the system has had time to start reflecting the long-term behavior of the new operating circumstances. Nonterminating systems begin with a warm-up (or transient) state and gradually move to a steady state. Once the initial transient phase has diminished to the point where the impact of the initial condition on the system's response is negligible, we consider it to have reached steady state.

9.5 Experimenting with Terminating Simulations

Experiments involving terminating simulations are usually conducted by making several simulation runs or replications of the period of interest using a different seed value for each run. This procedure enables statistically independent and unbiased observations to be made on the response variables of interest in the system (such as the number of customers in the system or the utilization of resources) over the period simulated. Statistics are also often gathered on performance measures for successive intervals during the period.

For terminating simulations, we are usually interested in final production counts and changing patterns of behavior over time (store opening until closing, for example) rather than the overall, or long-term, average behavior. It would be absurd, for example, to conclude that because two waitresses are busy an average of only 40 percent during the day that only one waitress is needed. This average measure reveals little about the utilization of the waitresses during peak periods. A more detailed report of how long customers wait for a waitress to serve them during successive intervals during the workday may reveal that three waitresses are needed to handle peak periods whereas only one waitress is necessary during off-peak hours. Therefore, statistics are also often gathered on performance measures for successive intervals during the period.

For terminating simulations, there are three important questions to answer in running the experiment:

1. What should be the initial state of the model?
2. What is the terminating event or time?
3. How many replications should be made?

9.5.1 Selecting the Initial Model State

The initial state of the model represents how the system looks (or is initialized) at the beginning of the simulation. The barbershop example presented in Section 9.2.2 is a terminating system. The barbershop opens at 8:00 A.M. and closes around 12:00 noon. The initial state of the shop at 8:00 A.M. is that there are no customers in the shop and all barbers are idle and reading the newspaper. Therefore, the initial state of the model at the beginning of the simulation is that no customers (entities) are in the barbershop (system) and all barbers (resources) are idle. The system is empty and idle until entities begin to arrive in the system as described by the logic in the model.

All terminating simulations do not necessarily begin with the system being empty and idle. For example, to measure the time that it takes for everyone to exit an arena after a sporting event, the simulation would be initialized with fans in their seats just before the sporting event ends. Therefore, the initial state of the model would reflect a large number of entities (fans) in the system.

9.5.2 Selecting a Terminating Event

A terminating event is an event that occurs during the simulation that causes the simulation run to end. This terminating event might occur when a certain time of day is reached (closing time of an airport's newspaper stand) or when certain conditions are met (the last passenger boards an aircraft). If the termination event is based on the satisfaction of conditions other than time, then the time that the simulation will end is not known in advance. Using the barbershop as an example, the simulation terminates sometime after 12:00 noon when the haircut for the last customer allowed in the shop is completed. Therefore, the termination condition would be when the simulation clock time is greater than or equal to 12:00 noon and when the last entity (customer) exits the system (barbershop).

9.5.3 Determining the Number of Replications

Given the initial or starting state of the simulation model and the terminating condition(s) for the simulation run, the analyst is ready to begin conducting experiments with the model. Experiments involving terminating simulations are usually conducted by making several simulation runs or replications of the period of interest using a different stream of random numbers for each run. This procedure enables statistically independent and unbiased observations to be made on response variables of interest in the system (the number of customers in the system, the utilization of resources) over the period simulated, as discussed in Section 9.3.

The answer to the question of how many replications are necessary is usually based on the analyst's desired half-width of a confidence interval. As a general guideline, begin by making 10 independent replications of the simulation and add more replications until the desired confidence interval half-width is reached. For the barbershop example, 18 independent replications of the simulation were required to achieve the desired confidence interval for the expected number of haircuts given on Saturday mornings.

9.6 Experimenting with Nonterminating Simulations

The issues associated with generating meaningful output for terminating simulations are somewhat different from those associated with nonterminating systems. However, the same statistical methods are used to analyze the output from both types of simulations, provided that the output observations are *unbiased, independent, and normally distributed.*

Nonterminating simulations are conducted when we are interested in the steady-state behavior of the simulation model. As such, the following topics must be addressed:

1. Determining and eliminating the initial warm-up bias.
2. Obtaining sample observations.
3. Determining run length.

9.6.1 Determining the Warm-up Period

In a steady-state simulation, we are interested in the steady-state behavior of the model. If a model starts out empty of entities, it usually takes some time before it reaches steady state. In a steady-state condition, the response variables in the system (like mean waiting time or throughput rate) exhibit statistical regularity (the distribution of a response variable is approximately the same from one time period to the next). Figure 9.3 illustrates the typical behavior of an output response measure as the simulation progresses through time. Note that the distribution of the output response changes very little once steady state is reached.

The time that it takes to reach steady state is a function of the activity times and the amount of activity taking place. For some models, steady state might be reached in a few hours of simulation time. For other models, it may take several hundred hours to reach steady state. In modeling steady-state behavior, we have the problem of determining when a model reaches steady state. This start-up period is usually referred to as the *warm-up period* (transient period). We want to wait until after the warm-up period before we start gathering any statistics. This way we eliminate any bias due to observations taken during the transient state of the model. For example, in Figure 9.3 any observations recorded before the warm-up ends would be biased to the low side. Therefore, we delete these observations

FIGURE 9.3

Behavior of model's output response as it reaches steady state.

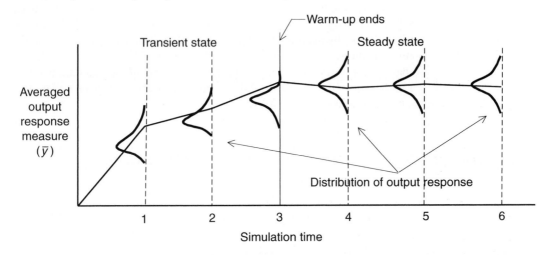

from the beginning of the run and use the remaining observations to estimate the true mean response of the model.

While several methods have been developed for estimating the warm-up time, the easiest and most straightforward approach is to run a preliminary simulation of the system, preferably with several (5 to 10) replications, average the output values at each time step across replications, and observe at what time the system reaches statistical stability. This usually occurs when the averaged output response begins to flatten out or a repeating pattern emerges. Plotting each data point (averaged output response) and connecting them with a line usually helps to identify when the averaged output response begins to flatten out or begins a repeating pattern. Sometimes, however, the variation of the output response is so large that it makes the plot erratic, and it becomes difficult to visually identify the end of the warm-up period. Such a case is illustrated in Figure 9.4. The raw data plot in Figure 9.4 was produced by recording a model's output response for a queue's average contents during 50-hour time periods (time slices) and averaging the output values from each time period across five replications. In this case, the model was initialized with several entities in the queue. Therefore, we need to eliminate this apparent upward bias before recording observations of the queue's average contents. Table 9.4 shows this model's output response for the 20 time periods (50 hours each) for each of the five replications. The raw data plot in Figure 9.4 was constructed using the 20 values under the $\bar{y}_i$ column in Table 9.4.

When the model's output response is erratic, as in Figure 9.4, it is useful to "smooth" it with a moving average. A moving average is constructed by calculating the arithmetic average of the w most recent data points (averaged output responses) in the data set. You have to select the value of w, which is called the *moving-average window*. As you increase the value of w, you increase the "smoothness" of the moving average plot. An indicator for the end of the warm-up

FIGURE 9.4

SimRunner uses the Welch moving-average method to help identify the end of the warm-up period
that occurs around the third or fourth period (150 to 200 hours) for this model.

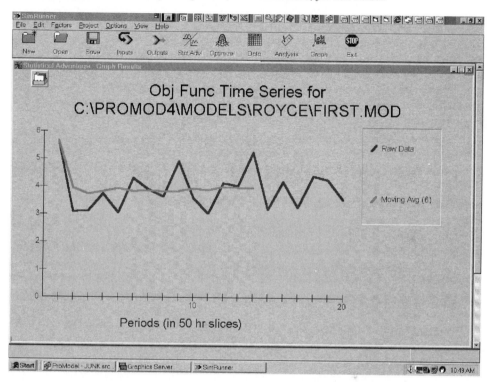

time is when the moving average plot appears to flatten out. Thus the routine is to
begin with a small value of w and increase it until the resulting moving average plot
begins to flatten out.

The moving-average plot in Figure 9.4 with a window of six ($w = 6$) helps
to identify the end of the warm-up period—when the moving average plot appears
to flatten out at around the third or fourth period (150 to 200 hours). Therefore,
we ignore the observations up to the 200th hour and record only those after the
200th hour when we run the simulation. The moving-average plot in Figure 9.4
was constructed using the 14 values under the $y_i(6)$ column in Table 9.4 that were
computed using

$$
\bar{y}_i(w) = \begin{cases} \dfrac{\sum_{s=-w}^{w} \bar{y}_{i+s}}{2w+1} & \text{if } i = w+1, \dots, m-w \\[2ex] \dfrac{\sum_{s=-(i-1)}^{i-1} \bar{y}_{i+s}}{2i-1} & \text{if } i = 1, \dots, w \end{cases}
$$

where m denotes the total number of periods and w denotes the window of the
moving average ($m = 20$ and $w = 6$ in this example).

TABLE 9.4   Welch Moving Average Based on Five Replications and 20 Periods

Period (i)	Time	Observed Mean Queue Contents (Replications)					Total Contents for Period, $Total_i$	Average Contents per Period, $\bar{y}_i = \dfrac{Total_i}{5}$	Welch Moving Average, $\bar{y}_i(6)$
		1	2	3	4	5			
1	50	4.41	4.06	6.37	11.72	1.71	28.27	5.65	5.65
2	100	3.15	2.95	2.94	3.52	2.86	15.42	3.08	3.95
3	150	2.58	2.50	3.07	3.14	4.32	15.61	3.12	3.73
4	200	2.92	3.07	4.48	4.79	3.47	18.73	3.75	3.84
5	250	3.13	3.28	2.34	3.32	3.19	15.26	3.05	3.94
6	300	2.51	3.07	5.45	5.15	5.34	21.52	4.30	3.82
7	350	5.09	3.77	4.44	3.58	2.63	19.51	3.90	3.86
8	400	3.15	2.89	3.63	4.43	4.15	18.25	3.65	3.83
9	450	2.79	3.40	9.78	4.13	4.48	24.58	4.92	3.84
10	500	4.07	3.62	4.50	2.38	3.42	17.99	3.60	3.93
11	550	3.42	3.74	2.46	3.08	2.49	15.19	3.04	3.89
12	600	3.90	1.47	5.75	5.34	4.19	20.65	4.13	3.99
13	650	3.63	3.77	2.14	4.24	6.36	20.14	4.03	3.99
14	700	9.12	4.25	3.83	3.19	5.91	26.30	5.26	3.96
15	750	3.88	3.54	3.08	2.94	2.51	15.95	3.19	
16	800	3.16	6.91	2.70	5.64	2.57	20.98	4.20	
17	850	5.34	3.17	2.47	2.73	2.68	16.39	3.28	
18	900	2.84	3.54	2.33	10.14	3.20	22.05	4.41	
19	950	2.65	4.64	5.18	4.97	3.93	21.37	4.27	
20	1000	3.27	4.68	3.39	2.95	3.46	17.75	3.55	

This is a somewhat busy-looking equation that turns out not to be so bad when you put some numbers into it, which we will now do to compute the Welch moving average, $\bar{y}_i(6)$, column in Table 9.4. We use the bottom part of the equation to compute the moving averages up to the wth moving average. Our $w = 6$, so we will compute $\bar{y}_1(6)$, $\bar{y}_2(6)$, $\bar{y}_3(6)$, $\bar{y}_4(6)$, $\bar{y}_5(6)$, and $\bar{y}_6(6)$ using the bottom part of the $\bar{y}_i(w)$ equation as follows:

$$\bar{y}_1(6) = \frac{\bar{y}_1}{1} = \frac{5.65}{1} = 5.65$$

$$\bar{y}_2(6) = \frac{\bar{y}_1 + \bar{y}_2 + \bar{y}_3}{3} = \frac{5.65 + 3.08 + 3.12}{3} = 3.95$$

$$\bar{y}_3(6) = \frac{\bar{y}_1 + \bar{y}_2 + \bar{y}_3 + \bar{y}_4 + \bar{y}_5}{5} = \frac{5.65 + 3.08 + 3.12 + 3.75 + 3.05}{5} = 3.73$$

$$\vdots$$

$$\bar{y}_6(6) = \frac{\bar{y}_1 + \bar{y}_2 + \bar{y}_3 + \bar{y}_4 + \bar{y}_5 + \bar{y}_6 + \bar{y}_7 + \bar{y}_8 + \bar{y}_9 + \bar{y}_{10} + \bar{y}_{11}}{11} = 3.82$$

Notice the pattern that as i increases, we average more data points together, with the ith data point appearing in the middle of the sum in the numerator (an equal number of data points are on each side of the center data point). This continues until we reach the $(w + 1)$th moving average, when we switch to the top part of the $\bar{y}_i(w)$ equation. For our example, the switch occurs at the 7th moving average because $w = 6$. From this point forward, we average the $2w + 1$ closest data points. For our example, we average the $2(6) + 1 = 13$ closest data points (the ith data point plus the $w = 6$ closest data points above it and the $w = 6$ closest data point below it in Table 9.4 as follows:

$$\bar{y}_7(6) = \frac{\bar{y}_1 + \bar{y}_2 + \bar{y}_3 + \bar{y}_4 + \bar{y}_5 + \bar{y}_6 + \bar{y}_7 + \bar{y}_8 + \bar{y}_9 + \bar{y}_{10} + \bar{y}_{11} + \bar{y}_{12} + \bar{y}_{13}}{13}$$

$$\bar{y}_7(6) = 3.86$$

$$\bar{y}_8(6) = \frac{\bar{y}_2 + \bar{y}_3 + \bar{y}_4 + \bar{y}_5 + \bar{y}_6 + \bar{y}_7 + \bar{y}_8 + \bar{y}_9 + \bar{y}_{10} + \bar{y}_{11} + \bar{y}_{12} + \bar{y}_{13} + \bar{y}_{14}}{13}$$

$$\bar{y}_8(6) = 3.83$$

$$\bar{y}_9(6) = \frac{\bar{y}_3 + \bar{y}_4 + \bar{y}_5 + \bar{y}_6 + \bar{y}_7 + \bar{y}_8 + \bar{y}_9 + \bar{y}_{10} + \bar{y}_{11} + \bar{y}_{12} + \bar{y}_{13} + \bar{y}_{14} + \bar{y}_{15}}{13}$$

$$\bar{y}_9(6) = 3.84$$

$$\vdots$$

$$\bar{y}_{14}(6) = \frac{\bar{y}_8 + \bar{y}_9 + \bar{y}_{10} + \bar{y}_{11} + \bar{y}_{12} + \bar{y}_{13} + \bar{y}_{14} + \bar{y}_{15} + \bar{y}_{16} + \bar{y}_{17} + \bar{y}_{18} + \bar{y}_{19} + \bar{y}_{20}}{13}$$

$$\bar{y}_{14}(6) = 3.96$$

Eventually we run out of data and have to stop. The stopping point occurs when $i = m - w$. In our case with $m = 20$ periods and $w = 6$, we stopped when $i = 20 - 6 = 14$.

The development of this graphical method for estimating the end of the warm-up time is attributed to Welch (Law and Kelton 2000). This method is sometimes referred to as the *Welch moving-average method* and is implemented in SimRunner. Note that when applying the method that the length of each replication should be relatively long and the replications should allow even rarely occurring events such as infrequent downtimes to occur many times. Law and Kelton (2000) recommend that w not exceed a value greater than about $m/4$. To determine a satisfactory warm-up time using Welch's method, one or more key output response variables, such as the average number of entities in a queue or the average utilization of a resource, should be monitored for successive periods. Once these variables begin to exhibit steady state, a good practice to follow would be to extend the warm-up period by 20 to 30 percent. This approach is simple, conservative, and usually satisfactory. The danger is in underestimating the warm-up period, not overestimating it.

9.6.2 Obtaining Sample Observations

It seems that for nonterminating simulations, you could simply run the simulation long enough for statistical output to stabilize and then analyze the results. In other words, you might use the results from a single long run after removing the bias created by the warm-up period. For computing confidence intervals, however, independent sample observations must be created. In a terminating simulation, simply running multiple replications creates independent sample observations. For steady-state simulations, we have several options for obtaining independent sample observations. Two approaches that are widely used are running multiple replications and interval batching.

Independent Observations via Replications

Running multiple replications for nonterminating simulations is similar to the way it is done for terminating simulations. The only differences are that (1) the initial warm-up bias must be determined and eliminated and (2) an appropriate run length must be determined. The warm-up issue was addressed in Section 9.6.1, and the run length issue is addressed in Section 9.6.3. Once the replications are made, confidence intervals can be computed as described earlier in Section 9.3. This was done for the ATM simulation of Lab Chapter 3 to produce Figure 9.2. Note the warm-up time of 1,000 hours and total simulation time of 1,500 hours at the top of the ProModel output report. For this simulation, statistics were collected from time 1,000 hours to 1,500 hours.

One advantage of running independent replications is that it ensures that samples are independent. On the negative side, running through the warm-up phase for each replication extends the length of time to perform the replications. Furthermore, there is a possibility that the length of the warm-up period is underestimated, causing biased results from each replication. However, because the replication method ensures that samples are independent, we strongly encourage its use when the warm-up period is properly detected and removed.

Independent Observations via Batch Means

Interval batching (also referred to as the *batch means technique*) is a method in which a single, long run is made and statistics are collected during separate periods. Although 5 to 10 replications are typically used to determine the warm-up period in preliminary experiments (see Section 9.6.1), the simulation is run only once to collect observations for statistical analysis purposes. Placing the output values from the simulation run into groups or batches forms the set of observations. For example, Figure 9.5 illustrates the output response from a simulation during a single run. Beyond the warm-up period, the run is divided into nonoverlapping intervals. Note that the intervals are of an equal amount of time (as in Figure 9.5) or an equal number of observations (every 1,000 entities observed). Next, the output responses that were generated during a common time interval are "batched" together. If the output measure of interest is a time-weighted statistic (such as the average number of entities in the system), then the batch interval

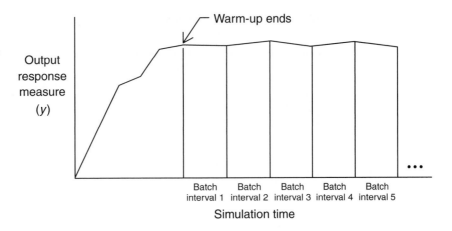

FIGURE 9.5

Individual statistics on the output measure are computed for each batch interval.

TABLE 9.5   A Single Replication Divided into Batch Intervals

Batch Interval (i)	Within Batch Interval Observations y_{ij}	Average x_i
1	$y_{11}, y_{12}, y_{13}, y_{14}, \ldots, y_{1m}$	x_1
2	$y_{21}, y_{22}, y_{23}, y_{24}, \ldots, y_{2m}$	x_2
3	$y_{31}, y_{32}, y_{33}, y_{34}, \ldots, y_{3m}$	x_3
4	$y_{41}, y_{42}, y_{43}, y_{44}, \ldots, y_{4m}$	x_4
5	$y_{51}, y_{52}, y_{53}, y_{54}, \ldots, y_{5m}$	x_5
$\vdots$	$\vdots$	$\vdots$
n	$y_{n1}, y_{n2}, y_{n3}, y_{n4}, \ldots, y_{nm}$	x_n

---

should be based on time. If the output measure is based on observations (like the waiting time of entities in a queue), then the batch interval is typically based on the number of observations.

Table 9.5 details how a single, long simulation run (one replication), like the one shown in Figure 9.5, is partitioned into batch intervals for the purpose of obtaining (approximately) independent and identically distributed observations of a simulation model's output response. Note the similarity between Table 9.3 of Section 9.3 and Table 9.5. As in Table 9.3, the observations represent the time an entity waited in queues during the simulation. The waiting time for an entity is denoted by y_{ij}, where the subscript i denotes the interval of time (batch interval) from which the observation came and the subscript j denotes the value of the counter used to count entities as they exit the system during a particular batch interval. For example, y_{23} is the waiting time for the third entity that was processed through the system during the second batch interval of time.

Because these observations are all from a single run (replication), they are not statistically independent. For example, the waiting time of the mth entity exiting the system during the first batch interval, denoted y_{1m}, is correlated with the waiting time of the first entity to exit the system during the second batch interval, denoted y_{21}. This is because if the waiting time observed for one entity is long, it is likely that the waiting time for the next entity observed is going to be long, and vice versa. Therefore, adjacent observations in Table 9.5 will usually be auto-correlated. However, the value of y_{14} can be somewhat uncorrelated with the value of y_{24} if they are spaced far enough apart such that the conditions that re-sulted in the waiting time value of y_{14} occurred so long ago that they have little or no influence on the waiting time value of y_{24}. Therefore, most of the observations within the interior of one batch interval can become relatively uncorrelated with most of the observations in the interior of other batch intervals, provided they are spaced sufficiently far apart. Thus the goal is to extend the batch interval length until there is very little correlation (you cannot totally eliminate it) between the observations appearing in different batch intervals. When this occurs, it is reason-able to assume that observations within one batch interval are independent of the observations within other batch intervals.

With the independence assumption in hand, the observations within a batch interval can be treated in the same manner as we treated observations within a replication in Table 9.3. Therefore, the values in the Average column in Table 9.5, which represent the average amount of time the entities processed during the ith batch interval waited in queues, are computed as follows:

$$x_i = \frac{\sum_{j=1}^{m} y_{ij}}{m}$$

where m is the number of entities processed through the system during the batch in-terval of time and y_{ij} is the time that the jth entity processed through the system waited in queues during the ith batch interval. The x_i values in the Average column are approximately independent and identically distributed and can be used to com-pute a sample mean for estimating the true average time an entity waits in queues:

$$\bar{x} = \frac{\sum_{i=1}^{n} x_i}{n}$$

The sample standard deviation is also computed as before:

$$s = \sqrt{\frac{\sum_{i=1}^{n} [x_i - \bar{x}]^2}{n - 1}}$$

And, if we assume that the observations (x_1 through x_n) are normally distributed (or at least approximately normally distributed), a confidence interval is computed by

$$\bar{x} \pm \frac{(t_{n-1,\alpha/2})s}{\sqrt{n}}$$

where $t_{n-1,\alpha/2}$ is a value from the Student's t table in Appendix B for an α level of significance.

ProModel automatically computes point estimates and confidence intervals using the above method. Figure 9.6 is a ProModel output report from a single,

FIGURE 9.6

Batch means technique used on ATM simulation in Lab Chapter 3. Warm-up period is from 0 to 1,000 hours. Each batch interval is 500 hours in length.

```
-----------------------------------------------------------------
General Report
Output from C:\Bowden\2nd Edition\ATM System Ch9.MOD [ATM System]
Date: Nov/26/2002   Time: 04:44:24 PM
-----------------------------------------------------------------
Scenario        : Normal Run
Replication     : 1 of 1
Period          : All
Warmup Time     : 1000 hr
Simulation Time : 3500 hr
-----------------------------------------------------------------

LOCATIONS
```

Location Name	Scheduled Hours	Capacity	Total Entries	Average Minutes Per Entry	Average Contents	Maximum Contents	Current Contents	% Util	
ATM Queue	500	999999	9903	9.457	3.122	26	18	0.0	(Batch 1)
ATM Queue	500	999999	10065	9.789	3.284	24	0	0.0	(Batch 2)
ATM Queue	500	999999	9815	8.630	2.823	23	16	0.0	(Batch 3)
ATM Queue	500	999999	9868	8.894	2.925	24	1	0.0	(Batch 4)
ATM Queue	500	999999	10090	12.615	4.242	34	0	0.0	(Batch 5)
ATM Queue	500	999999	9948.2	9.877	3.279	26.2	7	0.0	(Average)
ATM Queue	0	0	122.441	1.596	0.567	4.494	9.165	0.0	(Std. Dev.)
ATM Queue	500	999999	9796.19	7.895	2.575	20.620	-4.378	0.0	(95% C.I. Low)
ATM Queue	500	999999	10100.2	11.860	3.983	31.779	18.378	0.0	(95% C.I. High)

long run of the ATM simulation in Lab Chapter 3 with the output divided into five batch intervals of 500 hours each after a warm-up time of 1,000 hours. The column under the Locations section of the output report labeled "Average Minutes Per Entry" displays the average amount of time that customer entities waited in the queue during each of the five batch intervals. These values correspond to the x_i values of Table 9.5. Note that the 95 percent confidence interval automatically computed by ProModel in Figure 9.6 is comparable, though not identical, to the 95 percent confidence interval in Figure 9.2 for the same output statistic.

Establishing the batch interval length such that the observations $x_1, x_2, \ldots, x_n$ of Table 9.5 (assembled after the simulation reaches steady state) are approximately independent is difficult and time-consuming. There is no foolproof method for doing this. However, if you can generate a large number of observations, say $n \geq 100$, you can gain some insight about the independence of the observations by estimating the autocorrelation between adjacent observations (lag-1 autocorrelation). See Chapter 6 for an introduction to tests for independence and autocorrelation plots. The observations are treated as being independent if their lag-1 autocorrelation is zero. Unfortunately, current methods available for estimating the lag-1 autocorrelation are not very accurate. Thus we may be persuaded that our observations are *almost* independent if the estimated lag-1 autocorrelation computed from a large number of observations falls between −0.20 to +0.20. The word *almost* is emphasized because there is really no such thing as almost independent (the observations are independent or they are not). What we are indicating here is that we are leaning toward calling the observations independent when the estimate of the lag-1 autocorrelation is between -0.20 and $+0.20$. Recall that autocorrelation values fall between ± 1. Before we elaborate on this idea for determining if an acceptable batch interval length has been used to derive the observations, let's talk about positive autocorrelation versus negative autocorrelation.

Positive autocorrelation is a bigger enemy to us than negative autocorrelation because our sample standard deviation s will be biased low if the observations from which it is computed are positively correlated. This would result in a falsely narrow confidence interval (smaller half-width), leading us to believe that we have a better estimate of the mean μ than we actually have. Negatively correlated observations have the opposite effect, producing a falsely wide confidence interval (larger half-width), leading us to believe that we have a worse estimate of the mean μ than we actually have. A negative autocorrelation may lead us to waste time collecting additional observations to get a more precise (narrow) confidence interval but will not result in a hasty decision. Therefore, an emphasis is placed on guarding against positive autocorrelation. We present the following procedure adapted from Banks et al. (2001) for estimating an acceptable batch interval length.

For the observations $x_1, x_2, \ldots, x_n$ derived from n batches of data assembled after the simulation has researched steady state as outlined in Table 9.5, compute an estimate of the lag-1 autocorrelation $\hat{\rho}_1$ using

$$\hat{\rho}_1 = \frac{\sum_{i=1}^{n-1}(x_i - \bar{x})(x_{i+1} - \bar{x})}{s^2(n-1)}$$

where s^2 is the sample variance (sample standard deviation squared) and $\bar{x}$ is the sample mean of the n observations. Recall the recommendation that n should be *at least* 100 observations. If the estimated lag-1 autocorrelation is not between -0.20 and $+0.20$, then extend the original batch interval length by 50 to 100 percent, rerun the simulation to collect a new set of n observations, and check the lag-1 autocorrelation between the new x_1 to x_n observations. This would be repeated until the estimated lag-1 autocorrelation falls between -0.20 and $+0.20$ (or you give up trying to get within the range). Note that falling below -0.20 is less worrisome than exceeding $+0.20$.

Achieving $-0.20 \leq \hat{\rho}_1 \leq 0.20$: Upon achieving an estimated lag-1 autocorrelation within the desired range, "rebatch" the data by combining adjacent batch intervals of data into a larger batch. This produces a smaller number of observations that are based on a larger batch interval length. The lag-1 autocorrelation for this new set of observations will likely be closer to zero than the lag-1 autocorrelation of the previous set of observations because the new set of observations is based on a larger batch interval length. Note that at this point, you are not rerunning the simulation with the new, longer batch interval but are "rebatching" the output data from the last run. For example, if a batch interval length of 125 hours produced 100 observations (x_1 to x_{100}) with an estimated lag-1 autocorrelation $\hat{\rho}_1 = 0.15$, you might combine four contiguous batches of data into one batch with a length of 500 hours (125 hours $\times$ 4 = 500 hours). "Rebatching" the data with the 500-hour batch interval length would leave you 25 observations (x_1 to x_{25}) from which to construct the confidence interval. We recommend that you "rebatch" the data into 10 to 30 batch intervals.

Not Achieving $-0.20 \leq \hat{\rho}_1 \leq 0.20$: If obtaining an estimated lag-1 autocorrelation within the desired range becomes impossible because you cannot continue extending the length of the simulation run, then rebatch the data from your last run into no more than about 10 batch intervals and construct the confidence interval. Interpret the confidence interval with the apprehension that the observations may be significantly correlated.

Lab Chapter 9 provides an opportunity for you to gain experience applying these criteria to a ProModel simulation experiment. However, remember that there is no universally accepted method for determining an appropriate batch interval length. See Banks et al. (2001) for additional details and variations on this approach, such as a concluding hypothesis test for independence on the final set of observations. The danger is in setting the batch interval length too short, not too long. This is the reason for extending the length of the batch interval in the last step. A starting point for setting the initial batch interval length from which to begin the process of evaluating the lag-1 autocorrelation estimates is provided in Section 9.6.3.

In summary, the statistical methods in this chapter are applied to the data compiled from batch intervals in the same manner they were applied to the data compiled from replications. And, as before, the accuracy of point and interval estimates generally improves as the sample size (number of batch intervals) increases. In this case we increase the number of batch intervals by extending the length of the simulation run. As a general guideline, the simulation should be run long enough to create around 10 batch intervals and possibly more, depending on the desired confidence interval half-width. If a trade-off must be made between the number of batch intervals and the batch interval length, then err on the side of increasing the batch interval length. It is better to have a few independent observations than to have several autocorrelated observations when using the statistical methods presented in this text.

9.6.3 *Determining Run Length*

Determining run length for terminating simulations is quite simple because a natural event or time point defines it. Determining the run length for a steady-state simulation is more difficult because the simulation could be run indefinitely. Obviously, running extremely long simulations can be very time-consuming, so the objective is to determine an appropriate run length to ensure a representative sample of the steady-state response of the system.

The recommended length of the simulation run for a steady-state simulation depends on the interval between the least frequently occurring event and the type

Figure 9.7

*Batch intervals
compared with
replications.*

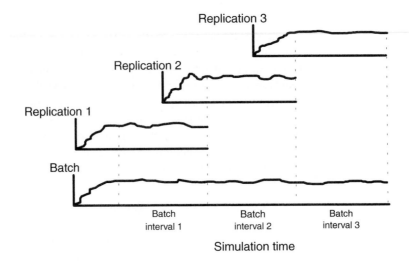

of sampling method (replication or interval batching) used. If running independent replications, it is usually a good idea to run the simulation long enough past the warm-up point to let every type of event (including rare ones) happen many times and, if practical, several hundred times. Remember, the longer the model is run, the more confident you can become that the results represent the steady-state behavior. A guideline for determining the initial run length for the interval batching method is much the same as for the replication method. Essentially, the starting point for selecting the length of each batch interval is the length of time chosen to run the simulation past the warm-up period when using the replication method. The total run length is the sum of the time for each batch interval plus the initial warm-up time. This guideline is illustrated in Figure 9.7.

The choice between running independent replications or batch intervals can be made based on how long it takes the simulation to reach steady state. If the warm-up period is short (that is, it doesn't take long to simulate through the warm-up period), then running replications is preferred over batch intervals because running replications guarantees that the observations are independent. However, if the warm-up period is long, then the batch interval method is preferred because the time required to get through the warm-up period is incurred only once, which reduces the amount of time to collect the necessary number of observations for a desired confidence interval. Also, because a single, long run is made, chances are that a better estimate of the model's true steady-state behavior is obtained.

9.7 Summary

Simulation experiments can range from a few replications (runs) to a large number of replications. While "ballpark" decisions require little analysis, more precise decision making requires more careful analysis and extensive experimentation.

Correctly configuring the model to produce valid output is essential to making sound decisions.

This chapter has identified the issues involved in producing and analyzing simulation output. Key issues include determining run length, number of replications, number of batch intervals, batch interval length, and warm-up times. When running multiple replications, the modeler needs to understand the concepts of random number streams and random number seeds. Several guidelines were presented throughout the chapter to help guide the analyst in interpreting simulation output data using point and interval estimates. Remember that the guidelines are not rigid rules that must be followed but rather are common practices that have worked well in the past.

9.8 Review Questions

1. Define *population, sample,* and *sample size.*
2. If you run the same simulation model twice, would you expect to see any difference in the results? Explain.
3. How are multiple simulation replications produced?
4. What do $\bar{x}$ and s define, and why are they called point estimators?
5. What does a confidence interval express?
6. What does the confidence interval on simulation output reveal about the validity of the model?
7. In order to apply statistical measures, such as standard deviation and confidence interval estimation, what assumptions must be made regarding simulation output?
8. What is the central limit theorem and how is it applied to simulation output?
9. Give an example of a system in a transient state; in a steady state.
10. How do you determine the length of a terminating simulation run?
11. How many replications should be made for terminating simulations?
12. Verify that the 95 percent confidence interval has been correctly computed by ProModel for the Average Minutes Per Entry statistic shown in Figure 9.2. The ProModel half-width of 1.408 minutes represents an approximate 14.3 percent error in the point estimate $\bar{x} = 9.820$ minutes. Determine the approximate number of additional replications needed to estimate the average time that customer entities wait in the ATM queue with a percentage error of 5 percent and a confidence level of 95 percent.
13. Describe two methods for obtaining sample observations for a nonterminating simulation.
14. When determining the batch interval length, what is the danger in making the interval too small?

15. Given the five batch means for the Average Minutes Per Entry of customer entities at the ATM queue location in Figure 9.6, estimate the lag-1 autocorrelation $\hat{\rho}_1$. Note that five observations are woefully inadequate for obtaining an accurate estimate of the lag-1 autocorrelation. Normally you will want to base the estimate on at least 100 observations. The question is designed to give you some experience using the $\hat{\rho}_1$ equation so that you will understand what the Stat::Fit software is doing when you use it in Lab Chapter 9 to crunch through an example with 100 observations.

16. Construct a Welch moving average with a window of 2 ($w = 2$) using the data in Table 9.4 and compare it to the Welch moving average with a window of 6 ($w = 6$) presented in Table 9.4.

References

Banks, Jerry; John S. Carson, II; Barry L. Nelson; and David M. Nicol. *Discrete-Event System Simulation.* New Jersey: Prentice-Hall, 2001.

Bateman, Robert E.; Royce O. Bowden; Thomas J. Gogg; Charles R. Harrell; and Jack R. A. Mott. *System Improvement Using Simulation.* Orem, UT: PROMODEL Corp., 1997.

Hines, William W., and Douglas C. Montgomery. *Probability and Statistics in Engineering and Management Science.* New York: John Wiley and Sons, 1990.

Law, Averill M., and David W. Kelton. *Simulation Modeling and Analysis.* New York: McGraw-Hill, 2000.

Montgomery, Douglas C. *Design and Analysis of Experiments.* New York: John Wiley & Sons, 1991.

Petersen, Roger G. *Design and Analysis of Experiments.* New York: Marcel Dekker, 1985.

10   COMPARING SYSTEMS

"The method that proceeds without analysis is like the groping of a blind man."
—Socrates

10.1 Introduction

In many cases, simulations are conducted to compare two or more alternative designs of a system with the goal of identifying the superior system relative to some performance measure. Comparing alternative system designs requires careful analysis to ensure that differences being observed are attributable to actual differences in performance and not to statistical variation. This is where running either multiple replications or batches is required. Suppose, for example, that method A for deploying resources yields a throughput of 100 entities for a given time period while method B results in 110 entities for the same period. Is it valid to conclude that method B is better than method A, or might additional replications actually lead to the opposite conclusion?

You can evaluate alternative configurations or operating policies by performing several replications of each alternative and comparing the average results from the replications. Statistical methods for making these types of comparisons are called *hypotheses tests*. For these tests, a hypothesis is first formulated (for example, that methods A and B both result in the same throughput) and then a test is made to see whether the results of the simulation lead us to reject the hypothesis. The outcome of the simulation runs may cause us to reject the hypothesis that methods A and B both result in equal throughput capabilities and conclude that the throughput does indeed depend on which method is used.

This chapter extends the material presented in Chapter 9 by providing statistical methods that can be used to compare the output of different simulation models that represent competing designs of a system. The concepts behind hypothesis testing are introduced in Section 10.2. Section 10.3 addresses the case when two alternative system designs are to be compared, and Section 10.4 considers the case when more than two alternative system designs are to be compared. Additionally, a technique called *common random numbers* is described in Section 10.5 that can sometimes improve the accuracy of the comparisons.

10.2 Hypothesis Testing

An inventory allocation example will be used to further explore the use of hypothesis testing for comparing the output of different simulation models. Suppose a production system consists of four machines and three buffer storage areas. Parts entering the system require processing by each of the four machines in a serial fashion (Figure 10.1). A part is always available for processing at the first machine. After a part is processed, it moves from the machine to the buffer storage area for the next machine, where it waits to be processed. However, if the buffer is full, the part cannot move forward and remains on the machine until a space becomes available in the buffer. Furthermore, the machine is blocked and no other parts can move to the machine for processing. The part exits the system after being processed by the fourth machine.

The question for this system is how best to allocate buffer storage between the machines to maximize the throughput of the system (number of parts completed per hour). The production control staff has identified two candidate strategies for allocating buffer capacity (number of parts that can be stored) between machines, and simulation models have been built to evaluate the proposed strategies.

FIGURE 10.1

Production system with four workstations and three buffer storage areas.

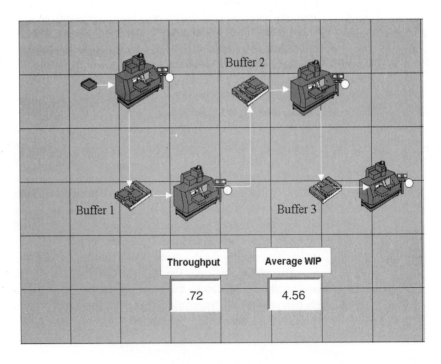

Throughput	Average WIP
.72	4.56

Suppose that Strategy 1 and Strategy 2 are the two buffer allocation strategies proposed by the production control staff. We wish to identify the strategy that maximizes the throughput of the production system (number of parts completed per hour). Of course, the possibility exists that there is no significant difference in the performance of the two candidate strategies. That is to say, the mean throughput of the two proposed strategies is equal. A starting point for our problem is to formulate our hypotheses concerning the mean throughput for the production system under the two buffer allocation strategies. Next we work out the details of setting up our experiments with the simulation models built to evaluate each strategy. For example, we may decide to estimate the true mean performance of each strategy (μ_1 and μ_2) by simulating each strategy for 16 days (24 hours per day) past the warm-up period and replicating the simulation 10 times. After we run experiments, we would use the simulation output to evaluate the hypotheses concerning the mean throughput for the production system under the two buffer allocation strategies.

In general, a null hypothesis, denoted H_0, is drafted to state that the value of μ_1 is not significantly different than the value of μ_2 at the α level of significance. An alternate hypothesis, denoted H_1, is drafted to oppose the null hypothesis H_0. For example, H_1 could state that μ_1 and μ_2 are different at the α level of significance. Stated more formally:

$$H_0: \mu_1 = \mu_2 \quad \text{or equivalently} \quad H_0: \mu_1 - \mu_2 = 0$$
$$H_1: \mu_1 \neq \mu_2 \quad \text{or equivalently} \quad H_1: \mu_1 - \mu_2 \neq 0$$

In the context of the example problem, the null hypothesis H_0 states that the mean throughputs of the system due to Strategy 1 and Strategy 2 do not differ. The alternate hypothesis H_1 states that the mean throughputs of the system due to Strategy 1 and Strategy 2 do differ. Hypothesis testing methods are designed such that the burden of proof is on us to demonstrate that H_0 is not true. Therefore, if our analysis of the data from our experiments leads us to reject H_0, we can be confident that there is a significant difference between the two population means. In our example problem, the output from the simulation model for Strategy 1 represents possible throughput observations from one population, and the output from the simulation model for Strategy 2 represents possible throughput observations from another population.

The α level of significance in these hypotheses refers to the probability of making a Type I error. A *Type I error* occurs when we reject H_0 in favor of H_1 when in fact H_0 is true. Typically α is set at a value of 0.05 or 0.01. However, the choice is yours, and it depends on how small you want the probability of making a Type I error to be. A *Type II error* occurs when we fail to reject H_0 in favor of H_1 when in fact H_1 is true. The probability of making a Type II error is denoted as β. Hypothesis testing methods are designed such that the probability of making a Type II error, β, is as small as possible for a given value of α. The relationship between α and β is that β increases as α decreases. Therefore, we should be careful not to make α too small.

We will test these hypotheses using a confidence interval approach to determine if we should reject or fail to reject the null hypothesis in favor of the alternative hypothesis. The reason for using the confidence interval method is that it is equivalent to conducting a two-tailed test of hypothesis with the added benefit of indicating the magnitude of the difference between μ_1 and μ_2 if they are in fact significantly different. The first step of this procedure is to construct a confidence interval to estimate the difference between the two means $(\mu_1 - \mu_2)$. This can be done in different ways depending on how the simulation experiments are conducted (we will discuss this later). For now, let's express the confidence interval on the difference between the two means as

$$P[(\bar{x}_1 - \bar{x}_2) - hw \leq \mu_1 - \mu_2 \leq (\bar{x}_1 - \bar{x}_2) + hw] = 1 - \alpha$$

where hw denotes the half-width of the confidence interval. Notice the similarities between this confidence interval expression and the one given on page 227 in Chapter 9. Here we have replaced $\bar{x}$ with $\bar{x}_1 - \bar{x}_2$ and μ with $\mu_1 - \mu_2$.

If the two population means are the same, then $\mu_1 - \mu_2 = 0$, which is our null hypothesis H_0. If H_0 is true, our confidence interval should include zero with a probability of $1 - \alpha$. This leads to the following rule for deciding whether to reject or fail to reject H_0. If the confidence interval includes zero, we fail to reject H_0 and conclude that the value of μ_1 is not significantly different than the value of μ_2 at the α level of significance (the mean throughput of Strategy 1 is not significantly different than the mean throughput of Strategy 2). However, if the confidence interval does not include zero, we reject H_0 and conclude that the value of μ_1 is significantly different than the value of μ_2 at the α level of significance (throughput values for Strategy 1 and Strategy 2 are significantly different).

Figure 10.2(a) illustrates the case when the confidence interval contains zero, leading us to fail to reject the null hypothesis H_0 and conclude that there is no significant difference between μ_1 and μ_2. The failure to obtain sufficient evidence to pick one alternative over another may be due to the fact that there really is no difference, or it may be a result of the variance in the observed outcomes being too high to be conclusive. At this point, either additional replications may be run or one of several variance reduction techniques might be employed (see Section 10.5). Figure 10.2(b) illustrates the case when the confidence interval is completely to the

FIGURE 10.2

Three possible positions of a confidence interval relative to zero.

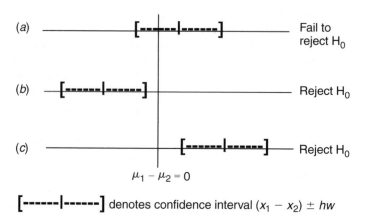

left of zero, leading us to reject H_0. This case suggests that $\mu_1 - \mu_2 < 0$ or, equivalently, $\mu_1 < \mu_2$. Figure 10.2(c) illustrates the case when the confidence interval is completely to the right of zero, leading us to also reject H_0. This case suggests that $\mu_1 - \mu_2 > 0$ or, equivalently, $\mu_1 > \mu_2$. These rules are commonly used in practice to make statements about how the population means differ ($\mu_1 > \mu_2$ or $\mu_1 < \mu_2$) when the confidence interval does not include zero (Banks et al. 2001; Hoover and Perry 1989).

10.3 Comparing Two Alternative System Designs

In this section, we will present two methods based on the confidence interval approach that are commonly used to compare two alternative system designs. To facilitate our understanding of the confidence interval methods, the presentation will relate to the buffer allocation example in Section 10.2, where the production control staff has identified two candidate strategies believed to maximize the throughput of the system. We seek to discover if the mean throughputs of the system due to Strategy 1 and Strategy 2 are significantly different. We begin by estimating the mean performance of the two proposed buffer allocation strategies (μ_1 and μ_2) by simulating each strategy for 16 days (24 hours per day) past the warm-up period. The simulation experiment was replicated 10 times for each strategy. Therefore, we obtained a sample size of 10 for each strategy ($n_1 = n_2 = 10$). The average hourly throughput achieved by each strategy is shown in Table 10.1.

As in Chapter 9, the methods for computing confidence intervals in this chapter require that our observations be independent and normally distributed. The 10 observations in column B (Strategy 1 Throughput) of Table 10.1 are independent

TABLE 10.1   Comparison of Two Buffer Allocation Strategies

(A) Replication	(B) Strategy 1 Throughput x_1	(C) Strategy 2 Throughput x_2
1	54.48	56.01
2	57.36	54.08
3	54.81	52.14
4	56.20	53.49
5	54.83	55.49
6	57.69	55.00
7	58.33	54.88
8	57.19	54.47
9	56.84	54.93
10	55.29	55.84
Sample mean $\bar{x}_i$, for $i = 1, 2$	56.30	54.63
Sample standard deviation s_i, for $i = 1, 2$	1.37	1.17
Sample variance s_i^2, for $i = 1, 2$	1.89	1.36

because a unique segment (stream) of random numbers from the random number generator was used for each replication. The same is true for the 10 observations in column C (Strategy 2 Throughput). The use of random number streams is discussed in Chapters 3 and 9 and later in this chapter. At this point we are assuming that the observations are also normally distributed. The reasonableness of assuming that the output produced by our simulation models is normally distributed is discussed at length in Chapter 9. For this data set, we should also point out that two different sets of random numbers were used to simulate the 10 replications of each strategy. Therefore, the observations in column B are independent of the observations in column C. Stated another way, the two columns of observations are not correlated. Therefore, the observations are independent within a population (strategy) and between populations (strategies). This is an important distinction and will be employed later to help us choose between different methods for computing the confidence intervals used to compare the two strategies.

From the observations in Table 10.1 of the throughput produced by each strategy, it is not obvious which strategy yields the higher throughput. Inspection of the summary statistics indicates that Strategy 1 produced a higher mean throughput for the system; however, the sample variance for Strategy 1 was higher than for Strategy 2. Recall that the variance provides a measure of the variability of the data and is obtained by squaring the standard deviation. Equations for computing the sample mean $\bar{x}$, sample variance s^2, and sample standard deviation s are given in Chapter 9. Because of this variation, we should be careful when making conclusions about the population of throughput values (μ_1 and μ_2) by only inspecting the point estimates ($\bar{x}_1$ and $\bar{x}_2$). We will avoid the temptation and use the output from the 10 replications of each simulation model along with a confidence interval to make a more informed decision.

We will use an $\alpha = 0.05$ level of significance to compare the two candidate strategies using the following hypotheses:

$$H_0: \; \mu_1 - \mu_2 = 0$$
$$H_1: \; \mu_1 - \mu_2 \neq 0$$

where the subscripts 1 and 2 denote Strategy 1 and Strategy 2, respectively. As stated earlier, there are two common methods for constructing a confidence interval for evaluating hypotheses. The first method is referred to as the Welch confidence interval (Law and Kelton 2000; Miller 1986) and is a modified *two-sample-t* confidence interval. The second method is the *paired-t* confidence interval (Miller et al. 1990). We've chosen to present these two methods because their statistical assumptions are more easily satisfied than are the assumptions for other confidence interval methods.

10.3.1 Welch Confidence Interval for Comparing Two Systems

The *Welch confidence interval method* requires that the observations drawn from each population (simulated system) be normally distributed and independent within a population and between populations. Recall that the observations in

Table 10.1 are independent and are assumed normal. However, the Welch confidence interval method does not require that the number of samples drawn from one population (n_1) equal the number of samples drawn from the other population (n_2) as we did in the buffer allocation example. Therefore, if you have more observations for one candidate system than for the other candidate system, then by all means use them. Additionally, this approach does not require that the two populations have equal variances ($\sigma_1^2 = \sigma_2^2 = \sigma^2$) as do other approaches. This is useful because we seldom know the true value of the variance of a population. Thus we are not required to judge the equality of the variances based on the sample variances we compute for each population (s_1^2 and s_2^2) before using the Welch confidence interval method.

The Welch confidence interval for an α level of significance is

$$P[(\bar{x}_1 - \bar{x}_2) - hw \leq \mu_1 - \mu_2 \leq (\bar{x}_1 - \bar{x}_2) + hw] = 1 - \alpha$$

where $\bar{x}_1$ and $\bar{x}_2$ represent the sample means used to estimate the population means μ_1 and μ_2; hw denotes the half-width of the confidence interval and is computed by

$$hw = t_{df,\alpha/2}\sqrt{\frac{s_1^2}{n_1} + \frac{s_2^2}{n_2}}$$

where df (degrees of freedom) is estimated by

$$df \approx \frac{\left[s_1^2/n_1 + s_2^2/n_2\right]^2}{\left[s_1^2/n_1\right]^2/(n_1 - 1) + \left[s_2^2/n_2\right]^2/(n_2 - 1)}$$

and $t_{df,\alpha/2}$ is a factor obtained from the Student's t table in Appendix B based on the value of $\alpha/2$ and the estimated degrees of freedom. Note that the degrees of freedom term in the Student's t table is an integer value. Given that the estimated degrees of freedom will seldom be an integer value, you will have to use interpolation to compute the $t_{df,\alpha/2}$ value.

For the example buffer allocation problem with an $\alpha = 0.05$ level of significance, we use these equations and data from Table 10.1 to compute

$$df \approx \frac{[1.89/10 + 1.36/10]^2}{[1.89/10]^2/(10 - 1) + [1.36/10]^2/(10 - 1)} \approx 17.5$$

and

$$hw = t_{17.5,0.025}\sqrt{\frac{1.89}{10} + \frac{1.36}{10}} = 2.106\sqrt{0.325} = 1.20 \text{ parts per hour}$$

where $t_{df,\alpha/2} = t_{17.5,0.025} = 2.106$ is determined from Student's t table in Appendix B by interpolation. Now the 95 percent confidence interval is

$$(\bar{x}_1 - \bar{x}_2) - hw \leq \mu_1 - \mu_2 \leq (\bar{x}_1 - \bar{x}_2) + hw$$
$$(56.30 - 54.63) - 1.20 \leq \mu_1 - \mu_2 \leq (56.30 - 54.63) + 1.20$$
$$0.47 \leq \mu_1 - \mu_2 \leq 2.87$$

With approximately 95 percent confidence, we conclude that there is a significant difference between the mean throughputs of the two strategies because the interval excludes zero. The confidence interval further suggests that the mean throughput μ_1 of Strategy 1 is higher than the mean throughput μ_2 of Strategy 2 (an estimated 0.47 to 2.87 parts per hour higher).

10.3.2 Paired-t Confidence Interval for Comparing Two Systems

Like the Welch confidence interval method, the *paired-t confidence interval method* requires that the observations drawn from each population (simulated system) be normally distributed and independent within a population. However, the paired-*t* confidence interval method does not require that the observations between populations be independent. This allows us to use a technique called *common random numbers* to force a positive correlation between the two populations of observations in order to reduce the half-width of our confidence interval without increasing the number of replications. Recall that the smaller the half-width, the better our estimate. The common random numbers technique is discussed in Section 10.5. Unlike the Welch method, the paired-*t* confidence interval method does require that the number of samples drawn from one population (n_1) equal the number of samples drawn from the other population (n_2) as we did in the buffer allocation example. And like the Welch method, the paired-*t* confidence interval method does not require that the populations have equal variances ($\sigma_1^2 = \sigma_2^2 = \sigma^2$). This is useful because we seldom know the true value of the variance of a population. Thus we are not required to judge the equality of the variances based on the sample variances we compute for each population (s_1^2 and s_2^2) before using the paired-*t* confidence interval method.

Given n observations ($n_1 = n_2 = n$), we pair the observations from each population (x_{1j} and x_{2j}) to define a new random variable $x_{(1-2)j} = x_{1j} - x_{2j}$, for $j = 1, 2, 3, \ldots, n$. The x_{1j} denotes the jth observation from the first population sampled (the output from the jth replication of the simulation model for the first alternative design), x_{2j} denotes the jth observation from the second population sampled (the output from the jth replication of the simulation model for the second alternative design), and $x_{(1-2)j}$ denotes the difference between the jth observations from the two populations. The point estimators for the new random variable are

$$\text{Sample mean} = \bar{x}_{(1-2)} = \frac{\sum_{j=1}^{n} x_{(1-2)j}}{n}$$

$$\text{Sample standard deviation} = s_{(1-2)} = \sqrt{\frac{\sum_{j=1}^{n} \left[x_{(1-2)j} - \bar{x}_{(1-2)} \right]^2}{n - 1}}$$

where $\bar{x}_{(1-2)}$ estimates $\mu_{(1-2)}$ and $s_{(1-2)}$ estimates $\sigma_{(1-2)}$.

The half-width equation for the paired-t confidence interval is

$$hw = \frac{(t_{n-1,\alpha/2})s_{(1-2)}}{\sqrt{n}}$$

where $t_{n-1,\alpha/2}$ is a factor that can be obtained from the Student's t table in Appendix B based on the value of $\alpha/2$ and the degrees of freedom $(n - 1)$. Thus the paired-t confidence interval for an α level of significance is

$$P\left(\bar{x}_{(1-2)} - hw \leq \mu_{(1-2)} \leq \bar{x}_{(1-2)} + hw\right) = 1 - \alpha$$

Notice that this is basically the same confidence interval expression presented in Chapter 9 with $\bar{x}_{(1-2)}$ replacing $\bar{x}$ and $\mu_{(1-2)}$ replacing μ.

Let's create Table 10.2 by restructuring Table 10.1 to conform to our new "paired" notation and paired-t method before computing the confidence interval necessary for testing the hypotheses

H_0: $\mu_1 - \mu_2 = 0$ or using the new "paired" notation $\mu_{(1-2)} = 0$
H_1: $\mu_1 - \mu_2 \neq 0$ or using the new "paired" notation $\mu_{(1-2)} \neq 0$

where the subscripts 1 and 2 denote Strategy 1 and Strategy 2, respectively. The observations in Table 10.2 are identical to the observations in Table 10.1. However, we added a fourth column. The values in the fourth column (column D) are computed by subtracting column C from column B. This fourth column represents the 10 independent observations ($n = 10$) of our new random variable $x_{(1-2)j}$.

TABLE 10.2 **Comparison of Two Buffer Allocation Strategies Based on the Paired Differences**

(A) *Replication (j)*	*(B)* *Strategy 1* *Throughput x_{1j}*	*(C)* *Strategy 2* *Throughput x_{2j}*	*(D)* *Throughput* *Difference ($B - C$)* *$x_{(1-2)j} = x_{1j} - x_{2j}$*
1	54.48	56.01	−1.53
2	57.36	54.08	3.28
3	54.81	52.14	2.67
4	56.20	53.49	2.71
5	54.83	55.49	−0.66
6	57.69	55.00	2.69
7	58.33	54.88	3.45
8	57.19	54.47	2.72
9	56.84	54.93	1.91
10	55.29	55.84	−0.55
Sample mean $\bar{x}_{(1-2)}$			1.67
Sample standard deviation $s_{(1-2)}$			1.85
Sample variance $s^2_{(1-2)}$			3.42

Now, for an $\alpha = 0.05$ level of significance, the confidence interval on the new random variable $x_{(1-2)j}$ in the Throughput Difference column of Table 10.2 is computed using our equations as follows:

$$\bar{x}_{(1-2)} = \frac{\sum_{j=1}^{10} x_{(1-2)j}}{10} = 1.67 \text{ parts per hour}$$

$$s_{(1-2)} = \sqrt{\frac{\sum_{j=1}^{10} \left[x_{(1-2)j} - 1.67 \right]^2}{10 - 1}} = 1.85 \text{ parts per hour}$$

$$hw = \frac{(t_{9,0.025})1.85}{\sqrt{10}} = \frac{(2.262)1.85}{\sqrt{10}} = 1.32 \text{ parts per hour}$$

where $t_{n-1,\alpha/2} = t_{9,0.025} = 2.262$ is determined from the Student's t table in Appendix B. The 95 percent confidence interval is

$$\bar{x}_{(1-2)} - hw \le \mu_{(1-2)} \le \bar{x}_{(1-2)} + hw$$
$$1.67 - 1.32 \le \mu_{(1-2)} \le 1.67 + 1.32$$
$$0.35 \le \mu_{(1-2)} \le 2.99$$

With approximately 95 percent confidence, we conclude that there is a significant difference between the mean throughputs of the two strategies given that the interval excludes zero. The confidence interval further suggests that the mean throughput μ_1 of Strategy 1 is higher than the mean throughput μ_2 of Strategy 2 (an estimated 0.35 to 2.99 parts per hour higher). This is basically the same conclusion reached using the Welch confidence interval method presented in Section 10.3.1.

Now let's walk back through the assumptions made when we used the paired-t method. The main requirement is that the observations in the Throughput Difference column be independent and normally distributed. Pairing the throughput observations for Strategy 1 with the throughput observations for Strategy 2 and subtracting the two values formed the Throughput Difference column of Table 10.2. The observations for Strategy 1 are independent because nonoverlapping streams of random numbers were used to drive each replication. The observations for Strategy 2 are independent for the same reason. The use of random number streams is discussed in Chapters 3 and 9 and later in this chapter. Therefore, there is no doubt that these observations meet the independence requirement. It then follows that the observations under the Throughput Difference column are also statistically independent. The assumption that has been made, without really knowing if it's true, is that the observations in the Throughput Difference column are normally distributed. The reasonableness of assuming that the output produced by our simulation models is normally distributed is discussed at length in Chapter 9.

10.3.3 Welch versus the Paired-t Confidence Interval

It is difficult to say beforehand which method would produce the smaller confidence interval half-width for a given comparison problem. If, however, the observations between populations (simulated systems) are not independent, then the Welch method cannot be used to compute the confidence interval. For this case, use the paired-t method. If the observations between populations are independent, the Welch method would be used to construct the confidence interval should you have an unequal number of observations from each population ($n_1 \neq n_2$) and you do not wish to discard any of the observations in order to pair them up as required by the paired-t method.

10.4 Comparing More Than Two Alternative System Designs

Sometimes we use simulation to compare more than two alternative designs of a system with respect to a given performance measure. And, as in the case of comparing two alternative designs of a system, several statistical methods can be used for the comparison. We will present two of the most popular methods used in simulation. The Bonferroni approach is presented in Section 10.4.1 and is useful for comparing three to about five designs. A class of linear statistical models useful for comparing any number of alternative designs of a system is presented in Section 10.4.2. A brief introduction to factorial design and optimization experiments is presented in Section 10.4.3.

10.4.1 The Bonferroni Approach for Comparing More Than Two Alternative Systems

The *Bonferroni approach* is useful when there are more than two alternative system designs to compare with respect to some performance measure. Given K alternative system designs to compare, the null hypothesis H_0 and alternative hypothesis H_1 become

$$H_0: \mu_1 = \mu_2 = \mu_3 = \cdots = \mu_K = \mu \qquad \text{for } K \text{ alternative systems}$$
$$H_1: \mu_i \neq \mu_{i'} \qquad \text{for at least one pair } i \neq i'$$

where i and i' are between 1 and K and $i < i'$. The null hypothesis H_0 states that the means from the K populations (mean output of the K different simulation models) are not different, and the alternative hypothesis H_1 states that at least one pair of the means are different.

The Bonferroni approach is very similar to the two confidence interval methods presented in Section 10.3 in that it is based on computing confidence intervals to determine if the true mean performance of one system (μ_i) is significantly different than the true mean performance of another system ($\mu_{i'}$). In fact, either the paired-t confidence interval or the Welch confidence interval can be used with the Bonferroni approach. However, we will describe it in the context of using paired-t confidence intervals, noting that the paired-t confidence interval method can be used when the observations across populations are either independent or correlated.

The Bonferroni method is implemented by constructing a series of confidence intervals to compare all system designs to each other (all pairwise comparisons).

The number of pairwise comparisons for K candidate designs is computed by $K(K-1)/2$. A paired-t confidence interval is constructed for each pairwise comparison. For example, four candidate designs, denoted D1, D2, D3, and D4, require the construction of six $[4(4-1)/2]$ confidence intervals to evaluate the differences $\mu_{(D1-D2)}$, $\mu_{(D1-D3)}$, $\mu_{(D1-D4)}$, $\mu_{(D2-D3)}$, $\mu_{(D2-D4)}$, and $\mu_{(D3-D4)}$. The six paired-t confidence intervals are

$$P\left(\bar{x}_{(D1-D2)} - hw \leq \mu_{(D1-D2)} \leq \bar{x}_{(D1-D2)} + hw\right) = 1 - \alpha_1$$
$$P\left(\bar{x}_{(D1-D3)} - hw \leq \mu_{(D1-D3)} \leq \bar{x}_{(D1-D3)} + hw\right) = 1 - \alpha_2$$
$$P\left(\bar{x}_{(D1-D4)} - hw \leq \mu_{(D1-D4)} \leq \bar{x}_{(D1-D4)} + hw\right) = 1 - \alpha_3$$
$$P\left(\bar{x}_{(D2-D3)} - hw \leq \mu_{(D2-D3)} \leq \bar{x}_{(D2-D3)} + hw\right) = 1 - \alpha_4$$
$$P\left(\bar{x}_{(D2-D4)} - hw \leq \mu_{(D2-D4)} \leq \bar{x}_{(D2-D4)} + hw\right) = 1 - \alpha_5$$
$$P\left(\bar{x}_{(D3-D4)} - hw \leq \mu_{(D3-D4)} \leq \bar{x}_{(D3-D4)} + hw\right) = 1 - \alpha_6$$

The rule for deciding if there is a significant difference between the true mean performance of two system designs is the same as before. Confidence intervals that exclude zero indicate a significant difference between the mean performance of the two systems being compared.

In a moment, we will gain some experience using the Bonferroni approach on our example problem. However, we should discuss an important issue about the approach first. Notice that the number of confidence intervals quickly grows as the number of candidate designs K increases [number of confidence intervals $= K(K-1)/2$]. This increases our computational workload, but, more importantly, it has a rather dramatic effect on the overall confidence we can place in our conclusions. Specifically, the overall confidence in the correctness of our conclusions goes down as the number of candidate designs increases. If we pick any one of our confidence intervals, say the sixth one, and evaluate it separately from the other five confidence intervals, the probability that the sixth confidence interval statement is correct is equal to $(1 - \alpha_6)$. Stated another way, we are $100(1 - \alpha_6)$ percent confident that the true but unknown mean ($\mu_{(D3-D4)}$) lies within the interval $(\bar{x}_{(D3-D4)} - hw)$ to $(\bar{x}_{(D3-D4)} + hw)$. Although each confidence interval is computed separately, it is the simultaneous interpretation of all the confidence intervals that allows us to compare the competing designs for a system. The Bonferroni inequality states that the probability of all six confidence intervals being simultaneously correct is at least equal to $(1 - \sum_{i=1}^{6} \alpha_i)$. Stated more generally,

$$P \text{ (all } m \text{ confidence interval statements are correct)} \geq (1 - \alpha) = \left(1 - \sum_{i=1}^{m} \alpha_i\right)$$

where $\alpha = \sum_{i=1}^{m} \alpha_i$ and is the overall level of significance and $m = \frac{K(K-1)}{2}$ and is the number of confidence interval statements.

If, in this example for comparing four candidate designs, we set $\alpha_1 = \alpha_2 = \alpha_3 = \alpha_4 = \alpha_5 = \alpha_6 = 0.05$, then the overall probability that all our conclusions are correct is as low as $(1 - 0.30)$, or 0.70. Being as low as 70 percent confident in our conclusions leaves much to be desired. To combat this, we simply lower the values of the individual significance levels ($\alpha_1 = \alpha_2 = \alpha_3 = \cdots = \alpha_m$) so their sum is not so large. However, this does not come without a price, as we shall see later.

One way to assign values to the individual significance levels is to first establish an overall level of significance α and then divide it by the number of pairwise comparisons. That is,

$$\alpha_i = \frac{\alpha}{K(K-1)/2} \qquad \text{for } i = 1, 2, 3, \ldots, K(K-1)/2$$

Note, however, that it is not required that the individual significance levels be assigned the same value. This is useful in cases where the decision maker wants to place different levels of significance on certain comparisons.

Practically speaking, the Bonferroni inequality limits the number of system designs that can be reasonably compared to about five designs or less. This is because controlling the overall significance level α for the test requires the assignment of small values to the individual significance levels ($\alpha_1 = \alpha_2 = \alpha_3 = \cdots = \alpha_m$) if more than five designs are compared. This presents a problem because the width of a confidence interval quickly increases as the level of significance is reduced. Recall that the width of a confidence interval provides a measure of the accuracy of the estimate. Therefore, we pay for gains in the overall confidence of our test by reducing the accuracy of our individual estimates (wide confidence intervals). When accurate estimates (tight confidence intervals) are desired, we recommend not using the Bonferroni approach when comparing more than five system designs. For comparing more than five system designs, we recommend that the analysis of variance technique be used in conjunction with perhaps the Fisher's least significant difference test. These methods are presented in Section 10.4.2.

Let's return to the buffer allocation example from the previous section and apply the Bonferroni approach using paired-t confidence intervals. In this case, the production control staff has devised three buffer allocation strategies to compare. And, as before, we wish to determine if there are significant differences between the throughput levels (number of parts completed per hour) achieved by the strategies. Although we will be working with individual confidence intervals, the hypotheses for the *overall* α level of significance are

H_0: $\mu_1 = \mu_2 = \mu_3 = \mu$
H_1: $\mu_1 \neq \mu_2$ or $\mu_1 \neq \mu_3$ or $\mu_2 \neq \mu_3$

where the subscripts 1, 2, and 3 denote Strategy 1, Strategy 2, and Strategy 3, respectively.

To evaluate these hypotheses, we estimated the performance of the three strategies by simulating the use of each strategy for 16 days (24 hours per day) past the warm-up period. And, as before, the simulation was replicated 10 times for each strategy. The average hourly throughput achieved by each strategy is shown in Table 10.3.

The evaluation of the three buffer allocation strategies ($K = 3$) requires that three $[3(3-1)/2]$ pairwise comparisons be made. The three pairwise

TABLE 10.3   **Comparison of Three Buffer Allocation Strategies ($K = 3$)**
Based on Paired Differences

(A) Rep. (j)	(B) Strategy 1 Throughput x_{1j}	(C) Strategy 2 Throughput x_{2j}	(D) Strategy 3 Throughput x_{3j}	(E) Difference (B − C) Strategy 1 − Strategy 2 $x_{(1-2)j}$	(F) Difference (B − D) Strategy 1 − Strategy 3 $x_{(1-3)j}$	(G) Difference (C − D) Strategy 2 − Strategy 3 $x_{(2-3)j}$
1	54.48	56.01	57.22	−1.53	−2.74	−1.21
2	57.36	54.08	56.95	3.28	0.41	−2.87
3	54.81	52.14	58.30	2.67	−3.49	−6.16
4	56.20	53.49	56.11	2.71	0.09	−2.62
5	54.83	55.49	57.00	−0.66	−2.17	−1.51
6	57.69	55.00	57.83	2.69	−0.14	−2.83
7	58.33	54.88	56.99	3.45	1.34	−2.11
8	57.19	54.47	57.64	2.72	−0.45	−3.17
9	56.84	54.93	58.07	1.91	−1.23	−3.14
10	55.29	55.84	57.81	−0.55	−2.52	−1.97
$\bar{x}_{(i-i')}$, for all i and i' between 1 and 3, with $i < i'$				1.67	−1.09	−2.76
$s_{(i-i')}$, for all i and i' between 1 and 3, with $i < i'$				1.85	1.58	1.37

comparisons are shown in columns E, F, and G of Table 10.3. Also shown in Table 10.3 are the sample means $\bar{x}_{(i-i')}$ and sample standard deviations $s_{(i-i')}$ for each pairwise comparison.

Let's say that we wish to use an overall significance level of $\alpha = 0.06$ to evaluate our hypotheses. For the individual levels of significance, let's set $\alpha_1 = \alpha_2 = \alpha_3 = 0.02$ by using the equation

$$\alpha_i = \frac{\alpha}{3} = \frac{0.06}{3} = 0.02 \qquad \text{for } i = 1, 2, 3$$

The computation of the three paired-t confidence intervals using the method outlined in Section 10.3.2 and data from Table 10.3 follows:

Comparing $\mu_{(1-2)}$: $\alpha_1 = 0.02$

$$t_{n-1,\alpha_1/2} = t_{9,0.01} = 2.821 \text{ from Appendix B}$$

$$hw = \frac{(t_{9,0.01})s_{(1-2)}}{\sqrt{n}} = \frac{(2.821)1.85}{\sqrt{10}}$$

$$hw = 1.65 \text{ parts per hour}$$

The approximate 98 percent confidence interval is

$$\bar{x}_{(1-2)} - hw \leq \mu_{(1-2)} \leq \bar{x}_{(1-2)} + hw$$
$$1.67 - 1.65 \leq \mu_{(1-2)} \leq 1.67 + 1.65$$
$$0.02 \leq \mu_{(1-2)} \leq 3.32$$

Comparing $\mu_{(1-3)}$: $\alpha_2 = 0.02$

$$t_{n-1,\alpha_2/2} = t_{9,0.01} = 2.821 \text{ from Appendix B}$$

$$hw = \frac{(t_{9,0.01})s_{(1-3)}}{\sqrt{n}} = \frac{(2.821)1.58}{\sqrt{10}}$$

$$hw = 1.41 \text{ parts per hour}$$

The approximate 98 percent confidence interval is

$$\bar{x}_{(1-3)} - hw \leq \mu_{(1-3)} \leq \bar{x}_{(1-3)} + hw$$
$$-1.09 - 1.41 \leq \mu_{(1-3)} \leq -1.09 + 1.41$$
$$-2.50 \leq \mu_{(1-3)} \leq 0.32$$

Comparing $\mu_{(2-3)}$: The approximate 98 percent confidence interval is

$$-3.98 \leq \mu_{(2-3)} \leq -1.54$$

Given that the confidence interval about $\mu_{(1-2)}$ excludes zero, we conclude that there is a significant difference in the mean throughput produced by Strategies 1 (μ_1) and 2 (μ_2). The confidence interval further suggests that the mean throughput μ_1 of Strategy 1 is higher than the mean throughput μ_2 of Strategy 2 (an estimated 0.02 parts per hour to 3.32 parts per hour higher). This conclusion should not be surprising because we concluded that Strategy 1 resulted in a higher throughput than Strategy 2 earlier in Sections 10.3.1 and 10.3.2. However, notice that this confidence interval is wider than the one computed in Section 10.3.2 for comparing Strategy 1 to Strategy 2 using the same data. This is because the earlier confidence interval was based on a significance level of 0.05 and this one is based on a significance level of 0.02. Notice that we went from using $t_{9,0.025} = 2.262$ for the paired-t confidence interval in Section 10.3.2 to $t_{9,0.01} = 2.821$ for this confidence interval, which increased the width of the interval to the point that it is very close to including zero.

Given that the confidence interval about $\mu_{(1-3)}$ includes zero, we conclude that there is no significant difference in the mean throughput produced by Strategies 1 (μ_1) and 3 (μ_3). And from the final confidence interval about $\mu_{(2-3)}$, we conclude that there is a significant difference in the mean throughput produced by Strategies 2 (μ_2) and 3 (μ_3). This confidence interval suggests that the throughput of Strategy 3 is higher than the throughput of Strategy 2 (an estimated 1.54 parts per hour to 3.98 parts per hour higher).

Recall that our *overall* confidence for these conclusions is approximately 94 percent. Based on these results, we may be inclined to believe that Strategy 2 is the least favorable with respect to mean throughput while Strategies 1 and 3 are the most favorable with respect to mean throughput. Additionally, the difference in the mean throughput of Strategy 1 and Strategy 3 is not significant. Therefore, we recommend that you implement Strategy 3 in place of your own Strategy 1 because Strategy 3 was the boss's idea.

In this case, the statistical assumptions for using the Bonferroni approach are the same as for the paired-t confidence interval. Because we used the Student's t

distribution to build the confidence intervals, the observations in the Throughput Difference columns of Table 10.3 must be independent and normally distributed. It is reasonable to assume that these two assumptions are satisfied here for the Bonferroni test using the same logic presented at the conclusion of Section 10.3.2 for the paired-t confidence interval.

10.4.2 Advanced Statistical Models for Comparing More Than Two Alternative Systems

Analysis of variance (ANOVA) in conjunction with a multiple comparison test provides a means for comparing a much larger number of alternative system designs than does the Welch confidence interval, paired-t confidence interval, or Bonferroni approach. The major benefit that the ANOVA procedure has over the Bonferroni approach is that the overall confidence level of the test of hypothesis does not decrease as the number of candidate system designs increases. There are entire textbooks devoted to ANOVA and multiple comparison tests used for a wide range of experimental designs. However, we will limit our focus to using these techniques for comparing the performance of multiple system designs. As with the Bonferroni approach, we are interested in evaluating the hypotheses

$$H_0: \mu_1 = \mu_2 = \mu_3 = \cdots = \mu_K = \mu \qquad \text{for } K \text{ alternative systems}$$
$$H_1: \mu_i \neq \mu_{i'} \qquad \text{for at least one pair } i \neq i'$$

where i and i' are between 1 and K and $i < i'$. After defining some new terminology, we will formulate the hypotheses differently to conform to the statistical model used in this section.

An *experimental unit* is the system to which treatments are applied. The simulation model of the production system is the experimental unit for the buffer allocation example. A *treatment* is a generic term for a variable of interest and a *factor* is a category of the treatment. We will consider only the single-factor case with K levels. Each factor level corresponds to a different system design. For the buffer allocation example, there are three factor levels—Strategy 1, Strategy 2, and Strategy 3. Treatments are applied to the experimental unit by running the simulation model with a specified factor level (strategy).

An *experimental design* is a plan that causes a systematic and efficient application of treatments to an experimental unit. We will consider the completely randomized (CR) design—the simplest experimental design. The primary assumption required for the CR design is that experimental units (simulation models) are homogeneous with respect to the response (model's output) before the treatment is applied. For simulation experiments, this is usually the case because a model's logic should remain constant except to change the level of the factor under investigation. We first specify a test of hypothesis and significance level, say an α value of 0.05, before running experiments. The null hypothesis for the buffer allocation problem would be that the mean throughputs due to the application of treatments (Strategies 1, 2, and 3) do not differ. The alternate hypothesis

TABLE 10.4   **Experimental Results and Summary Statistics for a Balanced Experimental Design**

Replication (j)	Strategy 1 Throughput (x_{1j})	Strategy 2 Throughput (x_{2j})	Strategy 3 Throughput (x_{3j})
1	54.48	56.01	57.22
2	57.36	54.08	56.95
3	54.81	52.14	58.30
4	56.20	53.49	56.11
5	54.83	55.49	57.00
6	57.69	55.00	57.83
7	58.33	54.88	56.99
8	57.19	54.47	57.64
9	56.84	54.93	58.07
10	55.29	55.84	57.81
Sum $x_i = \sum_{j=1}^{n} x_{ij} = \sum_{j=1}^{10} x_{ij}$, for $i = 1, 2, 3$	563.02	546.33	573.92
Sample mean $\bar{x}_i = \dfrac{\sum_{j=1}^{n} x_{ij}}{n} = \dfrac{\sum_{j=1}^{10} x_{ij}}{10}$, for $i = 1, 2, 3$	56.30	54.63	57.39

states that the mean throughputs due to the application of treatments (Strategies 1, 2, and 3) differ among at least one pair of strategies.

We will use a balanced CR design to help us conduct this test of hypothesis. In a balanced design, the same number of observations are collected for each factor level. Therefore, we executed 10 simulation runs to produce 10 observations of throughput for each strategy. Table 10.4 presents the experimental results and summary statistics for this problem. The response variable (x_{ij}) is the observed throughput for the treatment (strategy). The subscript i refers to the factor level (Strategy 1, 2, or 3) and j refers to an observation (output from replication j) for that factor level. For example, the mean throughput response of the simulation model for the seventh replication of Strategy 2 is 54.88 in Table 10.4. Parameters for this balanced CR design are

Number of factor levels = number of alternative system designs = $K = 3$
Number of observations for each factor level = $n = 10$
Total number of observations = $N = nK = (10)3 = 30$

Inspection of the summary statistics presented in Table 10.4 indicates that Strategy 3 produced the highest mean throughput and Strategy 2 the lowest. Again, we should not jump to conclusions without a careful analysis of the experimental data. Therefore, we will use analysis of variance (ANOVA) in conjunction with a multiple comparison test to guide our decision.

Analysis of Variance
Analysis of variance (ANOVA) allows us to partition the total variation in the output response from the simulated system into two components—variation due to

the effect of the treatments and variation due to experimental error (the inherent variability in the simulated system). For this problem case, we are interested in knowing if the variation due to the treatment is sufficient to conclude that the performance of one strategy is significantly different than the other with respect to mean throughput of the system. We assume that the observations are drawn from normally distributed populations and that they are independent within a strategy and between strategies. Therefore, the variance reduction technique based on common random numbers (CRN) presented in Section 10.5 cannot be used with this method.

The fixed-effects model is the underlying linear statistical model used for the analysis because the levels of the factor are fixed and we will consider each possible factor level. The fixed-effects model is written as

$$x_{ij} = \mu + \tau_i + \varepsilon_{ij} \begin{cases} \text{for } i = 1, 2, 3, \ldots, K \\ \text{for } j = 1, 2, 3, \ldots, n \end{cases}$$

where τ_i is the effect of the ith treatment (ith strategy in our example) as a deviation from the overall (common to all treatments) population mean μ and ε_{ij} is the error associated with this observation. In the context of simulation, the ε_{ij} term represents the random variation of the response x_{ij} that occurred during the jth replication of the ith treatment. Assumptions for the fixed-effects model are that the sum of all τ_i equals zero and that the error terms ε_{ij} are independent and normally distributed with a mean of zero and common variance. There are methods for testing the reasonableness of the normality and common variance assumptions. However, the procedure presented in this section is reported to be somewhat insensitive to small violations of these assumptions (Miller et al. 1990). Specifically, for the buffer allocation example, we are testing the equality of three treatment effects (Strategies 1, 2, and 3) to determine if there are statistically significant differences among them. Therefore, our hypotheses are written as

H_0: $\tau_1 = \tau_2 = \tau_3 = 0$
H_1: $\tau_i \neq 0$ for at least one i,   for $i = 1, 2, 3$

Basically, the previous null hypothesis that the K population means are all equal ($\mu_1 = \mu_2 = \mu_3 = \cdots = \mu_K = \mu$) is replaced by the null hypothesis $\tau_1 = \tau_2 = \tau_3 = \cdots = \tau_K = 0$ for the fixed-effects model. Likewise, the alternative hypothesis that at least two of the population means are unequal is replaced by $\tau_i \neq 0$ for at least one i. Because only one factor is considered in this problem, a simple one-way analysis of variance is used to determine F_{CALC}, the test statistic that will be used for the hypothesis test. If the computed F_{CALC} value exceeds a threshold value called the *critical value,* denoted F_{CRITICAL}, we shall reject the null hypothesis that states that the treatment effects do not differ and conclude that there are statistically significant differences among the treatments (strategies in our example problem).

To help us with the hypothesis test, let's summarize the experimental results shown in Table 10.4 for the example problem. The first summary statistic that we will compute is called the *sum of squares* (SS_i) and is calculated for the ANOVA for each factor level (Strategies 1, 2, and 3 in this case). In a balanced design where the number of observations n for each factor level is a constant, the sum of squares is calculated using the formula

$$\text{SS}_i = \left(\sum_{j=1}^{n} x_{ij}^2\right) - \frac{\left(\sum_{j=1}^{n} x_{ij}\right)^2}{n} \qquad \text{for } i = 1, 2, 3, \ldots, K$$

For this example, the sums of squares are

$$\text{SS}_1 = \left(\sum_{j=1}^{10} x_{1j}^2\right) - \frac{\left(\sum_{j=1}^{10} x_{1j}\right)^2}{10}$$

$$\text{SS}_1 = [(54.48)^2 + (57.36)^2 + \cdots + (55.29)^2] - \frac{(563.02)^2}{10} = 16.98$$

$$\text{SS}_2 = 12.23$$

$$\text{SS}_3 = 3.90$$

The grand total of the N observations ($N = nK$) collected from the output response of the simulated system is computed by

$$\text{Grand total} = x.. = \sum_{i=1}^{K} \sum_{j=1}^{n} x_{ij} = \sum_{i=1}^{K} x_i$$

The overall mean of the N observations collected from the output response of the simulated system is computed by

$$\text{Overall mean} = \bar{x}.. = \frac{\sum_{i=1}^{K} \sum_{j=1}^{n} x_{ij}}{N} = \frac{x..}{N}$$

Using the data in Table 10.4 for the buffer allocation example, these statistics are

$$\text{Grand total} = x.. = \sum_{i=1}^{3} x_i = 563.02 + 546.33 + 573.92 = 1{,}683.27$$

$$\text{Overall mean} = \bar{x}.. = \frac{x..}{N} = \frac{1{,}683.27}{30} = 56.11$$

Our analysis is simplified because a balanced design was used (equal observations for each factor level). We are now ready to define the computational formulas for the ANOVA table elements (for a balanced design) needed to conduct the hypothesis test. As we do, we will construct the ANOVA table for the buffer allocation example. The computational formulas for the ANOVA table elements are

Degrees of freedom total (corrected) = df(total corrected) = $N - 1$
$$= 30 - 1 = 29$$

Degrees of freedom treatment = df(treatment) = $K - 1 = 3 - 1 = 2$

Degrees of freedom error = df(error) = $N - K = 30 - 3 = 27$

and

$$\text{Sum of squares error} = \text{SSE} = \sum_{i=1}^{K} \text{SS}_i = 16.98 + 12.23 + 3.90 = 33.11$$

$$\text{Sum of squares treatment} = \text{SST} = \frac{1}{n}\left[\left(\sum_{i=1}^{K} x_i^2\right) - \frac{x_{..}^2}{K}\right]$$

$$\text{SST} = \frac{1}{10}\left[((563.02)^2 + (546.33)^2 + (573.92)^2) - \frac{(1{,}683.27)^2}{3}\right] = 38.62$$

$$\text{Sum of squares total (corrected)} = \text{SSTC} = \text{SST} + \text{SSE}$$
$$= 38.62 + 33.11 = 71.73$$

and

$$\text{Mean square treatment} = \text{MST} = \frac{\text{SST}}{\text{df(treatment)}} = \frac{38.62}{2} = 19.31$$

$$\text{Mean square error} = \text{MSE} = \frac{\text{SSE}}{\text{df(error)}} = \frac{33.11}{27} = 1.23$$

and finally

$$\text{Calculated } F \text{ statistic} = F_{\text{CALC}} = \frac{\text{MST}}{\text{MSE}} = \frac{19.31}{1.23} = 15.70$$

Table 10.5 presents the ANOVA table for this problem. We will compare the value of F_{CALC} with a value from the F table in Appendix C to determine whether to reject or fail to reject the null hypothesis $H_0: \tau_1 = \tau_2 = \tau_3 = 0$. The values obtained from the F table in Appendix C are referred to as *critical values* and are determined by $F_{(\text{df(treatment), df(error)}; \alpha)}$. For this problem, $F_{(2,27; 0.05)} = 3.35 = F_{\text{CRITICAL}}$, using a significance level (α) of 0.05. Therefore, we will reject H_0 since $F_{\text{CALC}} > F_{\text{CRITICAL}}$ at the $\alpha = 0.05$ level of significance. If we believe the data in Table 10.4 satisfy the assumptions of the fixed-effects model, then we would conclude that the buffer allocation strategy (treatment) significantly affects the mean

TABLE 10.5   Analysis of Variance Table

Source of Variation	Degrees of Freedom	Sum of Squares	Mean Square	F_{CALC}
Total (corrected)	$N - 1 = 29$	$\text{SSTC} = 71.73$		
Treatment (strategies)	$K - 1 = 2$	$\text{SST} = 38.62$	$\text{MST} = 19.31$	15.70
Error	$N - K = 27$	$\text{SSE} = 33.11$	$\text{MSE} = 1.23$	

throughput of the system. We now have evidence that at least one strategy produces better results than the other strategies. Next, a multiple comparison test will be conducted to determine which strategy (or strategies) causes the significance.

Multiple Comparison Test

Our final task is to conduct a multiple comparison test. The hypothesis test suggested that not all strategies are the same with respect to throughput, but it did not identify which strategies performed differently. We will use Fisher's least significant difference (LSD) test to identify which strategies performed differently. It is generally recommended to conduct a hypothesis test prior to the LSD test to determine if one or more pairs of treatments are significantly different. If the hypothesis test failed to reject the null hypothesis, suggesting that all μ_i were the same, then the LSD test would not be performed. Likewise, if we reject the null hypothesis, we should then perform the LSD test. Because we first performed a hypothesis test, the subsequent LSD test is often called a *protected* LSD test.

The LSD test requires the calculation of a test statistic used to evaluate all pairwise comparisons of the sample mean from each population ($\bar{x}_1, \bar{x}_2, \bar{x}_3, \ldots, \bar{x}_K$). In our example buffer allocation problem, we are dealing with the sample mean throughput computed from the output of our simulation models for the three strategies ($\bar{x}_1, \bar{x}_2, \bar{x}_3$). Therefore, we will make three pairwise comparisons of the sample means for our example, recalling that the number of pairwise comparisons for K candidate designs is computed by $K(K-1)/2$. The LSD test statistic is calculated as

$$\text{LSD}(\alpha) = t_{(\text{df(error)},\alpha/2)}\sqrt{\frac{2(\text{MSE})}{n}}$$

The decision rule states that if the difference in the sample mean response values exceeds the LSD test statistic, then the population mean response values are significantly different at a given level of significance. Mathematically, the decision rule is written as

If $|\bar{x}_i - \bar{x}_{i'}| > \text{LSD}(\alpha)$, then μ_i and $\mu_{i'}$ are significantly different at the α level of significance.

For this problem, the LSD test statistic is determined at the $\alpha = 0.05$ level of significance:

$$\text{LSD}_{(0.05)} = t_{27,0.025}\sqrt{\frac{2(\text{MSE})}{n}} = 2.052\sqrt{\frac{2(1.23)}{10}} = 1.02$$

Table 10.6 presents the results of the three pairwise comparisons for the LSD analysis. With 95 percent confidence, we conclude that each pair of means is different ($\mu_1 \neq \mu_2$, $\mu_1 \neq \mu_3$, and $\mu_2 \neq \mu_3$). We may be inclined to believe that the best strategy is Strategy 3, the second best strategy is Strategy 1, and the worst strategy is Strategy 2.

Recall that the Bonferroni approach in Section 10.4.1 did not detect a significant difference between Strategy 1 (μ_1) and Strategy 3 (μ_3). One possible

TABLE 10.6   LSD Analysis

	Strategy 2 $\bar{x}_2 = 54.63$	Strategy 1 $\bar{x}_1 = 56.30$
Strategy 3 $\bar{x}_3 = 57.39$	$\|\bar{x}_2 - \bar{x}_3\| = 2.76$ Significant $(2.76 > 1.02)$	$\|\bar{x}_1 - \bar{x}_3\| = 1.09$ Significant $(1.09 > 1.02)$
Strategy 1 $\bar{x}_1 = 56.30$	$\|\bar{x}_1 - \bar{x}_2\| = 1.67$ Significant $(1.67 > 1.02)$	

explanation is that the LSD test is considered to be more liberal in that it will in-dicate a difference before the more conservative Bonferroni approach. Perhaps if the paired-t confidence intervals had been used in conjunction with common random numbers (which is perfectly acceptable because the paired-t method does not require that observations be independent between populations), then the Bonferroni approach would have also indicated a difference. We are not sug-gesting here that the Bonferroni approach is in error (or that the LSD test is in error). It could be that there really is no difference between the performances of Strategy 1 and Strategy 3 or that we have not collected enough observations to be conclusive.

There are several multiple comparison tests from which to choose. Other tests include Tukey's honestly significant difference (HSD) test, Bayes LSD (BLSD) test, and a test by Scheffe. The LSD and BLSD tests are considered to be liberal in that they will indicate a difference between μ_i and $\mu_{i'}$ before the more conservative Scheffe test. A book by Petersen (1985) provides more information on multiple comparison tests.

10.4.3 Factorial Design and Optimization

In simulation experiments, we are sometimes interested in finding out how differ-ent decision variable settings impact the response of the system rather than simply comparing one candidate system to another. For example, we may want to measure how the mean time a customer waits in a bank changes as the number of tellers (the decision variable) is increased from 1 through 10. There are often many decision variables of interest for complex systems. And rather than run hundreds of experi-ments for every possible variable setting, experimental design techniques can be used as a shortcut for finding those decision variables of greatest significance (the variables that significantly influence the output of the simulation model). Using ex-perimental design terminology, decision variables are referred to as *factors* and the output measures are referred to as *responses* (Figure 10.3). Once the response of in-terest has been identified and the factors that are suspected of having an influence on this response defined, we can use a *factorial design* method that prescribes how many runs to make and what level or value to use for each factor.

FIGURE 10.3

Relationship between factors (decision variables) and output responses.

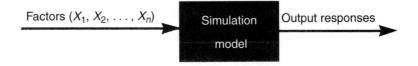

The natural inclination when experimenting with multiple factors is to test the impact that each individual factor has on system response. This is a simple and straightforward approach, but it gives the experimenter no knowledge of how factors interact with each other. It should be obvious that experimenting with two or more factors together can affect system response differently than experimenting with only one factor at a time and keeping all other factors the same.

One type of experiment that looks at the combined effect of multiple factors on system response is referred to as a *two-level, full-factorial design.* In this type of experiment, we simply define a high and a low setting for each factor and, since it is a full-factorial experiment, we try every combination of factor settings. This means that if there are five factors and we are testing two different levels for each factor, we would test each of the $2^5 = 32$ possible combinations of high and low factor levels. For factors that have no range of values from which a high and a low can be chosen, the high and low levels are arbitrarily selected. For example, if one of the factors being investigated is an operating policy (like first come, first served or last come, last served), we arbitrarily select one of the alternative policies as the high-level setting and a different one as the low-level setting.

For experiments in which a large number of factors are considered, a two-level, full-factorial design would result in an extremely large number of combinations to test. In this type of situation, a *fractional-factorial design* is used to strategically select a subset of combinations to test in order to "screen out" factors with little or no impact on system performance. With the remaining reduced number of factors, more detailed experimentation such as a full-factorial experiment can be conducted in a more manageable fashion.

After fractional-factorial experiments and even two-level, full-factorial experiments have been performed to identify the most significant factor level combinations, it is often desirable to conduct more detailed experiments, perhaps over the entire range of values, for those factors that have been identified as being the most significant. This provides more precise information for making decisions regarding the best, or optimal, factor values or variable settings for the system. For a more detailed treatment of factorial design in simulation experimentation, see Law and Kelton (2000).

In many cases, the number of factors of interest prohibits the use of even fractional-factorial designs because of the many combinations to test. If this is the case and you are seeking the best, or optimal, factor values for a system, an alternative is to employ an optimization technique to search for the best combination of values. Several optimization techniques are useful for searching for the combination that produces the most desirable response from the simulation model without evaluating all possible combinations. This is the subject of simulation optimization and is discussed in Chapter 11.

10.5 Variance Reduction Techniques

One luxury afforded to model builders is that the variance of a performance measure computed from the output of simulations can be reduced. This is a luxury because reducing the variance allows us to estimate the mean value of a random variable within a desired level of precision and confidence with fewer replications (independent observations). The reduction in the required number of replications is achieved by controlling how random numbers are used to "drive" the events in the simulation model. These time-saving techniques are called *variance reduction techniques*. The use of *common random numbers (CRN)* is perhaps one of the most popular variance reduction techniques. This section provides an introduction to the CRN technique, presents an example application of CRN, and discusses how CRN works. For additional details about CRN and a review of other variance reduction techniques, see Law and Kelton (2000).

10.5.1 Common Random Numbers

The common random numbers (CRN) technique was invented for comparing alternative system designs. Recall the proposed buffer allocation strategies for the production system presented in Section 10.2. The objective was to decide which strategy yielded the highest throughput for the production system. The CRN technique was not used to compare the performance of the strategies using the paired-*t* confidence interval method in Section 10.3.2. However, it would have been a good idea because the CRN technique provides a means for comparing alternative system designs under more equal experimental conditions. This is helpful in ensuring that the observed differences in the performance of two system designs are due to the differences in the designs and not to differences in experimental conditions. The goal is to evaluate each system under the exact same circumstances to ensure a fair comparison.

Suppose a system is simulated to measure the mean time that entities wait in a queue for service under different service policies. The mean time between arrivals of entities to the system is exponentially distributed with a mean of 5.5 minutes. The exponentially distributed variates are generated using a stream of numbers that are uniformly distributed between zero and 1, having been produced by the random number generator. (See Chapter 3 for a discussion on generating random numbers and random variates.) If a particular segment of the random number stream resulted in several small time values being drawn from the exponential distribution, then entities would arrive to the system faster. This would place a heavier workload on the workstation servicing the entities, which would tend to increase how long entities wait for service. Therefore, the simulation of each policy should be driven by the same stream of random numbers to ensure that the differences in the mean waiting times are due only to differences in the policies and not because some policies were simulated with a stream of random numbers that produced more extreme conditions.

FIGURE 10.4

*Unique seed value
assigned for each
replication.*

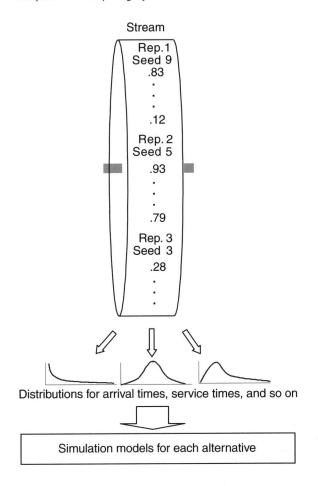

Distributions for arrival times, service times, and so on

Simulation models for each alternative

The goal is to use the exact random number from the stream for the exact purpose in each simulated system. To help achieve this goal, the random number stream can be seeded at the beginning of each independent replication to keep it synchronized across simulations of each system. For example, in Figure 10.4, the first replication starts with a seed value of 9, the second replication starts with a seed value of 5, the third with 3, and so on. If the same seed values for each replication are used to simulate each alternative system, then the same stream of random numbers will drive each of the systems. This seems simple enough. However, care has to be taken not to pick a seed value that places us in a location on the stream that has already been used to drive the simulation in a previous replication. If this were to happen, the results from replicating the simulation of a system would not be independent because segments of the random number stream would have been shared between replications, and this cannot be tolerated. Therefore, some simulation software provides a CRN option that, when selected, automatically assigns seed values to each replication to minimize the likelihood of this happening.

A common practice that helps to keep random numbers synchronized across systems is to assign a different random number stream to each stochastic element in the model. For this reason, most simulation software provides several unique streams of uniformly distributed random numbers to drive the simulation. This concept is illustrated in Figure 10.5 in that separate random number streams are used to generate service times at each of the four machines. If an alternative design for the system was the addition of a fifth machine, the effects of adding the fifth machine could be measured while holding the behavior of the original four machines constant.

In ProModel, up to 100 streams (1 through 100) are available to assign to any random variable specified in the model. Each stream can have one of 100 different initial seed values assigned to it. Each seed number starts generating random numbers at an offset of 100,000 from the previous seed number (seed 1 generates 100,000 random numbers before it catches up to the starting point in the cycle of seed 2). For most simulations, this ensures that streams do not overlap. If you do

FIGURE 10.5

Unique random number stream assigned to each stochastic element in system.

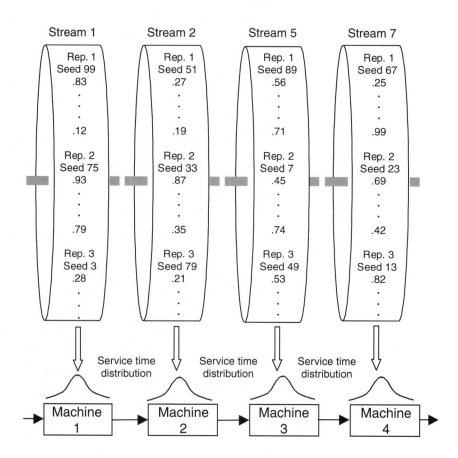

not specify an initial seed value for a stream that is used, ProModel will use the same seed number as the stream number (stream 3 uses the third seed). A detailed explanation of how random number generators work and how they produce unique streams of random numbers is provided in Chapter 3.

Complete synchronization of the random numbers across different models is sometimes difficult to achieve. Therefore, we often settle for partial synchronization. At the very least, it is a good idea to set up two streams with one stream of random numbers used to generate an entity's arrival pattern and the other stream of random numbers used to generate all other activities in the model. That way, activities added to the model will not inadvertently alter the arrival pattern because they do not affect the sample values generated from the arrival distribution.

10.5.2 *Example Use of Common Random Numbers*

In this section the buffer allocation problem of Section 10.2 will be simulated using CRN. Recall that the amount of time to process an entity at each machine is a random variable. Therefore, a unique random number stream will be used to draw samples from the processing time distributions for each machine. There is no need to generate an arrival pattern for parts (entities) because it is assumed that a part (entity) is always available for processing at the first machine. Therefore, the assignment of random number streams is similar to that depicted in Figure 10.5. Next, we will use the paired-*t* confidence interval to compare the two competing buffer allocation strategies identified by the production control staff. Also, note that the common random numbers technique can be used with the Bonferroni approach when paired-*t* confidence intervals are used.

There are two buffer allocation strategies, and the objective is to determine if one strategy results in a significantly different average throughput (number of parts completed per hour). Specifically, we will test at an $\alpha = 0.05$ level of significance the following hypotheses:

$$H_0:\ \mu_1 - \mu_2 = 0 \quad \text{or using the paired notation } \mu_{(1-2)} = 0$$
$$H_1:\ \mu_1 - \mu_2 \neq 0 \quad \text{or using the paired notation } \mu_{(1-2)} \neq 0$$

where the subscripts 1 and 2 denote Strategy 1 and Strategy 2, respectively. As before, each strategy is evaluated by simulating the system for 16 days (24 hours per day) past the warm-up period. For each strategy, the experiment is replicated 10 times. The only difference is that we have assigned individual random number streams to each process and have selected seed values for each replication. This way, both buffer allocation strategies will be simulated using identical streams of random numbers. The average hourly throughput achieved by each strategy is paired by replication as shown in Table 10.7.

Using the equations of Section 10.3.2 and an $\alpha = 0.05$ level of significance, a paired-*t* confidence interval using the data from Table 10.7 can be constructed:

$$\bar{x}_{(1-2)} = 2.67 \text{ parts per hour}$$
$$s_{(1-2)} = 1.16 \text{ parts per hour}$$

TABLE 10.7   Comparison of Two Buffer Allocation Strategies Using Common Random Numbers

(A) Replication (j)	(B) Strategy 1 Throughput x_{1j}	(C) Strategy 2 Throughput x_{2j}	(D) Throughput Difference $(B-C)$ $x_{(1-2)j} = x_{1j} - x_{2j}$
1	79.05	75.09	3.96
2	54.96	51.09	3.87
3	51.23	49.09	2.14
4	88.74	88.01	0.73
5	56.43	53.34	3.09
6	70.42	67.54	2.88
7	35.71	34.87	0.84
8	58.12	54.24	3.88
9	57.77	55.03	2.74
10	45.08	42.55	2.53
Sample mean $\bar{x}_{(1-2)}$			2.67
Sample standard deviation $s_{(1-2)}$			1.16
Sample variance $s^2_{(1-2)}$			1.35

and

$$hw = \frac{(t_{9,0.025})s_{(1-2)}}{\sqrt{n}} = \frac{(2.262)1.16}{\sqrt{10}} = 0.83 \text{ parts per hour}$$

where $t_{n-1,\alpha/2} = t_{9,0.025} = 2.262$ is determined from the Student's t table in Appendix B. The 95 percent confidence interval is

$$\bar{x}_{(1-2)} - hw \leq \mu_{(1-2)} \leq \bar{x}_{(1-2)} + hw$$
$$2.67 - 0.83 \leq \mu_{(1-2)} \leq 2.67 + 0.83$$
$$1.84 \leq \mu_{(1-2)} \leq 3.50$$

With approximately 95 percent confidence, we conclude that there is a significant difference between the mean throughputs of the two strategies because the interval excludes zero. The confidence interval further suggests that the mean throughput μ_1 of Strategy 1 is higher than the mean throughput μ_2 of Strategy 2 (an estimated 1.84 to 3.50 parts per hour higher).

The interesting point here is that the half-width of the confidence interval computed from the CRN observations is considerably shorter than the half-width computed in Section 10.3.2 without using CRN. In fact, the half-width using CRN is approximately 37 percent shorter. This is due to the reduction in the sample standard deviation $s_{(1-2)}$. Thus we have a more precise estimate of the true mean difference without making additional replications. This is the benefit of using variance reduction techniques.

10.5.3 Why Common Random Numbers Work

Using CRN does not actually reduce the variance of the output from the simulation model. It is the variance of the observations in the Throughput Difference column (column D) in Table 10.7 that is reduced. This happened because the observations in the Strategy 1 column are positively correlated with the observations in the Strategy 2 column. This resulted from "driving" the two simulated systems with exactly the same (or as close as possible) stream of random numbers. The effect of the positive correlation is that the variance of the observations in the Throughput Difference column will be reduced.

Although the observations between the strategy columns are correlated, the observations down a particular strategy's column are independent. Therefore, the observations in the Throughput Difference column are also independent. This is because each replication is based on a different segment of the random number stream. Thus the independence assumption for the paired-*t* confidence interval still holds. Note, however, that you cannot use data produced by using CRN to calculate the Welch confidence interval or to conduct an analysis of variance. These procedures require that the observations between populations (the strategy columns in this case) be independent, which they are not when the CRN technique is used.

As the old saying goes, "there is no such thing as a free lunch," and there is a hitch with using CRN. Sometimes the use of CRN can produce the opposite effect and increase the sample standard deviation of the observations in the Throughput Difference column. Without working through the mathematics, know that this occurs when a negative correlation is created between the observations from each system instead of a positive correlation. Unfortunately, there is really no way of knowing beforehand if this will happen. However, the likelihood of realizing a negative correlation in practice is low, so the ticket to success lies in your ability to synchronize the random numbers. If good synchronization is achieved, then the desired result of reducing the standard deviation of the observations in the Difference column will likely be realized.

10.6 Summary

An important point to make here is that simulation, by itself, does not solve a problem. Simulation merely provides a means to evaluate proposed solutions by estimating how they behave. The user of the simulation model has the responsibility to generate candidate solutions either manually or by use of automatic optimization techniques and to correctly measure the utility of the solutions based on the output from the simulation. This chapter presented several statistical methods for comparing the output produced by simulation models representing candidate solutions or designs.

When comparing two candidate system designs, we recommend using either the Welch confidence interval method or the paired-*t* confidence interval. Also, a variance reduction technique based on common random numbers can be used in conjunction with the paired-t confidence interval to improve the precision of the confidence interval. When comparing between three and five candidate system designs, the Bonferroni approach is useful. For more than five designs, the ANOVA procedure in conjunction with Fisher's least significance difference test is a good choice, assuming that the population variances are approximately equal. Additional methods useful for comparing the output produced by different simulation models can be found in Goldsman and Nelson (1998).

10.7 Review Questions

1. The following simulation output was generated to compare four candidate designs of a system.

   a. Use the paired-t confidence interval method to compare Design 1 with Design 3 using a 0.05 level of significance. What is your conclusion? What statistical assumptions did you make to use the paired-t confidence interval method?

   b. Use the Bonferroni approach with Welch confidence intervals to compare all four designs using a 0.06 overall level of significance. What are your conclusions? What statistical assumptions did you make to use the Bonferroni approach?

   c. Use a one-way analysis of variance (ANOVA) with $\alpha = 0.05$ to determine if there is a significant difference between the designs. What is your conclusion? What statistical assumptions did you make to use the one-way ANOVA?

	Waiting Time in System			
Replication	*Design 1*	*Design 2*	*Design 3*	*Design 4*
1	53.9872	58.1365	58.5438	60.1208
2	58.4636	57.6060	57.3973	59.6515
3	55.5300	58.5968	57.1040	60.5279
4	56.3602	55.9631	58.7105	58.1981
5	53.8864	58.3555	58.0406	60.3144
6	57.2620	57.0748	56.9654	59.1815
7	56.9196	56.0899	57.2882	58.3103
8	55.7004	59.8942	57.3548	61.6756
9	55.3685	57.5491	58.2188	59.6011
10	56.9589	58.0945	59.5975	60.0836
11	55.0892	59.2632	60.5354	61.1175
12	55.4580	57.4509	57.9982	59.5142

2. Why is the Bonferroni approach to be avoided when comparing more than about five alternative designs of a system?
3. What is a Type I error in hypothesis testing?
4. What is the relationship between a Type I and a Type II error?
5. How do common random numbers reduce variation when comparing two models?
6. Analysis of variance (ANOVA) allows us to partition the total variation in the output from a simulation model into two components. What are they?
7. Why can common random numbers not be used if the Welch confidence interval method is used?

References

Banks, Jerry; John S. Carson; Berry L. Nelson; and David M. Nicol. *Discrete-Event System Simulation*. Englewood Cliffs, NJ: Prentice Hall, 2001.

Bateman, Robert E.; Royce O. Bowden; Thomas J. Gogg; Charles R. Harrell; and Jack R. A. Mott. *System Improvement Using Simulation*. Orem, UT: PROMODEL Corp., 1997.

Goldsman, David, and Berry L. Nelson. "Comparing Systems via Simulation." Chapter 8 in *Handbook of Simulation*. New York: John Wiley and Sons, 1998.

Hines, William W., and Douglas C. Montgomery. *Probability and Statistics in Engineering and Management Science*. New York: John Wiley & Sons, 1990.

Hoover, Stewart V., and Ronald F. Perry. *Simulation: A Problem-Solving Approach*. Reading, MA: Addison-Wesley, 1989.

Law, Averill M., and David W. Kelton. *Simulation Modeling and Analysis*. New York: McGraw-Hill, 2000.

Miller, Irwin R.; John E. Freund; and Richard Johnson. *Probability and Statistics for Engineers*. Englewood Cliffs, NJ: Prentice Hall, 1990.

Miller, Rupert G. *Beyond ANOVA, Basics of Applied Statistics,* New York: Wiley, 1986.

Montgomery, Douglas C. *Design and Analysis of Experiments*. New York: John Wiley & Sons, 1991.

Petersen, Roger G. *Design and Analysis of Experiments*. New York: Marcel Dekker, 1985.

11   SIMULATION OPTIMIZATION

"Man is a goal seeking animal. His life only has meaning if he is reaching out and striving for his goals."
—Aristotle

11.1 Introduction

Simulation models of systems are built for many reasons. Some models are built to gain a better understanding of a system, to forecast the output of a system, or to compare one system to another. If the reason for building simulation models is to find answers to questions like "What are the optimal settings for _____ to minimize (or maximize) _____ ?" then optimization is the appropriate technology to combine with simulation. *Optimization* is the process of trying different combinations of values for the variables that can be controlled to seek the combination of values that provides the most desirable output from the simulation model.

For convenience, let us think of the simulation model as a black box that imitates the actual system. When inputs are presented to the black box, it produces output that estimates how the actual system responds. In our question, the first blank represents the inputs to the simulation model that are controllable by the decision maker. These inputs are often called *decision variables* or *factors*. The second blank represents the performance measures of interest that are computed from the stochastic output of the simulation model when the decision variables are set to specific values (Figure 11.1). In the question, "What is the optimal number of material handling devices needed to minimize the time that workstations are starved for material?" the decision variable is the number of material handling devices and the performance measure computed from the output of the simulation model is the amount of time that workstations are starved. The objective, then, is to seek the optimal value for each decision variable that minimizes, or maximizes, the expected value of the performance measure(s) of interest. The performance

FIGURE 11.1

Relationship between optimization algorithm and simulation model.

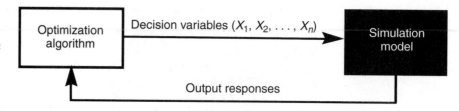

measure is traditionally called the *objective function*. Note that the expected value of the objective function is estimated by averaging the model's output over multiple replications or batch intervals. The simulation optimization problem is more formally stated as

$$\text{Min or Max } E[f(X_1, X_2, \ldots, X_n)]$$

Subject to

$$\text{Lower Bound}_i \leq X_i \leq \text{Upper Bound}_i \qquad \text{for } i = 1, 2, \ldots, n$$

where $E[f(X_1, X_2, \ldots, X_n)]$ denotes the expected value of the objective function, which is estimated.

The search for the optimal solution can be conducted manually or automated with algorithms specifically designed to seek the optimal solution without evaluating all possible solutions. Interfacing optimization algorithms that can automatically generate solutions and evaluate them in simulation models is a worthwhile endeavor because

- It automates part of the analysis process, saving the analyst time.
- A logical method is used to efficiently explore the realm of possible solutions, seeking the best.
- The method often finds several exemplary solutions for the analyst to consider.

The latter is particularly important because, within the list of optimized solutions, there may be solutions that the decision maker may have otherwise overlooked.

In 1995, PROMODEL Corporation and Decision Science, Incorporated, developed SimRunner based on the research of Bowden (1992) on the use of modern optimization algorithms for simulation-based optimization and machine learning. SimRunner helps those wishing to use advanced optimization concepts to seek better solutions from their simulation models. SimRunner uses an optimization method based on evolutionary algorithms. It is the first widely used, commercially available simulation optimization package designed for major simulation packages (ProModel, MedModel, ServiceModel, and ProcessModel). Although SimRunner is relatively easy to use, it can be more effectively used with a basic understanding of how it seeks optimal solutions to a problem. Therefore, the purpose of this chapter is fourfold:

- To provide an introduction to simulation optimization, focusing on the latest developments in integrating simulation and a class of direct optimization techniques called *evolutionary algorithms*.

- To give the reader an appreciation of the advantages and disadvantages of using evolutionary algorithms for optimizing simulated systems.
- To discuss tactical issues involved in the use of evolutionary algorithms for simulation optimization.
- To present examples highlighting how simulation optimization can help analysts in their quest to identify optimal solutions.

There are many approaches to optimizing simulated systems other than those based on evolutionary algorithms. For a more complete review of other approaches to simulation optimization, see Azadivar (1992), Jacobson and Schruben (1989), and Safizadeh (1990).

11.2 In Search of the Optimum

Finding the so-called optimal solution is not an easy task. In fact, it is somewhat like finding a needle in a haystack. The surest way to find the optimal solution is to follow these steps (Akbay 1996):

Step 1. Identify all possible decision variables that affect the output of the system.

Step 2. Based on the possible values of each decision variable, identify all possible solutions.

Step 3. Evaluate each of these solutions accurately.

Step 4. Compare each solution fairly.

Step 5. Record the best answer.

If the output from the simulation model for all possible values of the decision variables is recorded and plotted, the resulting plot would be called the *response surface*. For problems involving only one or two decision variables, the optimal solution can be identified quickly on the response surface. For example, the optimal solution would be located at the highest peak on the response surface for a maximization problem (Figure 11.2). However, it becomes difficult to view the response surface when there are more than two decision variables. Additionally, in many cases, there are simply too many solutions to evaluate (search through) in a reasonable amount of time. Therefore, a balance must be struck between finding the optimal solution and the amount of time to allocate to the search.

Fortunately, researchers have developed several optimization methods that can find the optimal solution for well-posed problems quickly without complete enumeration of all possible alternatives. Example techniques include the Newton-Raphson method and linear programming. Unfortunately, most realistic problems are not well posed and do not lend themselves to being solved by these optimization methods. For example, the response surfaces produced by stochastic simulation models can be highly nonlinear, multimodal, and noisy; can contain both discrete and continuous decision variables; and may not provide independent and identically distributed observations. However, many good heuristic optimization techniques relax the requirement that problems be well posed.

FIGURE 11.2

SimRunner plots the output responses generated by a ProModel simulation model as it seeks the optimal solution, which occurred at the highest peak.

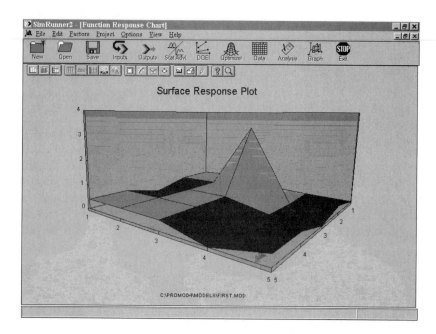

Heuristic techniques consistently provide good, or near optimal, solutions within a reasonable amount of search time. In fact, the use of a heuristic technique does not preclude the chance of finding the optimal solution. It may find the optimal solution, but there is no guarantee that it always will do so. For example, the Nelder Mead Simplex Search technique (Nelder and Mead 1965) has been extensively, and successfully, used for solving simulation optimization problems (Barton and Ivey 1996). This technique often finds the optimal solution even though there is no mathematical proof that it will converge to the optimal solution for these types of problems.

11.3 Combining Direct Search Techniques with Simulation

Direct search techniques are a class of search techniques designed to seek optimal values for the decision variables for a system so as to minimize, or maximize, the output measure of interest from the system. These techniques work directly with the output generated from the system of interest and require no additional information about the function that generates the output. They are ideally suited for optimization tasks when mathematical models do not exist from which gradient information can be computed to guide the search for the optimal solution or if the cost of estimating the gradient is computationally prohibitive. Such is often the case when dealing with stochastic simulation models.

Early on, researchers realized the benefits of combining simulation and direct search techniques. For example, Pegden and Gately (1977) developed an optimization module for the GASP IV simulation software package. Pegden and Gately (1980) later developed another optimization module for use with the SLAM simulation software package. Their optimization packages were based on a variant of a direct search method developed by Hooke and Jeeves (1961). After solving several problems, Pegden and Gately concluded that their packages extended the capabilities of the simulation language by "providing for automatic optimization of decision."

The direct search algorithms available today for simulation optimization are much better than those available in the late 1970s. Using these newer algorithms, the SimRunner simulation optimization tool was developed in 1995. Following SimRunner, two other simulation software vendors soon added an optimization feature to their products. These products are OptQuest96, which was introduced in 1996 to be used with simulation models built with Micro Saint software, and WITNESS Optimizer, which was introduced in 1997 to be used with simulation models built with Witness software. The optimization module in OptQuest96 is based on scatter search, which has links to Tabu Search and the popular evolutionary algorithm called the Genetic Algorithm (Glover 1994; Glover et al. 1996). WITNESS Optimizer is based on a search algorithm called Simulated Annealing (Markt and Mayer 1997). Today most major simulation software packages include an optimization feature.

SimRunner has an optimization module and a module for determining the required sample size (replications) and a model's warm-up period (in the case of a steady-state analysis). The optimization module can optimize integer and real decision variables. The design of the optimization module in SimRunner was influenced by optima-seeking techniques such as Tabu Search (Glover 1990) and evolutionary algorithms (Fogel 1992; Goldberg 1989; Schwefel 1981), though it most closely resembles an evolutionary algorithm (*SimRunner* 1996b).

11.4 Evolutionary Algorithms

Evolutionary algorithms (EAs) are a class of direct search techniques that are based on concepts from the theory of evolution. They mimic the underlying evolutionary process in that entities adapt to their environment in order to survive. EAs manipulate a population of solutions to a problem in such a way that poor solutions fade away and good solutions continually evolve in their search for the optimum (Bowden and Bullington 1996). Search techniques based on this concept have proved robust in that they have been successfully used to solve a wide variety of difficult problems (Biethahn and Nissen 1994; Bowden and Bullington 1995; Usher and Bowden 1996).

Evolutionary algorithms differ from traditional nonlinear optimization techniques in many ways. The most significant difference is that they search the response surface using a population of solutions as opposed to a single solution. This allows an EA to collect information about the response surface from many different points simultaneously. The EA uses the information reported back from multiple locations on the response surface, as opposed to a single point, to guide the search for the optimal solution. This population-based approach increases the chance of finding the global optimal solution (Biethahn and Nissen 1994). Research has been conducted to demonstrate the convergence characteristics of EAs and to develop mathematical proofs for global convergence (for example, see Bäck and Schwefel 1993).

The most popular EAs are genetic algorithms (Goldberg 1989), evolutionary programming (Fogel 1992), and evolution strategies (Schwefel 1981). Although these three EAs have some significant differences in the way they implement ideas from evolution, they share the same basic methodology. Four major algorithmic steps are needed to apply an EA:

Step 1. Generate an initial population of solutions to the problem by distributing them throughout the solution space.

Step 2. Accurately compute the fitness (response) of each solution.

Step 3. Based on the fitness of the solutions, select the best solutions and apply idealized-type genetic operators to produce a new generation of offspring solutions.

Step 4. Return to Step 2 as long as the algorithm is locating improved solutions.

11.4.1 Combining Evolutionary Algorithms with Simulation

The deviations mentioned previously from traditional nonlinear optimization techniques give EAs several advantages for solving simulation optimization problems. For example, Biethahn and Nissen (1994), Bäck et al. (1995), and Bäck and Schwefel (1993) report that EAs are very suitable for simulation optimization because

- They require no restrictive assumptions or prior knowledge about the topology of the response surface being searched, thereby making them broadly applicable techniques.
- They are well suited for problems with response surfaces that are highly dimensional, multimodal, discontinuous, nondifferentiable, and stochastic, and even for problems with moving response surfaces.
- They are very reliable search techniques and are relatively easy to use.

A potential disadvantage of using EAs for simulation optimization is that they sometimes require the evaluation of many solutions (calls to the simulation model). This could be prohibitive if computational time is limited. However, there are procedures that can be used to reduce the number of times the simulation model is called. For example, Hall and Bowden (1996) developed a method that combines an EA with a more locally oriented direct search technique that results in substantially fewer calls to the simulation model.

Several examples of using EAs for simulation optimization have appeared in the research literature. Lacksonen and Anussornnitisarn (1995) compared a genetic algorithm (GA) with three other direct search techniques (Hooke Jeeves Pattern Search, Nelder Mead Simplex Search, and Simulated Annealing) on 20 simulation optimization test problems and ranked GA as best. Stuckman, Evans, and Mollaghasemi (1991) suggest that GAs and Simulated Annealing are the algorithms of choice when dealing with a large number of decision variables. Tompkins and Azadivar (1995) recommend using GAs when the optimization problem involves qualitative (logical) decision variables. The authors have extensively researched the use of genetic algorithms, evolutionary programming, and evolution strategies for solving manufacturing simulation optimization problems and simulation-based machine learning problems (Bowden 1995; Bowden, Neppalli, and Calvert 1995; Bowden, Hall, and Usher 1996; Hall, Bowden, and Usher 1996).

Reports have also appeared in the trade journal literature on how the EA-based optimizer in SimRunner helped to solve real-world problems. For example, IBM; Sverdrup Facilities, Inc.; and Baystate Health Systems report benefits from using SimRunner as a decision support tool (Akbay 1996). The simulation group at Lockheed Martin used SimRunner to help determine the most efficient lot sizes for parts and when the parts should be released to the system to meet schedules (Anonymous 1996).

11.4.2 Illustration of an Evolutionary Algorithm's Search of a Response Surface

In this section, we demonstrate how EAs perform when applied to problems that are characterized by multimodal and stochastic response surfaces that can be produced by a simulation model's output. For convenience of presentation, a one-decision-variable version of a function developed by Ackley (1987) is used to gain insight on how EAs seek the optimal solution. First Ackley's function is transformed from a deterministic function to a stochastic function. This is done in the usual way by adding stochastic perturbations (noise) to the response generated by Ackley's function. Figure 11.3 illustrates Ackley's function when noise is added to

FIGURE 11.3

Ackley's function with noise and the ES's progress over eight generations.

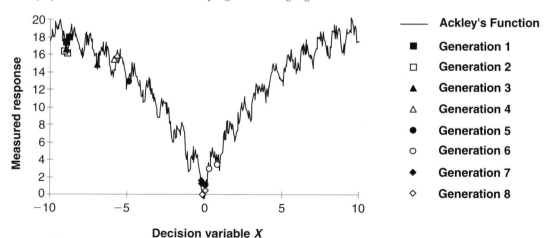

it by sampling from a uniform $(-1, 1)$ distribution. This simulates the variation that can occur in the output from a stochastic simulation model. The function is shown with a single decision variable X that takes on real values between -10 and 10. The response surface has a minimum expected value of zero, occurring when X is equal to zero, and a maximum expected value of 19.37 when X is equal to -9.54 or $+9.54$. Ackley's function is multimodal and thus has several *locally optimal* solutions (optima) that occur at each of the low points (assuming a minimization problem) on the response surface. However, the local optimum that occurs when X is equal to zero is the *global optimum* (the lowest possible response). This is a useful test function because search techniques can prematurely converge and end their search at one of the many optima before finding the global optimum.

According to Step 1 of the four-step process outlined in Section 11.4, an initial population of solutions to the problem is generated by distributing them throughout the solution space. Using a variant of Schwefel's (1981) Evolution Strategy (ES) with two parent solutions and 10 offspring solutions, 10 different values for the decision variable between -10 and 10 are randomly picked to represent an initial offspring population of 10 solutions. However, to make the search for the optimal solution more challenging, the 10 solutions in the initial offspring population are placed far from the global optimum to see if the algorithm can avoid being trapped by one of the many local optima. Therefore, the 10 solutions for the first generation were randomly picked between -10 and -8. So the test is to see if the population of 10 solutions can evolve from one generation to the next to find the global optimal solution without being trapped by one of the many local optima.

Figure 11.3 illustrates the progress that the ES made by following the four-step process from Section 11.4. To avoid complicating the graph, only the responses for the two best solutions (parents) in each generation are plotted on the response surface. Clearly, the process of selecting the best solutions and applying idealized genetic operators allows the algorithm to focus its search toward the optimal solution from one generation to the next. Although the ES samples many of the local optima, it quickly identifies the region of the global optimum and is beginning to hone in on the optimal solution by the eighth generation. Notice that in the sixth generation, the ES has placed solutions to the right side of the search space ($X > 0$) even though it was forced to start its search at the far left side of the solution space. This ability of an evolutionary algorithm allows it to conduct a more globally oriented search.

When a search for the optimal solution is conducted in a noisy simulation environment, care should be taken in measuring the response generated for a given input (solution) to the model. This means that to get an estimate of the expected value of the response, multiple observations (replications) of a solution's performance should be averaged. But to test the idea that EAs can deal with noisy response surfaces, the previous search was conducted using only one observation of a solution's performance. Therefore, the potential existed for the algorithm to become confused and prematurely converge to one of the many local optima because of the noisy response surface. Obviously, this was not the case with the EA in this example.

The authors are *not* advocating that analysts can forget about determining the number of replications needed to satisfactorily estimate the expected value of the response. However, to effectively conduct a search for the optimal solution, an algorithm must be able to deal with noisy response surfaces and the resulting uncertainties that exist even when several observations (replications) are used to estimate a solution's true performance.

11.5 Strategic and Tactical Issues of Simulation Optimization

When optimizing a simulated system for the purpose of applying the solution to an actual system, the analyst must first ensure that the simulation model accurately mimics the real system. This subject is discussed in Chapter 8. Additionally, the analyst must determine the appropriate amount of time to simulate the system, the warm-up period for the model (if applicable), and the required number of independent observations (model replications). A detailed treatment of these topics can be found in Chapter 9.

Another issue very relevant to simulation optimization is the efficiency of the simulation study. Here efficiency is measured in terms of the amount of time necessary to conduct experiments with the simulation model. In the context of an optimization project, an experiment is the evaluation of a solution to estimate its utility. Optimization projects often require the evaluation of many solutions, and a few seconds saved per experiment can translate into a significant reduction in time during an optimization project. Therefore, the amount of time it takes to run the model for one replication (operational efficiency) and the number of replications required to satisfactorily estimate the expected value of the model's response for a given input solution to the model (statistical efficiency) are important considerations.

11.5.1 *Operational Efficiency*

When a general-purpose language like C is used to build a simulation model, the programmer has a great deal of control over how efficiently, or fast, the program operates. For example, minimizing the number of lines of code in a program is one way to reduce the time required to execute the program. When using a simulation language or a simulator, the analyst has less control over the program's efficiency. However, the analyst still has control over other factors that affect the time required to run a model such as the level of detail included in the model, the number of different statistics collected during model execution, and the use of animation.

A good rule is not to take the approach of including every possible element of a system in a model, as a way of making sure something important is not left out of the model. Each element added to the model usually increases the time required to run the simulation. Therefore, include only those elements of the system that have a direct bearing on the problem being studied. For these essential elements, select the minimum number of statistics necessary to estimate the utility of a solution. ProModel products allow users to turn off statistical tracking for locations, entities, resources, and variables that are not needed to estimate the performance of a solution. If, for example, a model is to estimate an entity's time in the system, then only entity-based statistics need be recorded. Also, turning the animation feature off generally reduces the time to run the model.

11.5.2 Statistical Efficiency

One luxury afforded to model builders is that the variance of a performance measure computed from the output of simulations can be reduced. This is a luxury because reducing the variance allows us to estimate the mean value of a random variable within a desired level of precision and confidence with fewer replications (independent observations). The reduction in the required number of replications is achieved by controlling how random variates are used to "drive" the events in the simulation model. These time-saving techniques are called *variance reduction techniques*.

One popular variance reduction technique called *common random numbers (CRN)* was covered in Chapter 10. The CRN technique was invented to compare different system configurations, or solutions, in the context of simulation optimization. The CRN technique is intuitively appealing because it provides a means for comparing solutions under more equal experimental conditions than if CRN were not used. This is helpful in ensuring that the differences in the performance of two solutions are due to the differences in the solutions themselves and not to differences in experimental conditions. Therefore, the use of CRN is generally recommended.

11.5.3 General Optimization Procedure

After a credible and efficient (both operationally and statistically) simulation model has been built, it is time to begin the optimization project. The general procedure and rules of thumb for carrying out an optimization project with the EA-based optimization module in SimRunner follow.

Step 1.   The decision variables believed to affect the output of the simulation model are first programmed into the model as variables whose values can be quickly changed by the EA. Decision variables are typically the parameters whose values can be adjusted by management, such as the number of nurses assigned to a shift or the number of machines to be placed in a work cell.

Step 2.   For each decision variable, define its numeric data type (integer or real) and its lower bound (lowest possible value) and upper bound (highest possible value). During the search, the EA will generate solutions by varying the values of decision variables according to their data types, lower bounds, and upper bounds. The number of decision variables and the range of possible values affect the size of the search space (number of possible solutions to the problem). Increasing the number of decision variables or their range of values increases the size of the search space, which can make it more difficult and time-consuming to identify the optimal solution. As a rule, include only those decision variables known to significantly affect the output of the simulation model and judiciously define the range of possible values for each decision variable. Also, care should be taken when defining the lower and upper bounds of the decision variables to ensure that a combination of values will not be created that lead to a solution not envisioned when the model was built.

Step 3. After selecting the decision variables, construct the objective function to measure the utility of the solutions tested by the EA. Actually, the foundation for the objective function would have already been established when the goals for the simulation project were set. For example, if the goal of the modeling project is to find ways to minimize a customer's waiting time in a bank, then the objective function should measure an entity's (customer's) waiting time in the bank. The objective function is built using terms taken from the output report generated at the end of the simulation run. Objective function terms can be based on entity statistics, location statistics, resource statistics, variable statistics, and so on. The user specifies whether a term is to be minimized or maximized as well as the overall weighting of that term in the objective function. Some terms may be more or less important to the user than other terms. Remember that as terms are added to the objective function, the complexity of the search space may increase, which makes a more difficult optimization problem. From a statistical point of view, single-term objective functions are also preferable to multiterm objective functions. Therefore, strive to keep the objective function as specific as possible.

The objective function is a random variable, and a set of initial experiments should be conducted to estimate its variability (standard deviation). Note that there is a possibility that the objective function's standard deviation differs from one solution to the next. Therefore, the required number of replications necessary to estimate the expected value of the objective function may change from one solution to the next. Thus the objective function's standard deviation should be measured for several different solutions and the highest standard deviation recorded used to compute the number of replications necessary to estimate the expected value of the objective function. When selecting the set of test solutions, choose solutions that are very different from one another. For example, form solutions by setting the decision variables to their lower bounds, middle values, or upper bounds.

A better approach for controlling the number of replications used to estimate the expected value of the objective function for a given solution would be to incorporate a rule into the model that schedules additional replications until the estimate reaches a desired level of precision (confidence interval half-width). Using this technique can help to avoid running too many replications for some solutions and too few replications for others.

Step 4. Select the size of the EA's population (number of solutions) and begin the search. The size of the population of solutions used to conduct the search affects both the likelihood that the algorithm will locate the optimal solution and the time required to conduct the search. In general, as the population size is increased, the algorithm finds better solutions. However, increasing the population size generally increases the time required to conduct the search. Therefore, a balance must be struck between finding the optimum and the amount of available time to conduct the search.

Step 5. After the EA's search has concluded (or halted due to time constraints), the analyst should study the solutions found by the algorithm. In addition to the best solution discovered, the algorithm usually finds many other competitive solutions. A good practice is to rank each solution evaluated based on its utility as measured by the objective function. Next, select the most highly competitive solutions and, if necessary, make additional model replications of those solutions to get better estimates of their true utility. And, if necessary, refer to Chapter 10 for background on statistical techniques that can help you make a final decision between competing solutions. Also, keep in mind that the database of solutions evaluated by the EA represents a rich source of information about the behavior, or response surface, of the simulation model. Sorting and graphing the solutions can help you interpret the "meaning" of the data and gain a better understanding of how the system behaves.

If the general procedure presented is followed, chances are that a good course of action will be identified. This general procedure is easily carried out using ProModel simulation products. Analysts can use SimRunner to help

- Determine the length of time and warm-up period (if applicable) for running a model.
- Determine the required number of replications for obtaining estimates with a specified level of precision and confidence.
- Search for the optimal values for the important decision variables.

Even though it is easy to use SimRunner and other modern optimizers, do not fall into the trap of letting them become the decision maker. Study the top solutions found by the optimizers as you might study the performance records of different cars for a possible purchase. Kick their tires, look under their hoods, and drive them around the block before buying. Always remember that the optimizer is not the decision maker. It only suggest a possible course of action. It is the user's responsibility to make the final decision.

11.6 Formulating an Example Optimization Problem

The optimal allocation of buffer storage between workstations is a common industrial problem and has been extensively studied by academic researchers. If a production line consists of a network of tightly linked workstations, an inventory of parts (work in progress) between workstations can be used to minimize the chance that disruptions (machine failures, line imbalances, quality problems, or the like) will shut down the production line. Several strategies have been developed for determining the amount of buffer storage needed between workstations. However, these strategies are often developed based on simplifying assumptions, made for mathematical convenience, that rarely hold true for real production systems.

One way to avoid oversimplifying a problem for the sake of mathematical convenience is to build a simulation model of the production system and use it to help identify the amount of buffer storage space needed between workstations. However, the number of possible solutions to the buffer allocation problem grows rapidly as the size of the production system (number of possible buffer storage areas and their possible sizes) increases, making it impractical to evaluate all solutions. In such cases, it is helpful to use simulation optimization software like SimRunner to identify a set of candidate solutions.

This example is loosely based on the example production system presented in Chapter 10. It gives readers insight into how to formulate simulation optimization problems when using SimRunner. The example is not fully solved. Its completion is left as an exercise in Lab Chapter 11.

11.6.1 Problem Description

The production system consists of four machines and three buffers. Parts entering the system require processing by each of the four machines in a serial fashion (Figure 11.4). A part is always available for processing at the first machine. After a part is processed, it moves from the machine to the buffer storage area for the

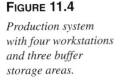

FIGURE 11.4

Production system with four workstations and three buffer storage areas.

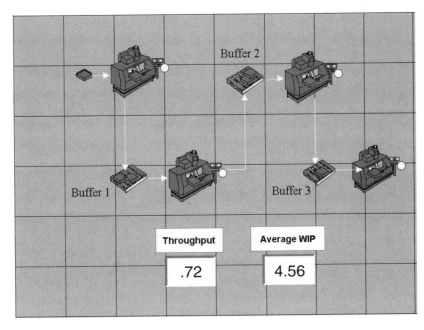

next machine, where it waits to be processed. However, if the buffer is full, the part cannot move forward and remains on the machine until a space becomes available in the buffer. Furthermore, the machine is blocked and no other parts can move to the machine for processing. The part exits the system after being processed by the fourth machine. Note that parts are selected from the buffers to be processed by a machine in a first-in, first-out order. The processing time at each machine is exponentially distributed with a mean of 1.0 minute, 1.3 minutes, 0.7 minute, and 1.0 minute for machines one, two, three, and four, respectively. The time to move parts from one location to the next is negligible.

For this problem, three decision variables describe how buffer space is allocated (one decision variable for each buffer to signify the number of parts that can be stored in the buffer). The Goal is to find the optimal value for each decision variable to maximize the profit made from the sale of the parts. The manufacturer collects $10 per part produced. The limitation is that each unit of space provided for a part in a buffer costs $1,000. So the buffer storage has to be strategically allocated to maximize the throughput of the system. Throughput will be measured as the number of parts completed during a 30-day period.

11.6.2 *Demonstration of the General Optimization Procedure*

The general optimization procedure outlined in Section 11.5.3 will be used to guide the formulation of the problem as a SimRunner simulation optimization project. The assumption is that a credible and efficient simulation model has been built of the production system.

Step 1.   In this step, the decision variables for the problem are identified and defined. Three decision variables are needed to represent the number of parts that can be stored in each buffer (buffer capacity). Let Q_1, Q_2, and Q_3 represent the number of parts that can be stored in buffers 1, 2, and 3, respectively.

Step 2.   The numeric data type for each Q_i is integer. If it is assumed that each buffer will hold a minimum of one part, then the lower bound for each Q_i is 1. The upper bound for each decision variable could arbitrarily be set to, say, 20. However, keep in mind that as the range of values for a decision variable is increased, the size of the search space also increases and will likely increase the time to conduct the search. Therefore, this is a good place to apply existing knowledge about the performance and design of the system.

For example, physical constraints may limit the maximum capacity of each buffer to no greater than 15 parts. If so, there is no need to conduct a search with a range of values that produce infeasible solutions (buffer capacities greater than 15 parts). Considering that the fourth machine's processing time is larger than the third machine's processing time, parts will tend to queue up at the third buffer. Therefore, it might be a good idea to set the upper bound for Q_3 to a larger value than the other two decision variables. However, be careful not to assume too much because it could prevent the optimization algorithm from exploring regions in the search space that may contain good solutions.

With this information, it is decided to set the upper bound for the capacity of each buffer to 15 parts. Therefore, the bounds for each decision variable are

$$1 \le Q_1 \le 15 \qquad 1 \le Q_2 \le 15 \qquad 1 \le Q_3 \le 15$$

Given that each of the three decision variable has 15 different values, there are 15^3, or 3,375, unique solutions to the problem.

Step 3. Here the objective function is formulized. The model was built to investigate buffer allocation strategies to maximize the throughput of the system. Given that the manufacturer collects $10 per part produced and that each unit of space provided for a part in a buffer costs $1,000, the objective function for the optimization could be stated as

$$\text{Maximize } [\$10(\text{Throughput}) - \$1,000(Q_1 + Q_2 + Q_3)]$$

where Throughput is the total number of parts produced during a 30-day period.

Next, initial experiments are conducted to estimate the variability of the objective function in order to determine the number of replications the EA-based optimization algorithm will use to estimate the expected value of the objective function for each solution it evaluates. While doing this, it was also noticed that the throughput level increased very little for buffer capacities beyond a value of nine. Therefore, it was decided to change the upper bound for each decision variable to nine. This resulted in a search space of 9^3, or 729, unique solutions, a reduction of 2,646 solutions from the original formulation. This will likely reduce the search time.

Step 4. Select the size of the population that the EA-based optimization algorithm will use to conduct its search. SimRunner allows the user to select an optimization profile that influences the degree of thoroughness used to search for the optimal solution. The three optimization profiles are aggressive, moderate, and cautious, which correspond to EA population sizes of small, medium, and large. The aggressive profile generally results in a quick search for locally optimal solutions and is used when computer time is limited. The cautious profile specifies that a more thorough search for the global optimum be conducted and is used when computer time is plentiful. At this point, the analyst knows the amount of time required to evaluate a solution. Only one more piece of information is needed to determine how long it will take the algorithm to conduct its search. That is the fraction of the 729 solutions the algorithm will evaluate before converging to a final solution. Unfortunately, there is no way of knowing this in advance. With time running out before a recommendation for the system must be given to management, the analyst elects to use a small population size by selecting SimRunner's aggressive optimization profile.

Step 5. After the search concludes, the analyst selects for further evaluation some of the top solutions found by the optimization algorithm. Note that this does not necessarily mean that only those solutions with the best objective function values are chosen, because the analyst should conduct both a quantitative and a qualitative analysis. On the quantitative side, statistical procedures presented in Chapter 10 are used to gain a better understanding of the relative differences in performance between the candidate solutions. On the qualitative side, one solution may be preferred over another based on factors such as ease of implementation.

Figure 11.5 illustrates the results from a SimRunner optimization of the buffer allocation problem using an aggressive optimization profile. The warm-up time for the simulation was set to 10 days, with each day representing a 24-hour production period. After the warm-up time of 10 days (240 hours), the system is simulated for an additional 30 days (720 hours) to determine throughput. The estimate for the expected value of the objective function was based on five replications of the simulation. The smoother line that is always at the top of the Performance Measures Plot in Figure 11.5 represents the value of the objective function for the best solution identified by SimRunner during the optimization. Notice the rapid improvement in the value of the objective function during the early part of the search as SimRunner identifies better buffer capacities. The other, more irregular line represents the value of the objective function for all the solutions that SimRunner tried.

SimRunner's best solution to the problem specifies a Buffer 1 capacity of nine, a Buffer 2 capacity of seven, and a Buffer 3 capacity of three and was the 33rd solution (experiment) evaluated by SimRunner. The best solution is located at the top of the table in Figure 11.5. SimRunner sorts the solutions in the table

FIGURE 11.5

SimRunner results for the buffer allocation problem.

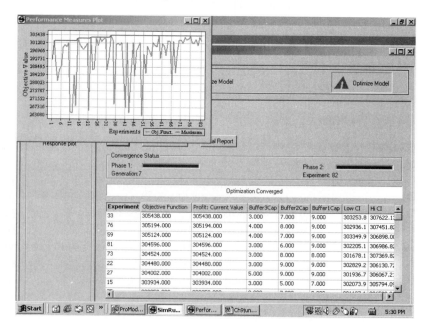

from best to worst. SimRunner evaluated 82 out of the possible 729 solutions. Note that there is no guarantee that SimRunner's best solution is in fact the optimum solution to the problem. However, it is likely to be one of the better solutions to the problem and could be the optimum one.

The last two columns in the SimRunner table shown in Figure 11.5 display the lower and upper bounds of a 95 percent confidence interval for each solution evaluated. Notice that there is significant overlap between the confidence intervals. Although this is not a formal hypothesis-testing procedure, the overlapping confidence intervals suggest the possibility that there is not a significant difference in the performance of the top solutions displayed in the table. Therefore, it would be wise to run additional replications of the favorite solutions from the list and/or use one of the hypothesis-testing procedures in Chapter 10 before selecting a particular solution as the best. The real value here is that SimRunner automatically conducted the search, without the analyst having to hover over it, and reported back several good solutions to the problem for the analyst to consider.

11.7 Real-World Simulation Optimization Project

In this section, we provide an account of how simulation combined with optimization was used to solve a kanban sizing problem. The objective for this problem is to find the minimum amount of inventory required to support production in a just-in-time (JIT) environment where parts are pulled through the system using kanban cards. This presentation is based on a problem faced by a major appliance factory in the United States. Two different techniques are used to solve the problem. One approach uses a technique developed by Toyota. The other approach uses an EA-based simulation optimization algorithm similar to the one used in SimRunner. The presentation concludes with a comparative analysis of the two solutions generated by these techniques.

11.7.1 Problem Description

A *kanban* is a card that authorizes a workstation to begin producing a part. Kanban cards are passed from downstream workstations to upstream workstations, signaling the upstream workstation to begin producing a part. This results in parts being *pulled* through the system as downstream workstations issue kanban cards to upstream workstations.

In a pull production system, the number of kanban cards directly influences the number of material-handling containers required for transporting (pulling) parts through the system and thus the maximum amount of work-in-progress (WIP) inventory. *Trigger values* represent the number of kanban cards that must accumulate before a workstation is authorized to begin production. Therefore, trigger values control the frequency of machine setups. While the ideal minimum trigger value equals 1, many processes do not allow for this due to the time required for machine setup or maintenance. Solving for the optimal number of kanban cards and corresponding trigger values presents a difficult task for production planners.

FIGURE 11.6

The two-stage pull production system.

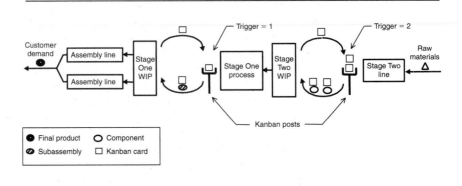

Figure 11.6 illustrates the relationship of the processes in the two-stage pull production system of interest that produces several different types of parts. Customer demand for the final product causes containers of subassemblies to be pulled from the Stage One WIP location to the assembly lines. As each container is withdrawn from the Stage One WIP location, a production-ordering kanban card representing the number of subassemblies in a container is sent to the kanban post for the Stage One processing system. When the number of kanban cards for a given subassembly meets its trigger value, the necessary component parts are pulled from the Stage Two WIP to create the subassemblies. Upon completing the Stage One process, subassemblies are loaded into containers, the corresponding kanban card is attached to the container, and both are sent to Stage One WIP. The container and card remain in the Stage One WIP location until pulled to an assembly line.

In Stage Two, workers process raw materials to fill the Stage Two WIP location as component parts are pulled from it by Stage One. As component parts are withdrawn from the Stage Two WIP location and placed into the Stage One process, a production-ordering kanban card representing the quantity of component parts in a container is sent to the kanban post for the Stage Two line. When the number of kanban cards for a given component part meets a trigger value, production orders equal to the trigger value are issued to the Stage Two line. As workers move completed orders of component parts from the Stage Two line to WIP, the corresponding kanban cards follow the component parts to the Stage Two WIP location.

While an overall reduction in WIP is sought, production planners desire a solution (kanban cards and trigger values for each stage) that gives preference to minimizing the containers in the Stage One WIP location. This requirement is due to space limitations.

11.7.2 Simulation Model and Performance Measure

Without sacrificing the presentation of the simulation optimization subject, only a sketch of the model is given. The pull production system was represented in a discrete-event simulation model. The model assumes constant interarrival times for customer orders to the assembly lines. However, workers at the assembly lines may find defects in the subassemblies, which increases subassembly demand above that needed for scheduled customer demand. The number of defective subassemblies found on the assembly lines follows a stochastic function. Production times for the Stage One process and Stage Two line are constant because they are automated processes. Setup time, time between failures, and time to repair for the Stage Two line are triangularly distributed. Each simulation run consists of a 10-day (two-week) warm-up period followed by a 20-day (one-month) steady-state period.

The performance measure to be maximized is

$$f(a) = W_1(\text{AvgPct} + \text{MinPct}) - W_2(\text{TK}_1) - W_3(\text{TK}_2)$$

where AvgPct is the average throughput percentage (percentage of demand satisfied) for all product types coming off the assembly lines; MinPct is the minimum throughput percentage for all product types coming off the assembly lines; TK_1 is the total number of kanban cards in Stage One; TK_2 is the total number of kanban cards in Stage Two; and W_1, W_2, and W_3 are coefficients that reflect the level of importance (weighting) assigned to each term by production planners. The weighting levels assigned to W_1, W_2, and W_3 are 20.0, 0.025, and 0.015, respectively. Using this performance function, the objective is to find a solution that will maximize throughput while minimizing the total number of kanban cards in the system.

The production planners decided to fix the trigger values in the Stage One process to 1; therefore, the optimal number of kanban cards in the Stage One process and the optimal number of kanban cards and corresponding trigger values in the Stage Two line for each of the 11 different part types being produced need to be determined. Thus the problem contains 33 integer decision variables. Each solution is a vector $a = (K1_1, K1_2, K1_3, \dots, K1_{11}; K2_1, K2_2, K2_3, \dots, K2_{11}; T2_1, T2_2, T2_3, \dots, T2_{11})$ where $K1_i$ represents the number of kanban cards for part type i circulating in the Stage One process, $K2_i$ represents the number of kanban cards for part type i circulating in the Stage Two line, and $T2_i$ represents the corresponding trigger quantities for part type i used in the Stage Two line. Considering the range of possible values for each of the decision variables (omitted for brevity), there are 4.81×10^{41} unique solutions to this problem.

11.7.3 Toyota Solution Technique

Because of their ease of use, the kanban sizing equations developed by Toyota are often used by manufacturers to estimate the number of kanban cards needed. For this problem, the constant-quantity, nonconstant-cycle withdrawal system as described by Monden (1993) was used. The number of kanbans required is determined by

$$\text{Kanbans} = \frac{\left[\dfrac{\text{Monthly demand}}{\text{Monthly setups}}\right] + \left[\text{Daily demand} \times \text{Safety coefficient}\right]}{\text{Container capacity}}$$

where the safety coefficient represents the amount of WIP needed in the system. Production planners assumed a safety coefficient that resulted in one day of WIP for each part type. Additionally, they decided to use one setup per day for each part type. Although this equation provides an estimate of the minimum number of kanban cards, it does not address trigger values. Therefore, trigger values for the Stage Two line were set at the expected number of containers consumed for each part type in one day. The Toyota equation recommended using a total of 243 kanban cards. The details of the calculation are omitted for brevity. When evaluated in the simulation model, this solution yielded a performance score of 35.140 (using the performance function defined in Section 11.7.2) based on four independent simulation runs (replications).

11.7.4 Simulation Optimization Technique

Here an EA-based optimization module like the one in SimRunner is used to search for an optimal solution to the problem. Given that it is not practical to assume that there is an unlimited amount of time for conducting a search, the software allows the user to select an optimization profile that influences the degree of thoroughness used to search for the optimal solution. The three optimization profiles are aggressive, moderate, and cautious, which correspond to EA population sizes of small, medium, and large. The aggressive profile generally results in a quick search and is used when computer time is limited. The cautious profile specifies that a more thorough search be conducted. The moderate profile was used to search for an optimal number of kanban cards and trigger values.

Each time the simulation model is called to evaluate a solution, the model runs four replications to get four independent observations of a solution's performance. The performance of a solution is computed at the end of each simulation run using the performance function defined in Section 11.7.2. The final performance score assigned to each solution is the average of the four independent observations.

Figure 11.7 illustrates the progress of the optimization module's search. Because it is difficult to visualize a response surface with 33 dimension variables, only the performance score of the best solution found per generation is plotted. The points are connected with a line to highlight the rate at which the algorithm finds solutions with improved performance scores. Note that the optimization module quickly finds solutions that are better (based on the four independent observations) than the solution found using the Toyota technique, and the optimization module steadily finds better solutions after that point. The search was halted at 150 generations because the solution was not improving significantly. The solution recommended using a total of 110 kanban cards and yielded a performance score of 37.945 based on four replications.

11.7.5 Comparison of Results

Final solutions generated by the optimization module and Toyota equation were run in the simulation model for 20 independent replications to get a better estimate of the true mean performance of the solutions. Table 11.1 presents a

FIGURE 11.7

*After the second
generation, the
solutions found by the
optimization module
are better than the
solutions generated
using the Toyota
method.*

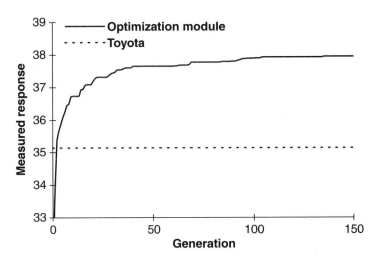

TABLE 11.1 **The Optimization Module's Solution Meets Production Throughput Goals with 54 Percent Fewer Kanban Cards Than Toyota**

Technique	Kanban Cards	Performance Scores
Optimization module	110	37.922
Toyota solution	243	35.115

comparison of the solutions. Both solutions achieved the required throughput using a substantially different number of kanban cards.

An application of Fisher's LSD statistical multiple comparison test presented in Chapter 10 suggests that the performance scores for the methods in Table 11.1 are different at a 0.05 level of significance. Results indicate that the optimization module produced a significantly better solution to the kanban sizing problem than did the production planners using the Toyota technique. The solution that the optimization module found requires approximately 54 percent fewer kanban cards (54 percent less WIP) than the solution generated using the Toyota technique—a considerable reduction in inventory carrying costs.

Considering the time required to run the simulation model on a personal computer with a P5-100 processor and the fact that there are 4.81×10^{41} unique solutions to the problem, it would take approximately 1.15×10^{38} days to evaluate all solutions to the problem. Obviously, this is not a practical alternative. The optimization module required only 24 hours (one day) to complete its search. It evaluated only a small fraction of the total number of solutions and provided a much better starting point than the Toyota solution for making the final decision regarding the actual pull production system, a noteworthy contribution to the decision-making process whether or not the solution found is the global optimum.

11.8 Summary

In recent years, major advances have been made in the development of user-friendly simulation software packages. However, progress in developing simulation output analysis tools has been especially slow in the area of simulation optimization because conducting simulation optimization with traditional techniques has been as much an art as a science (Greenwood, Rees, and Crouch 1993). There are such an overwhelming number of traditional techniques that only individuals with extensive backgrounds in statistics and optimization theory have realized the benefits of integrating simulation and optimization concepts. Using newer optimization techniques, it is now possible to narrow the gap with user-friendly, yet powerful, tools that allow analysts to combine simulation and optimization for improved decision support. SimRunner is one such tool.

Our purpose is not to argue that evolutionary algorithms are the panacea for solving simulation optimization problems. Rather, our purpose is to introduce the reader to evolutionary algorithms by illustrating how they work and how to use them for simulation optimization and to expose the reader to the wealth of literature that demonstrates that evolutionary algorithms are a viable choice for reliable optimization.

There will always be debate on what are the best techniques to use for simulation optimization. Debate is welcome as it results in better understanding of the real issues and leads to the development of better solution procedures. It must be remembered, however, that the practical issue is not that the optimization technique guarantees that it locates the optimal solution in the shortest amount of time for all possible problems that it may encounter; rather, it is that the optimization technique consistently finds good solutions to problems that are better than the solutions analysts are finding on their own. Newer techniques such as evolutionary algorithms and scatter search meet this requirement because they have proved robust in their ability to solve a wide variety of problems, and their ease of use makes them a practical choice for simulation optimization today (Boesel et al. 2001; Brady and Bowden 2001).

11.9 Review Questions

1. What is a response surface?
2. What does it mean if an optimization algorithm is classified as a heuristic technique?
3. What is the most significant difference between the way evolutionary algorithms (EAs) search a response surface and the way many traditional nonlinear optimization algorithms conduct a search? What advantage does this difference give to an EA over many traditional nonlinear optimization algorithms?
4. Assume you are working on a project with five decision variables (A, B, C, D, E). Each decision variable's numeric data type is integer with the

following values: 1, 2, 3, 4, 5, 6. How many different solutions can be generated from these five decision variables? If it takes one minute of computer time to evaluate a solution (run the simulation model for the required number of replications), how many hours would it take to evaluate all solutions?

5. Define *operational and statistical efficiency* in the context of simulation optimization.

References

Ackley, D. *A Connectionist Machine for Genetic Hill Climbing.* Boston, MA: Kluwer, 1987.

Akbay, K. "Using Simulation Optimization to Find the Best Solution." *IIE Solutions,* May 1996, pp. 24–29.

Anonymous, "Lockheed Martin." *IIE Solutions,* December, 1996, pp. SS48–SS49.

Azadivar, F. "A Tutorial on Simulation Optimization." *1992 Winter Simulation Conference.* Arlington Virginia, ed. Swain, J., D. Goldsman, R. Crain and J. Wilson. Institute of Electrical and Electronics Engineers, Piscataway, NJ: 1992, pp. 198–204.

Bäck, T.; T. Beielstein; B. Naujoks; and J. Heistermann. "Evolutionary Algorithms for the Optimization of Simulation Models Using PVM." *Euro PVM 1995—Second European PVM 1995, User's Group Meeting.* Hermes, Paris, ed. Dongarra, J., M. Gengler, B. Tourancheau and X. Vigouroux, 1995, pp. 277–282.

Bäck, T., and H.-P. Schwefel. "An Overview of Evolutionary Algorithms for Parameter Optimization." *Evolutionary Computation* 1, no. 1 (1993), pp. 1–23.

Barton, R., and J. Ivey. "Nelder-Mead Simplex Modifications for Simulation Optimization." *Management Science* 42, no. 7 (1996), pp. 954–73.

Biethahn, J., and V. Nissen. "Combinations of Simulation and Evolutionary Algorithms in Management Science and Economics." *Annals of Operations Research* 52 (1994), pp. 183–208.

Boesel, J.; Bowden, R. O.; Glover, F.; and J. P. Kelly. "Future of Simulation Optimization." *Proceedings of the 2001 Winter Simulation Conference,* 2001, pp. 1466–69.

Bowden, R. O. "Genetic Algorithm Based Machine Learning Applied to the Dynamic Routing of Discrete Parts." Ph.D. Dissertation, Department of Industrial Engineering, Mississippi State University, 1992.

Bowden, R. "The Evolution of Manufacturing Control Knowledge Using Reinforcement Learning." *1995 Annual International Conference on Industry, Engineering, and Management Systems,* Cocoa Beach, FL, ed. G. Lee. 1995, pp. 410–15.

Bowden, R., and S. Bullington. "An Evolutionary Algorithm for Discovering Manufacturing Control Strategies." In *Evolutionary Algorithms in Management Applications,* ed. Biethahn, J. and V. Nissen. Berlin: Springer, 1995, pp. 124–38.

Bowden, R., and S. Bullington. "Development of Manufacturing Control Strategies Using Unsupervised Learning." *IIE Transactions* 28 (1996), pp. 319–31.

Bowden, R.; J. Hall; and J. Usher. "Integration of Evolutionary Programming and Simulation to Optimize a Pull Production System." *Computers and Industrial Engineering* 31, no. 1/2 (1996), pp. 217–20.

Bowden, R.; R. Neppalli; and A. Calvert. "A Robust Method for Determining Good Combinations of Queue Priority Rules." *Fourth International Industrial Engineering Research Conference,* Nashville, TN, ed. Schmeiser, B. and R. Uzsoy. Norcross, GA: IIE, 1995, pp. 874–80.

Brady, T. F., and R. O. Bowden. "The Effectiveness of Generic Optimization Modules in Computer Simulation Languages." *Proceedings of the Tenth Annual Industrial Engineering Research Conference,* CD-ROM, 2001.

Fogel, D. "Evolving Artificial Intelligence." Ph.D. Thesis, University of California, 1992.

Glover, F. "Tabu Search: A Tutorial." *Interfaces,* 20, no. 4 (1990), pp. 41–52.

Glover, F. "Genetic Algorithms and Scatter Search: Unsuspected Potential," *Statistics and Computing* 4 (1994), pp. 131–40.

Glover, F.; P. Kelly, and M. Laguna. "New Advances and Applications of Combining Simulation and Optimization." *1996 Winter Simulation Conference,* Coronado, CA: ed. Charnes, J., Morrice, D., Brunner, D., and J. Swain. Piscataway, NJ: Institute of Electrical and Electronics Engineers, 1996, pp. 144–52.

Goldberg, D. *Genetic Algorithms in Search, Optimization, and Machine Learning.* Reading, MA: Addison Wesley, 1989.

Greenwood, A.; L. Rees; and I. Crouch. "Separating the Art and Science of Simulation Optimization: A Knowledge-Based Architecture Providing for Machine Learning." *IIE Transactions* 25, no. 6 (1993), pp. 70–83.

Hall, J., and R. Bowden. "Simulation Optimization for a Manufacturing Problem." *Proceedings of the Southeastern Simulation Conference,* Huntsville, AL, ed. J. Gauthier. San Diego, CA: The Society for Computer Simulation, 1996, pp. 135–40.

Hall, J.; R. Bowden; and J. Usher. "Using Evolution Strategies and Simulation to Optimize a Pull Production System." *Journal of Materials Processing Technology* 61 (1996), pp. 47–52.

Hooke, R., and T. Jeeves. "Direct Search Solution of Numerical Statistical Problems." *Journal of the Association of Computer Machines* 8 (1961), pp. 212–29.

Jacobson, S., and L. Schruben. "Techniques for Simulation Response Optimization." *Operations Research Letters* 8 (1989), pp. 1–9.

Lacksonen, T., and P. Anussornnitisarn. "Empirical Comparison of Discrete Event Simulation Optimization Techniques." *Proceedings of the Summer Computer Simulation Conference,* Ottawa, Canada. San Diego, CA: The Society for Computer Simulation, 1995, pp. 69–101.

Law, A., and W. Kelton. *Simulation Modeling and Analysis.* New York: McGraw-Hill, 2000.

Markt, P., and M. Mayer. "Witness Simulation Software: A Flexible Suite of Simulation Tools." *1997 Winter Simulation Conference,* Atlanta, GA, ed. Andradóttir, S., K. Healy, D. Withers and B. Nelson. Piscataway, NJ: Institute of Electrical and Electronics Engineers, 1997, pp. 127–33.

Monden, Y. *Toyota Production System.* Norcross, GA: IIE, 1993.

Nelder, J., and R. Mead. "A Simplex Method for Function Minimization." *Computer Journal* 7 (1965), pp. 308–13.

OptQuest 96 Information Sheet. Boulder, CO: Micro Analysis and Design, Inc., 1996.

Pegden, D., and M. Gately. "Decision Optimization for GASP IV Simulation Models." *1977 Winter Simulation Conference,* Gaitherburg, MD, ed. Highland, H. R. Sargent and J. Schmidt. Piscataway, NJ: Institute of Electrical and Electronics Engineers, 1977, pp. 127–33.

Pegden, D., and M. Gately. "A Decision-Optimization Module for SLAM." *Simulation* 34 (1980), pp. 18–25.

Safizadeh, M. "Optimization in Simulation: Current Issues and the Future Outlook." *Naval Research Logistics* 37 (1990), pp. 807–25.

Schwefel, H.-P. *Numerical Optimization of Computer Models.* Chichester: John Wiley & Sons, 1981.

SimRunner Online Software Help. Ypsilanti, MI: Decision Science, Inc., 1996b.

SimRunner User's Guide ProModel Edition. Ypsilanti, MI: Decision Science, Inc., 1996a.

Stuckman, B.; G. Evans; and M. Mollaghasemi. "Comparison of Global Search Methods for Design Optimization Using Simulation." *1991 Winter Simulation Conference,* Phoenix, AZ, ed. Nelson, B., W. Kelton and G. Clark. Piscataway, NJ: Institute of Electrical and Electronics Engineers, 1991, pp. 937–44.

Tompkins, G., and F. Azadivar. "Genetic Algorithms in Optimizing Simulated Systems." *1991 Winter Simulation Conference,* Arlington, VA, ed. Alexopoulos, C., K. Kang, W. Lilegdon, and D. Goldsman. Piscataway, NJ: Institute of Electrical and Electronics Engineers, 1995, pp. 757–62.

Usher, J., and R. Bowden. "The Application of Genetic Algorithms to Operation Sequencing for Use in Computer-Aided Process Planning." *Computers and Industrial Engineering Journal* 30, no. 4 (1996), pp. 999–1013.

12   MODELING MANUFACTURING SYSTEMS

"We no longer have the luxury of time to tune and debug new manufacturing systems on the floor, since the expected economic life of a new system, before major revision will be required, has become frighteningly short."

—Conway and Maxwell

12.1 Introduction

In Chapter 7 we discussed general procedures for modeling the basic operation of manufacturing and service systems. In this chapter we discuss design and operating issues that are more specific to manufacturing systems. Different applications of simulation in manufacturing are presented together with how specific manufacturing issues are addressed in a simulation model. Most manufacturing systems have material handling systems that, in some instances, have a major impact on overall system performance. We touch on a few general issues related to material handling systems in this chapter; however, a more thorough treatment of material handling systems is given in Chapter 13.

Manufacturing systems are processing systems in which raw materials are transformed into finished products through a series of operations performed at workstations. In the rush to get new manufacturing systems on line, engineers and planners often become overly preoccupied with the processes and methods without fully considering the overall coordination of system components.

Many layout and improvement decisions in manufacturing are left to chance or are driven by the latest management fad with little knowledge of how much improvement will result or whether a decision will result in any improvement at all. For example, work-in-process (WIP) reduction that is espoused by just-in-time (JIT) often disrupts operations because it merely uncovers the rocks (variability, long setups, or the like) hidden beneath the inventory water level that necessitated the WIP in the first place. To accurately predict the effect of lowering WIP levels requires sonar capability. Ideally you would like to identify and remove production rocks before arbitrarily lowering inventory levels and exposing production to these hidden problems. "Unfortunately," note Hopp and Spearman (2001), "JIT, as described in the American literature, offers neither sonar (models that predict the effects of system changes) nor a sense of the relative economics of level reduction versus rock removal."

Another popular management technique is the *theory of constraints*. In this approach, a constraint or bottleneck is identified and a best-guess solution is implemented, aimed at either eliminating that particular constraint or at least mitigating the effects of the constraint. The implemented solution is then evaluated and, if the impact was underestimated, another solution is attempted. As one manufacturing manager expressed, "Contraint-based management can't quantify investment justification or develop a remedial action plan" (Berdine 1993). It is merely a trial-and-error technique in which a best-guess solution is implemented with the hope that enough improvement is realized to justify the cost of the solution. Even Deming's plan–do–check–act (PDCA) cycle of process improvement implicitly prescribes checking performance *after* implementation. What the PDCA cycle lacks is an evaluation step to test or simulate the plan before it is implemented. While eventually leading to a better solution, this trial-and-error approach ends up being costly, time-consuming, and disruptive to implement.

In this chapter we address the following topics:

- What are the special characteristics of manufacturing systems?
- What terminology is used to describe manufacturing operations?
- How is simulation applied to manufacturing systems?
- What techniques are used to model manufacturing systems?

12.2 Characteristics of Manufacturing Systems

Manufacturing systems represent a class of processing systems in which entities (raw materials, components, subassemblies, pallet or container loads, and so on) are routed through a series of workstations, queues, and storage areas. As opposed to service systems where the entities are usually human, entities in manufacturing systems are inanimate objects that have a more controlled entry and routing sequence. Entity production is frequently performed according to schedules to fill predefined quotas. In other instances, production is triggered by low finished goods inventories (make-to-stock production) or by customer orders (make/assemble-to-order production). Manufacturing systems utilize varying degrees of mechanization and automation for entity processing and material movement. In the most highly automated facilities, there may be very little, if any, human involvement to be considered in the simulation study. These characteristics have the following implications for modeling manufacturing systems:

- Operation times often have little, if any, variability.
- Entity arrivals frequently occur at set times or conditions.

- Routing sequences are usually fixed from the start.
- Entities are often processed in batches.
- Equipment reliability is frequently a key factor.
- Material handling can significantly impact entity flow.
- We are usually interested in the steady-state behavior of the system.

Of course, actual modeling considerations will vary depending on the type of system being modeled and the purpose of the simulation.

12.3 Manufacturing Terminology

Manufacturing systems share many common elements and issues. They also share a vernacular that can seem foreign to those unfamiliar with manufacturing activities. Understanding how to model manufacturing systems requires a knowledge of the terminology that is used in manufacturing industries. Here are some of the terms that are useful to know when simulating manufacturing systems:

Operation—An operation is the activity performed on a product at a workstation. Operations are usually performed to produce some physical change to a part such as machining or molding. Assembly operations combine component parts. Nontransformational operations such as inspection or testing are also frequently performed.

Workstation—A workstation or work center is the place or location where an operation is performed. A workstation consists of one or more machines and/or personnel.

NC machine—An NC (numerically controlled) machine is a machine tool whose spindle and table action are automatically controlled by a computer.

Machining center—A machining center is essentially an NC machine with a tool exchanger that provides greater capability than a stand-alone NC machine. Machining centers have been defined as "multifunctional, numerically controlled machine tools with automatic tool changing capabilities and rotating cutting tools that permit the programmed, automated production of a wide variety of parts" (Wick 1987).

Master production schedule—The master production schedule defines what end products are to be produced for a given period, such as the number of office tables to produce for a week.

Production plan—The production plan is a detailed schedule of production for each individual component comprising each end product.

Bottleneck—The bottleneck is traditionally thought of as the workstation that has the highest utilization or largest ratio of required processing time to time available. In a broader sense, a bottleneck may be any constraint or limiting factor (an operator, a conveyor system) in the production of goods.

Setup—Setup is the activity required to prepare a workstation to produce a different part type.

Job—Depending on the industry, a job may be any of the following:
- In a general sense, a *job* refers to the activity or task being performed.
- In mass production, a *job* refers to each individual part.
- In a job shop, a *job* refers to a customer order to be produced.

Machine cycle time—Machine cycle time is the time required to perform a single operation. If load and unload times are included, it is sometimes referred to as *floor-to-floor time*.

Capacity—Capacity refers to either holding capacity, as in the case of a tank or storage bin, or production capacity, as in the case of a machine. Production capacity is usually expressed in one of the following four ways:
- *Theoretical capacity* is the rate at which a machine is capable of producing if it were in constant production, assuming 100 percent efficiency.
- *Effective capacity* is the actual capacity factored by the reliability of the machine with allowances for load and unload, chip removal, and so forth.
- *Expected capacity* is the effective capacity of the machine factored by anticipated allowances for system influences such as shortages and blockages (this is what simulation can tell you).
- *Actual capacity* is what the machine actually produces.

Scrap rate—Scrap rate is the percentage of defective parts that are removed from the system following an operation. The opposite of scrap rate is *yield*.

Reliability—Reliability is usually measured in terms of mean time between failures (MTBF), which is the average time a machine or piece of equipment is in operation before it fails.

Maintainability—Maintainability is usually measured in terms of mean time to repair (MTTR) and is the average time required to repair a machine or piece of equipment whenever it fails.

Availability—Availability is the percentage of total scheduled time that a resource is actually available for production. Availability is uptime, which is a function of reliability and maintainability.

Preventive or scheduled maintenance—Preventive or scheduled maintenance is periodic maintenance (lubrication, cleaning, or the like) performed on equipment to keep it in running condition.

Unit load—A unit load is a consolidated group of parts that is containerized or palletized for movement through the system. The idea of a unit load is to minimize handling through consolidation and provide a standardized pallet or container as the movement item.

12.4 Use of Simulation in Manufacturing

Simulation has proved effective in helping to sort through the complex issues surrounding manufacturing decisions. Kochan (1986) notes that, in manufacturing systems,

> The possible permutations and combinations of workpieces, tools, pallets, transport vehicles, transport routes, operations, etc., and their resulting performance, are almost endless. Computer simulation has become an absolute necessity in the design of practical systems, and the trend toward broadening its capabilities is continuing as systems move to encompass more and more of the factory.

As companies move toward greater vertical and horizontal integration and look for ways to improve the entire value chain, simulation will continue to be an essential tool for effectively planning the production and delivery processes.

Simulation in manufacturing covers the range from real-time cell control to long-range technology assessment, where it is used to assess the feasibility of new technologies prior to committing capital funds and corporate resources. Figure 12.1 illustrates this broad range of planning horizons.

Simulations used to make short-term decisions usually require more detailed models with a closer resemblance to current operations than what would be found in long-term decision-making models. Sometimes the model is an exact replica of the current system and even captures the current state of the system. This is true in real-time control and detailed scheduling applications. As simulation is used for more long-term decisions, the models may have little or no resemblance to current operations. The model resolution becomes coarser, usually because higher-level decisions are being made and data are too fuzzy and unreliable that far out into the future.

Simulation helps evaluate the performance of alternative designs and the effectiveness of alternative operating policies. A list of typical design and operational decisions for which simulation might be used in manufacturing includes the following:

Design Decisions

1. What type and quantity of machines, equipment, and tooling should be used?

2. How many operating personnel are needed?

3. What is the production capability (throughput rate) for a given configuration?

FIGURE 12.1

Decision range in manufacturing simulation.

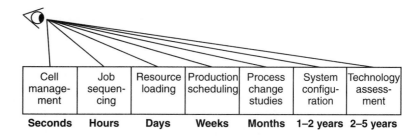

Cell manage-ment	Job sequen-cing	Resource loading	Production scheduling	Process change studies	System configu-ration	Technology assess-ment
Seconds	**Hours**	**Days**	**Weeks**	**Months**	**1–2 years**	**2–5 years**

4. What type and size of material handling system should be used?
5. How large should buffer and storage areas be?
6. What is the best layout of the factory?
7. What automation controls work the best?
8. What is the optimum unit load size?
9. What methods and levels of automation work best?
10. How balanced is a production line?
11. Where are the bottlenecks (bottleneck analysis)?
12. What is the impact of machine downtime on production (reliability analysis)?
13. What is the effect of setup time on production?
14. Should storage be centralized or localized?
15. What is the effect of vehicle or conveyor speed on part flow?

Operational Decisions

1. What is the best way to schedule preventive maintenance?
2. How many shifts are needed to meet production requirements?
3. What is the optimum production batch size?
4. What is the optimum sequence for producing a set of jobs?
5. What is the best way to allocate resources for a particular set of tasks?
6. What is the effect of a preventive maintenance policy as opposed to a corrective maintenance policy?
7. How much time does a particular job spend in a system (makespan or throughput time analysis)?
8. What is the best production control method (kanban, MRP, or other) to use?

In addition to being used for specific design or operational decisions, simulation has been used to derive general design rules so that the most cost-effective decisions can be made for a range of operating conditions. One study, for example, used simulation to estimate how the output capacity of a manufacturing system changes as the various components of the system are increased or improved for a known cost. These data were then used to develop cost versus output curves for a given part or product mix. By comparing these curves for different types of production systems, it was possible to identify break-even points where a particular type of system was economically preferred over other systems for particular products or product mixes (Renbold and Dillmann 1986).

12.5 Applications of Simulation in Manufacturing

There are numerous applications for simulation in manufacturing. Many of them are the same as those found in any type of system, such as capacity analysis and cycle time reduction. Applications that tend to be more specific to manufacturing include

- Methods analysis.
- Plant layout.
- Batch sizing.
- Production control.
- Inventory control.
- Supply chain planning.
- Production scheduling.
- Real-time control.
- Emulation.

Each of these application areas is discussed here.

12.5.1 Methods Analysis

Methods analysis looks at alternative ways of processing and handling material. An assembly process, for example, can be performed all at a single operation with several parallel stations or be broken down into multiple stages performed on a single assembly line. Production and assembly lines may be either paced, in which product movement occurs at a fixed rate with the operator keeping pace, or unpaced, in which the rate of flow is determined by the speed of the worker. The relative performance of these alternative methods of production can be easily demonstrated using simulation.

One of the decisions that addresses manufacturing methods pertains to the type and level of automation to use in the factory. Simulation has been especially helpful in designing automated systems to safeguard against suboptimal performance. As noted by Glenney and Mackulak (1985), "Computer simulation will permit evaluation of new automation alternatives and eliminate the problem of building a six-lane highway that terminates onto a single-lane dirt road." Often an unbalanced design occurs when the overall performance is overlooked because of preoccupation with a single element. "Islands of automation" that are not properly integrated and balanced lead to suboptimal performance. For example, an automated machine capable of producing 200 parts per hour fed by a manual handler at the rate of 50 parts per hour will produce only 50 parts per hour. The expression that "a chain is no stronger than its weakest link" has particular application to automated manufacturing systems.

There are three types of automation to consider in manufacturing: operation automation, material handling automation, and information automation. The degree of automation ranges from manual to fully automatic. Figure 12.2 shows the types and degrees of automation.

Although information flow is frequently overlooked in manufacturing simulations, it may be a contributing factor to the overall system performance and therefore may merit some analysis. As noted in a *Manufacturing Engineering* article (1993), "It's easier to make the product than to adjust the paper and data flow required to track the product."

FIGURE 12.2

Axes of automation.

Source: Adapted from
G. Boehlert and W. J.
Trybula, "Successful Factory
Automation," *IEEE
Transactions,* September
1984, p. 218.

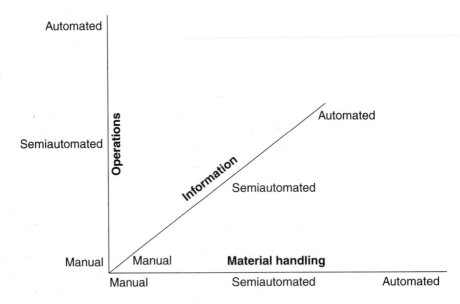

Automation itself does not solve performance problems. If automation is unsuited for the application, poorly designed, or improperly implemented and integrated, then the system is not likely to succeed. The best approach to automation is to first simplify, then systematize, and finally automate. Some companies find that after simplifying and systematizing their processes, the perceived need to automate disappears.

12.5.2 Plant Layout

With the equipment-intensive nature of manufacturing systems, plant layout is important to the smooth flow of product and movement of resources through the facility. A good layout results in a streamlined flow with minimal movement and thus minimizes material handling and storage costs. Even existing systems can benefit from a layout analysis. Layouts tend to evolve haphazardly over time with little thought given to work flow. This results in nonuniform, multidirectional flow patterns that cause management and material handling problems. Simulation helps to identify inefficient flow patterns and to create better layouts to economize the flow of material.

In planning the layout of a manufacturing system, it is important to have a total system perspective of material flow. Flow within the manufacturing operation is often hierarchical, beginning with the overall flow from receiving to manufacturing to shipping. At the lowest level, parts might flow from an input buffer to a machine (see Figure 12.3). Any or all of the levels shown in Figure 12.3 may be present within a plant. Interaction points occur between levels as well as within each level. When modeling material flow in a factory it is important to consider the level of flow and how flows interact.

FIGURE 12.3

Material flow system.

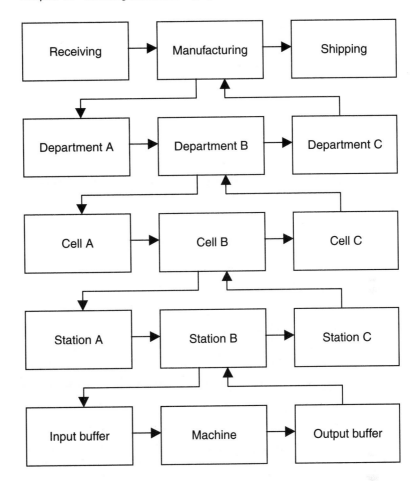

In laying out equipment in a facility, the designer must choose between a process layout, a product layout, and a part family layout of machines. In a *process layout,* machines are arranged together based on the process they perform. Machines performing like processes are grouped together. In a *product layout,* machines are arranged according to the sequence of operations that are performed on a product. This results in a serial flow of the product from machine to machine. In a *part family layout,* machines are arranged to process parts of the same family. Part family layouts are also called *group technology cells* because group technology is used to group parts into families. Figure 12.4 illustrates these three different types of arrangements and their implications on material flow.

Deciding which layout to use will depend largely on the variety and lot size of parts that are produced. If a wide variety of parts are produced in small lot sizes, a process layout may be the most appropriate. If the variety is similar enough in processing requirements to allow grouping into part families, a cell layout is best. If the variety is small enough and the volume is sufficiently large, a

FIGURE 12.4

Comparison between (a) process layout, (b) product layout, and (c) cell layout.

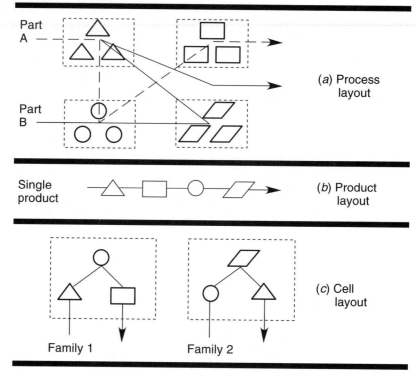

product layout is best. Often a combination of topologies is used within the same facility to accommodate mixed requirements. For example, a facility might be predominately a job shop with a few manufacturing cells. In general, the more the topology resembles a product layout where the flow can be streamlined, the greater the efficiencies that can be achieved.

Perhaps the greatest impact simulation can have on plant layout comes from designing the process itself, before the layout is even planned. Because the process plan and method selection provide the basis for designing the layout, having a well-designed process means that the layout will likely not require major changes once the best layout is found for the optimized process.

12.5.3 Batch Sizing

Batching decisions play a major role in meeting the flow and efficiency goals of a production facility. A *batch* or *lot* of parts refers to a quantity of parts grouped together for some particular purpose. Typically three different batch types are spoken of in manufacturing: production batch, move batch, and process batch.

The *production batch,* also referred to as the *production lot,* consists of all of the parts of one type that begin production before a new part type is introduced to the system. The *move* or *transfer batch* is the collection of parts that are grouped

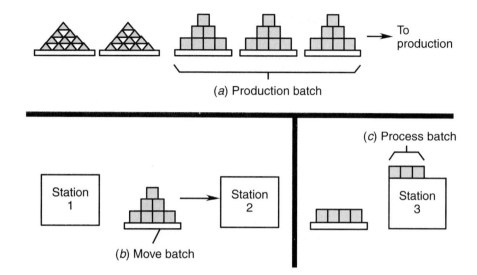

FIGURE 12.5

Illustration of
(a) production batch,
(b) move batch, and
(c) process batch.

and moved together from one workstation to another. A production batch is often broken down into smaller move batches. This practice is called *lot splitting*. The move batch need not be constant from location to location. In some batch manufacturing systems, for example, a technique called *overlapped production* is used to minimize machine idle time and reduce work in process. In overlapped production, a move batch arrives at a workstation where parts are individually processed. Then instead of accumulating the entire batch before moving on, parts are sent on individually or in smaller quantities to prevent the next workstation from being idle while waiting for the entire batch. The *process batch* is the quantity of parts that are processed simultaneously at a particular operation and usually consists of a single part. The relationship between these batch types is illustrated in Figure 12.5.

Deciding which size to use for each particular batch type is usually based on economic trade-offs between in-process inventory costs and economies of scale associated with larger batch sizes. Larger batch sizes usually result in lower setup costs, handling costs, and processing costs. Several commands are provided in ProModel for modeling batching operations such as GROUP, COMBINE, JOIN and LOAD.

12.5.4 Production Control

Production control governs the flow of material between individual workstations. Simulation helps plan the most effective and efficient method for controlling material flow. Several control methods exist, any of which may work well depending on the situation. Common methods include

- Push control.
- Pull control.
- Drum–buffer–rope (DBR) control.

Push Control

A *push system* is one in which production is driven by workstation capacity and material availability. Each workstation seeks to produce as much as it can, pushing finished work forward to the next workstation. In cases where the system is unconstrained by demand (the demand exceeds system capacity), material can be pushed without restraint. Usually, however, there is some synchronization of push with demand. In make-to-stock production, the triggering mechanism is a drop in finished goods inventory. In make-to-order or assemble-to-order production, a master production schedule drives production by scheduling through a material requirements planning (MRP) or other backward or forward scheduling system.

MRP systems determine how much each station should produce for a given period. Unfortunately, once planned, MRP is not designed to respond to disruptions and breakdowns that occur for that period. Consequently, stations continue to produce inventory as planned, regardless of whether downstream stations can absorb the inventory. Push systems such as those resulting from MRP tend to build up work in process (WIP), creating high inventory-carrying costs and long flow times.

Pull Control

At the other extreme of a push system is a *pull system,* in which downstream demand triggers each preceding station to produce a part with no more than one or two parts at a station at any given time (see Figure 12.6). Pull systems are often associated with the just-in-time (JIT) or lean manufacturing philosophy, which advocates the reduction of inventories to a minimum. The basic principle of JIT is to continuously reduce scrap, lot sizes, inventory, and throughput time as well as eliminate the waste associated with non–value added activities such as material handling, storage, machine setup, and rework.

FIGURE 12.6

Push versus pull system.

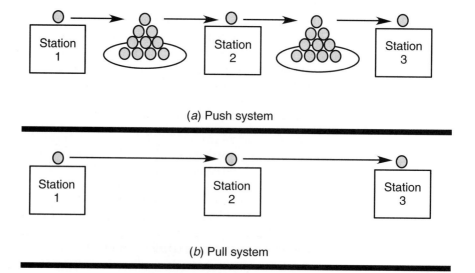

(*a*) Push system

(*b*) Pull system

A pull system is generally implemented using a method the Japanese call *kanban* (translated as *ticket*). When an operation is completed at the final location and exits the location, a new entity or part is pulled from the preceding WIP buffer. A kanban card accompanies each part bin in the buffer and is returned to the preceding station when the bin moves on. The returned card signals the preceding operation to pull a unit from its own WIP and replenish the kanban. A kanban system is implemented using either a *withdrawal* kanban, authorizing the retrieval of work items, or a *production* kanban, authorizing the production of more parts.

When modeling a kanban control system, it usually isn't necessary to explicitly model the cards or actual triggering mechanism used (some systems use golf balls, bins, or even a painted area on the floor). For most JIT systems where flow is serial, a kanban system is easily modeled by simply limiting the capacity of the workstation input and output buffers. A two-bin kanban system, for example, can be modeled by limiting the input buffer capacity of the workstation to two. Ultimately, a kanban system is just a way to limit queue capacities.

> Kanban is a queue limiter; that is, it tightly links a provider's service with a user's need. Thus, it limits the queue of work in front of the provider (the quantity/amount of items waiting and the time each item waits). Queue limitation replaces lax flow control with discipline, which results in customer satisfaction and competitive advantage (Schonberger and Knod 1994).

Because ProModel uses capacity limitations to control the flow of entities, entities will not advance to the next station until there is available capacity at that location. This effectively provides a built-in pull system in which the emptying of a buffer or workstation signals the next waiting entity to advance (push systems actually work the same way, only buffers effectively have infinite capacity). Several papers have been published (Mitra and Mitrani 1990; Wang and Wang 1990) showing the use of limited buffers to model JIT systems.

For certain pull systems, triggers signal the release of WIP further upstream than just one station. An example of this is another form of pull control called CONWIP (Hopp and Spearman 2001). A CONWIP (CONstant Work-In-Process) system is a form of pull system that pulls from the final operation. The completion of a part triggers the release of a new part to the beginning of the line. A CONWIP system limits work in process just like a kanban system. A new part is introduced to the system only when a part leaves the system, thus maintaining a constant inventory level. To simulate this process in ProModel, an ORDER statement should be used whenever an entity exits the system. This signals the creation of a new entity at the beginning of the line so that each part that exits the system releases a new part into the system. CONWIP has the advantage over kanban of being easier to administer while still achieving nearly the same, and sometimes even superior, results.

Drum–Buffer–Rope (DBR) Control

One approach to production control utilizes a drum–buffer–rope (DBR) metaphor. This technique is based on the theory of constraints and seeks to improve productivity by managing the bottleneck operation. The DBR approach basically lets the bottleneck operation pace the production rate by acting as a drum. The buffer is the inventory needed in front of the bottleneck operation to ensure that it is never starved for parts. The rope represents the tying or synchronizing of the production release rate to the rate of the bottleneck operation.

As noted earlier in this chapter, a weakness in constraint-based management is the inability to predict the level of improvement that can be achieved by eliminating a bottleneck. Simulation can provide this evaluative capability missing in constraint-based management. It can also help detect floating bottlenecks for situations where the bottleneck changes depending on product mix and production rates. Simulating DBR control can help determine where bottlenecks are and how much buffer is needed prior to bottleneck operations to achieve optimum results.

12.5.5 Inventory Control

Inventory control involves the planning, scheduling, and dispatching of inventory to support the production activity. Because of the high dollar impact of inventory decisions, developing and maintaining a good inventory control system are critical. Some common goals associated with inventory control at all stages in the supply chain include

- Responding promptly to customer demands.
- Minimizing order and setup costs associated with supplying inventory.
- Minimizing the amount of inventory on hand as well as all associated inventory carrying costs.

Decisions affecting inventory levels and costs include

- How much inventory should be ordered when inventory is replenished (the economic order quantity)?
- When should more inventory be ordered (the reorder point)?
- What are the optimum inventory levels to maintain?

There are two common approaches to deciding when to order more inventory and how much: the reorder point system and the MRP system. A reorder point system is an inventory control system where inventory for a particular item is replenished to a certain level whenever it falls below a defined level. MRP, on the other hand, is a bit more complicated in that it is time phased. This means that inventory is ordered to match the timing of the demand for products over the next planning horizon.

Inventory management must deal with random depletion or usage times and random lead times, both of which are easily handled in simulation. To model an inventory control system, one need only define the usage pattern, the demand pattern, and the usage policies (priorities of certain demands, batch or incremental commitment of inventories, and so on). The demand rate can be modeled using distributions of interarrival times. Replenishment of raw materials should be made using an ORDER command that is able to generate additional inventory when it drops below a particular level.

Running the simulation can reveal the detailed rises and drops in inventory levels as well as inventory summary statistics (average levels, total stockout time, and so forth). The modeler can experiment with different replenishment strategies and usage policies until the best plan is found that meets the established criteria.

Using simulation modeling over traditional, analytic modeling for inventory planning provides several benefits (Browne 1994):

- *Greater accuracy* because actual, observed demand patterns and irregular inventory replenishments can be modeled.
- *Greater flexibility* because the model can be tailored to fit the situation rather than forcing the situation to fit the limitations of an existing model.
- *Easier to model* because complex formulas that attempt to capture the entire problem are replaced with simple arithmetic expressions describing basic cause-and-effect relationships.
- *Easier to understand* because demand patterns and usage conditions are more descriptive of how the inventory control system actually works. Results reflect information similar to what would be observed from operating the actual inventory control system.
- *More informative output* that shows the dynamics of inventory conditions over time and provides a summary of supply, demand, and shortages.
- *More suitable for management* because it provides "what if" capability so alternative scenarios can be evaluated and compared. Charts and graphs are provided that management can readily understand.

Centralized versus Decentralized Storage.   Another inventory-related issue has to do with the positioning of inventory. Traditionally, inventory was placed in centralized locations for better control. Unfortunately, centralized inventory creates excessive handling of material and increased response times. More recent trends are toward decentralized, point-of-use storage, with parts kept where they are needed. This eliminates needless handling and dramatically reduces response time. Simulation can effectively assess which method works best and where to strategically place storage areas to support production requirements.

12.5.6 Supply Chain Management

Supply chain management is a key area for improving efficiencies in a corporation. A supply chain is made up of suppliers, manufacturers, distributors, and customers. There are two main goals with respect to supply chain management: to reduce costs and to improve customer service, which is achieved by reducing delivery time. Supply chain network analysis tools exist that can look at the problem in the aggregate and help identify optimum site locations and inventory levels given a general demand. Simulation, on the other hand, can look at the system on a more detailed level, analyzing the impact of interdependencies and variation over time. Simulation "accurately traces each transaction through the network at a detailed level, incorporating cost and service impacts of changes in inventory policies, shipment policies, uncertainty in demand, and other time-dependent factors" (Napolitano 1997).

Simulation of flow between facilities is similar to the simulation of flow within a facility. One of the most difficult challenges in modeling a supply chain is making the flow of entities in the model demand-driven. This is done by using the ORDER statement, with entities representing the orders waiting to be filled. Often order or customer consolidation must also be taken into account, which requires some data management using variables and attributes.

12.5.7 Production Scheduling

Production scheduling determines the start and finish times for jobs to be produced. The use of simulation for production scheduling is called *simulation-based scheduling*. Some scheduling is static and determined ahead of time. Other scheduling is performed dynamically, such as sequencing a set of jobs currently waiting to be processed at a particular machine. Simulation can help make both types of scheduling decisions and can assess any scheduling rules defined by the modeler. Once the simulation is run, a report can be generated showing the start and finish times of each job on each machine. Simulation does not set the schedule itself, but rather reports the schedule resulting from a given set of operating rules and conditions.

Job sequencing is a frequent challenge for job shops, which tend to be very due-date-driven in their scheduling and are often faced with the question "In what sequence should jobs be processed so as to minimize lateness or tardiness?" Suppose, for example, that three jobs are waiting to be processed, each requiring the same resources but in a different sequence. If the goal is to process the jobs in an order that will minimize total production time yet still permit on-time delivery, several simulation runs can be made just before production to find the optimal sequence in which the jobs should be processed.

In scheduling, we usually try to achieve four principal goals or objectives (Kiran and Smith 1984):

1. Minimize job lateness or tardiness.
2. Minimize the flow time or time jobs spend in production.
3. Maximize resource utilization.
4. Minimize production costs.

The decision of which job to process next is usually based one of the following rules:

Rule	Definition
Shortest processing time	Select the job having the least processing time.
Earliest due date	Select the job that is due the soonest.
First-come, first-served	Select the job that has been waiting the longest for this workstation.
First-in-system, first-served	Select the job that has been in the shop the longest.
Slack per remaining operations	Select the job with the smallest ratio of slack to operations remaining to be performed.

Traditional scheduling methods fail to account for fluctuations in quantities and resource capacity. This variation can be built into simulation models. However, in using simulation for dynamic scheduling, we are dealing with short time periods and are therefore less concerned about long-term statistical fluctuations in the system such as machine downtimes. In such instances, typically a deterministic simulation is used that is based on expected operation times. Models built for simulation-based scheduling have the following characteristics:

- The model captures an initial state.
- Operation times are usually based on expected times.
- Anomaly conditions (machine failures and the like) are ignored.
- The simulation is run only until the required production has been met.

Simulation can work with material requirements planning (MRP), advanced planning and scheduling (APS), and manufacturing execution systems (MES) to simulate the plans generated by these systems in order to test their viability. It can also produce a detailed schedule showing which resources were used and at what times. This information can then be used for resource scheduling.

It usually isn't practical to simulate an entire MRP system because the number of different entity types becomes too unmanageable. You should consider building a model based only on critical product components. You can forget the small inexpensive items like nuts and bolts, which are usually managed using a re-order point system, and focus on the few components that constitute the major scheduling challenge. This could provide a reasonably simple model that could be used regularly to test MRP-generated requirements. Such a model could then be used to achieve true, closed-loop MRP, in which production schedules could be iteratively generated and evaluated until the best schedule was found for the available capacity of the system.

12.5.8 Real-Time Control

During actual production, simulation has been integrated with manufacturing cells to perform real-time analysis such as next task selection and dynamic routing decisions. Initially, a model is built to operate with a simulated version of the cell, and then the simulated cell components are replaced with the actual cell components. Three benefits of using the simulation for cell management are

- Logic used in the simulator does not need to be recoded for the controller.
- Animation capabilities allow processes to be monitored on the layout.
- Statistical-gathering capabilities of simulation automatically provide statistics on selected measures of performance.

12.5.9 Emulation

A special use of simulation in manufacturing, particularly in automated systems, has been in the area of hardware emulation. As an emulator, simulation takes inputs from the actual control system (such as programmable controllers or microcomputers), mimics the behavior that would take place in the actual system, and then provides feedback signals to the control system. The feedback signals are synthetically created rather than coming from the actual hardware devices.

In using simulation for hardware emulation, the control system is essentially plugged into the model instead of the actual system. The hardware devices are then emulated in the simulation model. In this way simulation is used to test, debug, and even refine the actual control system before any or all of the hardware has been installed. Emulation can significantly reduce the time to start up new systems and implement changes to automation.

Emulation has the same real-time requirements as when simulation is used for real-time control. The simulation clock must be synchronized with the real clock to mimic actual machining and move times.

12.6 Manufacturing Modeling Techniques

In Chapter 7 we presented modeling techniques that apply to a broad range of systems, including manufacturing systems. Here we discuss a few modeling techniques that are more specific to manufacturing systems.

12.6.1 Modeling Machine Setup

Manufacturing systems often consist of processing equipment that has particular settings when working on a particular job. A machine setup or changeover occurs when tooling or other machine adjustments must be made to accommodate the next item to be processed. Machine setups usually occur only when there is a change of part type. For example, after running a lot of 200 parts of type A, the machine is retooled and adjusted for producing a lot of part type B. In some situations, adjustments must be made between processing parts of the same type. When setup is performed between different part types, the time and resources required to perform the setup may depend on both the part type of the new lot and the part type of the previous lot. For example, the setup time when changing over from part A to part B may be different than the setup time when changing over from part B to part C.

Setups are easily modeled in ProModel by selecting the Setup option under the Downtime menu for a location record. Setups can be defined to occur after every part or only after a change of part type.

12.6.2 Modeling Machine Load and Unload Time

Sometimes a key issue in modeling manufacturing systems is machine load and unload time. There are several approaches to dealing with load/unload activities:

- *Ignore them.* If the times are extremely short relative to the operation time, it may be safe to ignore load and unload times.
- *Combine them with the operation time.* If loading and unloading the machine are simply an extension of the operation itself, the load and

unload times can be tacked onto the operation time. A precaution here is in semiautomatic machines where an operator is required to load and/or unload the machine but is not required to operate the machine. If the operator is nearly always available, or if the load and unload activities are automated, this may not be a problem.

- *Model them as a movement or handling activity.* If, as just described, an operator is required to load and unload the machine but the operator is not always available, the load and unload activities should be modeled as a separate move activity to and from the location. To be accurate, the part should not enter or leave the machine until the operator is available. In ProModel, this would be defined as a routing from the input buffer to the machine and then from the machine to the output buffer using the operator as the movement resource.

12.6.3 Modeling Rework and Scrap

It is not uncommon in manufacturing systems to have defective items that require special treatment. A defective item is usually detected either at an inspection station or at the station that created the defect. Sometimes part defects are discovered at a later downstream operation where, for example, the part can't be assembled. In this case, the downstream workstation doubles as a sort of inspection station. When a defective item is encountered, a decision is usually made to

- *Reject the part* (in which case it exits the system). This is done by simply defining a probabilistic routing to EXIT.
- *Correct the defect at the current station.* To model this approach, simply add an additional WAIT statement to the operation logic that is executed the percentage of the time defects are expected to occur.
- *Route the part to an offline rework station.* This is modeled using a probabilistic routing to the rework station.
- *Recycle the part through the station that caused the defect.* This is also modeled using a probabilistic routing back to the station that caused the defect. Sometimes an entity attribute needs to be used to designate whether this is a first pass part or a rework part. This is needed when reworked parts require a different operation time or if they have a different routing.

The first consideration when modeling rework operations is to determine whether they should be included in the model at all. For example, if a part rejection rate is less than 1 percent, it may not be worth including in the model.

12.6.4 Modeling Transfer Machines

Transfer machines are machines that have multiple serial stations that all transfer their work forward in a synchronized fashion. Transfer machines may be either rotary or in-line. On a *rotary transfer machine,* also called a *dial-index machine,* the load and unload stations are located next to each other and parts are usually

FIGURE 12.7

Transfer line system.

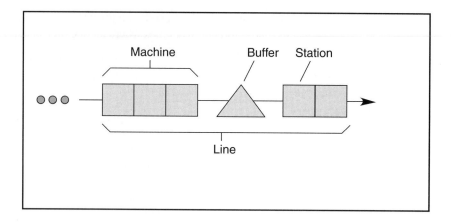

placed directly on the station fixture. *In-line transfer machines* have a load station at one end and an unload station at the other end. Some in-line transfer machines are coupled together to form a transfer line (see Figure 12.7) in which parts are placed on system pallets. This provides buffering (nonsynchronization) and even recirculation of pallets if the line is a closed loop.

Another issue is finding the optimum number of pallets in a closed, nonsynchronous pallet system. Such a system is characterized by a fixed number of pallets that continually recirculate through the system. Obviously, the system should have enough pallets to at least fill every workstation in the system but not so many that they fill every position in the line (this would result in a gridlock). Generally, productivity increases as the number of pallets increases up to a certain point, beyond which productivity levels off and then actually begins to decrease. Studies have shown that the optimal point tends to be close to the sum of all of the workstation positions plus one-half of the buffer pallet positions.

A typical analysis might be to find the necessary buffer sizes to ensure that the system is unaffected by individual failures at least 95 percent of the time. A similar study might find the necessary buffer sizes to protect the operation against the longest tool change time of a downstream operation.

Stations may be modeled individually or collectively, depending on the level of detail required in the model. Often a series of stations can be modeled as a single location. Operations in a transfer machine can be modeled as a simple operation time if an entire machine or block of synchronous stations is modeled as a single location. Otherwise, the operation time specification is a bit tricky because it depends on all stations finishing their operation at the same time. One might initially be inclined simply to assign the time of the slowest operation to every station. Unfortunately, this does not account for the synchronization of operations. Usually, synchronization requires a timer to be set up for each station that represents the operation for all stations. In ProModel this is done by defining an activated subroutine that increments a global variable representing the cycle completion after waiting for the cycle time. Entities at each station wait for the variable to be incremented using a WAIT UNTIL statement.

12.6.5 *Continuous Process Systems*

A continuous process system manufactures product, but in a different way than discrete part manufacturing systems. Continuous processing involves the production of bulk substances or materials such as chemicals, liquids, plastics, metals, textiles, and paper (see Figure 12.8).

Pritsker (1986) presents a simulation model in which the level of oil in a storage tank varies depending on whether oil is being pumped into or out of the tank. If the rate of flow into or out of the tank is constant, then the change in level is the rate of flow multiplied by the length of time. If, however, the rate of flow varies continuously, it is necessary to integrate the function defining the rate of flow over the time interval concerned (see Figure 12.9).

ProModel provides a tank construct and a set of routines for modeling systems in process industries. Both constant and variable flow rates can be defined. The routines provided are ideally suited for modeling combined discrete and continuous activities.

FIGURE 12.8

Continuous flow system.

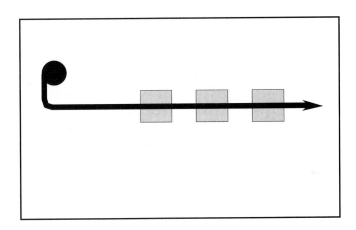

FIGURE 12.9

Quantity of flow when (a) rate is constant and (b) rate is changing.

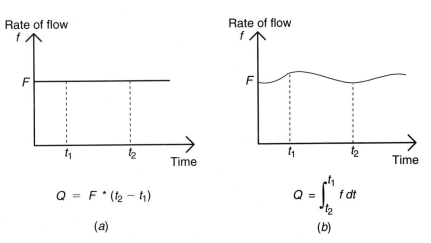

$$Q = F * (t_2 - t_1)$$

(a)

$$Q = \int_{t_2}^{t_1} f \, dt$$

(b)

12.7 Summary

In this chapter we focused on the issues and techniques for modeling manufacturing systems. Terminology common to manufacturing systems was presented and design issues were discussed. Different applications of simulation in manufacturing were described with examples of each. Issues related to modeling each type of system were explained and suggestions offered on how system elements for each system type might be represented in a model.

12.8 Review Questions

1. List four characteristics of manufacturing systems that make them different from most service systems in the way they are simulated.
2. What is the range of decision making for which simulation is used in manufacturing systems?
3. How can simulation help in planning the layout of a manufacturing facility?
4. In modeling automation alternatives, how can simulation reduce the risk of "building a six-lane highway that terminates into a single-lane dirt road"?
5. Conceptually, how would you model a pull system where the next downstream operation is pulling production? Where the completion of a final product triggers the start of a new product?
6. What are the characteristics of a simulation model that is used for production scheduling?
7. How can emulation help test an actual control system?
8. Why would it be invalid to model a synchronous transfer machine by simply giving all of the stations the same operation time?
9. In ProModel, how would you model an operation with an 85 percent yield where defective items are routed to a rework station?
10. In ProModel, how would you model a system where items are moved in batches of 50, but processed individually at each station?

References

Askin, Ronald G., and Charles R. Standridge. *Modeling and Analysis of Manufacturing Systems.* New York: John Wiley & Sons, 1993.

Berdine, Robert A. "FMS: Fumbled Manufacturing Startups?" *Manufacturing Engineering,* July 1993, p. 104.

Bevans, J. P. "First, Choose an FMS Simulator." *American Machinist,* May 1982, pp. 144–45.

Blackburn, J., and R. Millen. "Perspectives on Flexibility in Manufacturing: Hardware versus Software." In *Modelling and Design of Flexible Manufacturing Systems,* ed. Andrew Kusiak, pp. 99–109. Amsterdam: Elsevier, 1986.

Boehlert, G., and W. J. Trybula. "Successful Factory Automation." *IEEE Transactions,* September 1984, p. 218.

Bozer, Y. A., and J. A. White. "Travel-Time Models for Automated Storage/Retrieval Systems." *IIE Transactions,* December 1984, pp. 329–38.

Browne, Jim. "Analyzing the Dynamics of Supply and Demand for Goods and Services." *Industrial Engineering,* June 1994, pp. 18–19.

Cutkosky, M.; P. Fussell; and R. Milligan Jr. "The Design of a Flexible Machining Cell for Small Batch Production." *Journal of Manufacturing Systems* 3, no. 1 (1984), p. 39.

Fitzgerald, K. R. "How to Estimate the Number of AGVs You Need." *Modern Materials Handling,* October 1985, p. 79.

Gayman, David J. "Carving Out a Legend." *Manufacturing Engineering,* March 1988, pp. 75–79.

Glenney, N. E., and G. T. Mackulak. "Modeling and Simulation Provide Key to CIM Implementation Philosophy." *Industrial Engineering,* May 1985, p. 84.

Greene, James H. *Production and Inventory Control Handbook.* New York: McGraw-Hill, 1987.

Groover, Mikell P. *Automation, Production Systems, and Computer-Integrated Manufacturing.* Englewood Cliffs, NJ; Prentice Hall, 1987, pp. 119–29.

Hollier, R. H. "The Grouping Concept in Manufacture." *International Journal of Operations and Production Management,* 1980, pp. 1–71.

Hopp, Wallace J., and M. Spearman, *Factory Physics.* New York: Irwin/McGraw-Hill, 2001, p. 180.

Kiran, A. S., and M. L. Smith. "Simulation Studies in Job Shop Scheduling: A Survey." *Computers & Industrial Engineering,* vol 8(2). 1984, pp. 87–93.

Kochan, D. *CAM Developments in Computer Integrated Manufacturing.* Berlin: Springer-Verlag, 1986.

Langua, Leslie. "Simulation and Emulation." *Material Handling Engineering,* May 1997, pp. 41–43.

Law, A. M. "Introduction to Simulation: A Powerful Tool for Analyzing Complex Manufacturing Systems." *Industrial Engineering,* May 1986, pp. 57–58.

The Material Handling Institute. *Consideration for Planning and Installing an Automated Storage/Retrieval System.* Pittsburgh, PA: Material Handling Institute, 1977.

Miller, Richard Kendall. *Manufacturing Simulation.* Lilburn, GA: The Fairmont Press, 1990.

Miner, R. J. "Simulation as a CIM Planning Tool." In *Simulation,* ed. R. D. Hurrion, p. 73. Kempston, Bedford, UK: IFS Publications, 1987.

Mitra, Debasis, and Isi Mitrani. "Analysis of a Kanban Discipline for Cell Coordination in Production Lines." *Management Science* 36, no. 12 (1990), pp. 1548–66.

Napolitano, Maida. "Distribution Network Modeling" *IIE Solutions,* June 1997, pp. 20–25.

Noaker, Paula M. "Strategic Cell Design." *Manufacturing Engineering,* March 1993, pp. 81–84.

Pritsker, A. A. B. *Introduction to Simulation and Slam II.* West Lafayette, IN: Systems Publishing Corporation, 1986, p. 600.

Renbold, E. V., and R. Dillmann. *Computer-Aided Design and Manufacturing Methods and Tools.* Berlin: Springer-Verlag, 1986.

Schonberger, Richard J., and Edward M. Knod Jr. *Operations Management: Continuous Improvement.* 5th ed. Burr Ridge, IL: Richard D. Irwin, 1994.

Suri, R., and C. K. Whitney. "Decision Support Requirements in Flexible Manufacturing." *Journal of Manufacturing Systems* 3, no. 1 (1984).

Suzaki, Kiyoshi. *The New Manufacturing Challenge: Techniques for Continuous Improvement.* New York: Free Press, 1987.

Wang, Hunglin, and Hsu Pin Wang. "Determining the Number of Kanbans: A Step Toward Non-Stock Production." *International Journal of Production Research* 28, no. 11 (1990), pp. 2101–15.

Wick, C. "Advances in Machining Centers." *Manufacturing Engineering,* October 1987, p. 24.

Zisk, B. "Flexibility Is Key to Automated Material Transport System for Manufacturing Cells." *Industrial Engineering,* November 1983, p. 60.

13 MODELING MATERIAL HANDLING SYSTEMS

"Small changes can produce big results, but the areas of highest leverage are often the least obvious."
—Peter Senge

13.1 Introduction

Material handling systems utilize resources to move entities from one location to another. While material handling systems are not uncommon in service systems, they are found mainly in manufacturing systems. Apple (1977) notes that material handling can account for up to 80 percent of production activity. On average, 50 percent of a company's operation costs are material handling costs (Meyers 1993). Given the impact of material handling on productivity and operation costs, the importance of making the right material handling decisions should be clear.

This chapter examines simulation techniques for modeling material handling systems. Material handling systems represent one of the most complicated factors, yet, in many instances, the most important element, in a simulation model. Conveyor systems and automatic guided vehicle systems often constitute the backbone of the material flow process. Because a basic knowledge of material handling technologies and decision variables is essential to modeling material handling systems, we will briefly describe the operating characteristics of each type of material handling system.

13.2 Material Handling Principles

It is often felt that material handling is a necessary evil and that the ideal system is one in which no handling is needed. In reality, material handling can be a tremendous support to the production process if designed properly. Following is a list of 10 principles published by the Material Handling Institute as a guide to designing or modifying material handling systems:

1. *Planning principle:* The plan is the prescribed course of action and how to get there. At a minimum it should define what, where, and when so that the how and who can be determined.

2. *Standardization principle:* Material handling methods, equipment, controls, and software should be standardized to minimize variety and customization.

3. *Work principle:* Material handling work (volume × weight or count per unit of time × distance) should be minimized. Shorten distances and use gravity where possible.

4. *Ergonomic principle:* Human factors (physical and mental) and safety must be considered in the design of material handling tasks and equipment.

5. *Unit load principle:* Unit loads should be appropriately sized to achieve the material flow and inventory objectives.

6. *Space utilization principle:* Effective and efficient use should be made of all available (cubic) space. Look at overhead handling systems.

7. *System principle:* The movement and storage system should be fully integrated to form a coordinated, operational system that spans receiving, inspection, storage, production, assembly, packaging, unitizing, order selection, shipping, transportation, and the handling of returns.

8. *Automation principle:* Material handling operations should be mechanized and/or automated where feasible to improve operational efficiency, increase responsiveness, improve consistency and predictability, decrease operating costs, and eliminate repetitive or potentially unsafe manual labor.

9. *Environmental principle:* Environmental impact and energy consumption should be considered as criteria when designing or selecting alternative equipment and material handling systems.

10. *Life cycle cost principle:* A thorough economic analysis should account for the entire life cycle of all material handling equipment and resulting systems.

13.3 Material Handling Classification

With over 500 different types of material handling equipment to choose from, the selection process can be quite staggering. We shall not attempt here to describe or even list all of the possible types of equipment. Our primary interest is in modeling material handling systems, so we are interested only in the general categories of equipment having common operating characteristics. Material handling systems are traditionally classified into one of the following categories (Tompkins et al. 1996):

1. Conveyors.
2. Industrial vehicles.
3. Automated storage/retrieval systems.
4. Carousels.
5. Automatic guided vehicle systems.
6. Cranes and hoists.
7. Robots.

Missing from this list is "hand carrying," which still is practiced widely if for no other purpose than to load and unload machines.

13.4 Conveyors

A *conveyor* is a track, rail, chain, belt, or some other device that provides continuous movement of loads over a fixed path. Conveyors are generally used for high-volume movement over short to medium distances. Some overhead or towline systems move material over longer distances. Overhead systems most often move parts individually on carriers, especially if the parts are large. Floor-mounted conveyors usually move multiple items at a time in a box or container. Conveyor speeds generally range from 20 to 80 feet per minute, with high-speed sortation conveyors reaching speeds of up to 500 fpm in general merchandising operations.

Conveyors may be either gravity or powered conveyors. Gravity conveyors are easily modeled as queues because that is their principal function. The more challenging conveyors to model are powered conveyors, which come in a variety of types.

13.4.1 Conveyor Types

Conveyors are sometimes classified according to function or configuration:

- *Accumulation conveyor*—a conveyor that permits loads to queue up behind each other.
- *Transport conveyor*—a conveyor that transports loads from one point to another; usually nonaccumulating.
- *Sortation conveyor*—a conveyor that provides diverting capability to divert loads off the conveyor.
- *Recirculating conveyor*—a conveyor loop that causes loads to recirculate until they are ready for diverting.
- *Branching conveyor*—a conveyor that is characterized by a main line with either merging or diverging branches or spurs. Branches may, in turn, have connected spurs or branches.

There are many different types of conveyors based on physical characteristics. Among the most common types are the belt, roller, chain, trolley, power-and-free, tow, and monorail conveyors. A brief description will be given of each of these types.

Belt Conveyors.   Belt conveyors have a circulating belt on which loads are carried. Because loads move on a common span of fabric, if the conveyor stops, all loads must stop. Belt conveyors are often used for simple load transport.

Roller Conveyors.   Some roller conveyors utilize rollers that may be "dead" if the bed is placed on a grade to enable the effect of gravity on the load to actuate the rollers. However, many roller conveyors utilize "live" or powered rollers to convey loads. Roller conveyors are often used for simple load accumulation.

Chain Conveyors.   Chain conveyors are much like a belt conveyor in that one or more circulating chains convey loads. Chain conveyors are sometimes used in combination with roller conveyors to provide right-angle transfers for heavy loads. A pop-up section allows the chains running in one direction to "comb" through the rollers running in a perpendicular direction.

Trolley Conveyors.   A trolley conveyor consists of a series of trolleys (small wheels) that ride along a track and are connected by a chain, cable, or rod. Load carriers are typically suspended from the trolleys. The trolleys, and hence the load carriers, continually move so that the entire system must be stopped to bring a single load to a stop. No accumulation can take place on the conveyor itself. Parts are frequently loaded and unloaded onto trolley conveyors while the carriers are in motion (usually not faster than 30 fpm). However, because a trolley conveyor can be designed with a vertical dip, it can descend to automatically load a part. Automatic unloading is simpler. A carrier can be tipped to discharge a load or a cylinder can be actuated to push a load off a carrier onto an accumulation device. Discharge points can be programmed.

Power-and-Free Conveyors.   Power-and-free conveyors consist of two parallel trolley systems, one of which is powered while trolleys in the other system are nonpowered and free-moving. Loads are attached to the free trolleys and a "dog" or engaging pin on the powered trolleys engages and disengages the passive free trolleys to cause movement and stoppage. In some power-and-free systems, engaged load trolleys will automatically disengage themselves from the power line when they contact carriers ahead of them. This allows loads to be accumulated. In the free state, loads can be moved by hand or gravity; they can be switched, banked, or merely delayed for processing.

Tow Conveyors.   A tow conveyor or towline system is an endless chain running overhead in a track or on or beneath the ground in a track. The chain has a dog or other device for pulling (towing) wheeled carts on the floor. Towlines are best suited for applications where precise handling is unimportant. Tow carts are relatively inexpensive compared to powered vehicles; consequently, many of them can be added to a system to increase throughput and be used for accumulation.

An underfloor towline uses a tow chain in a trough under the floor. The chain moves continuously, and cart movement is controlled by extending a drive pin from the cart down into the chain. At specific points along the guideway, computer-operated stopping mechanisms raise the drive pins to halt cart movement. One advantage of this system is that it can provide some automatic buffering with stationary carts along the track.

Towline systems operate much like power-and-free systems, and, in fact, some towline systems are simply power-and-free systems that have been inverted. By using the floor to support the weight, heavier loads can be transported.

Monorail Conveyors.   Automatic monorail systems have self-powered carriers that can move at speeds of up to 300 fpm. In this respect, a monorail system is more of a discrete unit movement system than a conveyor system. Travel can be disengaged automatically in accumulation as the leading "beaver tail" contacts the limit switch on the carrier in front of it.

13.4.2 Operational Characteristics

The operational characteristics of conveyor systems are distinctly different from those of other material handling systems. Some of the main distinguishing features pertain to the nature of load transport, how capacity is determined, and how entity pickup and delivery occur.

Load Transport.   A conveyor spans the distance of travel along which loads move continuously, while other systems consist of mobile devices that move one or more loads as a batch move.

Capacity.   The capacity of a conveyor is determined by the speed and load spacing rather than by having a stated capacity. Specifically, it is a function of the minimum allowable interload spacing (which is the length of a queue position in the case of accumulation conveyors) and the length of the conveyor. In practice, however, this may never be reached because the conveyor speed may be so fast that loads are unable to transfer onto the conveyor at the minimum allowable spacing. Furthermore, intentional control may be imposed on the number of loads that are permitted on a particular conveyor section.

Entity Pickup and Delivery.   Conveyors usually don't pick up and drop off loads, as in the case of lift trucks, pallet jacks, or hand carrying, but rather loads are typically placed onto and removed from conveyors. Putting a load onto an input spur of a branching conveyor and identifying the load to the system so it can enter the conveyor system is commonly referred to as *load induction* (the spur is called an *induction conveyor*).

13.4.3 Modeling Conveyor Systems

Depending on the nature of the conveyor and its operation, modeling a conveyor can range from fairly straightforward to extremely complex. For single conveyor sections, modeling is very simple. Branching and recirculating conveyors, on the other hand, give rise to several complexities (recirculation, merging, and so on) that make it nearly impossible to predict how the system will perform. These types of conveyors make especially good problems for simulation.

From a modeling standpoint, a conveyor may be classified as being either accumulation or nonaccumulation (also referred to as accumulating or nonaccumulating), with either fixed or random load spacing (see Table 13.1).

Accumulation Conveyors.   *Accumulation conveyors* are a class of conveyors that provide queuing capability so that if the lead load stops, succeeding loads still advance along the conveyor and form a queue behind the stopped part. Accumulation conveyors permit independent or nonsynchronous part movement and therefore maximize part carrying capacity.

Nonaccumulation Conveyors.   *Nonaccumulation* or *simple transport conveyors* are usually powered conveyors having a single drive such that all movement on the conveyor occurs in unison, or synchronously. If a load must be brought to a stop, the entire conveyor and every load on it is also brought to a stop. When the conveyor starts again, all load movement resumes.

Fixed Spacing.   *Conveyors with fixed load spacing,* sometimes referred to as *segmented conveyors,* require that loads be introduced to the conveyor at fixed intervals. Tow conveyors and power-and-free conveyors, which have fixed spaced dogs that engage and disengage a part carrying cart or trolley, are examples of accumulation conveyors with fixed load spacing. Trolley conveyors with hanging carriers at fixed intervals or tray conveyors with permanently attached trays at fixed intervals are examples of nonaccumulation conveyors with fixed spacing.

Random Load Spacing.   *Random load spacing* permits parts to be placed at any distance from another load on the same conveyor. A powered roller conveyor is a common example of an accumulation conveyor with random load spacing. Powered conveyors typically provide noncontact accumulation, while gravity conveyors provide contact accumulation (that is, the loads accumulate against each

TABLE 13.1   Types of Conveyors for Modeling Purposes

	Accumulation	Nonaccumulation
Fixed Spacing	Power-and-free, towline	Trolley, sortation
Random Spacing	Roller, monorail	Belt, chain

other). A powered belt conveyor is a typical example of a nonaccumulation conveyor with random spacing. Parts may be interjected anywhere on the belt, and when the belt stops, all parts stop.

Performance Measures and Decision Variables.   Throughput capacity, delivery time, and queue lengths (for accumulation conveyors) are performance measures in conveyor system simulation. Several issues that are addressed in conveyor analysis include conveyor speed, accumulation capacity, and the number of carriers.

Questions to Be Answered.   Common questions that simulation can help answer in designing and operating a conveyor system are

- What is the minimum conveyor speed that still meets throughput requirements?
- What is the throughput capacity of the conveyor?
- What is the load delivery time for different activity levels?
- How much queuing is needed on accumulation conveyors?
- How many carriers are needed on a trolley or power-and-free conveyor?
- What is the optimum number of pallets that maximizes productivity in a recirculating conveyor?

13.4.4 Modeling Single-Section Conveyors

Conveyors used for simple queuing or buffering often consist of a single stretch or section of conveyor. Loads enter at one end and are removed at the other end. Load spacing is generally random. These types of conveyors are quite easy to model, especially using simulation products that provide conveyor constructs. In ProModel, a conveyor is simply a type of location. The modeler merely defines the length and speed and states whether the conveyor is an accumulation conveyor or nonaccumulation conveyor. Since a conveyor is just a location, routings can be defined to and from a conveyor. Operations can be defined at either the beginning or end of a conveyor.

If the simulation product being used does not have a conveyor construct, a simple accumulation conveyor can be modeled as a queue with a delay time equal to the time it takes to move from the beginning to the end. This can also be easily done in ProModel. For nonaccumulation conveyors the approach is more difficult. Henriksen and Schriber (1986) present a simple yet accurate approach to modeling single sections of accumulation and nonaccumulation conveyors with random load spacing and no branching or merging.

While these simplified approaches may be adequate for rough modeling of single conveyor sections in which parts enter at one end and exit at the other, they are inadequate for a conveyor system in which parts are continuously branching and merging onto conveyor sections. This type of system requires continuous tracking, which, in many discrete-oriented simulation languages, gets very complex or tends to be impractical.

13.4.5 Modeling Conveyor Networks

A *conveyor network* consists of two or more sections that are connected together to enable part merging, diverting, and recirculating. In such instances, a conveyor may have one or more entry points and one or more exit points. Furthermore, part entry and exit may not always be located at extreme opposite ends of the conveyor.

Conveyor networks are modeled in ProModel by placing conveyors next to each other and defining routings from one conveyor to the next. Branching is limited in ProModel in that entities can enter a conveyor only at the beginning and exit at the end. This may require breaking a conveyor into multiple sections to accommodate branching in the middle of a conveyor.

Conveyor networks may provide alternative routes to various destinations. Part movement time on a conveyor often is determinable only by modeling the combined effect of the conveying speed, the length of the conveyor, the load spacing requirements, the connecting conveyor logic, and any conveyor transfer times. The traffic and conveyor dynamics ultimately determine the move time.

13.5 Industrial Vehicles

Industrial vehicles include all push or powered carts and vehicles that generally have free movement. Powered vehicles such as lift trucks are usually utilized for medium-distance movement of batched parts in a container or pallet. For short moves, manual or semipowered carts are useful. Single-load transporters are capable of moving only one load at a time from one location to another. Such devices are fairly straightforward to model because they involve only a single source and a single destination for each move. Multiple-load transporters, on the other hand, can move more than one load at a time from one or more sources to one or more destinations. Defining the capacity and operation for multiple-load transporters can be extremely difficult because there are special rules defining when to retrieve additional loads and when to deliver the loads already on board. One example of a multiple-load transporter is a picking transporter, which simply may be a person pushing a cart through a part storage area loading several different part types onto the cart. These parts may then be delivered to a single destination, such as when filling a customer order, or to several destinations, such as the replenishment of stock along a production line. Towing vehicles are another example of this type of transporter. A towing vehicle pulls a train of carts behind it that may haul several loads at once. Towing vehicles may have a single pickup and drop point for each move similar to a unit-load transporter; however, they are also capable of making several stops, picking up multiple loads, and then delivering them to several remote locations. Such an operation can become extremely complex to incorporate into a generalized simulation language.

13.5.1 Modeling Industrial Vehicles

Modeling an industrial vehicle involves modeling a resource that moves along a path network. Paths are typically open aisles in which bidirectional movement is possible and passing is permitted. Some powered vehicles operate on a unidirectional track. Deployment strategies (work searches, idle vehicle parking, and so on) must be capable of being incorporated into the model, similar to that of an automated guided vehicle system (AGVS), described later.

Because industrial vehicles are often human-operated, they may take breaks and be available only during certain shifts.

Common performance measures that are of interest when modeling industrial vehicles include

- Vehicle utilization.
- Response time.
- Move rate capability.

Decision variables that are tested for industrial vehicles include

- Number of vehicles.
- Task prioritization.
- Empty vehicle positioning.

Typical questions to be answered regarding industrial vehicles are

- What is the required number of vehicles to handle the required activity?
- What is the best deployment of vehicles to maximize utilization?
- What is the best deployment of empty vehicles to minimize response time?

13.6 Automated Storage/Retrieval Systems

An *automated storage/retrieval system (AS/RS)* consists of one or more automated storage/retrieval machines that store and retrieve material to and from a storage facility such as a storage rack. Because it is automated, the system operates under computer control. The goal of an AS/RS is to provide random, high-density storage with quick load access, all under computer control.

An AS/RS is characterized by one or more aisles of rack storage having one or more *storage/retrieval (S/R) machines* (sometimes called *vertical* or *stacker cranes*) that store and retrieve material into and out of a rack storage system. Material is picked up for storage by the S/R machine at an input (pickup) station. Retrieved material is delivered by the S/R machine to an output (deposit) station to be taken away. Usually there is one S/R machine per aisle; however, there may also be one S/R machine assigned to two or more aisles or perhaps even two S/R machines assigned to a single aisle. For higher storage volume, some AS/RSs utilize double-deep storage, which may require load shuffling to access loads. Where even higher storage density is required and longer-term storage is needed, multidepth or deep lane storage systems are used. AS/RSs that use pallet storage racks are usually referred to as *unit-load systems,* while bin storage rack systems are known as *miniload systems.*

The throughput capacity of an AS/RS is a function of the rack configuration and speed of the S/R machine. Throughput is measured in terms of how many single or dual cycles can be performed per hour. A single cycle is measured as the average time required to pick up a load at the pickup station, store the load in a rack location, and return to the pickup station. A dual cycle is the time required to pick up a load at the input station, store the load in a rack location, retrieve a load from another rack location, and deliver the load to the output station. Obviously, there is considerable variation in cycle times due to the number of different possible rack locations that can be accessed. If the AS/RS is a stand-alone system with no critical interface with another system, average times are adequate for designing the system. If, however, the system interfaces with other systems such as front-end or remote-picking stations, it is important to take into account the variability in cycle times.

13.6.1 Configuring an AS/RS

Configuring an AS/RS depends primarily on the storage/retrieval or throughput activity that is required and the amount of storage space needed. Other considerations are building constraints in length, width, and height (especially if it is an existing facility); crane height limitations; and cost factors (a square building is less costly than a rectangular one).

Configuring an AS/RS is essentially an iterative process that converges on a solution that meets both total storage requirements and total throughput requirements. Once the number of rack locations is calculated based on storage requirements, an estimate of the number of aisles is made based on throughput requirements. This usually assumes a typical throughput rate of, say, 20 dual cycles per hour and assumes that one S/R machine is assigned to an aisle. If, for example, the required throughput is 100 loads in and 100 loads out per hour, an initial configuration would probably be made showing five aisles. If, however, it was shown using an AS/RS simulator or other analytical technique that the throughput capacity was only 15 cycles per hour due to the large number of rack openings required, the number of aisles may be increased to seven or eight and analyzed again. This process is repeated until a solution is found.

Occasionally it is discovered that the throughput requirement is extremely low relative to the number of storage locations in the system. In such situations, rather than have enormously long aisles, a person-aboard system will be used in which operator-driven vehicles will be shared among several aisles, or perhaps double-deep storage racks will be used.

Computerized Cycle Time Calculations

It is often thought necessary to simulate an AS/RS in order to determine throughput capacity. This is because the S/R machine accesses random bins or rack locations, making the cycle time extremely difficult to calculate. The problem is further compounded by mixed single (pickup and store or retrieve and deposit) and dual (pickup, storage, retrieval, and deposit) cycles. *Activity zoning,* in which items are stored in assigned zones based on frequency of use, also complicates cycle time calculations. The easiest way to accurately determine the cycle time for an AS/RS is by using a computer to enumerate the possible movements of the S/R machine from the pickup and delivery (P&D) stand to every rack or bin location. This produces an empirical distribution for the single cycle time. For dual cycle times, an intermediate cycle time—the time to go from any location to any other location—must be determined. For a rack 10 tiers high and 40 bays long, this can be 400×400, or 160,000, calculations! Because of the many calculations, sometimes a large sample size is used to develop the distribution. Most suppliers of automated storage/retrieval systems have computer programs for calculating cycle times that can be generated based on a defined configuration in a matter of minutes.

Analytical Cycle Time Calculations

Analytical solutions have been derived for calculating system throughput based on a given aisle configuration. Such solutions often rely on simplified assumptions about the operation of the system. Bozer and White (1984), for example, derive an equation for estimating single and dual cycle times assuming (1) randomized storage, (2) equal rack opening sizes, (3) P&D location at the base level on one end, (4) constant horizontal and vertical speeds, and (5) simultaneous horizontal and vertical rack movement. In actual practice, rack openings are seldom of equal size, and horizontal and vertical accelerations can have a significant influence on throughput capacity.

While analytical solutions to throughput estimation may provide a rough approximation for simple configurations, other configurations become extremely difficult to estimate. In addition, there are control strategies that may improve throughput rate, such as retrieving the load on the order list that is closest to the load just stored or storing a load in an opening that is closest to the next load to be retrieved. Finally, analytical solutions provide not a distribution of cycle times but merely a single expected time, which is inadequate for analyzing AS/RSs that interface with other systems.

AS/RS with Picking Operations

Whereas some AS/RSs (especially unit-load systems) have loads that are not captive to the system, many systems (particularly miniload systems) deliver bins or pallets either to the end of the aisle or to a remote area where material is picked from the bin or pallet, which is then returned for storage. Remote picking is usually achieved by linking a conveyor system to the AS/RS where loads are delivered to remote picking stations. In this way, containers stored in any aisle can be delivered to any workstation. This permits entire orders to be picked at a single station and eliminates the two-step process of picking followed by order consolidation.

Where picking takes place is an important issue achieving the highest productivity from both AS/RS and picker. Both are expensive resources, and it is undesirable to have either one waiting on the other.

13.6.2 Modeling AS/RSs

Simulation of AS/RSs has been a popular area of application for simulation. Ford developed a simulator called GENAWS as early as 1972. IBM developed one in 1973 called the IBM WAREHOUSE SIMULATOR. Currently, most simulation of AS/RSs is performed by vendors of the systems. Typical inputs for precision modeling of single-deep AS/RSs with each aisle having a captive S/R machine include

- Number of aisles.
- Number of S/R machines.
- Rack configuration (bays and tiers).
- Bay or column width.
- Tier or row height.
- Input point(s).
- Output point(s).
- Zone boundaries and activity profile if activity zoning is utilized.
- S/R machine speed and acceleration/deceleration.
- Pickup and deposit times.
- Downtime and repair time characteristics.

At a simple level, an AS/RS move time may be modeled by taking a time from a probability distribution that approximates the time to store or retrieve a load. More precise modeling incorporates the actual crane (horizontal) and lift (vertical) speeds. Each movement usually has a different speed and distance to travel, which means that movement along one axis is complete before movement along the other axis. From a modeling standpoint, it is usually necessary to calculate and model only the longest move time.

In modeling an AS/RS, the storage capacity is usually not a consideration and the actual inventory of the system is not modeled. It would require lots of overhead to model the complete inventory in a rack with 60,000 pallet locations. Because only the activity is of interest in the simulation, actual inventory is ignored. In fact, it is usually not even necessary to model specific stock keeping units (SKUs) being stored or retrieved, but only to distinguish between load types insofar as they affect routing and subsequent operations.

Common performance measures that are of interest in simulating an AS/RS include

- S/R machine utilization.
- Response time.
- Throughput capability.

Decision variables on which the modeler conducts experiments include

- Rack configuration and number of aisles.
- Storage and retrieval sequence and priorities.
- First-in, first-out or closest item retrieval.
- Empty S/R machine positioning.
- Aisle selection (random, round robin, or another method).

Typical questions addressed in simulation of an AS/RS include

- What is the required number of aisles to handle the required activity?
- Should storage activity be performed at separate times from retrieval activity, or should they be intermixed?
- How can dual cycling (combining a store with a retrieve) be maximized?
- What is the best stationing of empty S/R machines to minimize response time?
- How can activity zoning (that is, storing fast movers closer than slow movers) improve throughput?
- How is response time affected by peak periods?

13.7 Carousels

One class of storage and retrieval systems is a carousel. *Carousels* are essentially moving racks that bring the material to the retriever (operator or robot) rather than sending a retriever to the rack location.

13.7.1 Carousel Configurations

A carousel storage system consists of a collection of bins that revolve in either a horizontal or vertical direction. The typical front-end activity on a carousel is order picking. If carousels are used for WIP storage, bins might enter and leave the carousel. The average time to access a bin in a carousel is usually equal to one-half of the revolution time. If bidirectional movement is implemented, average access time is reduced to one-fourth of the revolution time. These times can be further reduced if special storing or ordering schemes are used to minimize access time.

A variation of the carousel is a *rotary rack* consisting of independently operating tiers. Rotary racks provide the added advantage of being able to position another bin into place while the operator is picking out of a bin from a different tier.

13.7.2 Modeling Carousels

Carousels are easily modeled by defining an appropriate response time representing the time for the carousel to bring a bin into position for picking. In addition to response times, carousels may have capacity considerations. The current contents may even affect response time, especially if the carousel is used to store multiple bins of the same item such as WIP storage. Unlike large AS/RSs, storage capacity may be an important issue in modeling carousel systems.

13.8 Automatic Guided Vehicle Systems

An *automatic guided vehicle system (AGVS)* is a path network along which computer-controlled, driverless vehicles transport loads. AGV systems are usually used for medium activity over medium distances. If parts are large, they are moved individually; otherwise parts are typically consolidated into a container or onto a pallet. Operationally, an AGVS is more flexible than conveyors, which provide fixed-path, fixed-point pickup and delivery. However, they cannot handle the high activity rates of a conveyor. On the other extreme, an AGVS is not as flexible as industrial vehicles, which provide open-path, any-point pickup and delivery. However, an AGVS can handle higher activity rates and eliminate the need for operators. AGVSs provide a flexible yet defined path of movement and allow pickup and delivery between any one of many fixed points, except for where certain points have been specifically designated as pickup or drop-off points.

AGVs are controlled by either a central computer, onboard computers, or a combination of the two. AGVs with onboard microprocessors are capable of having maps of the system in their memories. They can direct their own routing and blocking and can communicate with other vehicles if necessary. Some AGVs travel along an inductive guide path, which is a wire buried a few inches beneath the floor. Newer systems are non-wire guided. Travel is generally unidirectional to avoid traffic jams, although occasionally a path might be bidirectional.

The AGVS control system must make such decisions as what vehicle to dispatch to pick up a load, what route to take to get to the pickup point, what to do if the vehicle gets blocked by another vehicle, what to do with idle vehicles, and how to avoid collisions at intersections.

AGVs have acceleration and deceleration, and speeds often vary depending on whether vehicles are full or empty and whether they are on a turn or a stretch. There is also a time associated with loading and unloading parts. Furthermore, loaded vehicles may stop at points along the path for operations to be performed while the load is on the vehicle. One or more vehicles may be in a system, and more than one system may share common path segments. AGVs stop periodically for battery recharging and may experience unscheduled failures.

One of the modeling requirements of AGVSs is to accurately describe the method for controlling traffic and providing anticollision control. This is usually accomplished in one of two ways: (1) onboard vehicle sensing or (2) zone blocking.

Onboard vehicle sensing works by having a sensor detect the presence of a vehicle ahead and stop until it detects that the vehicle ahead has moved. This is usually accomplished using a light beam and receiver on the front of each vehicle with a reflective target on the rear of each.

In zone blocking, the guide path is divided into various zones (segments) that allow no more than one vehicle at a time. Zones can be set up using a variety of different sensing and communication techniques. When one vehicle enters a zone, other vehicles are prevented from entering until the current vehicle occupying the zone leaves. Once the vehicle leaves, any vehicle waiting for access to the freed zone can resume travel.

13.8.1 Designing an AGVS

When designing an AGV system, the first step is to identify all of the pickup and drop-off points. Next the path should be laid out. In laying out the guide path, several principles of good design should be followed (Askin and Standridge 1993):

- Travel should be unidirectional in the form of one or more recirculating loops unless traffic is very light. This avoids deadlock in traffic.
- Pickup stations should be downstream of drop-off stations. This enables an AGV to pick up a load, if available, whenever a load is dropped off.
- Pickup and deposit stations should generally be placed on spurs unless traffic is extremely light. This prevents AGVs from being blocked.
- Crossover paths should be used occasionally if measurable reductions in travel times can be achieved.

Once the path has been configured, a rough estimate is made of the number of AGVs required. This is easily done by looking at the total travel time required per hour, dividing it by the number of minutes in an hour (60), and multiplying it by 1 plus a traffic allowance factor:

$$AGV = \frac{\text{Total AGV minutes}}{60 \text{ minutes/hour}} \times (1 + \text{Traffic allowance})$$

The traffic allowance is a percentage factored into the vehicle requirements to reflect delays caused by waiting at intersections, rerouting due to congestion, or some other delay. This factor is a function of both poor path layout and quantity of vehicles and may vary anywhere from 0 in the case of a single-vehicle system to around 15 percent in congested layouts with numerous vehicles (Fitzgerald 1985).

The total vehicle minutes are determined by looking at the average required number of pickups and deposits per hour and converting this into the number of minutes required for one vehicle to perform these pickups and drop-offs. It is equal to the sum of the total empty move time, the total full move time, and the total pickup and deposit time for every delivery. Assuming I pickup points and J drop-off points, the total vehicle time can be estimated by

$$\text{Total vehicle minutes} = \sum_{i=1}^{I} \sum_{j=1}^{J} \left[2 \times \frac{\text{Distance}_{ij}}{\text{Avg. speed}_{ij}} + \text{Pickup}_i + \text{Deposit}_j \right] \\ \times \text{Deliveries}_{ij}$$

The expression

$$2 \times \frac{\text{Distance}_{ij}}{\text{Avg. speed}_{ij}}$$

implies that the time required to travel to the point of pickup (empty travel time) is the same as the time required to travel to the point of deposit (full travel time). This assumption provides only an estimate of the time required to travel to a point of pickup because it is uncertain where a vehicle will be coming from. In most cases, this should be a conservative estimate because (1) vehicles usually follow a work search routine in which the closest loads are picked up first and (2) vehicles frequently travel faster when empty than when full. A more accurate way of calculating empty load travel for complex systems is to use a compound weighted-averaging technique that considers all possible empty moves together with their probabilities (Fitzgerald 1985).

13.8.2 Controlling an AGVS

A multivehicle AGVS can be quite complex and expensive, so it is imperative that the system be well managed to achieve the utilization and level of productivity desired. Part of the management problem is determining the best deployment strategy or dispatching rules used. Some of these operating strategies are discussed next.

AGV Selection Rules.   When a load is ready to be moved by an AGV and more than one AGV is available for executing the move, a decision must be made as to which vehicle to use. The most common selection rule is a "closest rule" in which the nearest available vehicle is selected. This rule is intended to minimize empty vehicle travel time as well as to minimize the part waiting time. Other less frequently used vehicle selection rules include longest idle vehicle and least utilized vehicle.

Work Search Rules.   If a vehicle becomes available and two or more parts are waiting to be moved, a decision must be made as to which part should be moved first. A number of rules are used to make this decision, each of which can be effective depending on the production objective. Common rules used for dispatching an available vehicle include

- Longest-waiting load.
- Closest waiting load.
- Highest-priority load.
- Most loads waiting at a location.

Vehicle Parking Rules.   If a transporter delivers a part and no other parts are waiting for pickup, a decision must be made relative to the deployment of the transporter. For example, the vehicle can remain where it is, or it can be sent to a more strategic location where it is likely to be needed next. If several transport vehicles are idle, it may be desirable to have a prioritized list for a vehicle to follow for alternative parking preferences.

Work Zoning.   In some cases it may be desirable to keep a vehicle captive to a particular area of production and not allow it to leave this area unless it has work to deliver elsewhere. In this case, the transporter must be given a zone boundary within which it is capable of operating.

13.8.3 Modeling an AGVS

Modeling an AGVS is very similar to modeling an industrial vehicle (which it is in a sense) except that the operation is more controlled with less freedom of movement. Paths are generally unidirectional, and no vehicle passing is allowed.

One of the challenges in modeling an AGVS in the past has been finding the shortest routes between any two stops in a system. Current state-of-the-art simulation software, like ProModel, provides built-in capability to automatically determine the shortest routes between points in a complex network. Alternatively, the modeler may want to have the capability to explicitly define a set of paths for getting from one location to another.

Common performance measures that are of interest when modeling AGV systems include

- Resource utilization.
- Load throughput rate.
- Response time.
- Vehicle congestion.

In addition to the obvious purpose of simulating an AGVS to find out if the number of vehicles is sufficient or excessive, simulation can also be used to determine the following:

- Number of vehicles.
- Work search rules.
- Park search rules.
- Placement of crossover and bypass paths.

Typical questions that simulation can help answer include

- What is the best path layout to minimize travel time?
- Where are the potential bottleneck areas?
- How many vehicles are needed to meet activity requirements?
- What are the best scheduled maintenance/recharging strategies?
- Which task assignment rules maximize vehicle utilization?
- What is the best idle vehicle deployment that minimizes response time?
- Is there any possibility of deadlocks?
- Is there any possibility of collisions?

13.9 Cranes and Hoists

Cranes are floor-, ceiling-, or wall-mounted mechanical devices generally used for short- to medium-distance, discrete movement of material. Cranes utilize a hoist mechanism for attaching and positioning loads. Aside from bridge and gantry cranes, most other types of cranes and hoists can be modeled as an industrial vehicle. Since a gantry crane is just a simple type of bridge crane that is floor-mounted, our discussion here will be limited to bridge cranes. A bridge crane consists of a beam that bridges a bay (wall to wall). This beam or bridge moves on two tracks mounted on either wall. A retracting hoist unit travels back and forth under the bridge, providing a total of three axes of motion (gantry cranes have similar movement, only the runway is floor-mounted).

13.9.1 Crane Management

Managing crane movement requires an understanding of the priority of loads to be moved as well as the move characteristics of the crane. One must find the optimum balance of providing adequate response time to high-priority moves while maximizing the utilization of the crane. Crane utilization is maximized when drop-offs are always combined with a nearby pickup. Crane management becomes more complicated when (as is often the case) two or more cranes operate in the same bay or envelope of space. Pritsker (1986) identifies typical rules derived from practical experience for managing bridge cranes sharing the same runway:

1. Cranes moving to drop points have priority over cranes moving to pickup points.
2. If two cranes are performing the same type of function, the crane that was assigned its task earliest is given priority.
3. To break ties on rules 1 and 2, the crane closest to its drop-off point is given priority.
4. Idle cranes are moved out of the way of cranes that have been given assignments.

These rules tend to minimize the waiting time of loaded cranes and maximize crane throughput.

13.9.2 Modeling Bridge Cranes

For single cranes a simple resource can be defined that moves along a path that is the resultant of the *x* and *y* movement of the bridge and hoist. For more precise modeling or for managing multiple cranes in a runway, it is preferable to utilize built-in crane constructs. ProModel provides powerful bridge crane modeling capabilities that greatly simplify bridge crane modeling.

Common performance measures when simulating a crane include

- Crane utilization.
- Load movement rate.

- Response time.
- Percentage of time blocked by another crane.

Decision variables that become the basis for experimentation include

- Work search rules.
- Park search rules.
- Multiple-crane priority rules.

Typical questions addressed in a simulation model involving cranes include

- Which task assignment rules maximize crane utilization?
- What idle crane parking strategy minimizes response time?
- How much time are cranes blocked in a multicrane system?

13.10 Robots

Robots are programmable, multifunctional manipulators used for handling material or manipulating a tool such as a welder to process material. Robots are often classified based on the type of coordinate system on which they are based: cylindrical, cartesian, or revolute. The choice of robot coordinate system depends on the application. Cylindrical or polar coordinate robots are generally more appropriate for machine loading. Cartesian coordinate robots are easier to equip with tactile sensors for assembly work. Revolute or anthropomorphic coordinate robots have the most degrees of freedom (freedom of movement) and are especially suited for use as a processing tool, such as for welding or painting. Because cartesian or gantry robots can be modeled easily as cranes, our discussion will focus on cylindrical and revolute robots. When used for handling, cylindrical or revolute robots are generally used to handle a medium level of movement activity over very short distances, usually to perform pick-and-place or load/unload functions. Robots generally move parts individually rather than in a consolidated load.

One of the applications of simulation is in designing the cell control logic for a robotic work cell. A robotic work cell may be a machining, assembly, inspection, or a combination cell. Robotic cells are characterized by a robot with 3 to 5 degrees of freedom surrounded by workstations. The workstation is fed parts by an input conveyor or other accumulation device, and parts exit from the cell on a similar device. Each workstation usually has one or more buffer positions to which parts are brought if the workstation is busy. Like all cellular manufacturing, a robotic cell usually handles more than one part type, and each part type may not have to be routed through the same sequence of workstations. In addition to part handling, the robot may be required to handle tooling or fixtures.

13.10.1 Robot Control

Robotic cells are controlled by algorithms that determine the sequence of activities for the robot to perform. This control is developed offline using algorithms for selecting what activity to perform. The sequence of activities performed by the robot can be either a fixed sequence of moves that is continually repeated or a dynamically determined sequence of moves, depending on the status of the cell. For fixed-sequence logic, the robot control software consists of a sequence of commands that are executed sequentially. Dynamic sequence selection is programmed using if–then statements followed by branches dependent on the truth value of the conditional statement. System variables must also be included in the system to keep track of system status, or else the robot polls the sensors, and so on, of the system each time a status check is needed.

In developing an algorithm for a cell controller to make these decisions, the analyst must first decide if the control logic is going to control the robot, the part, or the workstation. In a part-oriented system, the events associated with the part determine when decisions are made. When a part finishes an operation, a routing selection is made and then the robot is requested. In a robot-oriented cell, the robot drives the decisions and actions so that when a part completes an operation, it merely sends a signal or sets a status indicator showing that it is finished. Then when the robot becomes available, it sees that the part is waiting and, at that moment, the routing decision for the part is made.

13.10.2 Modeling Robots

In modeling robots, it is sometimes presumed that the kinematics of the robot need to be simulated. Kinematic simulation is a special type of robot simulation used for cycle time analysis and offline programming. For discrete event simulation, it is necessary to know only the move times from every pickup point to every deposit point, which usually aren't very many. This makes modeling a robot no more difficult than modeling a simple resource. The advantage to having a specific robot construct in a simulation product is primarily for providing the graphic animation of the robot. ProModel doesn't have a built-in construct specifically for robots, although submodels have been defined that can be used in other models.

Common performance measures when modeling robots include

- Robot utilization.
- Response time.
- Throughput.

Decision variables that come into play when simulating robots include

- Pickup sequence.
- Task prioritization.
- Idle robot positioning.

Typical questions that simulation can help answer are

- What priority of tasks results in the highest productivity?
- Where is the best position for the robot after completing each particular drop-off?

13.11 Summary

Material handling systems can be one of the most difficult elements to model using simulation simply because of the sheer complexity. At the same time, the material handling system is often the critical element in the smooth operation of a manufacturing or warehousing system. The material handling system should be designed first using estimates of resource requirements and operating parameters (speed, move capacity, and so forth). Simulation can then help verify design decisions and fine-tune the design.

In modeling material handling systems, it is advisable to simplify wherever possible so models don't become too complex. A major challenge in modeling conveyor systems comes when multiple merging and branching occur. A challenging issue in modeling discrete part movement devices, such as AGVs, is how to manage their deployment in order to get maximum utilization and meet production goals.

13.12 Review Questions

1. What impact can the material handling system have on the performance of a manufacturing system?
2. What is the essential difference between a conveyor and other transport devices from a modeling standpoint?
3. In material handling simulation, should we be interested in modeling individual parts or unit loads? Explain.
4. How is the capacity of a conveyor determined?
5. What is the difference between an accumulation and a nonaccumulation conveyor?
6. What is the difference between fixed spacing and random spacing on a conveyor?
7. Under what conditions could you model an accumulation conveyor as a simple queue or buffer?
8. What are typical conveyor design decisions that simulation can help make?
9. What are typical simulation objectives when modeling an AS/RS?
10. When modeling an AS/RS, is there a benefit in modeling the actual storage positions? Explain.
11. Conceptually, how would you model the movement of a carousel?
12. Draw a logic diagram of what happens from the time an AGV deposits a load to the time it retrieves another load.
13. When modeling a robot in system simulation, is it necessary to model the actual kinematics of the robot? Explain.
14. What are the unique characteristics of a bridge crane from a modeling perspective?

References

Apple, J. M. *Plant Layout and Material Handling.* 3rd ed. N.Y.: Ronald Press, 1977.

Askin, Ronald G., and C. R. Standridge. *Modeling and Analysis of Manufacturing Systems.* New York: John Wiley & Sons, 1993.

"Automatic Monorail Systems." *Material Handling Engineering,* May 1988, p. 95.

Bozer, Y. A., and J. A. White. "Travel-Time Models for Automated Storage/Retrieval Systems." *IIE Transactions* 16, no. 4 (1984), pp. 329–38.

Fitzgerald, K. R. "How to Estimate the Number of AGVs You Need." *Modern Materials Handling,* October 1985, p. 79.

Henriksen, J., and T. Schriber. "Simplified Approaches to Modeling Accumulation and Non-Accumulating Conveyor Systems." In *Proceedings of the 1986 Winter Simulation Conference,* ed. J. Wilson, J. Henricksen, and S. Roberts. Piscataway, NJ: Institute of Electrical and Electronics Engineers, 1986.

Meyers, Fred E. *Plant Layout and Material Handling.* Englewood Cliffs, NJ: Regents/Prentice Hall, 1993.

Pritsker, A. A. B. *Introduction to Simulation and Slam II.* West Lafayette, IN: Systems Publishing Corporation, 1986, p. 600.

Tompkins, J. A.; J. A. White; Y. A. Bozer; E. H. Frazelle; J. M. A. Tanchoco; and J. Trevino. *Facilities Planning.* 2nd ed. New York: John Wiley & Sons, 1996.

Zisk, B. "Flexibility Is Key to Automated Material Transport System for Manufacturing Cells." *Industrial Engineering,* November 1983, p. 60.

14   MODELING SERVICE SYSTEMS

"No matter which line you move to, the other line always moves faster."
—Unknown

14.1 Introduction

A service system is a processing system in which one or more services are provided to customers. Entities (customers, patients, paperwork) are routed through a series of processing areas (check-in, order, service, payment) where resources (service agents, doctors, cashiers) provide some service. Service systems exhibit unique characteristics that are not found in manufacturing systems. Sasser, Olsen, and Wyckoff (1978) identify four distinct characteristics of services that distinguish them from products that are manufactured:

1. Services are *intangible;* they are not things.
2. Services are *perishable;* they cannot be inventoried.
3. Services provide *heterogeneous output;* output is varied.
4. Services involve *simultaneous production and consumption;* the service is produced and used at the same time.

These characteristics pose great challenges for service system design and management, particularly in the areas of process design and staffing. Having discussed general modeling procedures common to both manufacturing and service system simulation in Chapter 7, and specific modeling procedures unique to manufacturing systems in Chapter 12, in this chapter we discuss design and operating considerations that are more specific to service systems. A description is given of major classes of service systems. To provide an idea of how simulation might be performed in a service industry, a call center simulation example is presented.

14.2 Characteristics of Service Systems

Service systems represent a class of processing systems where entities (customers, orders, work, and so forth) are routed through a series of service stations and waiting areas. Although certain characteristics of service systems are similar to manufacturing systems, service systems have some unique characteristics. The aspects of service systems that involve work-flow processing (orders, paperwork, records, and the like) and product delivery are nearly identical to manufacturing and will not be repeated here. Those aspects of service systems that are most different from manufacturing systems are those involving customer processing. Many of the differences stem from the fact that in service systems, often both the entity being served and the resource performing the service are human. Humans have much more complex and unpredictable behavior than parts and machines. These special characteristics and their implications for modeling are described here:

- *Entities are capricious.* System conditions cause humans to change their minds about a particular decision once it has been made. Customer reactions to dissatisfactory circumstances include balking, jockeying, and reneging. *Balking* occurs when a customer attempts to enter a queue, sees that it is full, and leaves. *Jockeying* is where a customer moves to another queue that is shorter in hopes of being served sooner. *Reneging,* also called *abandonment,* is where a customer enters a waiting line or area, gets tired of waiting, and leaves. Modeling these types of situations can become complex and requires special modeling constructs or the use of programming logic to describe the behavior.

- *Entity arrivals are random and fluctuate over time.* Rather than scheduling arrivals as in production, customers enter most service systems randomly according to a Poisson process (interarrival times are exponentially distributed). Additionally, the rate of arrivals often changes depending on the time of day or the day of the week. The fluctuating pattern of arrivals usually repeats itself daily, weekly, or sometimes monthly. Accurate modeling of these arrival patterns and cycles is essential to accurate analysis. Having a fluctuating arrival rate makes most service simulations terminating simulations with focus on peak activities and other changes of interest as opposed to steady-state behavior.

- *Resource decisions are complex.* Typically resource allocation and task selection decisions are made according to some general rule (like first-come, first-served). In service systems, however, resources are intelligent and often make decisions based on more state-related criteria. An increase in the size of a waiting line at a cashier, for example, may prompt a new checkout lane to open. A change in the state of an entity (a patient in recovery requiring immediate assistance) may cause a resource (a nurse or doctor) to interrupt its current task to service the entity. The flexibility to respond to state changes is made possible because the resources are humans and therefore capable of making more complex decisions than machines. Because service personnel are often cross-trained and activity boundaries are flexible, resources can fill in for each other when needed. Modeling the complex behavior of human resources often requires the use of if–then logic to define decision rules.

- *Resource work pace fluctuates.* Another characteristic of human resources is that work pace tends to vary with time of day or work conditions. A change in the state of a queue (like the number of customers in line) may cause a resource (a cashier) to work faster, thereby reducing processing time. A change in the state of the entity (such as the length of waiting time) may cause a resource to work faster to complete the professional service. A change in the state of the resource (like fatigue or learning curve) may change the service time (slower in the case of fatigue, faster in the case of a learning curve). To model this variable behavior, tests must be continually made on state variables in the system so that resource behavior can be linked to the system state.

- *Processing times are highly variable.* Service processes vary considerably due to the nature of the process as well as the fact that the entity and the server are both human. Consequently, processing times tend to be highly variable. From a modeling standpoint, processing times usually need to be expressed using some probability distribution such as a normal or beta distribution.

- *Services have both front-room and back-room activities.* In front-room activities, customer service representatives meet with customers to take orders for a good or service. In the back room, the activities are carried out for producing the service or good. Once the service or good is produced, it is either brought to the customer who is waiting in the front room or, if the customer has gone, it is delivered to a remote site.

14.3 Performance Measures

The ultimate objectives of a service organization might include maximizing profits or maximizing customer satisfaction. However, these measures for success are considered *external performance criteria* because they are not completely determined by any single activity. Simulation modeling and analysis help evaluate those measures referred to as *internal performance criteria*—measures tied to a specific activity. Simulation performance measures are both quantitative and time-based and measure the efficiency and the effectiveness of a system configuration and operating logic. Examples of internal performance measures are waiting times, hours to process an application, cost per transaction, and percentage of time spent correcting transaction errors. Collier (1994) uses the term *interlinking* to define the process of establishing quantitative, causal relationships such as these to external performance measures. He argues that interlinking can be a powerful strategic and competitive weapon. Here are some typical internal performance measures that can be evaluated using simulation:

- Service time.
- Waiting time.
- Queue lengths.
- Resource utilization.
- Service level (the percentage of customers who can be promptly serviced, without any waiting).
- Abandonment rate (the percentage of impatient customers who leave the system).

14.4 Use of Simulation in Service Systems

The use of simulation in service industries has been relatively limited in the past, despite the many areas of application where simulation has proven beneficial: health care services (hospitals, clinics), food services (restaurants, cafeterias), and financial services (banks, credit unions) to name a few. In health care services, Zilm et al. (1983) studied the impact of staffing on utilization and cost; and Hancock et al. (1978) employed simulation to determine the optimal number of beds a facility would need to meet patient demand. In food services, Aran and Kang (1987) designed a model to determine the optimal seating configuration for a fast-food restaurant, and Kharwat (1991) used simulation to examine restaurant and delivery operations relative to staffing levels, equipment layout, work flow, customer service, and capacity. The use of simulation in the service sector continues to expand.

Even within manufacturing industries there are business or support activities similar to those found in traditional service industries. Edward J. Kane of IBM observed (Harrington 1991),

> Just taking a customer order, moving it through the plant, distributing these requirements out to the manufacturing floor—that activity alone has thirty sub-process steps to it. Accounts receivable has over twenty process steps. Information processing is a whole discipline in itself, with many challenging processes integrated into a single total activity. Obviously, we do manage some very complex processes separate from the manufacturing floor itself.

This entire realm of support processes presents a major area of potential application for simulation. Similar to the problem of dealing with excess inventory in manufacturing systems, customers, paperwork, and information often sit idle in service systems while waiting to be processed. In fact, the total waiting time for entities in service processes often exceeds 90 percent of the total flow time.

The types of questions that simulation helps answer in service systems can be categorized as being either design related or management related. Here are some typical design and management decisions that can be addressed by simulation:

Design Decisions
1. How much capacity should be provided in service and waiting areas?
2. What is the maximum throughput capability of the service system?
3. What are the equipment requirements to meet service demand?
4. How long does it take to service a customer?
5. How long do customers have to wait before being serviced?
6. Where should the service and waiting areas be located?
7. How can work flow and customer flow be streamlined?
8. What effect would information technology have on reducing non-value-added time?

Management Decisions
1. What is the best way to schedule personnel?
2. What is the best way to schedule appointments for customers?
3. What is the best way to schedule carriers or vehicles in transportation systems?
4. How should specific jobs be prioritized?
5. What is the best way to deal with emergency situations when a needed resource is unavailable?

Because service systems are nearly always in a state of transition, going from one activity level to another during different periods of the day or week, they rarely reach a steady state. Consequently, we are frequently interested in analyzing the transient behavior of service systems. Questions such as "How long is the transient cycle?" or "How many replications should be run?" become very important. Overall performance measures may not be as useful as the performance for each particular period of the transient cycle. An example where this is true is in the analysis of resource utilization statistics. In the types of service systems where arrival patterns and staff schedules fluctuate over the activity cycle (such as by day or week), the average utilization for the entire cycle is almost meaningless. It is much more informative to look at the resource utilization for each period of the activity cycle.

Most design and management decisions in service systems involve answering questions based on transient system conditions, so it is important that the results of the simulation measure transient behavior. Multiple replications should be run, with statistics gathered and analyzed for different periods of the transient cycle. In an attempt to simplify a simulation model, sometimes there is a temptation to model only the peak period, which is often the period of greatest interest. What is overlooked is the fact that the state of the system prior to each period and the length of each period significantly impact the performance measures for any particular period, including the peak period.

ProModel provides a periodic reporting option that allows statistics to be gathered by periods. This provides a more complete picture of system activity over the entire simulation.

14.5 Applications of Simulation in Service Industries

Some of the primary applications of simulation in service industries are process design, method selection, system layout, staff planning, and flow control.

14.5.1 Process Design

Because of the number of options in customer or paper processing as well as the flexibility of largely human resources, service systems can be quite fluid, allowing them to be quickly and easily reconfigured. Simulation helps identify configurations that are more efficient. Consider a process as simple as a car wash service that provides the following activities:

- Payment.
- Wash.
- Rinse.
- Dry.
- Vacuum.
- Interior cleaning.

How many ways can cars be processed through the car wash? Each activity could be done at separate places, or any and even all of them could be combined at a station. This is because of the flexibility of the resources that perform each operation and the imparticularity of the space requirements for most of the activities. Many of the activities could also be performed in any sequence. Payment, vacuuming, and interior cleaning could be done in almost any order. The only order that possibly could not change easily is washing, rinsing, and drying. The other consideration with many service processes is that not all entities receive the same services. A car wash customer, for example, may forgo getting vacuum service or interior cleaning. Thus it is apparent that the mix of activities in service processes can vary greatly.

Simulation helps in process design by allowing different processing sequences and combinations to be tried to find the best process flow. Modeling process flow is relatively simple in most service industries. It is only when shifts, resource pools, and preemptive tasks get involved that it starts to become more challenging.

14.5.2 Method Selection

Method selection usually accompanies process design. When determining what activities will be performed and how the process should be broken up, alternative methods for performing the activities are usually taken into consideration. Some of the greatest changes in the methods used in service industries have come about as a result of advances in information technology. Routine banking is being replaced by automated teller machines (ATMs). Shopping is being conducted over the Internet. Electronic document management systems are replacing paper document processing. Processing of orders and other documents that previously took weeks now takes only days or hours.

Automation of service processes presents challenges similar to the automation of manufacturing processes. If automating a service process speeds up a particular activity but does not minimize the overall processing time, it is not effective and may even be creating waste (such as large pools of waiting entities).

14.5.3 System Layout

An important consideration in designing a service system is the layout of the facility, especially if the building construction is going to be based on the system layout. For many service operations, however, the layout is easily changed as it may involve only rearranging desks and changing offices. Human resources and desktop computers are much easier to move around than an anchored 200-ton press used in sheet-metal processing. Still, a good layout can help provide a more streamlined flow and minimize the amount of movement for both customers and personnel.

The way in which work areas are laid out can have a significant impact on customer satisfaction and processing efficiency. In some systems, for example, multiple servers have individual queues for customers to wait before being served. This can cause customers to be served out of order of arrival and result in jockeying and customer discontent. Other systems provide a single input queue that feeds multiple servers (queuing for a bank teller is usually designed this way). This ensures customers will be serviced in the order of arrival. It may cause some reneging, however, if grouping all customers in a single queue creates the perception to an incoming customer that the waiting time is long.

14.5.4 Staff Planning

A major decision in nearly every service operation pertains to the level and type of staffing to meet customer demand. Understaffing can lead to excessive waiting times and lost or dissatisfied customers. Overstaffing can result in needless costs for resources that are inadequately utilized.

Related to staffing levels is the staff type. In service operations, personnel are frequently cross-trained so that alternative resources may perform any given task. One objective in this regard is to make the best use of resources by allowing higher-skilled resources to fill in occasionally when lower-skilled workers are unavailable.

Modeling staffing requirements is done by defining the pattern of incoming customers, specifying the servicing policies and procedures, and setting a trial staffing level. If more than one skill level is capable of performing a task, the list and priority for using these alternative resources need to be specified.

After running the simulation, waiting times, abandonment counts, and resource utilization rates can be evaluated to determine if the optimum conditions have been achieved. If results are unacceptable, either the incoming pattern, the service force, or the servicing policies and procedures can be modified to run additional experiments.

14.5.5 Flow Control

Flow control is to service systems what production control is to manufacturing systems. Service system operations planners must decide how to allow customers, documents, and so on to flow through the system. As in manufacturing systems, customers and information may be pushed or pulled through the system. By limiting queue capacities, a pull system can reduce the total number of customers or items waiting in the system. It also reduces the average waiting time of customers or items and results in greater overall efficiency.

Fast-food restaurants practice pull methods when they keep two or three food items (like burgers and fries) queued in anticipation of upcoming orders. When the queue limit is reached, further replenishment of that item is suspended. When a server withdraws an item, that is the pull signal to the cook to replenish it with another one. The lower the inventory of prepared food items, the more tightly linked the pull system is to customer demand. Excessive inventory results in deteriorating quality (cold food) and waste at the end of the day.

14.6 Types of Service Systems

Service systems cover a much broader range than manufacturing systems. They range from a pure service system to almost a pure production system. Customers may be directly involved in the service process or they may not be involved at all. The classification of service systems presented here is based on operational characteristics and is adapted, in part, from the classification given by Schmenner (1994):

- Service factory.
- Pure service shop.
- Retail service store.
- Professional service.
- Telephonic service.
- Delivery service.
- Transportation service.

A description of each in terms of its operating characteristics is given next.

14.6.1 Service Factory

Service factories are systems in which customers are provided services using equipment and facilities requiring low labor involvement. Consequently, labor costs are low while equipment and facility costs are high. Service factories usually have both front-room and back-room activities with total service being provided in a matter of minutes. Customization is done by selecting from a menu of options previously defined by the provider. Waiting time and service time are two primary factors in selecting the provider. Convenience of location is another important consideration. Customer commitment to the provider is low because there are usually alternative providers just as conveniently located.

Examples include banks (branch operations), restaurants, copy centers, barbers, check-in counters of airlines, hotels, and car rental agencies.

14.6.2 Pure Service Shop

In a pure service shop, service times are longer than for a service factory. Service customization is also greater. Customer needs must be identified before service can be provided. Customers may leave the location and return later for pickup, checking on an order, payment, or additional service. Price is often determined after the service is provided. Although front-room activity times may be short, back-room activity times may be long, typically measured in hours or days. The primary consideration is quality of service. Delivery time and price are of secondary importance. The customer's ability to describe the symptoms and possible service requirements are helpful in minimizing service and waiting times.

When customers arrive, they usually all go through some type of check-in activity. At this time, a record (paperwork or computer file) is generated for the customer and a sequence of service or care is prescribed. The duration of the service or the type of resources required may change during the process of providing service because of a change in the status of the entity. After the service is provided, tests may be performed to make sure that the service is acceptable before releasing the entity from the facility. If the results are acceptable, the customer and the record are matched and the customer leaves the system.

Examples include hospitals, repair shops (automobiles), equipment rental shops, banking (loan processing), Department of Motor Vehicles, Social Security offices, courtrooms, and prisons.

14.6.3 Retail Service Store

In retail services, the size of the facility is large in order to accommodate many customers at the same time. Customers have many product options from which to choose. Retail services require a high degree of labor intensity but a low degree of customization or interaction with the customer. Customers are influenced by price more than service quality or delivery time. Customers are interested in convenient location, assistance with finding the products in the store, and quick checkout. Total service time is usually measured in minutes.

When customers arrive in a retail shop, they often get a cart and use that cart as a carrier throughout the purchasing process. Customers may need assistance from customer service representatives during the shopping process. Once the customer has obtained the merchandise, he or she must get in line for the checkout process. For large items such as furniture or appliances, the customer may have to order and pay for the merchandise first. The delivery of the product may take place later.

Examples include department stores, grocery stores, hardware stores, and convenience stores.

14.6.4 Professional Service

Professional services are usually provided by a single person or a small group of experts in a particular field. The service is highly customized and provided by expensive resources. Duration of the service is long, extremely variable, and difficult to predict because customer involvement during the process is highly variable.

Processing may be performed by a single resource or multiple resources. When the customer arrives, the first process is diagnostic. Usually an expert resource evaluates the service needed by the customer and determines the type of service, the estimated service time, and the cost. This diagnosis then dictates what resources are used to process the order. The duration of the service or the type of resources required may change during the process of providing service. This is usually a result of the customer's review of the work. After the service is provided, a final review with the customer may be done to make sure that the service is acceptable. If the results are acceptable, the customer and the record are matched and the customer leaves the system.

Examples include auditing services, tax preparation, legal services, architectural services, construction services, and tailor services.

14.6.5 Telephonic Service

Telephonic services or teleservicing are services provided over the telephone. They are unique in that the service is provided without face-to-face contact with the customer. The service may be making reservations, ordering from a catalog, or providing customer support. In a telephonic service system, there are a number of issues to address, including

- *Overflow calls*—the caller receives a busy signal.
- *Reneges*—the customer gets in but hangs up after a certain amount of time if no assistance is received.
- *Redials*—a customer who hangs up or fails to get through calls again.

The most important criterion for measuring effectiveness is delivery time. The customer is interested in getting the service or ordering the product as quickly as possible. The customer's ability to communicate the need is critical to the service time.

Calls usually arrive in the incoming call queue and are serviced based on the FIFO rule. Some advanced telephone systems allow routing of calls into multiple queues for quicker service. Processing of a call is done by a single resource. Service duration depends on the nature of the service. If the service is an ordering process, then the service time is short. If the service is a technical support process, then the service time may be long or the call may require a callback after some research.

Examples include technical support services (hotlines) for software or hardware, mail-order services, and airline and hotel reservations.

14.6.6 Delivery Service

Delivery services involve the ordering, shipping, and delivery of goods, raw materials, or finished products to points of use or sale. Customers may accept deliveries only within certain time schedules. In practice, there are often other constraints besides time windows. Certain sequences in deliveries may be inflexible. Customers are interested in convenient, fast delivery. If the products that are delivered are perishable or fragile goods, the quality of the products delivered is also important to the customer.

Deliveries begin with the preparation of the product and loading of the product on the delivery resources. Determination of the best routing decisions for drivers may depend on the number or proximity of customers waiting for the product.

14.6.7 Transportation Service

Transportation services involve the movement of people from one place to another. A fundamental difference between transportation and delivery systems is that people are being transported rather than goods. Another important difference is that the routes in transportation services tend to be fixed, whereas the routes in delivery services are somewhat flexible. Customers are interested in convenient, fast transportation. Cost of transportation plays a significant role in the selection of the service. Because set schedules and routes are used in transportation, customers expect reliable service.

Two types of systems are used in transportation: (1) multiple pickup and drop-off points and (2) single pickup and drop-off points. In multiple pickup and drop-off point systems, customers enter and leave the transportation vehicle independently. In single pickup and drop-off transportation, customers all enter at one place and are dropped off at the same destination.

Examples include airlines, railroads, cruise lines, mass transit systems, and limousine services.

14.7 Simulation Example: A Help Desk Operation

This section, based on an article by Singer and Gasparatos (1994), presents an example of a simulation project performed to improve a help desk operation. This project illustrates the flexibility of processing and how variable some of the data can be. Even for existing service systems, radical changes can be made with little disruption that can dramatically improve performance.

14.7.1 Background

Society Bank's Information Technology and Operations (ITO) group offers a "help desk" service to customers of the ITO function. This service is offered to both internal and external customers, handling over 12,000 calls per month. The client services help desk provides technical support and information on a variety of technical topics including resetting passwords, ordering PC equipment, requesting phone installation, ordering extra copies of internal reports, and reporting mainframe and network problems. The help desk acts as the primary source of communication between ITO and its customers. It interacts with authority groups within ITO by providing work and support when requested by a customer.

The old client services help desk process consisted of (1) a mainframe help desk, (2) a phone/local area network help desk, and (3) a PC help desk. Each of the three operated separately with separate phone numbers, operators, and facilities. All calls were received by their respective help desk operators, who manually logged all information about the problem and the customer, and then proceeded to pass the problem on to an authority group or expert for resolution.

Because of acquisitions, the increased use of information technologies, and the passing of time, Society's help desk process had become fragmented and layered with bureaucracy. This made the help desk a good candidate for a process redesign. It was determined that the current operation did not have a set of clearly defined goals, other than to provide a help desk service. The organizational boundaries of the current process were often obscured by the fact that much of the different help desks' work overlapped and was consistently being handed off. There were no process performance measures in the old process, only measures of call volume. A proposal was made to consolidate the help desk functions. The proposal also called for the introduction of automation to enhance the speed and accuracy of the services.

14.7.2 Model Description

The situation suggested a process redesign to be supported by simulation of the business processes to select and validate different operational alternatives.

Detailed historical information was gathered on call frequencies, call arrival patterns, and length of calls to the help desk. This information was obtained from the help desk's database, ASIM. Table 14.1 summarizes the call breakdown by

TABLE 14.1   Types of Calls for Help Desk

Time Period	Password Reset	Device Reset	Inquiries	Percent Level 1	Percent Level 1A	Percent Level 2
7 A.M.–11 A.M.	11.7%	25.7%	8.2%	45.6%	4.6%	47.3%
11 A.M.–2 P.M.	8.8	29.0	10.9	48.7	3.6	44.3
2 P.M.–5 P.M.	7.7	27.8	11.1	46.6	4.4	45.8
5 P.M.–8 P.M.	8.6	36.5	17.8	62.9	3.7	32.2
Average	9.9%	27.5%	9.9%	47.3%	4.3%	48.4%

call level. Level 1 calls are resolved immediately by the help desk, Level 1A calls are resolved later by the help desk, and Level 2 calls are handed off to an authority group for resolution.

Historically, calls averaged 2.5 minutes, lasting anywhere from 30 seconds to 25 minutes. Periodically, follow-up work is required after calls that ranges from 1 to 10 minutes. Overall, the help desk service abandonment rate was 4 to 12 percent (as measured by the percentage of calls abandoned), depending on staffing levels.

The help desk process was broken down into its individual work steps and owners of each work step were identified. Then a flowchart that described the process was developed (Figure 14.1). From the flowchart, a computer simulation model was developed of the old operation, which was validated by comparing actual performance of the help desk with that of the simulation's output. During the 10-day test period, the simulation model produced results consistent with those of the actual performance. The user of the model was able to define such model parameters as daily call volume and staffing levels through the use of the model's Interact Box, which provided sensitivity analysis.

Joint requirements planning (JRP) sessions allowed the project team to collect information about likes, dislikes, needs, and improvement suggestions from users, customers, and executives. This information clarified the target goals of the process along with its operational scope. Suggestions were collected and prioritized from the JRP sessions for improving the help desk process. Internal benchmarking was also performed using Society's customer service help desk as a reference for performance and operational ideas.

FIGURE 14.1

Flow diagram of client services.

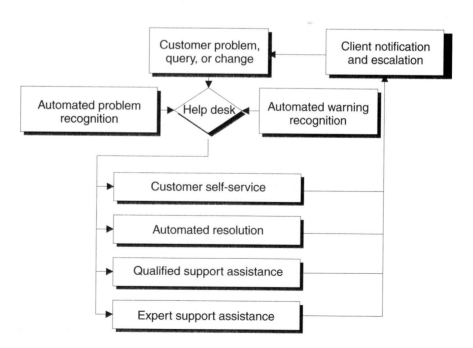

A target process was defined as providing a single-source help desk ("one-stop shopping" approach) for ITO customers with performance targets of

- 90 percent of calls to be answered by five rings.
- Less than 2 percent abandonment rate.

Other goals of the target process were to enhance the user's perception of the help desk and to significantly reduce the time required to resolve a customer's request. A combination of radical redesign ideas (reengineering) and incremental change ideas (TQM) formed the nucleus of a target help desk process. The redesigned process implemented the following changes:

- Consolidate three help desks into one central help desk.
- Create a consistent means of problem/request logging and resolution.
- Introduce automation for receiving, queuing, and categorizing calls for resetting terminals.
- Capture information pertaining to the call once at the source, and, if the call is handed off, have the information passed along also.
- Use existing technologies to create a hierarchy of problem resolution where approximately 60 percent of problems can be resolved immediately without using the operators and approximately 15 percent of the calls can be resolved immediately by the operators.
- Create an automated warning and problem recognition system that detects and corrects mainframe problems before they occur.

The original simulation model was revisited to better understand the current customer service level and what potential impact software changes, automation, and consolidation would have on the staffing and equipment needs and operational capacity. Simulation results could also be used to manage the expectations for potential outcomes of the target process implementation.

Immediate benefit was gained from the use of this application of simulation to better understand the old operational interrelationships between staffing, call volume, and customer service. Figure 14.2 shows how much the abandonment rate

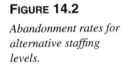

FIGURE 14.2

Abandonment rates for alternative staffing levels.

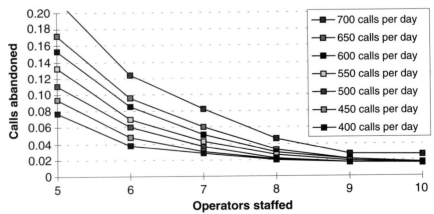

will change when the average daily call volume or the number of operators varies. The importance of this graph is realized when one notices that it becomes increasingly harder to lower the abandonment rate once the number of operators increases above seven. Above this point, the help desk can easily handle substantial increases in average daily call volume while maintaining approximately the same abandonment rate.

After modeling and analyzing the current process, the project team evaluated the following operational alternatives using the simulation model:

- The option to select from a number of different shift schedules so that staffing can easily be varied from current levels.
- The introduction of the automated voice response unit and its ability to both receive and place calls automatically.
- The ability of the automated voice response unit to handle device resets, password resets, and system inquiries.
- The incorporation of the PC and LAN help desks so that clients with PC-related problems can have their calls routed directly to an available expert via the automated voice response unit.
- The ability to change the response time of the ASIM problem-logging system.

Additionally, two alternative staffing schedules were proposed. The alternative schedules attempt to better match the time at which operators are available for answering calls to the time the calls are arriving. The two alternative schedules reduce effort hours by up to 8 percent while maintaining current service levels.

Additional results related to the Alternative Operations simulation model were

- The automated voice response unit will permit approximately 75 percent of PC-related calls to be answered immediately by a PC expert directly.
- Using Figure 14.2, the automated voice response unit's ability to aid in reducing the abandonment rate can be ascertained simply by estimating the reduction in the number of calls routed to help desk operators and finding the appropriate point on the chart for a given number of operators.
- Improving the response time of ASIM will noticeably affect the operation when staffing levels are low and call volumes are high. For example, with five operators on staff and average call volume of 650 calls per day, a 25 percent improvement in the response time of ASIM resulted in a reduction in the abandonment rate of approximately 2 percent.

14.7.3 Results

The nonlinear relationship between the abandonment rate and the number of operators on duty (see Figure 14.2) indicates the difficulty in greatly improving performance once the abandonment rate drops below 5 percent. Results generated from the validated simulation model compare the impact of the proposed staffing changes with that of the current staffing levels. In addition, the analysis of the effect of the automated voice response unit can be predicted before implementation so that the best alternative can be identified.

The introduction of simulation to help desk operations has shown that it can be a powerful and effective management tool that should be utilized to better achieve operational goals and to understand the impact of changes. As the automation project continues to be implemented, the simulation model can greatly aid management and the project team members by allowing them to intelligently predict how each new phase will affect the help desk.

14.8 Summary

Service systems provide a unique challenge in simulation modeling, largely due to the human element involved. Service systems have a high human content in the process. The customer is often involved in the process and, in many cases, is the actual entity being processed. In this chapter we discussed the aspects that should be considered when modeling service systems and suggested ways in which different situations might be modeled. We also discussed the different types of service systems and addressed the modeling issues associated with each. The example case study showed how fluid service systems can be.

14.9 Review Questions

1. What characteristics of service systems make them different from manufacturing systems?
2. What internal performance measures are linked to customer satisfaction?
3. Why are service systems more reconfigurable than manufacturing systems?
4. Identify a service system where the customer is not part of the process.
5. What is abandonment or reneging in simulation terminology?
6. Explain how simulation was used to create the abandonment rate chart shown in Figure 14.2.
7. What are three questions that simulation might help answer in designing bus routes for a city bus line?
8. Is a transportation system entity-driven or resource-driven? Explain.
9. Describe conceptually how you would model jockeying in a simulation model.

References

Aran, M. M., and K. Kang. "Design of a Fast Food Restaurant Simulation Model." *Simulation.* Norcross, GA: Industrial Engineering and Management Press, 1987.

Collier, D. A. *The Service/Quality Solution.* Milwaukee: ASQC Quality Press, 1994.

Hancock, W.; D. C. Magerlein; R. H. Stores; and J. B. Martin. "Parameters Affecting Hospital Occupancy and Implications for Facility Sizing." *Health Services Research* 13, 1978: pp. 276–89.

Harrington, J. H. *Business Process Improvement: The Breakthrough Strategy for Total Quality, Productivity, and Competitiveness.* New York: McGraw-Hill, 1991.

Harrington, J. H. *Business Process Improvement.* New York: McGraw-Hill, 1994.

Kharwat, A. K. "Computer Simulation: An Important Tool in the Fast Food Industry." *Proceedings of the 1991 Winter Simulation Conference,* eds. B. Nelson, W. D. Kelton, and G. M. Clark. Piscataway, NJ: Institute of Electrical and Electronics Engineers, 1991.

Sasser, W. E.; R. P. Olsen; and D. D. Wyckoff. *Management of Service Operations.* Boston: Allyn & Bacon, 1978, pp. 8–21.

Schmenner, Roger W. *Plant and Service Tours in Operations Management.* 4th ed. New York: MacMillan, 1994.

Singer, R., and A. Gasparatos. "Help Desks Hear Voice." *Software Magazine,* February 1994, pp. 24–26.

Zilm, F., and R. Hollis. "An Application of Simulation Modeling to Surgical Intensive Care Bed Needs Analysis." *Hospital and Health Services Administration,* Sept/Oct. 1983, pp. 82–101.

P A R T

II LABS

1   INTRODUCTION TO PROMODEL 6.0

Imagination is the beginning of creation. You imagine what you desire, you will what you imagine and at last you create what you will.

—George Bernard Shaw

ProModel (Production Modeler) by PROMODEL Corporation is a simulation tool for modeling various manufacturing and service systems. Manufacturing systems such as job shops, conveyors, transfer lines, mass production, assembly lines, flexible manufacturing systems, cranes, just-in-time systems, kanban systems, and so forth can be modeled by ProModel. Service systems such as hospitals, call centers, warehouse operations, transportation systems, grocery/department stores, information systems, customer service management, supply chains, logistic systems, and other business processes also can be modeled efficiently and quickly with ProModel.

ProModel is a powerful tool in the hands of engineers and managers to test various alternative designs, ideas, and process maps before actual implementation. Improvements in existing systems or the design of new systems can be modeled and tested before committing any money, time, or other resources. Various operating strategies and control alternatives can be compared and analyzed. Typically, most people use simulation tools to accurately predict and improve system performance by modeling the actual location (such as a plant floor, a bank lobby, or an emergency room) or an abstract process. Through testing various what-if scenarios, one can determine the best (optimum) way to conduct operations.

ProModel concentrates on resource utilization, production capacity, productivity, inventory levels, bottlenecks, throughput times, and other performance measures.

ProModel is a discrete event simulator and is intended to model discrete systems. Also, it is designed to model systems where events happen at definite points in time. The time resolution is controllable and ranges from 0.01 hours to 0.00001 seconds.

ProModel uses a graphical user interface (GUI). It is a true Windows (XP, Me, 2000, 98, 95, or NT) based simulation tool and utilizes all the Windows features such as a standard user interface, multitasking, built-in printer drivers, and "point and click" operation. ProModel has an online help system and a trainer.

Background graphics in .BMP, .PCX, .WMF, and .GIF formats can be imported into ProModel. Input data from spreadsheets can be seamlessly read into ProModel for quick and easy updates. Also, multiple-scenario analysis is possible using various input data files with the same model. Other model data may be read from or written to general text files.

L1.1 ProModel 6.0 Opening Screen

ProModel can be installed in one of the following ways:

    a. Runtime/evaluation package.

    b. Standard package.

    c. Student package.

    d. Network package.

Figure L1.1

ProModel opening screen (student package).

ProModel's opening screen (student package) is shown in Figure L1.1. There are six items (buttons) in the opening menu:

1. *Open a model:* Allows models created earlier to be opened.
2. *Install model package:* Copies to the specified destination directory all of the files contained in a model package file.
3. *Run demo model:* Allows one of several example models packed with the software to be run.
4. *www.promodel.com:* Allows the user to connect to the PROMODEL Corporation home page on the World Wide Web.
5. *SimRunner:* This new addition to the ProModel product line evaluates your existing simulation models and performs tests to find better ways to achieve desired results. A design of experiment methodology is used in SimRunner. For a detailed description of SimRunner, please refer to Lab 11.
6. *Stat::Fit:* This module allows continuous and/or discrete distributions to be fitted to a set of input data automatically. For a detailed discussion on the modeling of input data distribution, please refer to Lab 6.

L1.2 Simulation in Decision Making

ProModel is an excellent decision support tool and is used to help plan and make improvements in many areas of manufacturing and service industries. Click on Run demo model from the shortcut panel and select any model that looks interesting to run. You can also use the main menu to select File → Open and then choose a model from the Demos subdirectory. To run a model, select Simulation → Run. To stop a simulation prematurely, select Simulation → End Simulation.

Problem Statement

At a call center for **California Cellular,** customer service associates are employed to respond to customer calls and complaints. On average, 10 customers call per hour. The time between two calls is exponentially distributed with a mean of six minutes. Responding to each call takes a time that varies from a low of 2 minutes to a high of 10 minutes, with a mean of 6 minutes. If the company had a policy that

a. The average time to respond to a customer call should not be any more than six minutes, how many customer service associates should be employed by the company?

b. The maximum number of calls waiting should be no more than five, how many customer service associates should be employed by the company?

L1.2.1 *Average Waiting Time*

First we build a simulation model with one customer service associate answering all customer calls (Figure L1.2). The average time a customer has to wait before reaching the customer service associate is 18.92 minutes (Figure L1.3). This is a much longer wait than the company policy of six minutes (average).

If we employ two customer service associates (Figure L1.4), we would like to figure out what the average wait would be. The average wait drops down to 5.9 minutes (Figure L1.5), which is clearly a much more acceptable result. Hence the decision recommended to management will be to hire two associates.

FIGURE L1.2

California Cellular with one customer service agent.

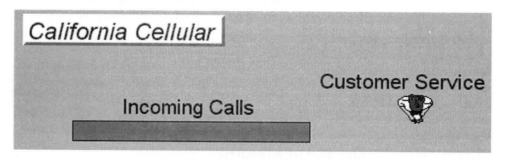

FIGURE L1.3

Customer waiting time statistics with one customer service agent.

Entity States	Variables	Location Costing	Resource Costing	Entity Costing	Logs

Logs for lab_l_1_2_one server. **Normal Run**				
Name	**Number Observations**	**Minimum Value**	**Maximum Value**	**Avg Value**
Wait Time=	5054	2.25	94.45	18.92

FIGURE L1.4

California Cellular with two customer service agents.

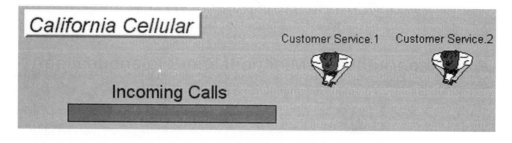

FIGURE L1.5

Customer waiting time statistics with two customer service agents.

Entity States	Variables	Location Costing	Resource Costing	Entity Costing	Logs

	Logs for lab_I_1_2_two server, Normal Run			
Name	**Number Observations**	**Minimum Value**	**Maximum Value**	**Avg Value**
Wait Time =	4945	2.23	17.65	5.90

FIGURE L1.6

Number of calls waiting with one customer service agent.

eneral	Locations	Location States Multi	Location States Single/Tank	Resources	Resource States	Node Entries	Failed Arrivals	E

	Locations for lab_I_1_2_one server, Normal Run							
Name	**Scheduled Time (MIN)**	**Capacity**	**Total Entries**	**Avg Time Per Entry (MIN)**	**Avg Contents**	**Maximum Contents**	**Current Contents**	**Pct Utilization**
Customer Ser...	30000.00	1	5055	5.00	0.84	1	1	84.19
Incoming Calls	30000.00	999999	5055	13.92	2.35	18	0	0.00

FIGURE L1.7

Number of calls waiting with two customer service agents.

General	Locations	Location States Multi	Location States Single/Tank	Resources	Resource States	Node Entries	Failed Arrivals	E

	Locations for lab_I_1_2_two server, Normal Run							
Name	**Scheduled Time (MIN)**	**Capacity**	**Total Entries**	**Avg Time Per Entry (MIN)**	**Avg Contents**	**Maximum Contents**	**Current Contents**	**Pct Utilization**
Customer Ser...	60000.00	2	4945	5.01	0.41	2	0	41.30
Customer Ser...	30000.00	1	3050	5.01	0.51	1	0	50.90
Customer Ser...	30000.00	1	1895	5.02	0.32	1	0	31.69
Incoming Calls	30000.00	999999	4945	0.88	0.15	5	0	0.00

L1.2.2 Maximum Queue Length

Now we want to evaluate the policy of no more than five calls waiting for response at any time. First we look at the results (Figure L1.6) of having one customer service associate working at a time. The maximum number of customers waiting for response is 18. This is clearly not an acceptable situation.

If we change the number of associates to two, the results obtained are given in Figure L1.7. The maximum number of calls waiting is five. This is acceptable according to the company policy of no more than five calls waiting. When we graph the number of incoming calls waiting for the duration of the simulation run of 500 hours (Figure L1.8), there were only three occasions of five calls waiting. Hence, we recommend to management to hire two associates.

FIGURE L1.8

Graph of number of customers waiting versus simulation run time.

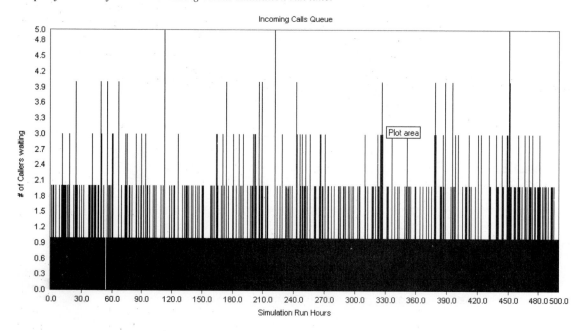

L1.3 Exercises

1. How do you open an existing simulation model?
2. What is SimRunner? How can you use it in your simulation analysis?
3. What does the Stat::Fit package do? Do you need it when building a simulation model?
4. At the most, how many locations, entities, and types of resources can be modeled using the student version of ProModel?
5. Open the Manufacturing Cost model (Mfg_cost) from the Demos subdirectory and run the model three different times to find out whether one, two, or three operators are optimal for minimizing the cost per part (the cost per part is displayed on the scoreboard during the simulation). Selecting Model Parameters, you can change the number of operators from the Simulation menu by double-clicking on the first parameter (number of operators) and entering 1, 2, or 3. Then select Run from the Model Parameters dialog. Each simulation will run for 15 hours.
6. Without knowing how the model was constructed, can you give a rational explanation for the number of operators that resulted in the least cost?
7. Go to the ProModel website on the Internet (www.promodel.com). What are some of the successful real-world applications of the ProModel software? Is ProModel applied only to manufacturing problems?

2   PROMODEL WORLD VIEW, MENU, AND TUTORIAL

I only wish that ordinary people had an unlimited capacity for doing harm; then they might have an unlimited power for doing good.
—Socrates (469–399 B.C.)

In this lab, Section L2.1 introduces you to various commands in the ProModel menu. In Section L2.2 we discuss the basic modeling elements in a ProModel model file. Section L2.3 discusses some of the innovative features of ProModel. Section L2.4 refers to a short tutorial on ProModel in a PowerPoint presentation format. Some of the material describing the use and features of ProModel has been taken from the ProModel User Guide as well as ProModel's online help system.

L2.1 Introduction to the ProModel Menu

In this section we introduce you to the title bar and various actions and commands in the ProModel menu bar. In subsequent chapters we will use many of these commands in various examples and you will have an opportunity to master them. We encourage you to use the ProModel's online help to learn more about all the menu bar commands.

L2.1.1 The Title and the Menu Bars

The title bar at the top of the screen (Figure L2.1) holds the name of the model document currently being created. Until the document is given a name, the title bar shows "ProModel". Once the document is given a title (like Paula's Production Shop) in the general information dialog box, the title bar shows "ProModel— (Paula's Production Shop)."

FIGURE L2.1

The title and the menu bars.

FIGURE L2.2

The File menu.

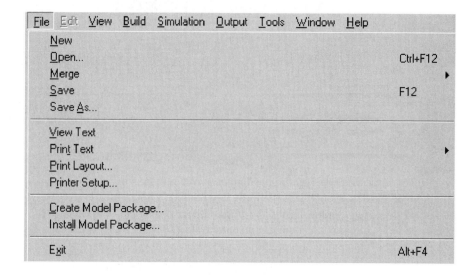

The menu bar, just below the title bar (Figure L2.1), is used to call up menus, or lists of tasks. The menu bar of the ProModel screen displays the commands you use to work with ProModel. Some of the items in the menu, like File, Edit, View, Tools, Window, and Help, are common to most Windows applications. Others such as Build, Simulation, and Output provide commands specific to programming in ProModel. In the following sections we describe all the menu commands and the tasks within each menu.

L2.1.2 File Menu

The File menu (Figure L2.2) allows the user to open a new model, open an existing model, merge two or more submodels into one, save the current model, and import models created in earlier versions of ProModel. It also allows you to view a text version of the model and print either the model text file or the graphic layout of the model. The printer setup can be modified from this menu. Model packages can be created for distribution to others and can also be installed using this menu. The last item on this menu allows the user to exit the ProModel environment and go back to the Windows (XP, Me, 2000, 98, 95, or NT) environment.

L2.1.3 Edit Menu

The Edit menu (Figure L2.3) contains selections for editing the contents of edit tables and logic windows. The selections available from this menu will change according to the module from which the Edit menu is selected. The selections also

FIGURE L2.3

The Edit menu.

FIGURE L2.4

The Build menu.

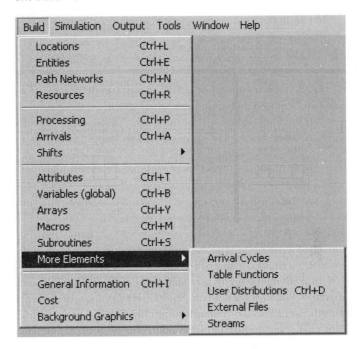

vary according to the currently selected window. The Edit menu is active only when a model file is open.

L2.1.4 Build Menu

The Build menu (Figure L2.4) consists of all of the modules for creating and editing a model, which include the following basic and optional modules:

Basic Modules

Locations	Processing
Entities	Arrivals

Optional Modules

Path Networks	Macros
Resources	Subroutines
Shifts	Arrival Cycles
Cost	Table Functions
Attributes	User Distributions
Variables	External Files
Arrays Streams	

FIGURE L2.5

General Information dialog box.

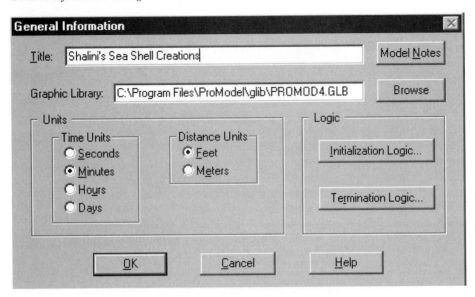

In addition, two more modules are available in the Build menu: General Information and Background Graphics.

General Information.   This dialog box (Figure L2.5) allows the user to specify the name of the model, the default time unit, the distance unit, and the graphic library to be used. The model's initialization and termination logic can also be specified using this dialog box. A Notes window allows the user to save information such as the analyst's name, the revision date, any assumptions made about the model, and so forth. These notes can also be displayed at the beginning of a simulation run.

Background Graphics.   The Background Graphics module (Figure L2.6) allows the user to create a unique background for the model using the tools in the graphics editor. An existing background can also be imported from another application such as AutoCAD.

Generally, most graphics objects are laid out in front of the grid. Large objects as well as imported backgrounds are placed behind the grid.

L2.1.5 Simulation Menu

The Simulation menu (Figure L2.7) controls the execution of a simulation and contains options for running a model, defining model parameters, and defining and running scenarios.

ProModel also has an add-on feature: SimRunner.

FIGURE L2.6

Background Graphics dialog box.

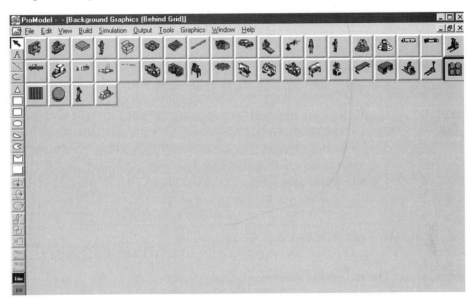

FIGURE L2.7

The Simulation menu.

FIGURE L2.8

The Output menu.

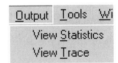

FIGURE L2.9

The Tools menu.

L2.1.6 Output Menu

The Output menu (Figure L2.8) starts the ProModel Output Processor for viewing model statistics. It also allows the user to view the trace, which was generated during simulation model runtime.

L2.1.7 Tools Menu

The Tools menu (Figure L2.9) contains various utilities as follows:

- *Graphics Editor:* For creating, editing, and modifying graphic icons.
- *Stat::Fit:* For fitting empirical distributions to a set of user data.

- *Expression Search:* A search-and-replace feature for finding or replacing expressions throughout a model.
- *Options:* The paths for model files, the graphic library, output results, and auto saves are specified in this dialog box. The time between auto saves is also specified here.
- *Customize:* With this utility one can add any tools (viz. Dashboard, MPA, ProClare, Prosetter, ProStat, or Quick Bar), or any other Windows executable files to the Tools Menu.
- *QuickBar:* QuickBar is a versatile tool that can make your work in ProModel much faster. It allows you to remove buttons, add custom ProModel buttons, or add custom buttons launching other applications.

L2.1.8 View Menu

The View menu (Figure L2.10) contains selections for setting up the model viewing environment. These settings are used whenever ProModel is started.

The Zoom submenu (Figure L2.11) allows the model to be displayed from 25 percent magnification to 400 percent magnification or any other user-specified custom magnification.

The Layout Settings submenu is shown in Figure L2.12. The grid settings allow the properties of the grids to be changed. The background and routing path colors can also be changed using this submenu. The various editing preferences are shown in the Edit Tables submenu (Figure L2.13).

FIGURE L2.10

The View menu.

FIGURE L2.11

The Zoom menu.

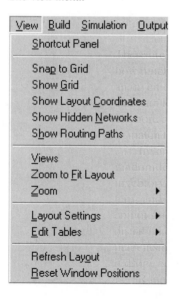

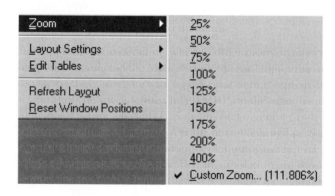

FIGURE L2.12

The Layout Settings submenu.

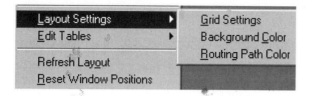

FIGURE L2.13

Editing preferences in ProModel.

FIGURE L2.14

The Window menu.

FIGURE L2.15

The Help menu.

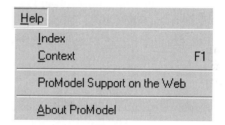

L2.1.9 Window Menu

The Window menu (Figure L2.14) allows you to arrange the windows (or iconized windows) that are currently displayed on the screen so that all windows are visible at once. It also lets you bring any individual window to the front of the display.

L2.1.10 Help Menu

The Help menu (Figure L2.15) accesses the ProModel Online Help System and provides access to the ProModel Tutorial (getting started). Access to the PROMODEL Corporation website is also provided here. The index is an alphabetic list of all the help topics in the online help system. The main help topics are categorized as follows:

- Common procedures.
- Overview of ProModel.
- Main menus.
- Shortcut menu.
- Model elements (entities, resources, locations, path networks, and so on).
- Logic elements (variables, attributes, arrays, expressions, and so forth).
- Statements (GET, JOIN, and the like).

- Functions.
- Customer support—telephone, pager, fax, e-mail, online file transfer, and so on.

To quickly learn what is new in ProModel Version 6.0, go to the Help → Index menu and type "new features" for a description of the latest features of the product.

L2.2 Basic Modeling Elements

In this section we introduce the various basic modeling elements in ProModel. In Lab 3 you will actually build your first ProModel simulation model. The basic modeling elements in ProModel are

    a. Locations

    b. Entities

    c. Arrivals

    d. Processing

Click Build from the menu bar to access these modeling elements (Figure L2.4).

The ProModel Student Version 6.0 limits the user to no more than 20 locations, five entity types, five resource types, five attributes, and 10 RTI parameters in a simulation model. If more capability is required for special projects, ask your instructor to contact the PROMODEL corporation about faculty or network versions of the software.

L2.2.1 Locations

Locations represent fixed places in the system where entities are routed for processing, delay, storage, decision making, or some other activity. We need some type of receiving locations to hold incoming entities. We also need processing locations where entities have value added to them. To build locations:

    a. Left-click on the desired location icon in the Graphics toolbox. Left-click in the layout window where you want the location to appear.

    b. A record is automatically created for the location in the Locations edit table (Figure L2.16).

    c. Clicking in the appropriate box and typing in the desired changes can now change the name, units, capacity, and so on. Note that in Lab 3 we will actually fill in this information for an example model.

L2.2.2 Entities

Anything that a model can process is called an *entity*. Some examples are parts or widgets in a factory, patients in a hospital, customers in a bank or a grocery store, and travelers calling in for airline reservations.

FIGURE L2.16

The Locations edit screen.

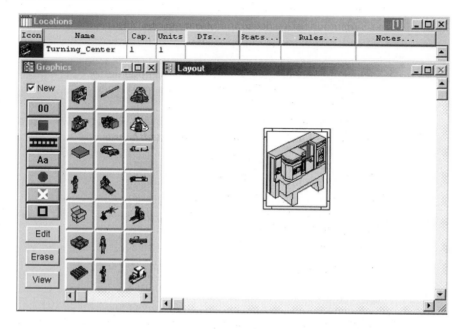

To build entities:

a. Left-click on the desired entity graphic in the Entity Graphics toolbox.

b. A record will automatically be created in the Entities edit table (Figure L2.17).

c. Moving the slide bar in the toolbox can then change the name. Note that in Lab 3 we will actually fill in this information for an example model.

L2.2.3 Arrivals

The mechanism for defining how entities enter the system is called arrivals. Entities can arrive singly or in batches. The number of entities arriving at a time is called the batch size (Qty each). The time between the arrivals of successive entities is called interarrival time (Frequency). The total number of batches of arrivals is termed Occurrences. The batch size, time between successive arrivals, and total number of batches can be either constants or random (statistical distributions). Also, the first time that the arrival pattern is to begin is termed First Time.

To create arrivals:

a. Left-click on the entity name in the toolbox and left-click on the location where you would like the entities to arrive (Figure L2.18).

b. Enter the various required data about the arrival process. Note that in Lab 3 we will actually fill in this information for an example model.

FIGURE L2.17

The Entities edit table.

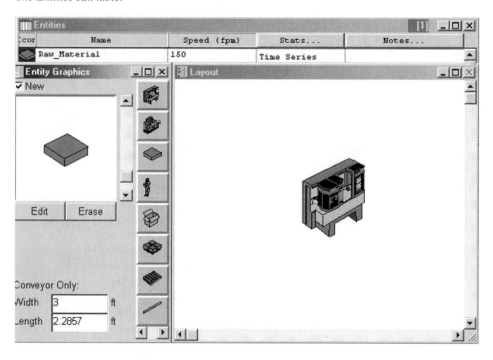

FIGURE L2.18

The Arrivals edit table.

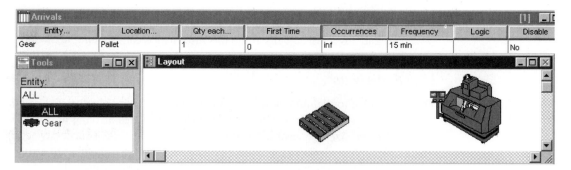

L2.2.4 Processing

Processing describes the operations that take place at a location, such as the amount of time an entity spends there, the resources it needs to complete processing, and anything else that happens at the location, including selecting an entity's next destination.

FIGURE L2.19

The Process edit table.

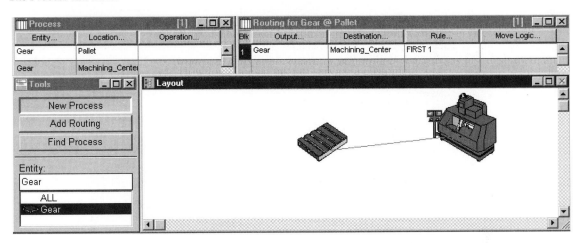

To create processing we need to do the following:

a. Left-click on the entity name in the toolbar and then left-click on the desired beginning location (Figure L2.19).

b. Left-click on the destination location. A processing record is created.

c. To add multiple routing lines to the same record, left-click on the Add Routing button in the toolbox.

d. To route the part to exit, left-click on the Route to Exit button in the toolbox. Note that in Lab 3 we will actually fill in this information for an example model.

L2.3 Innovative Features in ProModel

In this section we describe some of the innovative features in ProModel. These features help automate the model-building process, eliminate the need to remember command syntax, and all but eliminate unintentional errors.

L2.3.1 Logic Builder

The Logic Builder (Figure L2.20) is a tool that makes it easier to create valid logic statements without having to remember keywords, syntax, required arguments, or model element names. It takes you through the process of creating statements or expressions, as well as providing point-and-click access to every element defined in your model. The Logic Builder knows the syntax of every statement and function, allowing you to define logic simply by filling in the blanks.

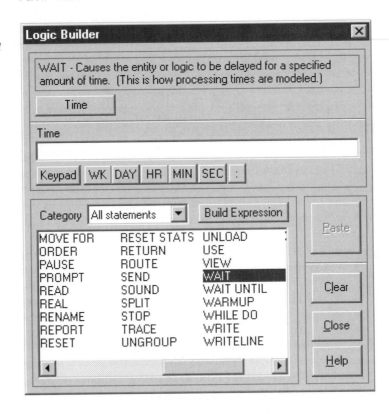

When the Logic Builder is open from a logic window, it remains on the screen until you click the Close button or close the logic window or table from which it was invoked. This allows you to enter multiple statements in a logic window and even move around to other logic windows without having to constantly close and reopen the Logic Builder. However, the Logic Builder closes automatically after pasting to a field in a dialog box or edit table or to an expression field because you must right-click anyway to use the Logic Builder in another field.

You can move to another logic window or field while the Logic Builder is still up by right clicking in that field or logic window. The Logic Builder is then reset with only valid statements and elements for that field or window, and it will paste the logic you build into that field or window. Some of the commonly used logic statements available in ProModel are as follows:

- WAIT: Used for delaying an entity for a specified duration at a location, possibly for processing it.
- STOP: Terminates the current replication and optionally displays a message.
- GROUP: Temporarily consolidates a specified quantity of similar entities together.
- LOAD: Temporarily attaches a specified quantity of entities to the current entity.

- INC: Increments a variable, array element, or attribute by the value of a specified numeric expression.
- UNGROUP: Separates entities that were grouped using the GROUP statement.
- MOVE: Moves the entity to the end of the queue in the specified time.
- VIEW: Changes the view of the layout window.
- GRAPHIC: Changes the entity's current picture (graphic).
- IF THEN ELSE: Executes a block of statements if the condition is true.
- WAIT UNTIL: Delays the processing of the current logic until the condition is true.
- MOVE FOR: Moves the entity to the next location, in the specified time.
- SPLIT: Splits the entity into a specified number of entities.
- PAUSE: Causes the execution of simulation to pause until the user selects the Resume option.

L2.3.2 Dynamic Plots

Dynamic plots enable you to graphically observe and record statistical information about the performance of model elements during run time. When the simulation model is running, it needs to be paused (Simulation → Pause Simulation). With the model in the pause mode select Dynamic Plots → New from the Information menu as in Figure L2.21.

The dynamic plots dialog contains a factors list, four panels, and several button controls. The factors list (Figure L2.22) provides each of the available model factors from your model, and the panels (Figure L2.23) display performance graphs for the factors you select (ProModel Help, 2001 PROMODEL Corporation).

Button Controls

- *Save:* Saves a copy of all model data—from the time you start the graphic display to when you click save—to an Excel spreadsheet.
- *Snapshot:* Saves a copy of currently displayed, graphed model data to an Excel spreadsheet.
- *Grid:* Turns the main panel grid lines on and off.
- *Multi-line:* Displays a combined graph of panels 1, 2, and 3.
- *Refresh:* Redraws the graph.

FIGURE L2.21

Dynamic Plots menu.

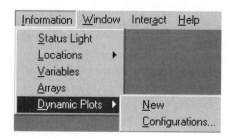

FIGURE L2.22

Dynamic Plot edit table.

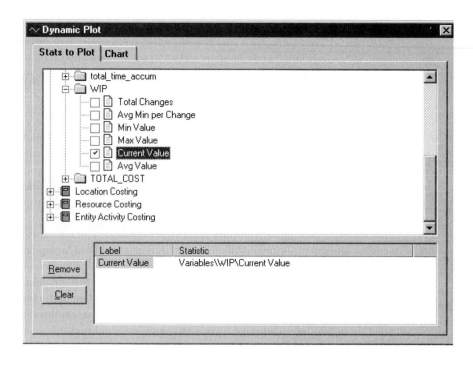

FIGURE L2.23

Dynamic Plot of the current value of WIP.

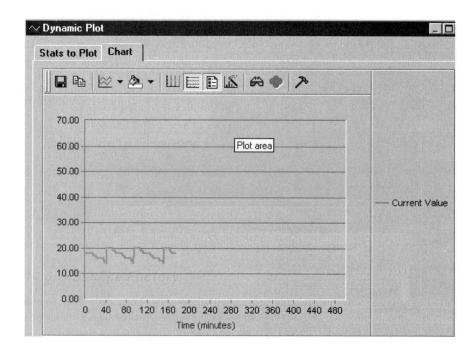

Right-Click Menus

The right-click menu for the graphic display is available for panels 1, 2, and 3, and the main panel. When you right-click in any of these panels, the right-click menu appears.

Panels 1, 2, and 3

- *Move Up:* Places the graph in the main panel.
- *Clear Data:* Removes the factor and its graph from panel 1, 2, or 3 and the main panel. If you created a multi-line graph, Clear Data removes the selected line from the graph and does not disturb the remaining graph lines.
- *Line Color:* Allows you to assign a specific line color to the graph.
- *Background Color:* Allows you to define a specific background color for panels 1, 2, and 3.

Main Panel

- *Clear All Data:* Removes all factors and graphs from panels 1, 2, 3, and the main panel.
- *Remove Line 1, 2, 3:* Deletes a specific line from the main panel.
- *Line Color:* Allows you to assign a specific line color to the graph.
- *Background Color:* Allows you to define a specific background color for panels 1, 2, and 3.
- *Grid Color:* Allows you to assign a specific line color to the grid.

L2.3.3 Customize

Customize

You can add direct links to applications and files right on your ProModel toolbar. Create a link to open a spreadsheet, a text document, or your favorite calculator (Figure L2.24).

To create or modify your Custom Tools menu, select Tools → Customize from your ProModel menu bar. This will pull up the Custom Tools dialog window. The Custom Tools dialog window allows you to add, delete, edit, or rearrange the menu items that appear on the Tools drop-down menu in ProModel.

L2.3.4 Quick Bar

Overview

QuickBar is a versatile tool that can make your work in ProModel much faster (Figure L2.25). This tool is similar to the Windows Taskbar. You can move the bar around your screen or dock it on one side. Each button on the toolbar resembles an associated action in ProModel. You can jump straight to the Build Locations module, or open a new model with the click of a button.

FIGURE L2.24

Adding Calculator to the Customized Tools menu.

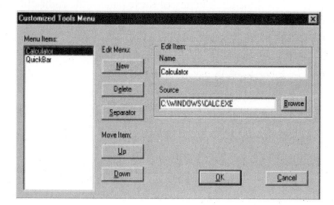 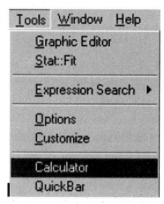

FIGURE L2.25

The QuickBar task bar.

QuickBar is fully customizable. QuickBar allows you to remove buttons, add custom ProModel buttons, or add custom buttons launching other applications (ProModel Help, 2001 PROMODEL Corporation).

System Button

Selecting the System button brings up a menu with the following options:

- *Auto-Hide:* The Auto-Hide feature is used when QuickBar is in the docked state. If docked, and the Auto-Hide is on, QuickBar will shrink, so that only a couple of pixels show. When your pointer touches these visible pixels, the QuickBar will appear again.
- *Docking Position:* The Docking Position prompt allows you to "dock" the QuickBar along one side of your screen, or leave it floating.
- *Customize:* Access the customize screen by clicking the System button, then choose "Customize. . . ." Once in the Customize dialog, you can add or remove toolbars, add or remove buttons from the toolbars, show or hide toolbars, or move buttons up or down on the toolbar.

- *About:* The About option will display information about QuickBar.
- *Close QuickBar:* The Close QuickBar option will close QuickBar, but will not change any customizing you may have made to QuickBar.

L2.4 A Tutorial on ProModel 6.0

A tutorial on ProModel is included in the accompanying CD along with the ProModel software (Figure L2.26). Click on the lower right button to run the ProModel tutorial. Microsoft PowerPoint is needed to view the tutorial.

This tutorial is meant to familiarize the user with various features of ProModel and its strengths. The file named tutorial.ppt is in a PowerPoint presentation format and will execute by itself. The tutorial takes approximately 20 minutes to run and is divided into 10 major steps. Slides for each of the 10 steps also include animation that shows the model-building process clearly:

1. Start new model
2. Background graphic
3. Locations
4. Entities
5. Path networks

FIGURE L2.26

Tutorial on ProModel.

6. Resources
7. Processing
8. Arrivals
9. Run simulation
10. View output

L2.5 Exercises

1. Identify the ProModel menu where you will find the following items:
   a. Save As
   b. Delete
   c. View Trace
   d. Shifts
   e. Index
   f. General Information
   g. Options
   h. Printer Setup
   i. Processing
   j. Scenarios
   k. Tile
   l. Zoom

2. Which of the following is not a valid ProModel menu or submenu item?
   a. AutoBuild
   b. What's This?
   c. Merge
   d. Merge Documents
   e. Snap to Grid
   f. Normal
   g. Paste
   h. Print Preview
   i. View Text

3. Some of the following are not valid ProModel element names. Which ones?
   a. Activities
   b. Locations
   c. Conveyors
   d. Queues
   e. Shifts
   f. Station
   g. Server
   h. Schedules
   i. Arrivals
   j. Expressions

    k. Variables

    l. Create

4. What are some of the valid logic statements used in ProModel?

5. What are some of the differences between the following logic statements:

    a. Wait **versus** Wait Until.

    b. Move **versus** Move For.

    c. Pause **versus** Stop.

    d. View **versus** Graphic.

    e. Split **versus** Ungroup.

6. Describe the functions of the following items in the ProModel Edit menu:

    a. Delete

    b. Insert

    c. Append

    d. Move

    e. Move to

    f. Copy Record

    g. Paste Record

7. Describe the differences between the following items in the ProModel View menu:

    a. Zoom vs. Zoom to Fit Layout

    b. Show Grid vs. Snap to Grid

3   RUNNING A PROMODEL SIMULATION

As far as the laws of mathematics refer to reality, they are not certain; and as far as they are certain, they do not refer to reality.
—Albert Einstein

Lab 3 provides some experience running ProModel simulations before building your first simulation model using ProModel in Lab 4. The example model used in this lab is of the ATM system that was simulated in Chapter 3 with a spreadsheet. Lab 3 demonstrates the efficiency with which systems can be simulated with ProModel.

L3.1 ATM System Specifications and Problem Statement

In a moment you will load and run a ProModel simulation of an automatic teller machine (ATM) system described as follows:

> Customers arrive to use an ATM at an average interarrival time of 3.0 minutes exponentially distributed. When customers arrive to the system, they join a queue to wait for their turn on the ATM. The queue has the capacity to hold an infinite number of customers. Customers spend an average of 2.4 minutes, exponentially distributed, at the ATM to complete their transactions, which is called the service time at the ATM.

The objective is to simulate the system to estimate the expected waiting time for customers in the queue (the average time customers wait in line for the ATM) and the expected time in the system (the average time customers wait in the queue plus the average time it takes them to complete their transaction at the ATM).

This is the same ATM system simulated by spreadsheet in Chapter 3 but with a different objective. The Chapter 3 objective was to simulate only the first 25 customers arriving to the system. Now no such restriction has been applied. This new

403

objective provides an opportunity for comparing the simulated results with those computed using queuing theory, which was presented in Section 2.9.3.

Queuing theory allows us to compute the exact values for the expected time that customers wait in the queue and in the system. Given that queuing theory can be used to get exact answers, why are we using simulation to estimate the two expected values? There are two parts to the answer. First, it gives us an opportunity to measure the accuracy of simulation by comparing the simulation output with the exact results produced using queuing theory. Second, most systems of interest are too complex to be modeled with the mathematical equations of queuing theory. In those cases, good estimates from simulation are valuable commodities when faced with expensive decisions.

L3.1.1 Queuing Theory's Answer to the ATM System

We learned in Chapter 2 that queuing theory can be used to calculate the steady-state or long-term expected values of particular performance measures. A system reaches steady state after it has been running awhile with no changes in the conditions influencing the system. After a stream of customers arrives to the ATM over time, the system will begin to reach steady state, which means that the statistical distributions that describe the output of the system stabilize. The steady-state topic is covered in detail in Chapter 9.

Using the queuing theory equations from Section 2.9.3, the expected time customers wait in the queue is 9.6 minutes and the expected amount of time customers spend in the system is 12.0 minutes. Let's use a ProModel simulation to estimate these values.

L3.1.2 ProModel's Answer to the ATM System

The ProModel model of the ATM system is provided with the Student ProModel CD accompanying the textbook. Load the ATM system simulation model using the File option on the ProModel menu bar across the top of the screen. Click on File and Open, and then select the Lab3_1_2 ATM System.mod file from the CD.

Begin the simulation by pressing the F10 function key. The animation is very informative and is more interesting to watch than was the Excel spreadsheet simulation of Chapter 3 (Figure L3.1).

The ProModel simulation is programmed to pause the simulation and display progress messages. To continue the simulation after reading a message, press the OK button at the bottom of the message (Figures L3.2 and L3.3).

The simulation results of an average 9.47 minutes in the queue and an average of 11.86 minutes in the system (Figure L3.4) are close to the steady-state expected values of 9.6 minutes in the queue and 12.0 minutes in the system computed using queuing theory. You can get closer to the exact values if needed. Chapter 9 discusses how to achieve a desired level of accuracy.

ProModel displays a prompt at the end of the simulation asking if you want to collect statistics; select "No" for now (Figure L3.5). ProModel automatically

FIGURE L3.1

ATM simulation in progress. System events are animated and key performance measures are dynamically updated.

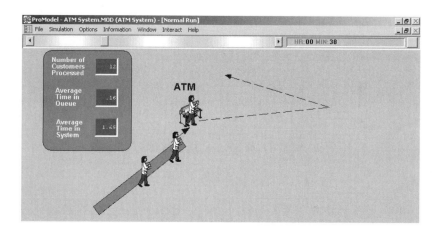

FIGURE L3.2

The simulation clock shows 1 hour and 20 minutes (80 minutes) to process the first 25 customers arriving to the system. The simulation takes only a second of your time to process 25 customers.

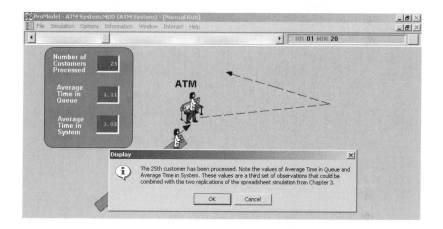

FIGURE L3.3

ATM simulation on its 15,000th customer and approaching steady state.

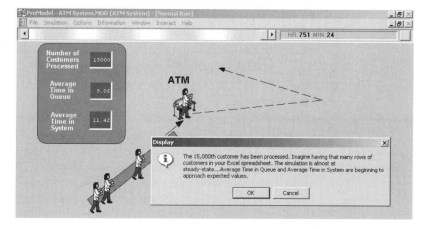

FIGURE L3.4

ATM simulation after reaching steady state.

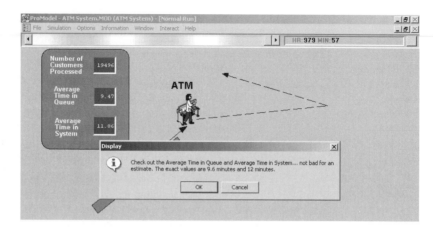

FIGURE L3.5

ATM simulation at its end.

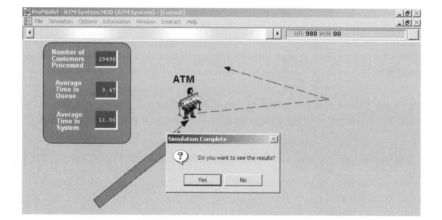

collects a multitude of statistics over the course of the simulation. You will learn this feature in Lab 4.

The simulation required only a minute of your time to process 19,496 customers. In Lab 4, you will begin to see how easy it is to build models using ProModel as compared to building them with spreadsheets.

L3.2 Exercises

1. The values obtained for average time in queue and average time in system from the ATM ProModel simulation of the first 25 customers processed represent a third set of observations that can be combined with the observations for the same performance measures presented in Table 3.3 of Section 3.5.3 in Chapter 3 that were derived from two

replications of the spreadsheet simulation. The ProModel output represents a third replication of the simulation.

   a. What must be true in order for the ProModel output to be counted as a third replication?

   b. Compute new estimates for the average time in queue and average time in system for the first 25 customers processed by combining the appropriate observations from the three replications (two spreadsheet replications and one ProModel replication). Do you think your new estimates are better than the previous estimates of 1.39 minutes for the average time in queue and 3.31 minutes for average time in system from Table 3.3? Why?

2. Select Locations from the ProModel Build menu with the Lab3_1_2 ATM System.mod file loaded. Notice that the Capacity (Cap.) field for the ATM_Queue location is set to INFINITY in the Locations Edit Table. The Cap. field defines the maximum number of entities that can simultaneously occupy the location. In this case, an infinite number of entities can occupy the location. Also, notice that FIFO (first-in, first-out) appears in the Rules field for the ATM_Queue location. FIFO indicates that entities waiting for the ATM are processed through the ATM_Queue in the order that they arrived to the location. Are these assumptions of queuing theory? Which assumption is the least realistic? What would you program into the ProModel model to make it a more realistic simulation of an ATM system?

3. Select Arrivals from the ProModel Build menu with the Lab3_1_2 ATM System.mod file loaded. The first two fields in the Arrivals Edit Table declare which entities are to arrive into the simulation and where they are to arrive. In this case, the entities are named ATM_Customer. They arrive to the ATM_Queue location. Notice that the Frequency field is set to E(3.0). The Frequency field controls the interarrival time of entities, which is exponentially distributed with a mean of 3.0 minutes, E(3.0), for the ATM system. Which column in Table 3.2 of the spreadsheet simulation in Chapter 3 corresponds to this Frequency field?

4. Select Processing from the ProModel Build menu with the Lab3_1_2 ATM System.mod file loaded. After the ATM_Customer arrives to the ATM_Queue and waits for its turn at the automatic teller machine, it is routed to the ATM location. The customer's transaction time at the ATM is modeled with the `Wait E(2.4)` statement under the Operation field of the processing record for the ATM_Customer at the ATM location. Which column in Table 3.2 of the spreadsheet simulation in Chapter 3 corresponds to this `Wait E(2.4)` statement in the Operation field?

4   BUILDING YOUR FIRST MODEL

Knowing is not enough; we must apply. Willing is not enough; we must do.
—Johann von Goethe

In this lab we build our first simulation model using ProModel. In Section L4.1 we describe some of the basic concepts of building your first ProModel simulation model. Section L4.2 introduces the concept of queue in ProModel. Section L4.3 lets us build a model with multiple locations and multiple entities. In Section L4.4 we show how to modify an existing model and add more locations to it. Finally, in Section L4.5 we show how variability in arrival time and customer service time affect the performance of the system.

L4.1 Building Your First Simulation Model

In this section we describe how you can build your very first simulation model using ProModel software. We introduce the concepts of locations, entities, entity arrivals, processes, and routing.

Customers visit the neighborhood barbershop **Fantastic Dan** for a haircut. The customer interarrival time is exponentially distributed with an average of 10 minutes. Dan (the barber) takes anywhere from 8 to 10 minutes, uniformly distributed (mean and half-width of 9 and 1 minute respectively) for each haircut. This time also includes the initial greetings and the transaction of money at the end of the haircut. Run the simulation model for one day (480 minutes). Find these answers:

    a. About how many customers does Dan process per day?

    b. What is the average number of customers waiting to get a haircut? What is the maximum?

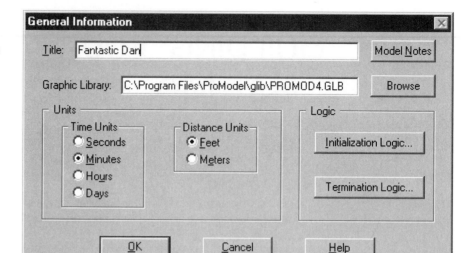

c. What is the average time spent by a customer in the salon?

d. What is the utilization of Barber Dan?

e. What is the maximum and average number of customers waiting for a haircut?

From the menu bar select File → New. In the General Information panel (Figure L4.1) fill in the title of the simulation model as "Fantastic Dan." Fill in some of other general information about the model like the time and distance units. Click OK to proceed to define the locations.

From the menu bar select Build → Locations. Define two locations— Waiting_for_Barber and Barber_Dan (Figure L4.2). Note that the first location is actually a region (it is the icon that looks like a square box). A region is a boundary used to represent a location's area. When the model is running the region is not visible. The icon selected for the second location is actually called operator. We changed its name to Barber_Dan (Name column in the Location table).

Check off the New button on the Graphics panel (Figure L4.3) and click the button marked Aa. Click on the location icon in the Layout panel. The name of the location (Barber_Dan) appears on the location icon.

Define the entity (Figure L4.4) and change its name to Customer. Define the processes and the routings (Figures L4.5 and L4.6) the customers go through at the barbershop. All customers arrive and wait at the location Waiting_for_Barber. Then they are routed to the location Barber_Dan. At this location the barber performs the haircut, which takes an amount of time uniformly distributed between 8 and 10 minutes or Uniform (9,1). Use the step-by-step procedure detailed in section L2.2.4 to create the process and routing tables graphically.

FIGURE L4.2

Defining locations Waiting_for_Barber and BarberDan.

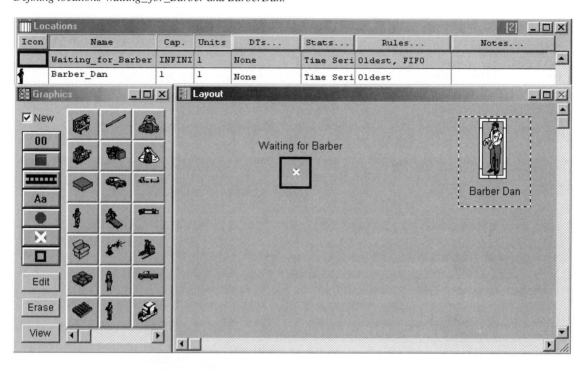

FIGURE L4.3

The Graphics panel.

FIGURE L4.4

Define the entity—Customer.

Icon	Name	Speed (fpm)	Stats...
	Customer	150	Time Series

Entities

FIGURE L4.5

Process and Routing tables for Fantastic Dan model.

Process [1]

Entity...	Location...	Operation...
Customer	iting_for_Barber	
Customer	Barber_Dan	Wait U(9, 1)

Routing for Customer @ Waiting_for_Barber [1]

Blk	Output...	Destination...	Rule...	Move Logic...
1	Customer	Barber_Dan	FIRST 1	

Process [2]

Entity...	Location...	Operation...
Customer	Waiting_for_Barb	
Customer	Barber_Dan	Wait U(9, 1)

Routing for Customer @ Barber_Dan [1]

Blk	Output...	Destination...	Rule...	Move Logic...
1	Customer	EXIT	FIRST 1	

FIGURE L4.6

Process and Routing tables for Fantastic Dan model in text format.

			Process		Routing		
Entity	Location	Operation	Blk	Output	Destination	Rule	Move Logic
Customer	Waiting_for_Barber		1	Customer	Barber_Dan	FIRST 1	
Customer	Barber_Dan	Wait U(9, 1)	1	Customer	EXIT	FIRST 1	

To define the haircut time, click Operation in the Process table. Click the button with the hammer symbol. A new window named Logic Builder opens up. Select the command Wait. The ProModel expression Wait causes the customer (entity) to be delayed for a specified amount of time. This is how processing times are modeled.

Click Build Expression. In the Logic window, select Distribution Function (Figure L4.7). In the Distribution Function window, select Uniform distribution. Click Mean and select 9. Click Half-Range and select 1. Click Return. Click Paste. Close the Logic Builder window. Close the Operation window.

Finally the customers leave the barbershop. They are routed to a default location called EXIT in ProModel. When entities (or customers) are routed to the EXIT location, they are in effect disposed from the system. All the information associated with the disposed entity is deleted from the computer's memory to conserve space.

FIGURE L4.7

*The Logic Builder
menu.*

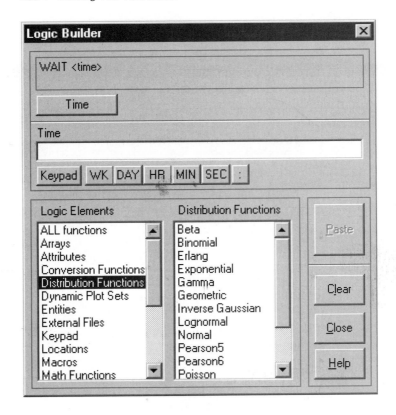

TABLE L4.1   Commonly Used Distribution Functions

Distribution	*ProModel Expression*
Uniform	U (mean, half-range)
Triangular	T (minimum, mode, maximum)
Exponential	E (mean)
Normal	N (mean, std. dev.)

The distribution functions are built into ProModel and generate random values based on the specified distribution. Some of the commonly used distribution functions are shown in Table 4.1.

Now we will define the entity arrival process, as in Figure L4.8.

Next we will define some of the simulation options—that is, run time, number of replications, warm-up time, unit of time, and clock precision (Figure L4.9). The run time is the number of hours the simulation model will be run. The number of replications refers to number of times the simulation model will be run (each time the model will run for an amount of time specified by run hours). The

FIGURE L4.8

Customer arrival table.

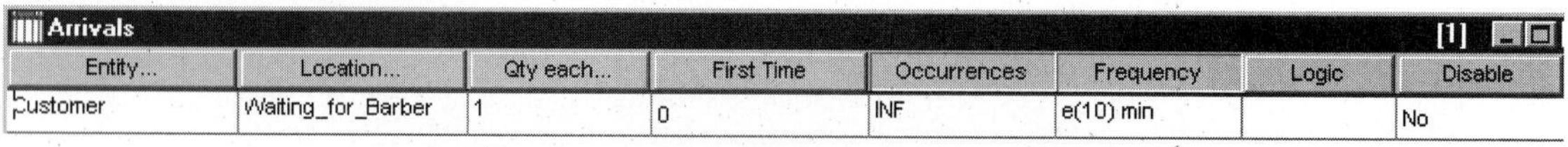

Entity...	Location...	Qty each...	First Time	Occurrences	Frequency	Logic	Disable
Customer	Waiting_for_Barber	1	0	INF	e(10) min		No

Arrivals [1]

FIGURE L4.9

Definition of simulation run options.

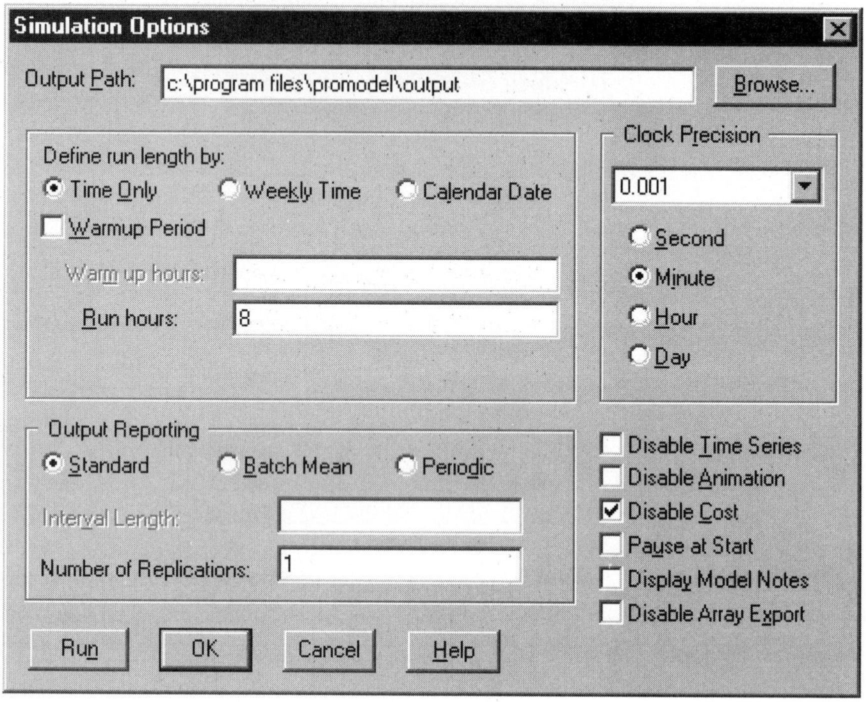

Simulation Options

Output Path: c:\program files\promodel\output   Browse...

Define run length by:
- ⦿ Time Only   ○ Weekly Time   ○ Calendar Date
- ☐ Warmup Period

Warm up hours: [   ]

Run hours: [8   ]

Clock Precision
[0.001   ▼]
- ○ Second
- ⦿ Minute
- ○ Hour
- ○ Day

Output Reporting
- ⦿ Standard   ○ Batch Mean   ○ Periodic

Interval Length: [   ]

Number of Replications: [1   ]

- ☐ Disable Time Series
- ☐ Disable Animation
- ☑ Disable Cost
- ☐ Pause at Start
- ☐ Display Model Notes
- ☐ Disable Array Export

[Run]   [OK]   [Cancel]   [Help]

warm-up time refers to the amount of time to let the simulation model run to achieve steady-state behavior. Statistics are usually collected after the warm-up period is over. The run time begins at the end of the warm-up period. The unit of time used in the model can be seconds, minutes, or hours. The clock precision refers to the precision in the time unit used to measure all simulation event timings.

Let us select the Run option from the Simulation Options menu (or click F10). Figure L4.10 shows a screen shot during run time. The button in the middle of the scroll bar at the top controls the speed of the simulation run. Pull it right to increase the speed and left to decrease the speed.

After the simulation runs to its completion, the user is prompted, "Do you want to see the results?" (Figure L4.11). Click Yes. Figures L4.12 and L4.13 are part of the results that are automatically generated by ProModel in the 3DR (three-dimensional report) Output Viewer.

FIGURE L4.10

Screen shot at run time.

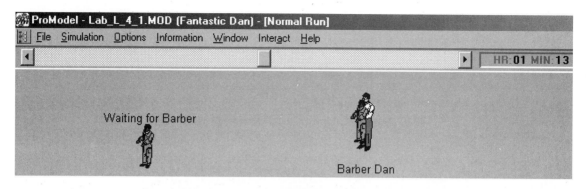

FIGURE L4.11

Simulation complete prompt.

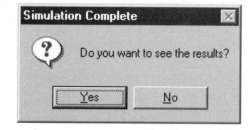

FIGURE L4.12

The 3DR Output Viewer for the Fantastic Dan model.

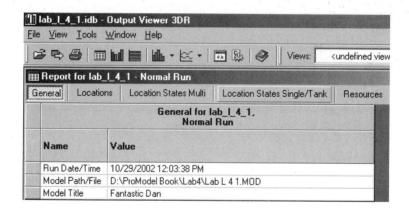

Note that the average time a customer spends waiting for Barber Dan is 22.95 minutes. The average time spent by a customer in the barbershop is 32.28 minutes. The utilization of the Barber is 89.15 percent. The number of customers served in 480 minutes (or 8 hours) is 47. On average 5.875 customers are served per hour. The maximum number of customers waiting for a haircut is 8, although the average number of customers waiting is only 2.3.

FIGURE L4.13

FIGURE **L4.13**

Results of the Fantastic Dan simulation model.

| General | Locations | Location States Multi | Location States Single/Tank | Resources | Resource States | Node Entries | Failed Arrivals | E |

Locations for lab_l_4_1,
Normal Run

Name	Scheduled Time (MIN)	Capacity	Total Entries	Avg Time Per Entry (MIN)	Avg Contents	Maximum Contents	Current Contents	Pct Utilization
Barber Dan	480.00	1	48	8.92	0.89	1	1	89.15
Waiting for Ba...	480.00	999999	48	22.95	2.30	8	0	0.00

| General | Locations | Location States Multi | Location States Single/Tank | Resources | Resou |

Location States Multi for lab_l_4_1,
Normal Run

Name	Scheduled Time (MIN)	Pct Empty	Pct Part Occupied	Pct Full	Pct Down
Waiting for Ba...	480.00	36.11	63.89	0.00	0.00

| General | Locations | Location States Multi | Location States Single/Tank | Resources | Resource States | Node Entries | Fail |

Location States Single/Tank for lab_l_4_1,
Normal Run

Name	Scheduled Time (MIN)	Pct Operation	Pct Setup	Pct Idle	Pct Waiting	Pct Blocked	Pct Down
Barber Dan	480.00	89.15	0.00	10.85	0.00	0.00	0.00

| Failed Arrivals | Entity Activity | Entity States | Variables | Location Costing | Resource Costing | Entity Costing | Logs |

Entity Activity for lab_l_4_1,
Normal Run

Name	Total Exits	Current Qty In System	Avg Time In System (MIN)	Avg Time In Move Logic (MIN)	Avg Time Wait For Res (MIN)	Avg Time In Operation (MIN)	Avg Time Blocked (MIN)
Customer	47	1	32.28	0.00	16.91	8.92	6.45

| Failed Arrivals | Entity Activity | Entity States | Variables | Location Costing | R |

Entity States for lab_l_4_1,
Normal Run

Name	Pct In Move Logic	Pct Wait For Res	Pct In Operation	Pct Blocked
Customer	0.00	52.40	27.62	19.98

L4.2 Building the Bank of USA ATM Model

In this section we describe how you can build a simplified version of the automatic teller machine (ATM) system model described in Lab 3, Section L3.1, using ProModel software. We also introduce the concept of a queue.

Customers arrive to use a **Bank of USA ATM.** The average customer inter-arrival time is 3.0 minutes exponentially distributed. When customers arrive to the system they join a queue to wait for their turn on the ATM. Customers spend an average of 2.4 minutes exponentially distributed at the ATM to complete their transactions, which is called the service time at the ATM. Build a simulation model of the Bank of USA ATM. Run the simulation model for 980 hours.

 a. About how many customers are served per hour?
 b. What is the average customer waiting time in the queue?
 c. What is the average time spent by a customer in the system?
 d. What is the utilization of the ATM?
 e. What are the maximum and average numbers of customers waiting in the ATM queue?

From the menu bar select File → New. In the General Information panel (Figure L4.14) fill in the title of the simulation model as "Bank of USA ATM." Fill in some of the other general information about the model like the time and distance units. Click OK to proceed to define the locations.

From the menu bar select Build → Locations. Define two locations—ATM and ATM_Queue (Figure L4.15). The icon selected for the first location is actually called brake. We changed its name to ATM (Name column in the Location table). The icon for the second location (a queue) is selected from the graphics panel. The icon (third from top) originally looks like a "ladder on its side." To place it in our model layout first, left-click the mouse at the start location of the queue. Then drag the mouse pointer to the end of the queue and right-click. Change the name of this queue location from Loc1 → ATM_Queue. Now double-click the ATM_Queue icon on the layout. This opens another window as follows (Figure L4.16). Make sure to click on the Queue option in the Conveyor/Queue options window. Change the length of the queue to be exactly 31.413 feet.

FIGURE L4.14

General Information for the Bank of USA ATM simulation model.

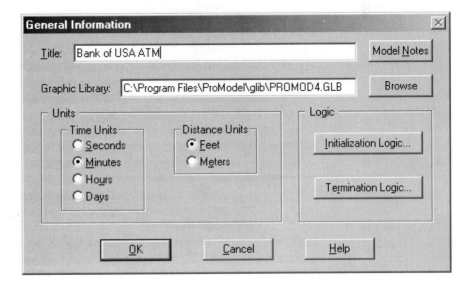

FIGURE L4.15

Defining locations ATM_Queue and ATM.

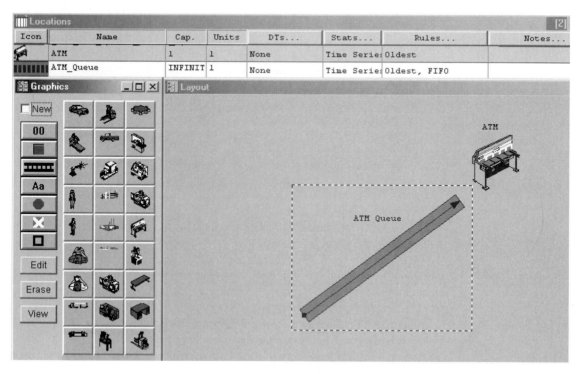

FIGURE L4.16

Click on the Queue option in the Conveyor/Queue options window.

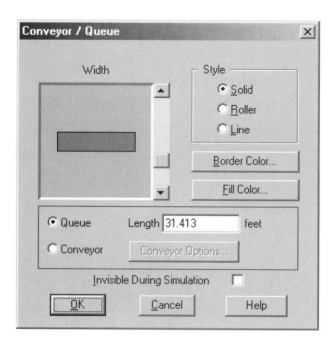

Check off the New button on the graphics panel (Figure L4.15) and click the button marked Aa (fourth icon from top). Click on the location icon in the layout panel. The name of the location (ATM) appears on the location icon. Do the same for the ATM_Queue location.

Define the entity (Figure L4.17) and change its name to ATM_Customer. Define the processes and the routings (Figures L4.18 and L4.19) the customers go through at the ATM system. All customers arrive and wait at the location ATM_Queue. Then they are routed to the location ATM. At this location the customers deposit or withdraw money or check their balances, which takes an average of 2.4 minutes exponentially distributed. Use the step-by-step procedure detailed in section L2.2.4 to create the process and routing tables graphically.

To define the service time at the ATM, click Operation in the Process table. Click the button with the hammer symbol. A new window named Logic Builder opens up. Select the command Wait. The ProModel expression Wait causes the ATM customer (entity) to be delayed for a specified amount of time. This is how processing times are modeled.

FIGURE L4.17

Define the entity—ATM_Customer.

Icon	Name	Speed (fpm)	Stats...
👤	ATM_Customer	150	Time Series

FIGURE L4.18

Process and Routing tables for Bank of USA ATM model.

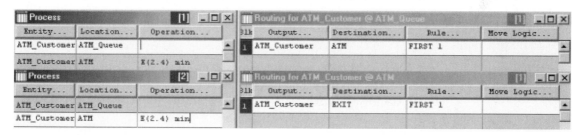

FIGURE L4.19

Process and Routing tables for Bank of USA ATM model in text format.

```
*******************************************************************************
*                            Processing                                      *
*******************************************************************************

                        Process                        Routing

Entity        Location   Operation        Blk  Output        Destination  Rule     Move Logic
--------      --------   ---------        ---  ------        -----------  ----     ----------
ATM_Customer  ATM_Queue                   1    ATM_Customer  ATM          FIRST 1
ATM_Customer  ATM        E(2.4) min       1    ATM_Customer  EXIT         FIRST 1
```

FIGURE L4.20

The Logic Builder menu.

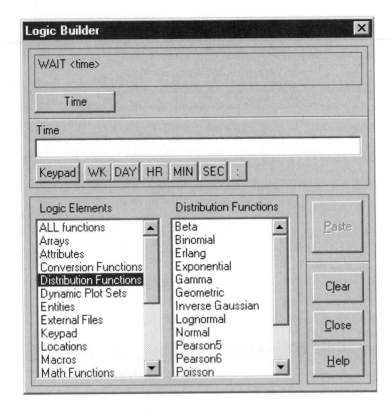

FIGURE L4.21

Customer arrival table.

Entity...	Location...	Qty each...	First Time	Occurrences	Frequency	Logic	Disable
ATM_Customer	ATM_Queue	1		INF	E(3.0)		No

Click Build Expression. In the Logic window, select Distribution Functions (Figure L4.20). In the Distribution Functions window, select Exponential distribution. Click Mean and select 2.4. Click Return. Click MIN. Click Paste. Close the Logic Builder window. Close the Operation window.

Finally the customers leave the Bank of USA ATM. They are routed to a default location called EXIT in ProModel. When entities (or customers) are routed to the EXIT location, they are in effect disposed from the system. All the information associated with the disposed entity is deleted from the computer's memory to conserve space.

Now we will define the entity arrival process, as in Figure L4.21.

Next we will define some of the simulation options—that is, run time, number of replications, warm-up time, unit of time, and clock precision (Figure L4.22). The run time is the number of hours the simulation model will be run. In this example

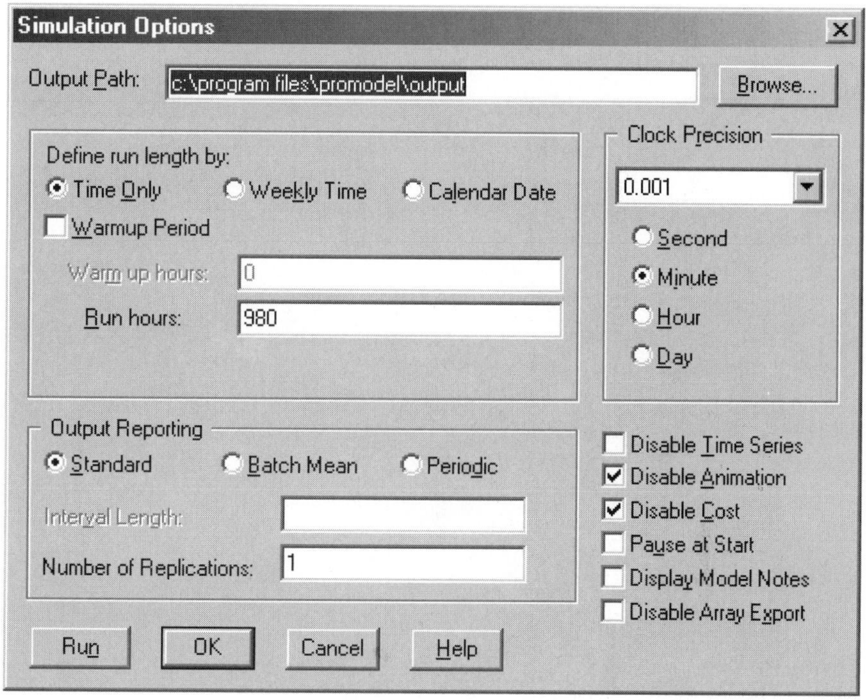

FIGURE L4.22

*Definition of
simulation run options.*

we are going to model 980 hours of operation of the ATM system. The number of
replications refers to the number of times the simulation model will be run (each
time the model will run for an amount of time specified by run hours). The warm-up
time refers to the amount of time to let the simulation model run to achieve steady-
state behavior. Statistics are usually collected after the warm-up period is over. The
run time begins at the end of the warm-up period. For a more detailed discussion on
warm-up time, please refer to Chapter 9, Section 9.6.1 and Lab 9. The unit of time
used in the model can be seconds, minutes, or hours. The clock precision refers to the
precision in the time unit used to measure all simulation event timings.

Let us select the Run option from the Simulation Options menu (or click
F10). Figure L4.23 shows a screen shot during run time. The button in the middle
of the scroll bar at the top controls the speed of the simulation run. Pull it right to
increase the speed and left to decrease the simulation execution speed.

After the simulation runs to its completion, the user is prompted, "Do you want
to see the results?" (Figure L4.24). Click Yes. Figures L4.25 and L4.26 are part of
the results that are automatically generated by ProModel in the Output Viewer.

Note that the average time a customer spends waiting in the ATM Queue
is 9.62 minutes. The average time spent by a customer in the ATM system is
12.02 minutes. The utilization of the ATM is 79.52 percent. Also, 20,000 customers
are served in 60,265.64 minutes or 19.91 customers per hour. The maximum number of customers waiting in the ATM Queue is 31, although the average number of

FIGURE L4.23

Screen shot at run time.

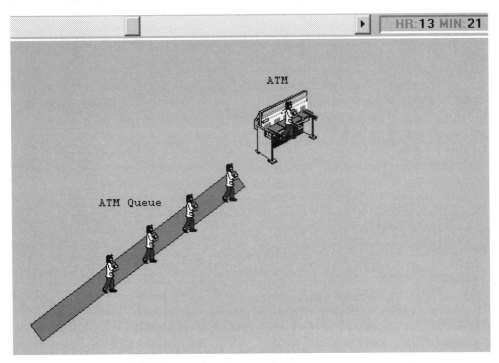

FIGURE L4.24

Simulation complete prompt.

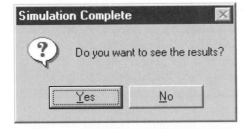

FIGURE L4.25

The output viewer for the Bank of USA ATM model.

Report for bankofusa_atm - Normal Run			
General	Locations	Location States Multi	Location States Single/Tank

	General for bankofusa_atm, Normal Run	
Name	**Value**
Run Date/Time	10/29/2002 9:38:28 AM
Model Path/File	C:\WINDOWS\Desktop\BankofUSA ATM.MOD
Model Title	Bank of USA ATM

FIGURE L4.26

Results of the Bank of USA ATM simulation model.

| General | Locations | Location States Multi | Location States Single/Tank | Resources | Resource States | Node Entries | Failed Arrivals | E |

**Locations for lab_l_4_2,
Normal Run**

Name	Scheduled Time [MIN]	Capacity	Total Entries	Avg Time Per Entry [MIN]	Avg Contents	Maximum Contents	Current Contents	Pct Utilization
ATM	60265.64	1	20000	2.40	0.80	1	0	79.52
ATM Queue	60265.64	999999	20000	9.62	3.19	31	0	0.00

| General | Locations | Location States Multi | Location States Single/Tank | Resources | Resou |

**Location States Multi for lab_l_4_2,
Normal Run**

Name	Scheduled Time [MIN]	Pct Empty	Pct Part Occupied	Pct Full	Pct Down
ATM Queue	60265.64	34.67	65.33	0.00	0.00

| General | Locations | Location States Multi | Location States Single/Tank | Resources | Resource States | Node Entries | Fai |

**Location States Single/Tank for lab_l_4_2,
Normal Run**

Name	Scheduled Time [MIN]	Pct Operation	Pct Setup	Pct Idle	Pct Waiting	Pct Blocked	Pct Down
ATM	60265.64	79.52	0.00	20.48	0.00	0.00	0.00

| Failed Arrivals | Entity Activity | Entity States | Variables | Location Costing | Resource Costing | Entity Costing | Logs |

**Entity Activity for lab_l_4_2,
Normal Run**

Name	Total Exits	Current Qty In System	Avg Time In System [MIN]	Avg Time In Move Logic [MIN]	Avg Time Wait For Res [MIN]	Avg Time In Operation [MIN]	Avg Time Blocked [MIN]
ATM Customer	20000	0	12.02	0.00	7.52	2.61	1.89

| Failed Arrivals | Entity Activity | Entity States | Variables | Location Costing | R |

**Entity States for lab_l_4_2,
Normal Run**

Name	Pct In Move Logic	Pct Wait For Res	Pct In Operation	Pct Blocked
ATM Customer	0.00	62.57	21.68	15.75

customers waiting is only 3.19. This model is an enhancement of the ATM model in Lab 3, Section L3.1.2. Results will not match exactly as some realism has been added to the model that cannot be addressed in queuing theory.

L4.3 Locations, Entities, Processing, and Arrivals

In Sections L4.1 and L4.2 we have used the basic elements of ProModel. In this section we incorporate locations, entities, arrivals, processing, and routing logic into another simulation model example.

Problem Statement—Multiple Locations, Multiple Entities

Wooden logs are received at the receiving dock of the **Poly Furniture Factory** at the rate of one every 10 minutes. Logs go to the splitter, where four pieces are made from each log. The splitting time is Normal (4,1) minutes. The individual pieces go to the lathe, where they are turned for another Triangular (3,6,9) minutes and made into rounds. The rounds go on to a paint booth, where they are converted into painted logs. Painting takes Exponential (5) minutes. Painted logs go to the store. Consider a material handling time of one minute between each process. Make a simulation model and run the simulation for 10 hours.

To build this model first select General Information under the Build menu. Fill in some of the general information about the model (Figure L4.27). Define the locations (Figure L4.28)—receiving dock, splitter saw, lathe, paint booth, and painted logs store. Note that the capacity of the receiving dock location has been changed to infinity (INF) to make sure that all incoming entities (raw material) are allowed into the manufacturing facility.

Define the entities (Figure L4.29)—logs, piece, rounds, and painted logs. Define the entity arrival process (Figure L4.30)—logs arriving to the location receiving dock at the rate of one every 10 minutes.

Define the processes and the routings (Figure L4.31). All logs arrive at the receiving dock. From there they go to the splitter. At the splitter each log is made into four pieces. To be able to model this, click on the Rule button in the routing table. Change the quantity to four. This models the process of one log going into the location splitter saw and four pieces coming out. Pieces are routed to the lathe, where they become rounds. Rounds are routed to the paint booth, where they become painted logs. Painted logs are sent to the painted log store. Finally, the painted logs are sent to the default location EXIT for disposal.

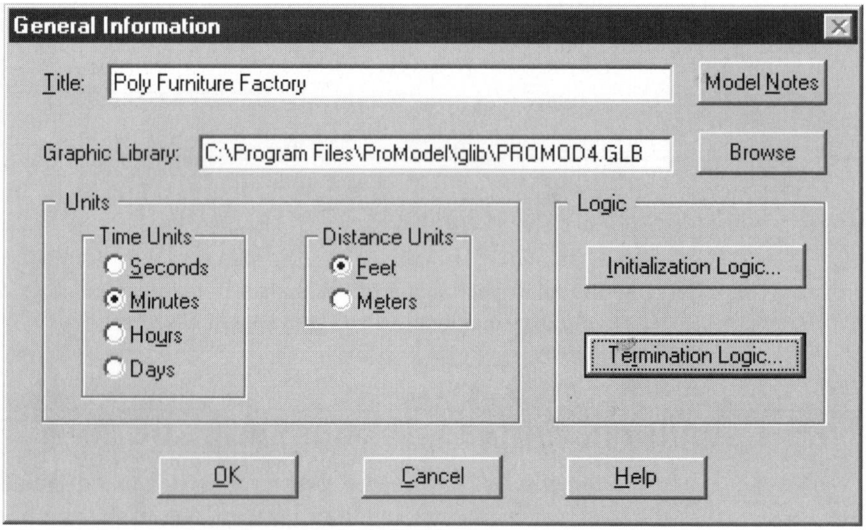

FIGURE L4.27

General information for the Poly Furniture Factory simulation.

FIGURE L4.28

Locations in the Poly Furniture Factory.

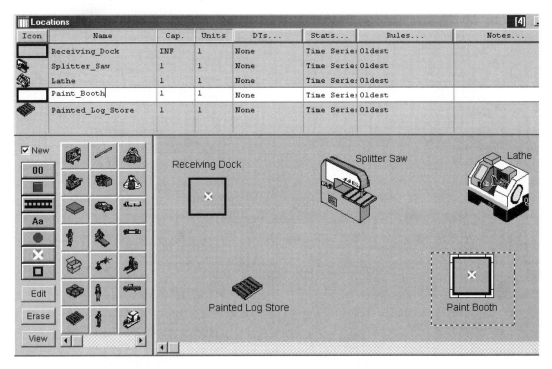

FIGURE L4.29

Entities in the Poly Furniture Factory.

Icon	Name	Speed (fpm)	Stats...	Notes...
	Logs	150	Time Series	
	Piece	150	Time Series	
	Rounds	150	Time Series	
	Painted_Logs	150	Time Series	

FIGURE L4.30

Entity arrivals in the Poly Furniture Factory.

Entity...	Location...	Qty each...	First Time	Occurrences	Frequency	Logic	Disable
Logs	Receiving_Dock	1	0	INF	10		No

FIGURE L4.31

Processes and routings in the Poly Furniture Factory.

Entity	Location	Operation	Blk	Output	Destination	Rule	Move Logic
Logs	Receiving_Dock		1	Logs	Splitter_Saw	FIRST 1	MOVE FOR 1
Logs	Splitter_Saw	Wait N(4,1)	1	Piece	Lathe	FIRST 4	MOVE FOR 1
Piece	Lathe	Wait T(3,6,9)	1	Rounds	Paint_Booth	FIRST 1	MOVE FOR 1
Rounds	Paint_Booth	Wait E(5)	1	Painted_Logs	Painted_Log_Store	FIRST 1	MOVE FOR 1
Painted_Logs	Painted_Log_Store		1	Painted_Logs	EXIT	FIRST 1	

FIGURE L4.32

Simulation options in the Poly Furniture Factory.

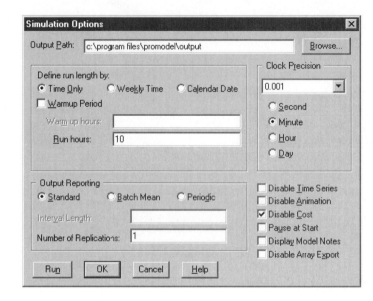

The time to move material between processes is modeled in the Move Logic field of the Routing table. Four choices of constructs are available in the Move Logic field:

- MOVE—to move the entity to the end of a queue or conveyor.
- MOVE FOR—to move the entity to the next location in a specific time.
- MOVE ON—to move the entity to the next location using a specific path network.
- MOVE WITH—to move the entity to the next location using a specific resource (forklift, crane).

Define some of the simulation options: the simulation run time (in hours), the number of replications, the warm-up time (in hours), and the clock precision (Figure L4.32).

Now we go on to the Simulation menu. Select Save & Run. This will save the model we have built so far, compile it, and also run it. When the simulation model finishes running, we will be asked if we would like to view the results. Select Yes. A sample of the results is shown in Figure L4.33.

FIGURE L4.33

Sample of the results of the simulation run for the Poly Furniture Factory.

lab_l_4_3.idb - Output Viewer 3DR - [Report for lab_l_4_3 - Normal Run]

File   View   Tools   Window   Help

Views: <undefined view>

General | Locations | Location States Multi | Location States Single/Tank | Resources | Resource States | Node Entries | Failed Arrivals

Locations for lab_l_4_3, Normal Run

Name	Scheduled Time [MIN]	Capacity	Total Entries	Avg Time Per Entry [MIN]	Avg Contents	Maximum Contents	Current Contents	Pct Utilization
Lathe	600.00	1	69	7.63	0.88	1	1	87.72
Paint Booth	600.00	1	68	5.33	0.60	1	0	60.43
Painted Log S...	600.00	1	68	0.00	0.00	1	0	0.00
Receiving Dock	600.00	999999	61	209.52	21.30	43	43	0.00
Splitter Saw	600.00	1	18	32.33	0.97	1	1	97.00

General | Locations | Location States Multi | Location States Single/Tank | Resources | Resou

Location States Multi for lab_l_4_3, Normal Run

Name	Scheduled Time [MIN]	Pct Empty	Pct Part Occupied	Pct Full	Pct Down
Receiving Dock	600.00	1.67	98.33	0.00	0.00

General | Locations | Location States Multi | Location States Single/Tank | Resources | Resource States | Node Entries | Fai

Location States Single/Tank for lab_l_4_3, Normal Run

Name	Scheduled Time [MIN]	Pct Operation	Pct Setup	Pct Idle	Pct Waiting	Pct Blocked	Pct Down
Lathe	600.00	68.99	0.00	12.28	0.00	18.73	0.00
Paint Booth	600.00	60.43	0.00	39.57	0.00	0.00	0.00
Painted Log S...	600.00	0.00	0.00	100.00	0.00	0.00	0.00
Splitter Saw	600.00	12.34	0.00	3.00	0.00	84.66	0.00

Node Entries | Failed Arrivals | Entity Activity | Entity States | Variables | Location Costing | Resource Costing | Entity Costing

Entity Activity for lab_l_4_3, Normal Run

Name	Total Exits	Current Qty In System	Avg Time In System [MIN]	Avg Time In Move Logic [MIN]	Avg Time Wait For Res [MIN]	Avg Time In Operation [MIN]	Avg Time Blocked [MIN]
Logs	0	44	0.00	0.00	0.00	0.00	0.00
Painted Logs	68	0	233.32	4.00	0.00	15.53	213.80
Piece	0	1	0.00	0.00	0.00	0.00	0.00
Rounds	0	0	0.00	0.00	0.00	0.00	0.00

Node Entries | Failed Arrivals | Entity Activity | Entity States | Variables | Local

Entity States for lab_l_4_3, Normal Run

Name	Pct In Move Logic	Pct Wait For Res	Pct In Operation	Pct Blocked
Logs	0.00	0.00	0.00	0.00
Painted Logs	1.71	0.00	6.65	91.63
Piece	0.00	0.00	0.00	0.00
Rounds	0.00	0.00	0.00	0.00

L4.4 Add Location

Now let us add a location to an existing model. For the Poly Furniture Factory example in Section L4.3, we will add an oven after the painting booth for drying the painted logs individually. The drying takes a time that is normally distributed with a mean of 20 minutes and standard deviation of 2 minutes. After drying, the painted logs go on to the painted logs store.

To add an additional location to the model, select Locations from the Build menu. With the mouse, first select the appropriate resource icon from the Graphics toolbar (left-click), then left-click again in the Layout window. Change the name of the location (in the Locations table) from Loc1 to Oven (Figure L4.34).

Locations can also be added from the Edit menu. Select Append from the Edit menu. A new location called Loc1 is appended to the end of the Locations table. Change the name to Oven. Deselect the New option in the Graphics toolbar. Select an appropriate icon from the Graphics toolbar. Click in the Layout window.

To display the names of each location in the Layout window, first select a location (left-click). Next deselect the New option in the graphics toolbar. Left-click on the command button Aa in the Graphics toolbar (left column, fourth from top). Finally, left-click on the location selected. The name of the location will appear in the Layout window. Now it can be repositioned and its fonts and color changed.

The complete simulation model layout is shown in Figure L4.35. Note that the capacity of the oven as well as the paint booth has been changed to 10. The processes and routings are created as shown in Figure L4.36. At this point it is worthwhile to note the differences between the capacity and the units of a location.

- *Capacity:* This is the number of units of entities that the location can hold simultaneously. The default capacity of a location is one.
- *Units:* A multiunit location consists of several identical units that are referenced as a single location for processing and routing purposes. A multiunit location eliminates the need to create multiple locations and multiple processes for locations that do the same thing. The default number of units of a location is one.

FIGURE L4.34

Locations at the Poly Furniture Factory with oven.

Icon	Name	Cap.	Units	DTs...	Stats...	Rules...
	Receiving_Dock	INF	1	None	Time Series	Oldest
	Splitter_Saw	1	1	None	Time Series	Oldest
	Lathe	1	1	None	Time Series	Oldest
	Paint_Booth	10	1	None	Time Series	Oldest
	Painted_Log_Store	1	1	None	Time Series	Oldest
	Oven	10	1	None	Time Series	Oldest

FIGURE L4.35

Simulation model layout of the Poly Furniture Factory.

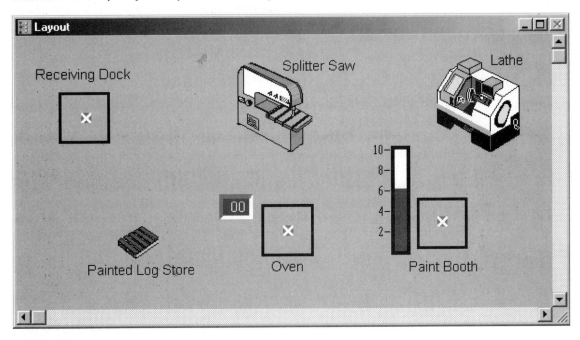

FIGURE L4.36

Processes and routings at the Poly Furniture Factory.

		Process			Routing		
Entity	Location	Operation	Blk	Output	Destination	Rule	Move Logic
Logs	Receiving_Dock		1	Logs	Splitter_Saw	FIRST 1	MOVE FOR 1
Logs	Splitter_Saw	Wait N(4,1)	1	Piece	Lathe	FIRST 4	MOVE FOR 1
Piece	Lathe	Wait T(3,6,9)	1	Rounds	Paint_Booth	FIRST 1	MOVE FOR 1
Rounds	Paint_Booth	Wait E(5)	1	Painted_Logs	Oven	FIRST 1	MOVE FOR 1
Painted_Logs	Oven	Wait N(20,2)	1	Painted_Logs	Painted_Log_Store	FIRST 1	MOVE FOR 1
Painted_Logs	Painted_Log_Store		1	Painted_Logs	EXIT	FIRST 1	

The contents of a location can be displayed in one of the following two alternative ways:

    a. To show the contents of a location as a counter, first deselect the New option from the Graphics toolbar. Left-click on the command button 00 in the Graphics toolbar (left column, top). Finally, left-click on the location selected (Oven). The location counter will appear in the Layout window next to the location Oven (Figure L4.35).

    b. To show the contents of a location (Paint Booth) as a gauge, first deselect the New option from the Graphics toolbar. Left-click on the second command button from the top in the left column in the Graphics toolbar.

The gauge icon will appear in the Layout window next to the location Paint Booth (Figure L4.35). The fill color and fill direction of the gauge can now be changed if needed.

L4.5 Effect of Variability on Model Performance

Variability in the data makes a big impact on the performance of any system. Let us take the example of Fantastic Dan, the barbershop in Section L4.1. Assume that one customer arrives for getting a haircut every 10 minutes. Also, Dan takes exactly nine minutes for each haircut.

Let us modify the model of Section L4.1 to reflect the above changes. The arrival table is changed as shown in Figure L4.37 and the process table is changed as shown in Figure L4.38. The results of the model with and without variability are compared and shown in Table L4.2.

FIGURE L4.37

Customer arrival for haircut.

Entity...	Location...	Qty each...	First Time	Occurrences	Frequency	Logic	Disable
Customer	Waiting_for_Barb	1	0	INF	10 min		No

FIGURE L4.38

Processing of customers at the barbershop.

Process				Routing for Customer @ Waiting_for_Barber			
Entity...	Location...	Operation...	Blk	Output...	Destination...	Rule...	Move Logic...
Customer	Waiting_for_			Customer	Barber_Dan	FIRST 1	
Customer	Barber_Dan	Wait 9					

TABLE L4.2 **Comparison of the Barbershop Model with and without Variability**

	With Variability	Without Variability
Average customer time at the barbershop	32.27 min.	9 min.
Average waiting in line for barber	22.95 min.	0 min.

L4.6 Blocking

With respect to the way statistics are gathered, here are the rules that are used in ProModel (see the ProModel Users Manual, p. 636):

1. *Average <time> in system:* The average total time the entity spends in the system, from the time it arrives till it exits the system.

2. *Average <time> in operation:* The average time the entity spends in processing at a location (due to a WAIT statement) or traveling on a conveyor or queue.

3. *Average <time> in transit:* The average time the entity spends traveling to the next location, either in or out of a queue or with a resource. The move time in a queue is decided by the length of the queue (defined in the queue dialog, Figure L4.16) and the speed of the entity (defined in the entity dialog, Figure L4.4 or L4.17).

4. *Average <time> wait for resource, etc.:* The average time the entity spends waiting for a resource or another entity to join, combine, or the like.

5. *Average <time> blocked:* The average time the entity spends waiting for a destination location to become available. Any entities held up behind another blocked entity are actually waiting on the blocked entity, so they are reported as "time waiting for resource, etc."

Example

At the **SoCal Machine Shop** (Figure L4.39) gear blanks arriving to the shop wait in a queue (Incoming_Q) for processing on a turning center and a mill, in that

FIGURE L4.39

The Layout of the SoCal Machine Shop.

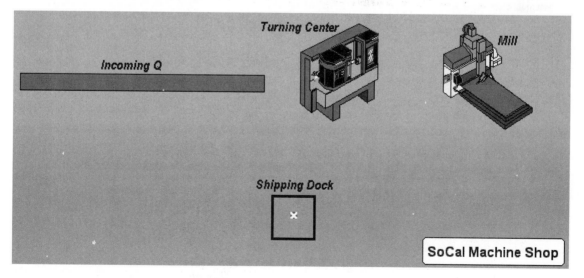

order. A total of 100 gear blanks arrive at the rate of one every eight minutes. The processing times on the turning center and mill are eight minutes and nine minutes, respectively. Develop a simulation model and run it.

To figure out the time the "gear blanks" are blocked in the machine shop, waiting for a processing location, we have entered "Move for 0" in the operation logic (Figure L4.40) of the Incoming_Q. Also, the decision rule for the queue has been changed to "No Queuing" in place of FIFO (Figure L4.41). This way all the entities waiting in the queue for the turning center to be freed up are reported as blocked. When you specify FIFO as the queuing rule for a location, only the lead entity is ever blocked (other entities in the location are waiting for the lead entity and are reported as "wait for resource, etc.").

FIGURE L4.40

Process and Routing tables for SoCal Machine Shop.

```
***********************************************************************************
*                                Processing                                      *
***********************************************************************************

                                   Process                          Routing

Entity      Location        Operation        Blk  Output      Destination      Rule
----------  --------------  ---------------   ---  ----------  ---------------  -------
Gear_Blank  Incoming_Q      Move for 0         1   Gear_Blank  Turning_Center  FIRST 1
Gear_Blank  Turning_Center  WAIT 8 min
                                               1   Gear_Blank  Mill            FIRST 1
Gear_Blank  Mill            WAIT 9 min         1   Gear_Blank  Shipping_Dock   FIRST 1
Gear_Blank  Shipping_Dock                      1   Gear_Blank  EXIT            FIRST 1
```

FIGURE L4.41

Decision rules for Incoming_Q.

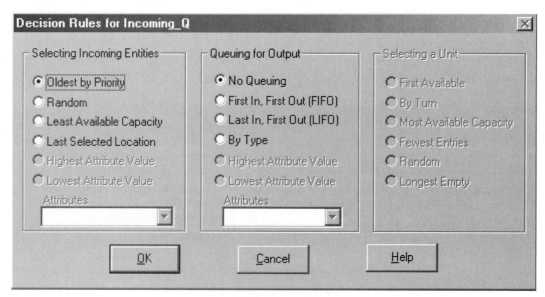

FIGURE L4.42

Entity activity statistics at the SoCal Machine Shop.

ENTITY ACTIVITY

Entity Name	Total Exits	Current Quantity In System	Average Minutes In System	Average Minutes In Move Logic	Average Minutes Wait For Res, etc.	Average Minutes In Operation	Average Minutes Blocked
Gear Blank	100	0	66.50	0.0	0.0	17.00	49.50

FIGURE L4.43

Entity activity statistics at the SoCal Machine Shop with two mills.

ENTITY ACTIVITY

Entity Name	Total Exits	Current Quantity In System	Average Minutes In System	Average Minutes In Move Logic	Average Minutes Wait For Res, etc.	Average Minutes In Operation	Average Minutes Blocked
Gear Blank	100	0	17.00	0.0	0.0	17.00	0.0

From the entity activity statistics (Figure L4.42) in the output report we can see the entities spend on average or 66.5 minutes in the system, of which 49.5 minutes are blocked (waiting for another process location) and 17 minutes are spent in operation (eight minutes at the turning center and nine minutes at the mill). Blocking as a percentage of the average time in system is 74.44 percent. The utilization of the turning center and the mill are 98.14 percent and 98.25 percent, respectively.

In general, the blocking time as a percentage of the time in system increases as the utilization of the processing locations increases. To reduce blocking in the machine shop, let us install a second mill. In the location table, change the number of units of mill to 2. The entity activity statistics from the resulting output report are shown in Figure L4.43. As expected, the blocking time has been reduced to zero.

L4.7 Exercises

1. Run the Receiving Dock simulation example model (*Receive.mod*) from the Demos subdirectory for 80 hours. Please answer the following questions:

   a. Average number of pallets put away?

   b. Average number of items put away?

   c. Average pallets per load?

   d. Average items per load?

    e. What are the cycle times for semi-trucks and small trucks?

    f. What are the utilizations of the three docks?

    g. What are the utilizations of the three forklift trucks?

    h. What is the utilization of the supervisor?

    i. Average time (in minutes) a truck spends in the system? Average time (in minutes) a semi spends in the system?

    j. How many pallets are used in the Receiving faility?

2. Run the Semiconductor Cluster Tool model (*Semicon.mod*) from the Demos subdirectory. Select Simulation → Scenarios → scen1. Run this scenario and answer the following:

    a. What is the number of wafers processed per hour?

    b. What is the steady state throughput of wafers per hour?

    c. What is the utilization of the robot?

    d. What is the average value of wip?

    e. The percentage utilization of Soft Etch and VCE1?

3. Run the Order Processing simulation model (*Orders.mod*) from the Demos subdirectory. This is a model of a proposed order processing system. The objectives of the model were to assist in communicating concepts involved with the proposed order processing system, focusing on the flow of information and the actual operating procedures. Please answer the following questions:

    a. How many dockworkers are there?

    b. What is the average utilization of the dockworkers?

    c. What is the average and maximum processing time (current value column in results file)?

    d. What are the utilizations of the forklift trucks—ShipHiLo and PickHiLo?

    e. What is the throughput volume (current value column in results file)?

4. Run the Runoff simulation model (Run-off.mod) from the Demos subdirectory for 20 hours. In this example, six different machine configurations are run-off on the same simulation screen. Each machining center is connected by conveyor sections. In each case, the final product is identical. Please answer the following questions:

    a. Amonst options 1–6 which option is better in terms of throughput?

    b. What are the average times a part spends in the system for each of the six optional machine configurations? Which configuration is better?

    c. How does the cost of the part compare for the six options?

5. For the example in Section L4.1 (Fantastic Dan), run the simulation model for a whole year (250 days, eight hours each day) and answer the following questions:

    a. On average, how many customers does Dan serve each day?

    b. What is the average number of customers waiting to get a haircut? What is the maximum?

    c. What is the average time spent by a customer in the salon?

  d. How busy is Dan on average? Is this satisfactory?

  e. How many chairs should Dan have for customers waiting for a haircut?

6. If Dan could take exactly nine minutes for each haircut, will it improve the situation? Rework Question 1 and answer parts *a* through *e*.

7. For the example in Section L4.1, what will happen if we change the capacity of the location Waiting_for_Barber to one? Is this change desirable? What are the implications? Make the appropriate change in the model and show the results to support your argument.

8. For the Bank of USA ATM example in Section L4.2, run the model with and without any variability in the customer interarrival time and service time. Compare the average time in system and the average time in queue based on 1000 hours of simulation runs.

9 . For the SoCal Machine Shop in Section L4.6 assume the length of the Incoming_Q is 30 feet, and the speed of the arriving Gear_Blanks is 10 feet/minute. Delete the statement `Move for 0` in the operation logic of the Incoming_Q. Change the queuing rule to FIFO. Answer the following questions and explain why.

  a. What is the average time in operation?

  b. What is the average time waiting for resource?

  c. What is the average time blocked?

10. In the Arrivals element (table), what are the differences between Occurrences and Frequency? What is the significance of the Qty each column?

11. In the Locations element (table), what are the differences between Cap and Units?

12. What are the various time units that can be used while developing a model? Where is this information provided? If one forgets to specify the time unit in the model, what time unit is used by ProModel?

13. The Processing element has two tables that need to be edited. What are their names and their functions?

14. When an entity has completed all the processing in the system, what should we do with it? Where should it be routed?

15. Differentiate between the following:

  a. Entity versus locations.

  b. Locations versus resources.

  c. Attributes versus variables.

  d. Save versus Save As.

  e. `Move, Move For, Move On, Move With`.

  f. `Wait` versus `Wait Until`.

  g. `Stop` versus `End`.

16. Increased utilization of processing locations or resources lead to increased levels of blocking in the system. Explain why.

5   PROMODEL'S OUTPUT MODULE

I am not discouraged, because every wrong attempt discarded is another step forward.
—Thomas Edison

In this chapter we discuss ProModel's Output Program Manager in detail with examples.

L5.1 The Output Program Manager

The ProModel Output Program Manager, also called Output Viewer 3DR (three-dimensional report), can be run either as a stand-alone application by choosing View Statistics from the Output menu or from within the Windows Program Manager by choosing Results Viewer. This allows us to view ProModel output files at any time, with or without starting ProModel. The Results Viewer has two options—3DR, and Classic. The default output viewer can be selected in the Tools → Options menu in ProModel. The Classic output viewer uses a proprietary binary file format that works faster. However, the 3DR output viewer produces much nicer output. Some examples of the output view in the 3DR mode are shown in Figures L4.13, L4.26, and L4.33. The output viewer simplifies the process of generating and charting graphs and reports and analyzing the output data. We encourage you to use the ProModel's online help to learn more about the Output Program Manager.

The File menu (Figure L5.1) allows the user to open one or more output databases for review, analysis, and comparison. It also allows the user to export raw data to a common delimited format (*.CSV), to keep reports and graphs for all subsequent runs, to print, and to set up printing.

FIGURE L5.1

File menu in the 3DR Output Viewer.

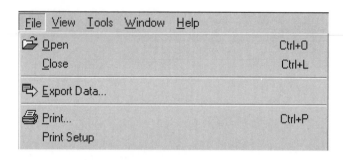

FIGURE L5.2

Output menu options in ProModel.

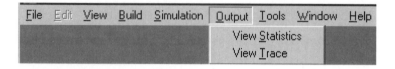

FIGURE L5.3

View menu in the 3DR Output Viewer.

The Output menu in ProModel (Figure L5.2) has the following options:

- *View Statistics:* Allows the user to view the statistics generated from running a simulation model. Selecting this option loads the Output Viewer 3DR.
- *View Trace:* Allows the user to view the trace file generated from running a simulation model. Sending a trace listing to a text file during runtime generates a trace. Please refer to Lab 8, section L8.2, for a more complete discussion of tracing a simulation model.

The View menu (Figure L5.3) allows the user to select the way in which output data, charts, and graphs can be displayed. The View menu has the following options:

a. Report	*d.* Histogram
b. Category Chart	*e.* Time Plot
c. State Chart	*f.* Sheet Properties

FIGURE L5.4

3DR Report view of the results of the ATM System in Lab3.

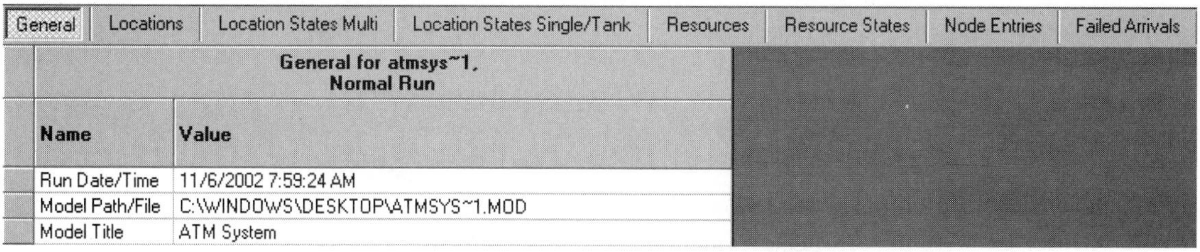

General	Locations	Location States Multi	Location States Single/Tank	Resources	Resource States	Node Entries	Failed Arrivals

	General for atmsys~1, Normal Run			
Name	**Value**			
Run Date/Time	11/6/2002 7:59:24 AM			
Model Path/File	C:\WINDOWS\DESKTOP\ATMSYS~1.MOD			
Model Title	ATM System			

L5.1.1 Report View

The report view (Figure L5.4) of the Output Viewer 3DR is similar to the Classic view of the output report. A major difference is that there are tabs for various parts of the report as follows:

a. General

b. Locations

c. Location States Multiunit

d. Location States Single/Tank

e. Resources

f. Resource States

g. Node Entries

h. Failed Arrivals

i. Entity Activity

j. Entity States

k. Variables

l. Location Costing

m. Resource Costing

n. Entity Costing

o. Logs

L5.1.2 Category Chart

The category chart (Figure L5.5) is a graphical representation of the status of the category of data selected at the end of the simulation run. Here are some of the data categories available for charting:

a. Entity Activity
  i. Total exits
  ii. Current quantity in system
  iii. Average time in system (Figure L5.6)

b. Location States Multi
  i. Pct. empty
  ii. Pct. full

c. Location States Single
  i. Pct. idle
  ii. Pct. waiting
  iii. Pct. blocked

d. Locations
  i. Total entries
  ii. Average contents

e. Logs
  i. Minimum value
  ii. Average value

f. Resource States
  i. Pct. idle
  ii. Pct. in use

g. Variables
  i. Minimum value
  ii. Average value
  iii. Maximum contents

Categories of charts available in the Category Chart Selection menu.

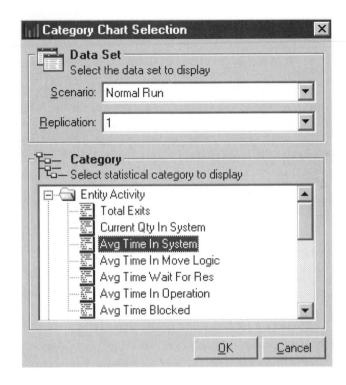

FIGURE L5.6

An example of a category chart presenting entity average time in system.

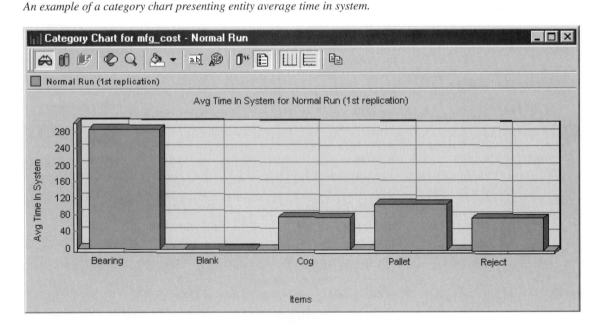

FIGURE L5.7

Categories of charts available in the State Chart Selection menu.

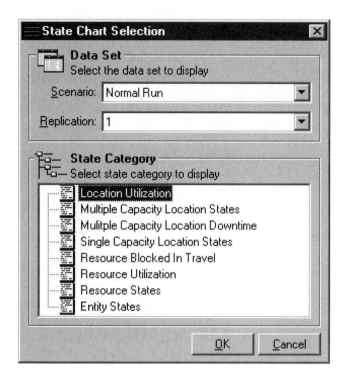

L5.1.3 State Chart

In ProModel the state and utilization graphs produce an averaged state summary for multiple replications or batches. By selecting State or Utilization Summary, it is possible to select the type of graph in the dialog shown in Figure L5.7.

The Output Program allows you to create seven different types of state and utilization graphs to illustrate the percentage of time that locations and resources were in a particular state: operation, waiting, blocked, down, or the like.

- *Location Utilization:* Location Utilization graphs show the percentage of time that each location in the system was utilized. ProModel allows you to see this information for all of the locations in the system at once (Figure L5.8).

- *Location State:* Single and Multiple Capacity Location State graphs show the percentage of time that each location in the system was in a particular state, such as idle, in operation, waiting for arrivals, blocked or down (Figure L5.9). ProModel shows this information for all of the locations in the system at once. For further enhancement of a particular location you can create a pie chart. To create a pie chart from a state graph double click on one of the state bars (alternatively you can right-click on a bar and select Create Pie Chart from the pop-up menu) for that location (Figure L5.10).

State chart representation of location utilization.

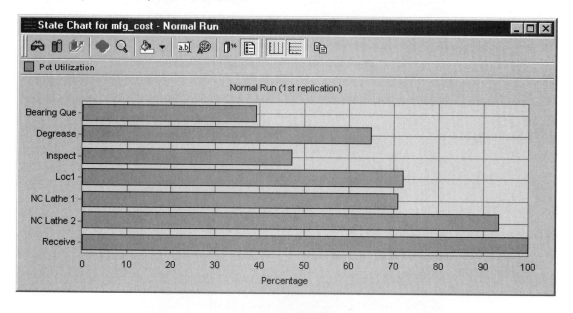

FIGURE L5.9

A state chart representation of all the locations' states.

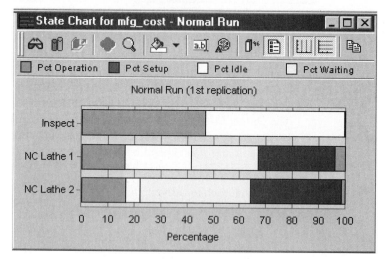

- *Multiple Capacity Location Downtime:* Multiple Capacity Location Downtime graphs show the percentage of time that each multicapacity location in the system was down. A pie chart can be created for any one of the locations.
- *Resource Blocked in Travel:* Resource Blocked in Travel graphs show the percentage of time a resource was blocked. A resource is blocked if it is unable to move to a destination because the next path node along the route of travel was blocked (occupied).

FIGURE L5.10

A pie chart representing the states of the location Inspect.

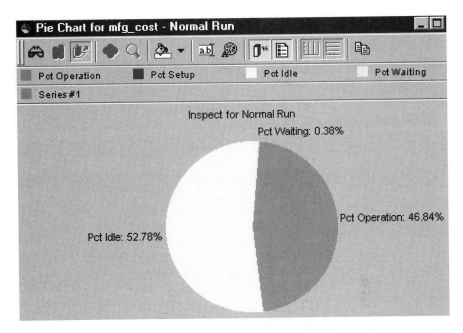

• *Resource Utilization:* Resource Utilization graphs show the percentage of time that each resource in the system was utilized. A resource is utilized when it is transporting or processing an entity or servicing a location or other resource. ProModel shows this information for all resources in the system at once.

• *Resource State:* The following information is contained in the report for each resource (Figure L5.11).

Resource Name	The name of the resource.
Scheduled Hours	The total number of hours the resource was scheduled to be available, which excludes off-shift time and scheduled downtime.
Pct. in Use	The percentage of time the resource spent transporting or processing an entity, or servicing a location or other resource.
Pct. Travel to Use	The percentage of time the resource spent traveling to a location or resource to transport, process, or service an entity, location, or resource. This also includes pickup and deposit time.
Pct. Travel to Park	The percentage of time the resource spent traveling to a path node to park.
Pct. Idle	The percentage of time the resource was available but not in use.
Pct. Down	The percentage of time the resource was unavailable due to unscheduled downtime.

FIGURE L5.11

State chart for the Cell Operator resource states.

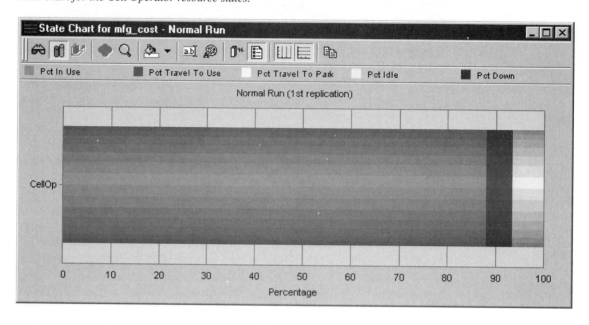

FIGURE L5.12

State chart representation of entity states.

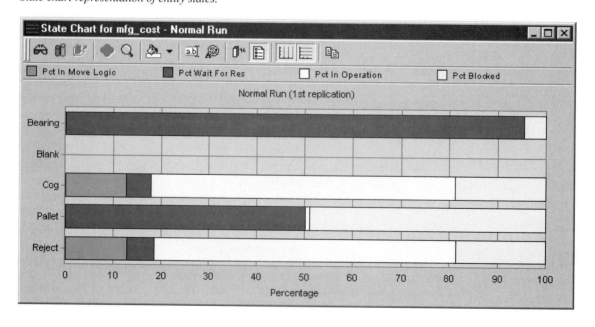

- *Entity State:* The following information is given for each entity type (Figure L5.12)

Pct. in Move Logic	The percentage of time the entity spent traveling to the next location, either in or out of a queue or with a resource.
Pct. Wait for Res	The percentage of time the entity spent waiting for a resource or another entity to join, combine, or the like.
Pct. in Operation	The percentage of time the entity spent in processing at a location or traveling on a conveyor.
Pct. Blocked	The percentage of time the entity spent waiting for a freed destination.

L5.1.4 Histogram and Time Plot

Time series statistics can be collected for locations, entities, resources, and variables. By selecting one of the Time Series menu item choices, one can select the values to graph and the options to customize the appearance of the graph (Figure L5.13).

- *Time series histogram* (Figure L5.14): Selecting Edit from the Options menu displays the Graph Options dialog that lets you select time units, bar width, and other options to modify the look of the graph.
- *Time plots:* Time plots (Figure L5.15) give a different perspective from histograms. One can track global variables such as total work-in-process inventory, contents of a queue or location, and total items produced. Figure L5.15 is a time plot of the status of the work-in-process inventory over 900 minutes of simulation of the *mfg_cost.mod* model in the Demo subdirectory. Figure L5.16 is a time plot of the contents of the Part_Conveyor over about 515 minutes of simulation of the *pracmod.mod* model in the Training subdirectory.

  Notice by clicking on the time-plot menu, there are actually three different types of values that can be plotted over time. One is Time-weighted values in which the values that are plotted are weighted by time. So if you want to display the contents of a location for each 15-minute interval of the simulation, it will plot the time-weighted average value for each 15-minute interval. The other two are Simple Values which takes a simple average when plotting by time interval, and Counts which plots the number of times a value changed per time interval.
- *Export plots/histograms:* When you plot one or more model elements using a plot or histogram, ProModel allows you to export these data to an Excel spreadsheet. When you export the plot or histogram data, ProModel creates a new spreadsheet and defines a new layer for each data element.

FIGURE L5.13

Dialog box for plotting a histogram.

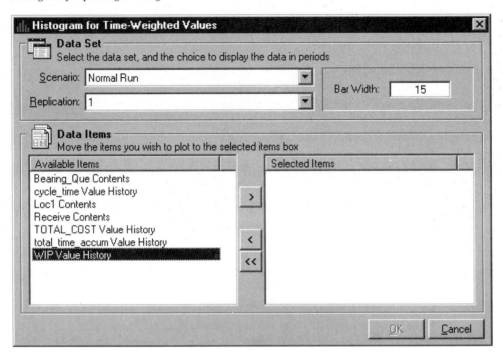

FIGURE L5.14

A time-weighted histogram of the contents of the Bearing Queue.

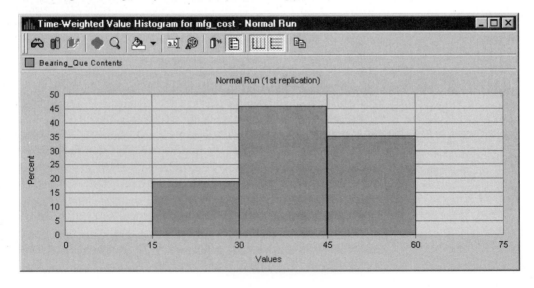

FIGURE L5.15

A time series plot of the Bearing Queue contents over time.

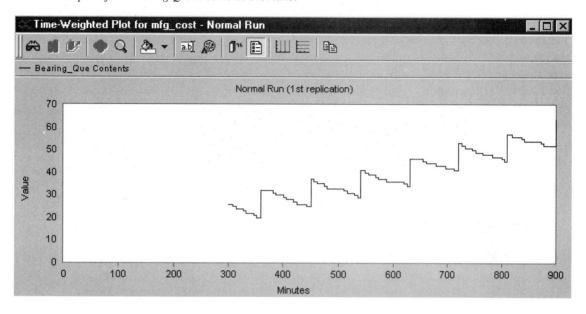

FIGURE L5.16

A time series plot of WIP.

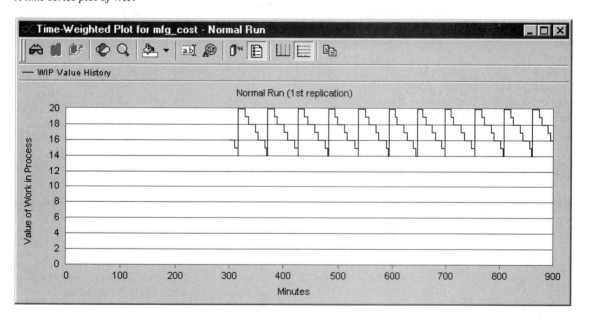

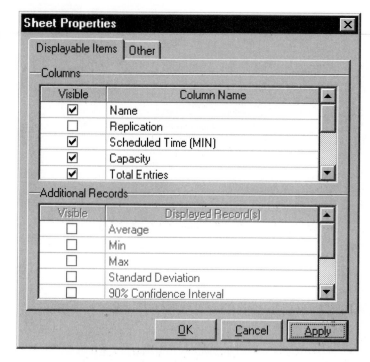

L5.1.5 Sheet Properties

The Sheet Properties menu in the View menu allows the user to customize the columns in the report view (Figure L5.17). The headings and gridlines can be turned on and off. The font type, size, and color of the column headings can be selected here. The time unit of the columns with time values can be selected as seconds, minutes, hours, days, or weeks.

L5.2 Classic View

The output view in the Classic mode for the example in Section L4.3 is shown in Figure L5.18. The View menu in the Classic output viewer consists of the following options (Figure L5.19):

a. *General Stats:* Creates a general summary report. If there are only one scenario, one replication, and one report period, the report will be generated automatically.

b. *Selected Stats:* The Selected Stats report allows you to create a replication, batch mean, or periodic report with only the specific statistics and elements you want in the report. The Selected Stats report is available only if there are multiple replications, periods, or batches.

FIGURE L5.18

The results of Poly Furniture Factory (with Oven) in Classic view.

```
General Report
Output from C:\My Documents\ProModelBook_New\Lab_L_3_3.MOD [Poly Furniture Fact
Date: May/07/2002   Time: 01:42:13 PM
--------------------------------------------------------------------------------
Scenario     : Normal Run
Replication  : 1 of 1
Simulation Time : 10 hr
--------------------------------------------------------------------------------
```

LOCATIONS

Location Name	Scheduled Hours	Capacity	Total Entries	Average Minutes Per Entry	Average Contents	Maximum Contents
Receiving Dock	10	999999	61	188.04	19.11	39
Splitter Saw	10	1	22	26.27	0.96	1
Lathe	10	1	86	5.90	0.84	1
Paint Booth	10	10	85	4.80	0.68	4
Painted Log Store	10	1	81	0.0	0	1
Oven	10	10	84	20.00	2.80	6

LOCATION STATES BY PERCENTAGE (Multiple Capacity)

Location Name	Scheduled Hours	% Empty	% Partially Occupied	% Full	% Down
Receiving Dock	10	1.67	98.33	0.0	0.0
Paint Booth	10	42.63	57.37	0.0	0.0
Oven	10	2.86	97.14	0.0	0.0

LOCATION STATES BY PERCENTAGE (Single Capacity/Tanks)

Location Name	Scheduled Hours	% Operation	% Setup	% Idle	% Waiting	% Blocked	% Down
Splitter Saw	10	11.38	0.0	3.67	0.0	81.95	0.0
Lathe	10	81.57	0.0	15.43	0.0	0.0	0.0
Painted Log Store	10	0.0	0.0	100.00	0.0	0.0	0.0

FAILED ARRIVALS

Entity Name	Location Name	Total Failed
Logs	Receiving Dock	0

ENTITY ACTIVITY

Entity Name	Total Exits	Current Quantity In System	Average Minutes In System	Average Minutes In Move Logic	Average Minutes Wait For Res, etc.	Average Minutes In Operation	Average Minutes Blocked
Logs	0	40	-	-	-	-	-
Piece	0	1	-	-	-	-	-
Rounds	0	1	-	-	-	-	-
Painted Logs	81	3	221.25	5.00	0.0	35.04	181.20

ENTITY STATES BY PERCENTAGE

Entity Name	% In Move Logic	% Wait For Res, etc.	% In Operation	% Blocked
Logs	-	-	-	-
Piece	-	-	-	-
Rounds	-	-	-	-
Painted Logs	2.26	0.0	15.84	81.90

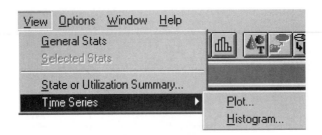

c. *State or Utilization Summary:* In ProModel, the state and utilization graphs produce an averaged state summary for multiple replications or batches.

d. *Time Series Plot/Histogram:* In previous versions, time series plots and histograms were known as throughput, content, value, and duration plots and histograms. These charts can now be combined on one chart for improved comparisons. You must specify Time Series statistics for the desired model element to view this information.

The Options menu in the Classic output viewer allows the user to display the dialog box for various options pertaining to the current report or graphs. Also, the most recently viewed reports and graphs can be automatically opened.

L5.2.1 Time Series Plot

The numbers of customers waiting for Barber Dan in Lab 4, Section L4.1, over the eight-hour simulation period are shown in Figure L5.20 as a time series plot. The number of customers waiting for a haircut varies between zero and one for the first hour. In the second hour, the number of customers waiting grows to as many as four. After dropping back to zero, the number of customers waiting grows again to as many as eight. However, it drops down to zero again toward the end of the day.

L5.2.2 Time Series Histogram

The number of customers waiting for a haircut at Fantastic Dan (Lab 4, Section L4.1) is shown in Figure L5.21 as a histogram. For about 63 percent of the time (simulation run time) the number of customers waiting varies from zero to three, and for 20 percent of the time this number varies from three to six. For about 17 percent of the time the number of customers waiting for haircuts varies from six to nine.

L5.2.3 Location State Graphs

The utilization of all the locations at the Poly Furniture Factory in Lab 4, Section L4.3, with unit capacity is shown in Figure L5.22. The different location

FIGURE L5.20

Time series plot of customers waiting for Barber Dan.

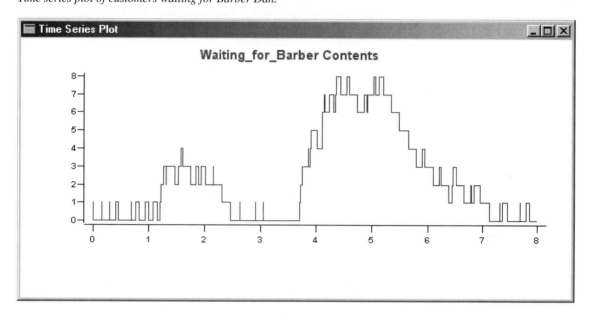

FIGURE L5.21

Time series histogram of customers waiting for Barber Dan.

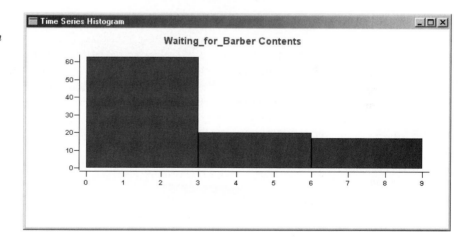

states are Operation, Setup, Idle, Waiting, Blocked, and Down. The location states for the Splitter Saw are shown in Figure L5.23 as a pie graph. The utilization of multiple capacity locations at Poly Furniture Factory is shown in Figure L5.24. All the states the entity (Painted_Logs) is in are shown in Figure L5.25 as a state graph and in Figure L5.26 as a pie chart. The different states are move, wait for resource, and operation.

FIGURE L5.22

State graphs for the utilization of single capacity locations.

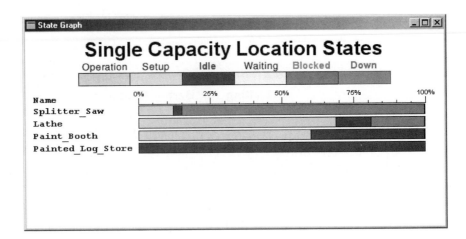

FIGURE L5.23

Pie chart for the utilization of the Splitter Saw.

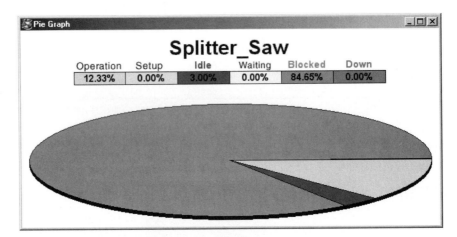

FIGURE L5.24

State graphs for the utilization of multiple capacity locations.

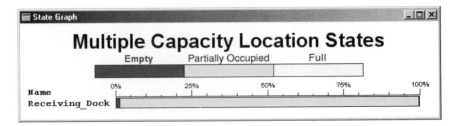

FIGURE L5.25

Graph of the states of the entity Painted_Logs.

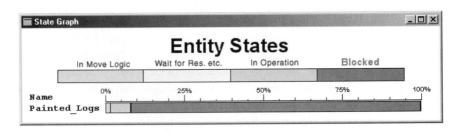

FIGURE L5.26

Pie graph of the states of the entity Painted_Logs.

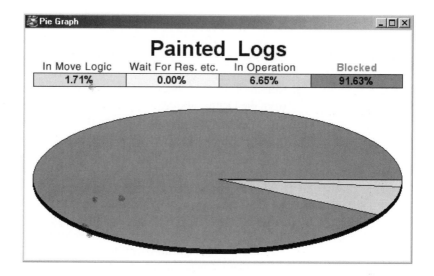

L5.3 Exercises

1. Customers arrive at the Lake Gardens post office for buying stamps, mailing letters and packages, and so forth. The interarrival time is exponentially distributed with a mean of 2 minutes. The time to process each customer is normally distributed with a mean of 10 minutes and a standard deviation of 2 minutes.

   a. Make a time series plot of the number of customers waiting in line at the post office in a typical eight-hour day.

   b. How many postal clerks are needed at the counter so that there are no more than 15 customers waiting in line at the post office at any time? There is only one line serving all the postal clerks. Change the number of postal clerks until you find the optimum number.

2. The Lake Gardens postmaster in Exercise 1 wants to serve his customers well. She would like to see that the average time spent by a postal customer at the post office is no more than 15 mins. How many postal clerks should she hire?

3. For the Poly Furniture Factory example in Lab 4, Section L4.3,

   a. Make a state graph and a pie graph for the splitter and the lathe.

   b. Find the percentage of time the splitter and the lathe are idle.

4. For the Poly Furniture Factory example in Lab 4, Section L4.4,

   a. Make histograms of the contents of the oven and the paint booth. Make sure the bar width is set equal to one. What information can you gather from these histograms?

   b. Plot a pie chart for the various states of the entity Painted_Logs. What percentage of time the Painted_Logs are in operation?

   c. Make a time series plot of the oven and the paint booth contents. How would you explain these plots?

5. For the Bank of USA ATM example in Lab 4, Section L4.2,
   a. Plot a pie chart for the various states of the ATM customer.
   b. What is the percentage of time the ATM customer is in process (using the ATM)?

6. Run the Tube Distribution Supply Chain example model (logistcs.mod from the Demo subdirectory) for 40 hours. What are the various entities modeled in this example? What are the various operations and processes modeled in this example?

   Look at the results and find
   a. The percentage utilization of the locations Mill and the Process Grades Threads.
   b. The capacities of Inventory and Inventory 2–6. The maximum contents of Inventory and Inventory 2–6.
   c. The idle time percent of the location Mill.
   d. What are the utilizations of the fork lift trucks 1–6? How well utilized are they? Can we reduce the fleet of fork trucks without much loss in the level of service?
   e. What are the various states (in percentages) of the location Process Grades Threads?
   f. What are the utilizations of the five transport resources Transport 1–5? Are they almost equally utilized? Why or why not? Please discuss.

7. Run the Warehouse model (deaerco.mod from the Demo subdirectory) for 100 hours. Run only Scenario 1. Go into the Simulation → Options menu and change the Run Hours from 10 to 100.
   a. What are the average values of inventory of Inventory Aisle 1–12?
   b. What is the average Time to Fill, Box, and Check?
   c. What is the Average Time in System?
   d. What are the cost of customer order, receiving order, and the sum total of cost per order?
   e. What are the percentage idle times for Checker 1 and Checker 2?
   f. What are the utilizations of Boxers 1–4 and the average utilization of all the Boxers?

6   FITTING STATISTICAL DISTRIBUTIONS TO INPUT DATA

There are three kinds of lies: lies, damned lies, and statistics.
—Benjamin Disraeli

Input data drive our simulation models. Input data can be for interarrival times, material handling times, setup and process times, demand rates, loading and unloading times, and so forth. The determination of what data to use and where to get the appropriate data is a complicated and time-consuming task. The quality of data is also very important. We have all heard the cliché "garbage in, garbage out." In Chapter 6 we discussed various issues about input data collection and analysis. We have also described various empirical discrete and continuous distributions and their characteristics. In this lab we describe how ProModel helps in fitting empirical statistical distributions to user input data.

L6.1 An Introduction to Stat::Fit

Stat::Fit is a utility packaged with the ProModel software, which is available from the opening screen of ProModel (Figure L1.1) and also from the Windows Start → Programs menu. The Stat::Fit opening screen is shown in Figure L6.1. Stat::Fit can be used for analyzing user-input data and fitting an appropriate empirical distribution. The empirical distribution can be either continuous or discrete. If there are enough data points, say 100 or more, it may be appropriate to fit an empirical distribution to the data using conventional methods. ProModel offers built-in capability to perform input data analysis with the tool Stat::Fit. Stat::Fit fits probability distributions to empirical data. It allows comparison among various distribution functions. It performs goodness-of-fit tests using chi-square, Kolmogorov–Smirnov, and Anderson–Darling procedures. It calculates appropriate parameters for distributions. It provides distribution expressions for use in the simulation model. When the amount of data is

FIGURE L6.1

*Stat::Fit opening
screen.*

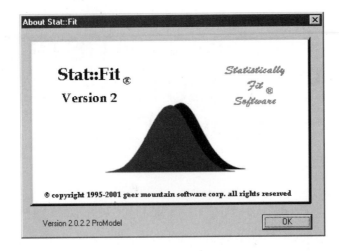

FIGURE L6.2

Stat::Fit opening menu.

small, the goodness-of-fit tests are of little use in selecting one distribution over another because it is inappropriate to fit one distribution over another in such a situation. Also, when conventional techniques have failed to fit a distribution, the empirical distribution is used directly as a user distribution (Chapter 6, Section 6.9).

The opening menu of Stat::Fit is shown in Figure L6.2. Various options are available in the opening menu:

1. *File:* File opens a new Stat::Fit project or an existing project or data file. The File menu is also used to save a project.

2. *Edit:*

3. *Input:*

4. *Statistics:*

5. *Fit:* The Fit menu provides a Fit Setup dialog and a Distribution Graph dialog. Other options are also available when a Stat::Fit project is opened. The Fit Setup dialog lists all the distributions supported by Stat::Fit and the relevant choices for goodness-of-fit tests. At least one distribution must be chosen before the estimate, test, and graphing commands become available. The Distribution Graph command uses the distribution and parameters provided in the Distribution Graph dialog to create a graph of any analytical distribution supported by Stat::Fit. This graph is not connected to any input data or document.

6. *Utilities:* The Replications command allows the user to calculate the number of independent data points or replications of an experiment necessary to provide a given range or confidence interval for the estimate of a parameter. The confidence interval is given for the confidence level specified. The default is a 95 percent confidence interval. The resulting number of replications is calculated using the *t* distribution.

7. *View:*

8. *Window:* The Window menu is used to either cascade or tile various windows opened while working on the Stat::Fit projects.

9. *Help:*

Figure L6.3 shows the data/document input screen. Figure L6.4 shows the various data input options available in Stat::Fit. The type of distribution is also specified

FIGURE L6.3

Document input screen.

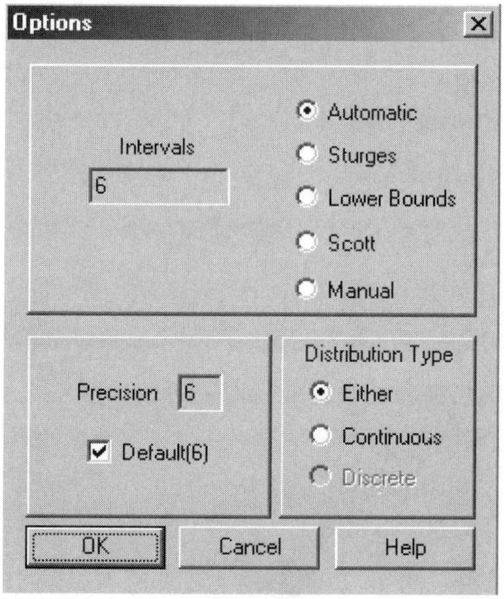

FIGURE L6.4

Stat::Fit data input options.

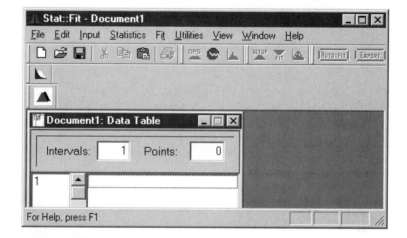

here. The Input Options command can be accessed from the Input menu as well as the Input Options button on the Speed Bar.

L6.2 An Example Problem

Problem Statement

The time between arrivals of cars at **San Dimas Gas Station** were collected as shown in Table L6.1. This information is also saved and available in an Excel worksheet named *L6.2_Gas Station Time Between Arrival*. Use Stat::Fit to analyze the data and fit an appropriate continuous distribution to the data. Figure L6.5 shows part of the actual data of times between arrival of cars in minutes. The sample data have 30 data points. Figure L6.6 shows the histogram of the input data, while Figure L6.7 shows some of the descriptive statistics generated by Stat::Fit.

TABLE L6.1   Times between Arrival of Cars at San Dimas Gas Station

Number	Times between Arrival, Minutes
1	12.36
2	5.71
3	16.79
4	18.01
5	5.12
6	7.69
7	19.41
8	8.58
9	13.42
10	15.56
11	10.
12	18.
13	16.75
14	14.13
15	17.46
16	10.72
17	11.53
18	18.03
19	13.45
20	10.54
21	12.53
22	8.91
23	6.78
24	8.54
25	11.23
26	10.1
27	9.34
28	6.53
29	14.35
30	18.45

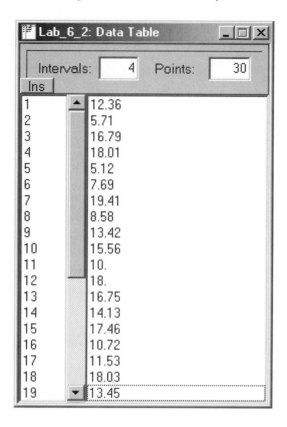

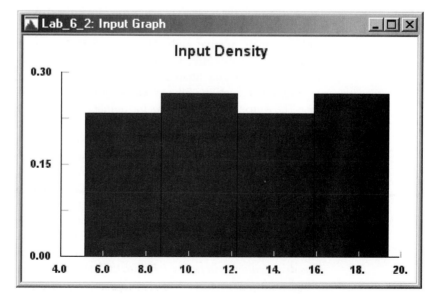

FIGURE L6.7

Descriptive statistics for the input data.

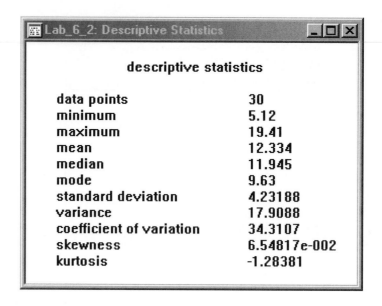

Lab_6_2: Descriptive Statistics

descriptive statistics

data points	30
minimum	5.12
maximum	19.41
mean	12.334
median	11.945
mode	9.63
standard deviation	4.23188
variance	17.9088
coefficient of variation	34.3107
skewness	6.54817e-002
kurtosis	-1.28381

L6.3 Auto::Fit Input Data

Automatic fitting of continuous distributions can be performed by clicking on the Auto::Fit icon or by selecting Fit from the Menu bar and then Auto::Fit from the Submenu (Figure L6.8).

This command follows the same procedure as discussed in Chapter 6 for manual fitting. Auto::Fit will automatically choose appropriate continuous distributions to fit to the input data, calculate Maximum Likelihood Estimates for those distributions, test the results for Goodness of Fit, and display the distributions in order of their relative rank. The relative rank is determined by an empirical method, which uses effective goodness-of-fit calculations. While a good rank usually indicates that the fitted distribution is a good representation of the input data, an absolute indication of the goodness of fit is also given.

The Auto::Fit dialog allows the number of continuous distributions to be limited by choosing only those distributions with a lower bound or by forcing a lower bound to a specific value as in Fit Setup. Also, the number of distributions will be limited if the skewness of the input data is negative; many continuous distributions with lower bounds do not have good parameter estimates in this situation.

The acceptance of fit usually reflects the results of the goodness-of-fit tests at the level of significance chosen by the user. However, the acceptance may be modified if the fitted distribution would generate significantly more data points in the tails of the distribution than are indicated by the input data.

The Auto::Fit function forces the setup of the document so that *only continuous distributions will be used*. Figure L6.9 shows various continuous distributions fitted to the input data of the San Dimas Gas Station (Section L6.2) and their

FIGURE L6.8

The Auto::Fit submenu.

FIGURE L6.9

Various distributions fitted to the input data.

Lab_6_2: Automatic Fitting

Auto::Fit of Distributions

distribution	rank	acceptance
Power Function(5., 19.5, 1.02)	100	do not reject
Uniform(5., 19.4)	98.1	do not reject
Johnson SB(5., 14.6, -1.26e-002, 0.555)	83.	do not reject
Weibull(5., 1.65, 8.09)	80.	do not reject
Beta(5., 19.4, 1.1, 1.17)	78.2	do not reject
Erlang(5., 2., 3.67)	75.6	do not reject
Pearson 6(5., 3.46e+005, 1.83, 8.64e+004)	67.	do not reject
Rayleigh(5., 5.96)	66.5	do not reject
Gamma(5., 1.82, 4.03)	66.2	do not reject
LogLogistic(5., 2.07, 6.35)	30.8	reject
Lognormal(5., 1.69, 1.)	12.1	reject
Triangular(5., 23.4, 5.)	3.94	do not reject
Chi Squared(5., 6.41)	3.19	do not reject
Exponential(5., 7.33)	2.78	reject
Inverse Weibull(5., 0.696, 0.329)	7.31e-002	reject
Pareto(5., 1.19)	2.94e-002	reject
Pearson 5(5., 0.62, 1.25)	1.13e-002	reject
Inverse Gaussian(5., 2.78, 7.33)	7.69e-004	reject

rank in terms of the amount of fit. Both the Kolmogorov–Smirnov and the Anderson–Darling goodness-of-fit tests will be performed on the input data as shown in Figure L6.10. The Maximum Likelihood Estimates will be used with an accuracy of at least 0.00003. The actual data and the fitted uniform distribution are compared and shown in Figure L6.11.

FIGURE L6.10

Goodness-of-fit tests performed on the input data.

goodness of fit

data points	30
estimates	maximum likelihood estimates
accuracy of fit	3.e-004
level of significance	5.e-002

summary

distribution	Kolmogorov Smirnov	Anderson Darling
Beta(5., 19.4, 1.1, 1.17)	0.114	2.23
Chi Squared(5., 6.41)	0.19	2.41
Erlang(5., 2., 3.67)	0.102	0.706
Exponential(5., 7.33)	0.216	1.97
Gamma(5., 1.82, 4.03)	0.109	0.75
Inverse Gaussian(5., 2.78, 7.33)	0.358	5.76
Inverse Weibull(5., 0.696, 0.329)	0.287	3.4
Johnson SB(5., 14.6, -1.26e-002, 0.555)	9.34e-002	0.26
LogLogistic(5., 2.07, 6.35)	0.155	0.875
Lognormal(5., 1.69, 1.)	0.168	1.61
Pareto(5., 1.19)	0.305	3.96
Pearson 5(5., 0.62, 1.25)	0.319	4.51
Pearson 6(5., 3.46e+005, 1.83, 8.64e+000	0.108	0.747
Power Function(5., 19.5, 1.02)	7.44e-002	0.199
Rayleigh(5., 5.96)	0.123	0.827
Triangular(5., 23.4, 5.)	0.182	1.62
Uniform(5., 19.4)	8.21e-002	2.07
Weibull(5., 1.65, 8.09)	0.109	0.522

FIGURE L6.11

Comparison of actual data and fitted uniform distribution.

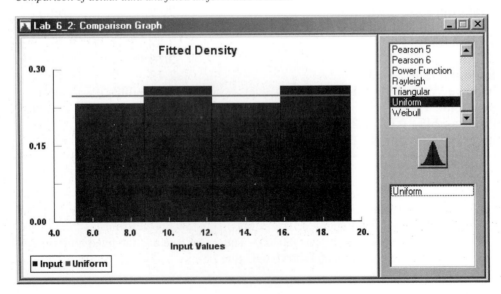

Because the Auto::Fit function requires a specific setup, the Auto::Fit view can be printed only as the active window or part of the active document, not as part of a report. The Auto::Fit function will not fit discrete distributions. The manual method, previously described, should be used instead.

L6.4 Exercises

1. Consider the operation of a fast-food restaurant where customers arrive for ordering lunch. The following is a log of the time (minutes) between arrivals of 40 successive customers. Use Stat::Fit to analyze the data and fit an appropriate continuous distribution. What are the parameters of this distribution?

11	11	12	8	15	14	15	13
9	13	14	9	14	9	13	7
12	12	7	13	12	16	7	10
8	8	17	15	10	7	16	11
11	10	16	10	11	12	14	15

2. The servers at the restaurant in Question 1 took the following time (minutes) to serve food to these 40 customers. Use Stat::Fit to analyze the data and fit an appropriate continuous distribution. What are the parameters of this distribution?

11	11	12	8	15	14	15	13
9	13	14	10	14	9	13	12
12	12	11	13	12	16	11	10
10	8	17	12	10	7	13	11
11	10	13	10	11	12	14	15

3. The following are the numbers of incoming calls (each hour for 80 successive hours) to a call center set up for serving customers of a certain Internet service provider. Use Stat::Fit to analyze the data and fit an appropriate discrete distribution. What are the parameters of this distribution?

12	12	11	13	12	16	11	10
9	13	14	10	14	9	13	12
12	12	11	13	12	16	11	10
10	8	17	12	10	7	13	11
11	11	12	8	15	14	15	13
9	13	14	10	14	9	13	12
12	12	11	13	12	16	11	10
10	8	17	12	10	7	13	11
11	10	13	10	11	12	14	15
10	8	17	12	10	7	13	11

4. Observations were taken on the times to serve online customers at a stockbroker's site (STOCK.com) on the Internet. The times (in seconds) are shown here, sorted in ascending order. Use Stat::Fit and fit an appropriate distribution to the data. What are the parameters of this distribution?

1.39	21.47	39.49	58.78	82.10
3.59	22.55	39.99	60.61	83.52
7.11	28.04	41.42	63.38	85.90
8.34	28.97	42.53	65.99	88.04
11.14	29.05	47.08	66.00	88.40
11.97	35.26	51.53	73.55	88.47
13.53	37.65	55.11	73.81	92.63
16.87	38.21	55.75	74.14	93.11
17.63	38.32	55.85	79.79	93.74
19.44	39.17	56.96	81.66	98.82

5. Forty observations for a bagging operation was shown in Chapter 6, Table 6.5. Use Stat::Fit to find out what distribution best fits the data. Compare your results with the results obtained in Chapter 6 (Suggestion: In Chapter 6 five equal probability cells were used for the chi-square goodness of fit test. Use same number of cells in Stat::Fit to match the results).

7 BASIC MODELING CONCEPTS

Imagination is more important than knowledge.
—Albert Einstein

In this chapter we continue to describe other basic features and modeling concepts of ProModel. In Section L7.1 we show an application with multiple locations and multiple entity types. Section L7.2 describes modeling of multiple parallel locations. Section L7.3 shows various routing rules. In Section L7.4 we introduce the concept of variables. Section L7.5 introduces the inspection process, tracking of defects, and rework. In Section L7.6 we show how to assemble nonidentical entities and produce assemblies. In Section L7.7 we show the process of making temporary entities through the process of loading and subsequent unloading. Section L7.8 describes how entities can be accumulated before processing. Section L7.9 shows the splitting of one entity into multiple entities. In Section L7.10 we introduce various decision statements with appropriate examples. Finally, in Section L7.11 we show you how to model a system that shuts down periodically.

L7.1 Multiple Locations, Multiple Entity Types

Problem Statement
In one department at **Pomona Electronics,** three different printed circuit boards are assembled. Each board is routed through three assembly areas. The routing order is different for each of the boards. Further, the time to assemble a board depends on the board type and the operation. The simulation model is intended to determine the time to complete 500 boards of each type. The assembly time for each board is exponentially distributed with the mean times shown in Table L7.1.

Define three locations (Area1, Area2, and Area3) where assembly work is done and another location, PCB_Receive, where all the printed circuit boards

465

FIGURE L7.1

The three processing locations and the receiving dock.

Icon	Name	Cap.	Units	DTs...	Stats...	Rules...
	PCB_Receive	inf	1	None	Time Series	Oldest
	Area1	inf	1	None	Time Series	Oldest
	Area2	inf	1	None	Time Series	Oldest
	Area3	inf	1	None	Time Series	Oldest

FIGURE L7.2

Layout of Pomona Electronics.

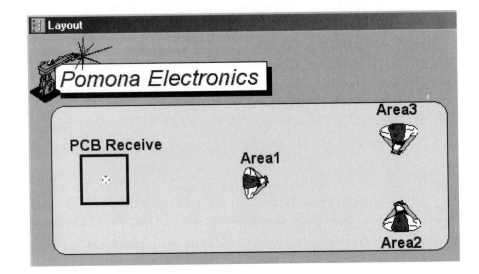

TABLE L7.1   Assembly Times for the Printed Circuit Boards

Printed Circuit Board 1		Printed Circuit Board 2		Printed Circuit Board 3	
Area	*Mean Time*	*Area*	*Mean Time*	*Area*	*Mean Time*
1	10	2	5	3	12
2	12	1	6	2	14
3	15	3	8	1	15

arrive (Figure L7.1). Assume each of the assembly areas has infinite capacity. The layout of Pomona Electronics is shown in Figure L7.2. Note that we used Background Graphics → Behind Grid, from the Build menu, to add the Pomona Electronics logo on the simulation model layout. Add the robot graphics (or something appropriate). Define three entities as PCB1, PCB2, and PCB3 (Figure L7.3).

FIGURE L7.3

The three types of circuit boards.

Icon	Name	Speed (fpm)	Stats...
▦ Entities			
◆	PCB1	150	Time Series
◆	PCB2	150	Time Series
◇	PCB3	150	Time Series

FIGURE L7.4

The arrival process for all circuit boards.

```
××××××××××××××××××××××××××××××××××××××××××××××××××××××××××××××××××××
                              Arrivals
××××××××××××××××××××××××××××××××××××××××××××××××××××××××××××××××××××

Entity    Location      Qty each   First Time  Occurrences  Frequency   Logic
--------  ------------  ---------   ----------  -----------  ----------  ------
PCB1      PCB_Receive   500                     1            0
PCB2      PCB_Receive   500                     1            0
PCB3      PCB_Receive   500                     1            0
```

FIGURE L7.5

Processes and routings for Pomona Electronics.

```
                              Process                       Routing

Entity    Location      Operation             Blk  Output   Destination  Rule
--------  ------------  --------------------  ---  -------   -----------  ------
PCB1      PCB_Receive                         1    PCB1      Area1        FIRST 1
PCB1      Area1         wait e(10)            1    PCB1      Area2        FIRST 1
PCB1      Area2         wait e(12)            1    PCB1      Area3        FIRST 1
PCB1      Area3         wait e(15)            1    PCB1      EXIT         FIRST 1
PCB2      PCB_Receive                         1    PCB2      Area2        FIRST 1
PCB2      Area2         wait e(5)             1    PCB2      Area1        FIRST 1
PCB2      Area1         wait e(6)             1    PCB2      Area3        FIRST 1
PCB2      Area3         wait e(8)             1    PCB2      EXIT         FIRST 1
PCB3      PCB_Receive                         1    PCB3      Area3        FIRST 1
PCB3      Area3         wait e(12)            1    PCB3      Area2        FIRST 1
PCB3      Area2         wait e(14)            1    PCB3      Area1        FIRST 1
PCB3      Area1         wait e(15)            1    PCB3      EXIT         FIRST 1
```

Define the arrival process (Figure L7.4). Assume all 1500 boards are in stock when the assembly operations begin. The process and routing tables are developed as in Figure L7.5.

Run the simulation model. Note that the whole batch of 1500 printed circuit boards (500 of each) takes a total of 2 hours and 27 minutes to be processed.

L7.2 Multiple Parallel Identical Locations

Location: Capacity versus Units

Each location in a model has associated with it some finite capacity. The *capacity* of a location refers to the maximum number of entities that the location can hold at any time. The reserved word INF or INFINITE sets the capacity to the maximum allowable value. The default capacity for a location is one unless a counter, conveyor, or queue is the first graphic assigned to it. The default capacity in that case is INFINITE.

The *number of units* of a location refers to the number of identical, interchangeable, and parallel locations referenced as a single location for routing and processing. A multiunit location eliminates the need to create multiple locations and multiple processes for locations that do the same thing. While routings to and from each unit are the same, each unit may have unique downtimes.

A multiunit location is created by entering a number greater than one as the number of units for a normal location. A corresponding number of locations will appear below the original location with a numeric extension appended to each copied location name designating the unit number of that location. The original location record becomes the prototype for the unit records. Each unit will inherit the prototype's characteristics unless the individual unit's characteristics are changed. Downtimes, graphic symbols, and, in the case of multicapacity locations, capacities may be assigned to each unit. As the number of units of a location is changed, the individual unit locations are automatically created or destroyed accordingly.

The following three situations describe the advantages and disadvantages of modeling multicapacity versus multiunit locations.

- *Multicapacity locations.* Locations are modeled as a single unit with multicapacity (Figure L7.6). All elements of the location perform identical operations. When one element of the location is unavailable due to downtime, all elements are unavailable. Only clock-based downtimes are allowed.

- *Multiunit locations.* Locations are modeled as single capacity but multiple units (Figure L7.7). These are locations consisting of two or more parallel and interchangeable processing units. Each unit shares the same sources of input and the same destinations for output, but each may have independent operating characteristics. This method provides more flexibility in the assignment of downtimes and in selecting an individual unit to process a particular entity.

- *Multiple, single-capacity locations.* Locations are modeled as individual and single capacity (Figure L7.8). Usually noninterchangeable locations are modeled as such. By modeling as individual locations, we gain the flexibility of modeling separate downtimes for each element. In addition, we now have complete flexibility to determine which element will process a particular entity.

FIGURE L7.6

Single unit of multicapacity location.

Icon	Name	Cap.	Units	DTs...	Stats...	Rules...
	Inspector	3	1	None	Time Series	Oldest

FIGURE L7.7

Multiple units of single-capacity locations.

Icon	Name	Cap.	Units	DTs...	Stats...	Rules...
	Inspector	1	4	None		
	Inspector.1	1	1	None	Time Series	Oldest
	Inspector.2	1	1	None	Time Series	Oldest
	Inspector.3	1	1	None	Time Series	Oldest
	Inspector.4	1	1	None	Time Series	Oldest

FIGURE L7.8

Multiple single-capacity locations.

Icon	Name	Cap.	Units	DTs...	Stats...	Rules...
	Inspector1	1	1	None	Time Series	Oldest
	Inspector2	1	1	None	Time Series	Oldest
	Inspector3	1	1	None	Time Series	Oldest

Problem Statement

At *San Dimas Electronics,* jobs arrive at three identical inspection machines according to an exponential distribution with a mean interarrival time of 12 minutes. The first available machine is selected. Processing on any of the parallel machines is normally distributed with a mean of 10 minutes and a standard deviation of 3 minutes. Upon completion, all jobs are sent to a fourth machine, where they queue up for date stamping and packing for shipment; this takes five minutes normally distributed with a standard deviation of two minutes. Completed jobs then leave the system. Run the simulation for one month (20 days, eight hours each). Calculate the average utilization of the four machines. Also, how many jobs are processed by each of the four machines?

Define a location called Inspect. Change its units to 3. Three identical parallel locations—that is, Inspect.1, Inspect.2, and Inspect.3—are thus created. Also, define a location for all the raw material to arrive (Material_Receiving). Change the capacity of this location to infinite. Define a location for Packing (Figure L7.9). Select Background Graphics from the Build menu. Make up a label "San Dimas Electronics." Add a rectangular border. Change the font and color appropriately.

FIGURE L7.9

The locations and the layout of San Dimas Electronics.

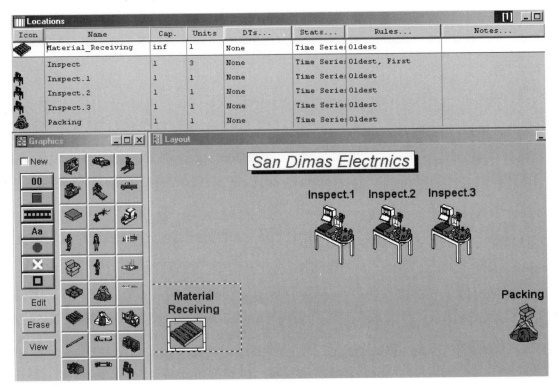

FIGURE L7.10

Arrivals of PCB at San Dimas Electronics.

	Arrivals						[1]
Entity...	Location...	Qty each...	First Time	Occurrences	Frequency	Logic	Disable
PC_Board	Material_Receiving	1	0	inf	e(12) min		No

Define an entity called PCB. Define the frequency of arrival of the entity PCB as exponential with a mean interarrival time of 12 minutes (Figure L7.10). Define the process and routing at San Dimas Electronics as shown in Figure L7.11.

In the Simulation menu select Options. Enter 160 in the Run Hours box. Run the simulation model. The average utilization and the number of jobs processed at the four locations are given in Table L7.2.

FIGURE L7.11

Process and routing tables at San Dimas Electronics.

```
*********************************************************************************
*                                 Processing                                   *
*********************************************************************************
                              Process                    Routing
Entity    Location          Operation          Blk  Output    Destination  Rule
────────  ─────────────────  ─────────────────  ───  ────────  ───────────  ──────
PC_Board  Material_Receiving                    1    PC_Board  Inspect      FIRST 1
PC_Board  Inspect           Wait N(10,3) min    1    PC_Board  Packing      FIRST 1
PC_Board  Packing           Wait N(5,2) min     1    PC_Board  EXIT         FIRST 1
```

TABLE L7.2   Summary of Results of the Simulation Run

	Average Utilization	*Number of Jobs Processed*
Inspector1	49.5	437
Inspector2	29.7	246
Inspector3	15.9	117
Packing	41.5	798

L7.3 Routing Rules

Routing rules are used in selecting among competing downstream processes for routing an entity. Sometimes customers prefer one downstream resource to another. The selection among several competing resources may follow one of these rules:

- FIRST *available:* Select the first location that has available capacity.
- MOST *available:* Select the location that has the most available capacity.
- *By* TURN: Rotate the selection among two or more available locations.
- RANDOM: Select randomly among two or more available locations.
- *If* JOIN *request:* Select the location that satisfies a JOIN request (see the JOIN statement).
- *If* LOAD *request:* Select the location that satisfies a LOAD request (see the LOAD statement).
- *If* SEND *request:* Select the location that satisfies a SEND request (see the SEND statement).
- *Longest:* Select the location that has been unoccupied the longest.
- *Unoccupied until* FULL: Continue to select the same location until it is full.
- *If* EMPTY: Select the location only when empty and continue to select until it is full.

- *Probabilistic:* Select based on the probability specified (such as .75).
- *User condition:* Select the location that satisfies the Boolean condition specified by the user (such as AT2>5). Conditions may include any numeric expression except for location attributes, resource-specific functions, and downtime-specific functions.
- CONTINUE: Continue at the same location for additional operations. CONTINUE is allowed only for blocks with a single routing.
- *As* ALTERNATE *to:* Select as an alternate if available and if none of the above rules can be satisfied.
- *As* BACKUP: Select as a backup if the location of the first preference is down.
- DEPENDENT: Select only if the immediately preceding routing was processed.

If only one routing is defined for a routing block, use the first available, join, load, send, if empty, or continue rule. The most available, by turn, random, longest un-occupied, until full, probabilistic, and user condition rules are generally only useful when a block has multiple routings.

Problem Statement
Amar, Akbar, and Anthony are three tellers in the local branch of **Bank of India.** Figure L7.12 shows the layout of the bank. Assume that customers arrive at the bank according to a uniform distribution (mean of five minutes and half-width of four minutes). All the tellers service the customers according to another uniform

FIGURE L7.12

Layout of the Bank of India.

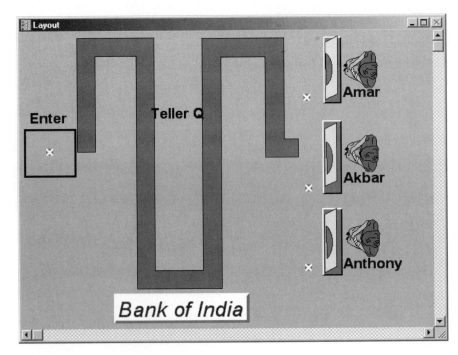

FIGURE L7.13

Locations at the Bank of India.

Icon	Name	Cap.	Units	DTs...	Stats...	Rules...
	Amar	1	1	None	Time Series	Oldest
	Akbar	1	1	None	Time Series	Oldest
	Anthony	1	1	None	Time Series	Oldest
	Enter	1	1	None	Time Series	Oldest
	Teller_Q	INFINITE	1	None	Time Series	Oldest, FIFO

FIGURE L7.14

Queue menu.

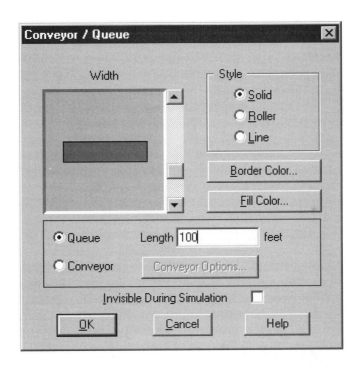

distribution (mean of 10 minutes and half-width of 6 minutes). However, the customers prefer Amar to Akbar, and Akbar over Anthony. If the teller of choice is busy, the customers choose the first available teller. Simulate the system for 200 customer service completions. Estimate the teller's utilization (percentage of time busy).

The locations are defined as Akbar, Anthony, Amar, Teller_Q, and Enter as shown in Figure L7.13. The Teller_Q is exactly 100 feet long. Note that we have checked the queue option in the Conveyor/Queue menu (Figure L7.14). The

FIGURE L7.15

Customer arrival at the Bank of India.

▥ Arrivals							[1] ▃▢
Entity...	Location...	Qty each...	First Time	Occurrences	Frequency	Logic	Disable
Customers	Enter	1		200	U(5,4) min		No

FIGURE L7.16

Process and routing tables at the Bank of India.

```
********************************************************************************
*                              Processing
********************************************************************************

                              Process                    Routing

Entity     Location  Operation              Blk  Output     Destination  Rule
--------   --------  ---------              ---  ------     -----------  ----
Customers  Enter                            1    Customers  Teller_Q     FIRST 1
Customers  Teller_Q                         1    Customers  Amar         FIRST 1
                                                 Customers  Akbar        FIRST
                                                 Customers  Anthony      FIRST
Customers  Amar      Wait U(10,6) min       1    Customers  EXIT         FIRST 1
Customers  Akbar     Wait U(10,6) min       1    Customers  EXIT         FIRST 1
Customers  Anthony   Wait U(10,6) min       1    Customers  EXIT         FIRST 1
```

TABLE L7.3   Utilization of Tellers at the Bank of India

	% Utilization	
Tellers	Selection in Order of Preference	Selection by Turn
Amar	79	63.9
Akbar	64.7	65.1
Anthony	46.9	61.5

customer arrival process is shown in Figure L7.15. The processes and routings are shown in Figure L7.16. Note that the customers go to the tellers Amar, Akbar, and Anthony in the order they are specified in the routing table.

The results of the simulation model are shown in Table L7.3. Note that Amar, being the favorite teller, is much more busy than Akbar and Anthony.

If the customers were routed to the three tellers in turn (selected in rotation), the process and routing tables would be as in Figure L7.17. Note that By Turn was selected from the Rule menu in the routing table. These results of the simulation model are also shown in Table L7.3. Note that Amar, Akbar, and Anthony are now utilized almost equally.

FIGURE L7.17

Process and routing tables for tellers selected by turn.

```
***********************************************************************
*                            Processing
***********************************************************************

                       Process                        Routing

Entity     Location  Operation          Blk  Output     Destination  Rule
--------   --------  ------------------  ---  --------   -----------  -------
Customers  Enter                         1   Customers  Teller_Q     FIRST 1
Customers  Teller_Q                      1   Customers  Amar         TURN 1
                                             Customers  Akbar        TURN
                                             Customers  Anthony      TURN
Customers  Amar      Wait U(10,6) min    1   Customers  EXIT         FIRST 1
Customers  Akbar     Wait U(10,6) min    1   Customers  EXIT         FIRST 1
Customers  Anthony   Wait U(10,6) min    1   Customers  EXIT         FIRST 1
```

L7.4 Variables

Variables are placeholders for either real or integer numbers that may change during the simulation. Variables are typically used for making decisions or for gathering data. Variables can be defined to track statistics and monitor other activities during a simulation run. This is useful when the built-in statistics don't capture a particular performance metric of interest. Variables might be defined to track

- The number of customers waiting in multiple queues.
- Customer waiting time during a specific time period.
- Customer time in the bank.
- Work-in-process inventory.
- Production quantity.

In ProModel two types of variables are used—local variables and global variables.

- *Global variables* are accessible from anywhere in the model and at any time. Global variables are defined through the Variables(global) editor in the Build menu. The value of a global variable may be displayed dynamically during the simulation. It can also be changed interactively. Global variables can be referenced anywhere a numeric expression is valid.
- *Local variables* are temporary variables that are used for quick convenience when a variable is needed only within a particular operation (in the Process table), move logic (in the Routing table), logic (in the Arrivals, Resources, or Subroutine tables), the initialization or termination logic (in the General Information dialog box), and so forth. Local variables are available only within the logic in which they are declared

and are not defined in the Variables edit table. They are created for each entity, downtime occurrence, or the like executing a particular section of logic. A new local variable is created for each entity that encounters an INT or REAL statement. It exists only while the entity processes the logic that declared the local variable. Local variables may be passed to subroutines as parameters and are available to macros.

A local variable must be declared before it is used. To declare a local variable, use the following syntax:

```
INT or REAL <name1>{= expression}, <name2>{= expression}
```

Examples:

```
INT HourOfDay, WIP
REAL const1 = 2.5, const2 = 5.0
INT Init_Inventory = 170
```

In Section L7.11 we show you how to use a local variable in your simulation model logic.

Problem Statement—Tracking Work in Process and Production

In the **Poly Casting Inc.** machine shop, raw castings arrive in batches of four every hour. From the raw material store they are sent to the mill, where they undergo a milling operation that takes an average of three minutes with a standard deviation of one minute (normally distributed). The milled castings go to the grinder, where they are ground for a duration that is uniformly distributed (minimum four minutes and maximum six minutes) or U(5,1). After grinding, the ground pieces go to the finished parts store. Run the simulation for 100 hours. Track the work-in-process inventory and the production quantity.

The complete simulation model layout is shown in Figure L7.18. The locations are defined as Receiving_Dock, Mill, Grinder, and Finish_Parts_Store (Figure L7.19). Castings (entity) are defined to arrive in batches of four (Qty each) every 60 minutes (Frequency) as shown in Figure L7.20. The processes and routings are shown in Figure L7.21.

Define a variable in your model to track the work-in-process inventory (WIP) of parts in the machine shop. Also define another variable to track the production (PROD_QTY) of finished parts (Figure L7.22). Note that both of these are integer type variables.

In the process table, add the following operation statement in the Receiving location (Figure L7.21).

```
                    WIP = WIP + 1
```

Add the following operation statements in the outgoing Finish_Parts_Store location (Figure L7.21):

```
                 WIP = WIP - 1
                 PROD_QTY = PROD_QTY + 1
```

Alternatively, these three statements could be written as INC WIP, DEC WIP, and INC PROD_QTY.

FIGURE L7.18

Layout of Poly Casting Inc.

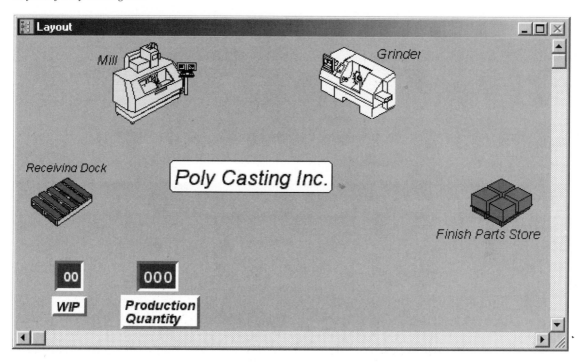

FIGURE L7.19

Locations at Poly Casting Inc.

Icon	Name	Cap.	Units	DTs...	Stats...	Rules...
	Mill	1	1	None	Time Series	Oldest
	Grinder	1	1	None	Time Series	Oldest
	Finish_Parts_Store	1	1	None	Time Series	Oldest
	Receiving_Dock	inf	1	None	Time Series	Oldest

FIGURE L7.20

Arrival of castings at Poly Casting Inc.

Entity...	Location...	Qty each...	First Time	Occurrences	Frequency	Logic	Disable
Casting	Receiving_Dock	4	0	inf	60 min		No

FIGURE L7.21

Processes and routings for the Poly Casting Inc. model.

					Routing		
		Process					
Entity	Location	Operation	Blk	Output	Destination	Rule	Move Logic
Casting	Receiving_Dock	WIP = WIP +1	1	Casting	Mill	FIRST 1	MOVE FOR 1 MIN
Casting	Mill	Wait N(3,1) min	1	Casting	Grinder	FIRST 1	MOVE FOR 1 MIN
Casting	Grinder	Wait U(5,1) min	1	Casting	Finish_Parts_Store	FIRST 1	MOVE FOR 1 MIN
Casting	Finish_Parts_Store	WIP = WIP -1					
		PROD_QTY = PROD_QTY + 1					
			1	Casting	EXIT	FIRST 1	

FIGURE L7.22

Variables for the Poly Casting Inc. model.

	Variables (global)				
Icon	ID	Type...	Initial value	Stats...	
Yes	WIP	Integer	0	Time Series, Tir	
Yes	PROD_QTY	Integer	0	Time Series, Tir	

L7.5 Uncertainty in Routing—Track Defects and Rework

Sometimes an entity needs to be routed to a destination from a list of available locations on a probabilistic basis. The location to which a specific entity is routed remains uncertain. A certain percentage of entities will be routed to a specific destination. The total of these percentages, however, should be 100 percent.

After processing is completed, a workpiece is frequently inspected and checked for correctness. If found to be good, the workpiece moves on to the next operation. If found to be defective, it is either scrapped or reworked. For rework, the workpiece is sent to one of the earlier operations. The inspection process is carried out either on all workpieces or on a sample of workpieces.

Problem Statement

Poly Casting Inc. in Section L7.4 decides to add an inspection station at the end of the machine shop, after the grinding operation. After inspection, 30 percent of the widgets are sent back to the mill for rework, 10 percent are sent back to the grinder for rework, and 5 percent are scrapped. The balance, 55 percent, pass inspection and go on to the finished parts store. The inspection takes a time that is triangularly distributed with a minimum of 4, mode of 5, and maximum of 6 minutes. The process times for rework are the same as those for new jobs. Track the amount of rework at the mill and the grinder. Also track the amount of scrapped parts and finished production. Run the simulation for 100 hours.

The locations are defined as mill, grinder, inspection, finish parts store, receiving dock, scrap parts, mill rework, and grind rework as shown in Figure L7.23.

FIGURE L7.23

Simulation model layout for Poly Castings Inc. with inspection.

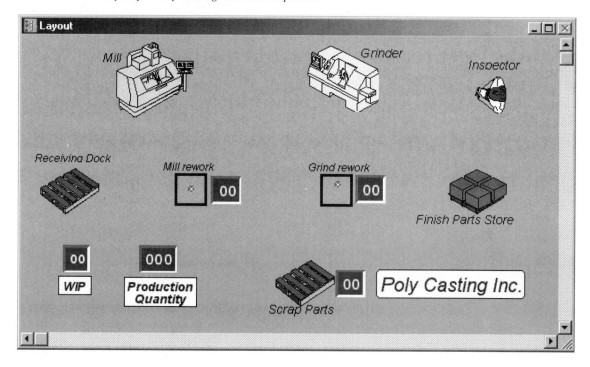

FIGURE L7.24

Variables for the Poly Castings Inc. with inspection model.

Icon	ID	Type...	Initial value	Stats...
Yes	WIP	Integer	0	Time Series, Tir
Yes	PROD_QTY	Integer	0	Time Series, Tir
Yes	M_rwrk	Integer	0	Time Series, Tir
Yes	G_rwrk	Integer	0	Time Series, Tir
Yes	Scrap	Integer	0	Time Series, Tir

The last four locations are defined with infinite capacity. The arrivals of castings are defined in batches of four every hour. Next we define five variables (Figure L7.24) to track work in process, production quantity, mill rework, grind rework, and scrap quantity. The processes and routings are defined as in Figure L7.25.

FIGURE L7.25

Processes and routings for the Poly Castings Inc. with inspection model.

		Process			Routing		
Entity	Location	Operation	Blk	Output	Destination	Rule	Move Logic
Casting	Receiving_Dock	WIP = WIP +1	1	Casting	Mill	FIRST 1	MOVE FOR 1 MIN
Casting	Mill	Wait N(3,1) min	1	Casting	Grinder	FIRST 1	MOVE FOR 1 MIN
Casting	Grinder	Wait U(5,1) min	1	Casting	Inspector	FIRST 1	MOVE FOR 1 MIN
Casting	Inspector	Wait T(4,5,6) min					
			1	Casting	Mill_rework	0.300000 1	MOVE FOR 1 MIN
				Casting	Grind_rework	0.100000	MOVE FOR 1 MIN
				Casting	Scrap_Parts	0.050000	MOVE FOR 1 MIN
				Casting	Finish_Parts_Store	0.550000	MOVE FOR 1 MIN
Casting	Mill_rework	M_rwrk = M_rwrk + 1	1	Casting	Mill	FIRST 1	
Casting	Grind_rework	G_rwrk = G_rwrk + 1	1	Casting	Grinder	FIRST 1	
Casting	Scrap_Parts	WIP = WIP -1 Scrap = Scrap + 1	1	Casting	EXIT	FIRST 1	
Casting	Finish_Parts_Store	WIP = WIP -1 PROD_QTY = PROD_QTY + 1	1	Casting	EXIT	FIRST 1	

L7.6 Batching Multiple Entities of Similar Type

L7.6.1 Temporary Batching—GROUP/UNGROUP

Frequently we encounter a situation where widgets are batched and processed together. After processing is over, the workpieces are unbatched again. For example, an autoclave or an oven is fired with a batch of jobs. After heating, curing, or bonding is done, the batch of jobs is separated and the individual jobs go their way. The individual pieces retain their properties during the batching and processing activities.

For such temporary batching, use the GROUP statement. For unbatching, use the UNGROUP statement. One may group entities by individual entity type by defining a process record for the type to group, or group them irrespective of entity type by defining an ALL process record. ProModel maintains all of the characteristics and properties of the individual entities of the grouped entities and allows them to remain with the individual entities after an UNGROUP command. Note that the capacity of the location where Grouping occurs must be at least as large as the group size.

Problem Statement

El Segundo Composites receive orders for aerospace parts that go through cutting, lay-up, and bonding operations. Cutting and lay-up take uniform (20,5) minutes and uniform (30,10) minutes. The bonding is done in an autoclave in batches of five parts and takes uniform (100,10) minutes. After bonding, the parts go to the shipment clerk individually. The shipment clerk takes normal (20,5) minutes to get each part ready for shipment. The orders are received on average once every 60 minutes, exponentially distributed. The time to transport these parts from one machine to another takes on average 15 minutes. Figure out the amount of WIP in the shop. Simulate for six months or 1000 working hours.

The locations are defined as Ply_Cutting, LayUp, Oven, Order_Q, Ship_Clerk, and Ship_Q. The layout of El Segundo Composites is shown in Figure L7.26. The processes and the routings are shown in Figure L7.27. A variable WIP is defined to keep track of the amount of work in process in the shop. The WIP value history over the simulated 1000 hours is shown in Figure L7.28. Note that the average WIP in the shop has been around 7, although on rare occasions the WIP has been as much as 23.

FIGURE L7.26

Layout of El Segundo Composites.

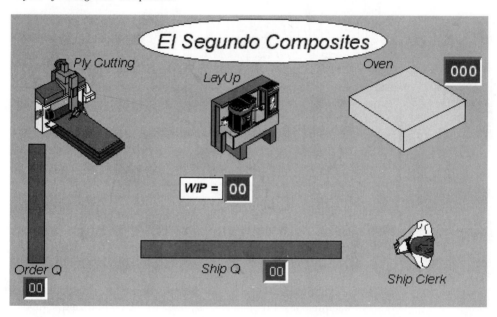

FIGURE L7.27

Process and routing tables for El Segundo Composites.

	Process			Routing			
Entity	Location	Operation	Blk	Output	Destination	Rule	Move Logic
Raw_Material	Order_Q	Inc WIP	1	Raw_Material	Ply_Cutting	FIRST 1	Move for 15 min
Raw_Material	Ply_Cutting	Wait U(20,5) min	1	Raw_Material	LayUp	FIRST 1	Move for 15 min
Raw_Material	LayUp	Wait U(30,10) min	1	Raw_Material	Oven	FIRST 1	Move for 15 min
Raw_Material	Oven	Group 5 as Batch					
Batch	Oven	Wait U(100,10) min UNGROUP					
Raw_Material	Oven		1	Raw_Material	Ship_Q	FIRST 1	Move for 15 min
Raw_Material	Ship_Q		1	Raw_Material	Ship_Clerk	FIRST 1	Move for 15 min
Raw_Material	Ship_Clerk	Wait N(20,5) min Dec WIP	1	Raw_Material	EXIT	FIRST 1	

FIGURE L7.28

Work in process value history.

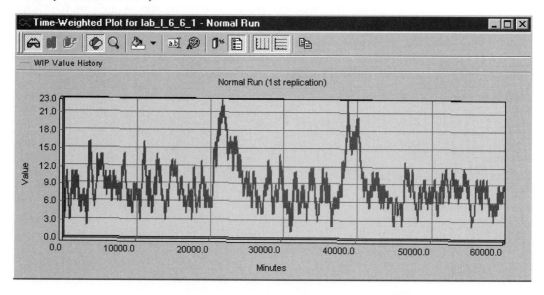

L7.6.2 Permanent Batching—COMBINE

In some situations, multiple entities of the same type or different types are batched together permanently. After batching, a single batch entity is formed, alternatively with a different name. In such permanent batching, the characteristics of the individual entities are lost; that is, the individual entities cannot be ungrouped later.

When defining the location, the capacity of the location where you use the COMBINE statement should be at least as large as the combined quantity.

Problem Statement

At the Garden Reach plant of the **Calcutta Tea Company,** the filling machine fills empty cans with 50 bags of the best Darjeeling tea at the rate of one can every 1±0.5 seconds uniformly distributed. The tea bags arrive to the packing line with a mean interarrival time of one second exponentially distributed. The filled cans go to a packing machine where 20 cans are combined into one box. The packing operation takes uniform (20±10) seconds. The boxes are shipped to the dealers. This facility runs 24 hours a day. Simulate for one day.

The various locations at the Calcutta Tea Company plant are shown in Figure L7.29. Three entities (Teabag, Can, and Box) are defined next. Teabags are defined to arrive with exponential interarrival time with a mean of one second. The processes and routing logic are shown in Figure L7.30. The layout of the Calcutta Tea Company plant is shown in Figure L7.31.

FIGURE L7.29

Locations at the Calcutta Tea Company.

Icon	Name	Cap.	Units	DTs...	Stats...	Rules...
	Incoming	1	1	None	Time Series	Oldest
	Filling_Machine	50	1	None	Time Series	Oldest
	Packing_Machine	20	1	None	Time Series	Oldest
	packing_Q	INFINITE	1	None	Time Series	Oldest, FIFO
	ship_Q	INFINITE	1	None	Time Series	Oldest, FIFO
	Shipping	1	1	None	Time Series	Oldest
	fill_Q	INFINITE	1	None	Time Series	Oldest

FIGURE L7.30

Process and routing tables at the Calcutta Tea Company.

```
**********************************************************************************
*                              Processing                                       *
**********************************************************************************

                               Process                        Routing

Entity   Location        Operation        Blk  Output  Destination       Rule
-------  --------------  --------------    ---  ------  ---------------   ------
TeaBag   Incoming                          1    TeaBag  fill_Q            FIRST 1
TeaBag   fill_Q                            1    TeaBag  Filling_Machine   FIRST 1
TeaBag   Filling_Machine COMBINE 50
                         Wait U(1,0.5) sec
                                           1    Can     packing_Q         FIRST 1
Can      packing_Q                         1    Can     Packing_Machine   FIRST 1
Can      Packing_Machine COMBINE 20
                         WAIT U(20,10) sec
                                           1    Box     ship_Q            FIRST 1
Box      ship_Q                            1    Box     Shipping          FIRST 1
Box      Shipping                          1    Box     EXIT              FIRST 1
```

FIGURE L7.31

Layout of the Calcutta Tea Company.

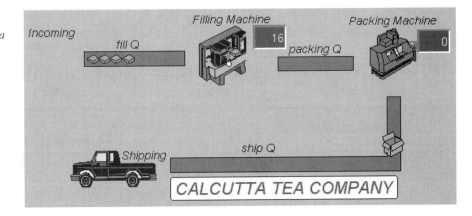

L7.7 Attaching One or More Entities to Another Entity

L7.7.1 Permanent Attachment—JOIN

Sometimes one or more entities are attached permanently to another entity, as in an assembly operation. The assembly process is a permanent bonding: the assembled entities lose their separate identities and properties. The individual entities that are attached cannot be separated again.

The *join* process is used to permanently assemble two or more individual entities together. For every JOIN statement, there must be a corresponding If Join Request rule. Joining is a two-step process:

1. Use the JOIN statement at the designated assembly location.
2. Use the join routing rule for all joining entities.

One of the joining entities, designated as the "base" entity, issues the JOIN statement. All other joining entities must travel to the assembly location on an If Join Request routing rule.

Problem Statement

At **Shipping Boxes Unlimited** computer monitors arrive at Monitor_Q at the rate of one every 15 minutes (exponential) and are moved to the packing table. Boxes arrive at Box_Q, at an average rate of one every 15 minutes (exponential) and are also moved to the packing table. At the packing table, monitors are packed into boxes. The packing operation takes normal (5,1) minutes. Packed boxes are sent to the inspector (Inspect_Q). The inspector checks the contents of the box and tallies with the packing slip. Inspection takes normal (4,2) minutes. After inspection, the boxes are loaded into trucks at the shipping dock (Shipping_Q). The loading takes uniform (5,1) minutes. Simulate for 100 hours. Track the number of monitors shipped and the WIP of monitors in the system.

The locations at Shipping Boxes Unlimited are defined as Monitor_Q, Box_Q, Shipping_Q, Inspect_Q, Shipping_Dock, Packing_Table, and Inspector. All the queue locations are defined with a capacity of infinity. Three entities (Monitor, Empty Box, and Full_Box) are defined next. The arrivals of monitors and empty boxes are shown in Figure L7.32. The processes and routings are shown in Figure L7.33. A snapshot of the simulation model is captured in

FIGURE L7.32

Arrival of monitors and empty boxes at Shipping Boxes Unlimited.

Entity...	Location...	Qty each...	First Time	Occurrences	Frequency	Logic	Disable
Monitor	monitor_Q	1		inf	e(5) min		No
Empty_Box	box_Q	1		inf	e(4) min		No

FIGURE L7.33

Processes and routings for Shipping Boxes Unlimited.

```
*************************************************************************
*                           Processing                                 *
*************************************************************************
```

		Process				Routing	
Entity	Location	Operation	Blk	Output	Destination	Rule	
----------	-------------	-------------------	---	----------	-------------	------	
Monitor	monitor_Q	Inc WIP	1	Monitor	Packing_Table	JOIN 1	
Empty_Box	box_Q		1	Empty_Box	Packing_Table	FIRST 1	
Empty_Box	Packing_Table	Join 1 Monitor					
		Wait N(5,1) min	1	Full_Box	Inspect_Q	FIRST 1	
Full_Box	Inspect_Q		1	Full_Box	Inspector	FIRST 1	
Full_Box	Inspector	Wait N(4,2) min	1	Full_Box	shipping_Q	FIRST 1	
Full_Box	shipping_Q		1	Full_Box	shipping	FIRST 1	
Full_Box	shipping	Wait U(5,1) min					
		Dec WIP					
		Inc monitor_shipped					
			1	Full_Box	EXIT	FIRST 1	

FIGURE L7.34

A snapshot of the simulation model for Shipping Boxes Unlimited.

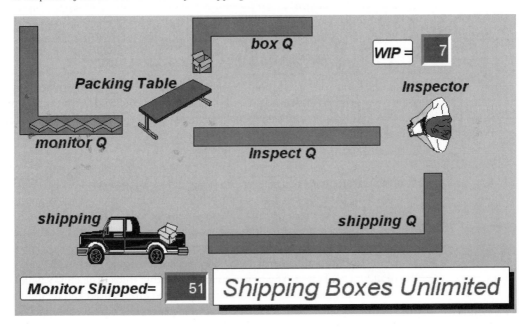

Figure L7.34. The plot of the work-in-process inventory for the 100 hours of simulation run is shown in Figure L7.35. Note that the work-in-process inventory rises to as much as 12 in the beginning. However, after achieving steady state, the WIP inventory stays mostly within the range of 0 to 3.

Time-weighted plot of the WIP inventory at Shipping Boxes Unlimited.

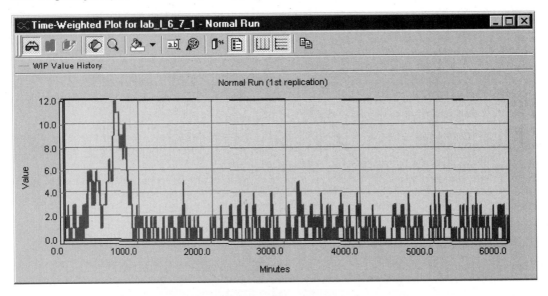

L7.7.2 Temporary Attachment—*LOAD/UNLOAD*

Sometimes one or more entities are attached to another entity temporarily, processed or moved together, and detached later in the system. In ProModel, we use the LOAD and UNLOAD statements to achieve that goal. The properties of the individual attached entities are preserved. This of course requires more memory space and hence should be used only when necessary.

Problem Statement

For the **Shipping Boxes Unlimited** problem in Section L7.7.1, assume the inspector places (loads) a packed box on an empty pallet. The loading takes anywhere from two to four minutes, uniformly distributed. The loaded pallet is sent to the shipping dock and waits in the shipping queue. At the shipping dock, the packed boxes are unloaded from the pallet. The unloading time is also uniformly distributed; U(3,1) min. The boxes go onto a waiting truck. The empty pallet is returned, via the pallet queue, to the inspector. One pallet is used and recirculated in the system. Simulate for 100 hours. Track the number of monitors shipped and the WIP of monitors.

 The locations defined in this model are Monitor_Q, Box_Q, Inspect_Q, Shipping_Q, Pallet_Q, Packing_Table, Inspector, and Shipping_Dock (Figure L7.36). All queue locations have infinite capacity. The layout of *Shipping Boxes Unlimited* is shown in Figure L7.37. Five entities are defined: Monitor, Box, Empty_Box, Empty_Pallet, and Full_Pallet as shown in Figure L7.38. The

FIGURE L7.36

The locations at Shipping Boxes Unlimited.

Icon	Name	Cap.	Units	DTs...	Stats...	Rules...
	monitor_Q	INFINITE	1	None	Time Series	Oldest, FIFO
	box_Q	INFINITE	1	None	Time Series	Oldest, FIFO
	Packing_Table	1	1	None	Time Series	Oldest
	Inspect_Q	INFINITE	1	None	Time Series	Oldest, FIFO
	Inspector	1	1	None	Time Series	Oldest
	shipping_Q	INFINITE	1	None	Time Series	Oldest, FIFO
	shipping	1	1	None	Time Series	Oldest
	Pallet_Q	INFINITE	1	None	Time Series	Oldest, FIFO

FIGURE L7.37

Boxes loaded on pallets at Shipping Boxes Unlimited.

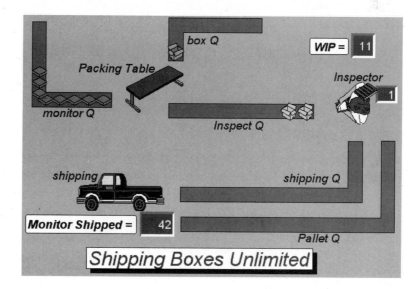

arrivals of monitors, empty boxes, and empty pallets are shown in Figure L7.39. The processes and routings are shown in Figure L7.40. Note that comments can be inserted in a line of code as follows (Figure L7.40):

```
/* inspection time */
```

The plot of the work-in-process inventory for the 100 hours of simulation run is presented in Figure L7.41. Note that after the initial transient period ($\cong 1200$ minutes), the work-in-process inventory drops and stays mostly in a range of 0–2.

FIGURE L7.38

Entities at Shipping Boxes Unlimited.

Icon	Name	Speed (fpm)	Stats...
	Empty_Box	150	Time Series
	Monitor	150	Time Series
	Full_Box	150	Time Series
	Empty_Pallet	150	Time Series
	Full_Pallet	150	Time Series

FIGURE L7.39

Arrival of monitors, empty boxes, and empty pallets at Shipping Boxes Unlimited.

Entity...	Location...	Qty each...	First Time	Occurrences	Frequency	Logic	Disable
Monitor	monitor_Q	1		inf	e(15) min		No
Empty_Box	box_Q	1		inf	e(15) min		No
Empty_Pallet	Pallet_Q	1		1	0		No

FIGURE L7.40

Process and routing tables at Shipping Boxes Unlimited.

```
**********************************************************************************
*                              Processing                                       *
**********************************************************************************

                              Process                        Routing

Entity         Location        Operation           Blk  Output         Destination    Rule
----------     -----------     ------------------   ---  ------------   ------------   -------
Monitor        monitor_Q       Inc WIP              1    Monitor        Packing_Table  JOIN 1
Empty_Box      box_Q                                1    Empty_Box      Packing_Table  FIRST 1
Empty_Box      Packing_Table   Join 1 Monitor
                               Wait N(5,1) min      1    Full_Box       Inspect_Q      FIRST 1
Full_Box       Inspect_Q                            1    Full_Box       Inspector      LOAD 1
Empty_Pallet   Pallet_Q                             1    Empty_Pallet   Inspector      FIRST 1
Empty_Pallet   Inspector       Wait N(4,2) min /*inspection time*/
                               Wait U(3,1) min /*load time*/
                               Load 1

                                                    1    Full_Pallet    shipping_Q     FIRST 1
                                                    1    Full_Pallet    shipping       FIRST 1
Full_Pallet    shipping_Q
Full_Pallet    shipping        Wait U(3,1) min /*unload time*/
                               Unload 1
                                                    1    Empty_Pallet   Pallet_Q       FIRST 1
Full_Box       shipping        Wait U(5,1) min /*loading in truck*/
                               Dec WIP
                               Inc monitor_shipped
                                                    1    Full_Box       EXIT           FIRST 1
```

FIGURE L7.41

Time-weighted plot of the WIP inventory at Shipping Boxes Unlimited.

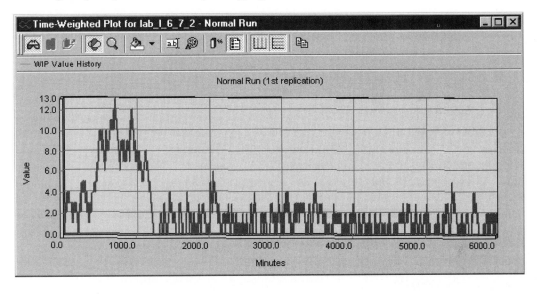

L7.8 Accumulation of Entities

Sometimes we need to hold entities at a location until a certain number accumulate. Once this critical limit is reached, the entities are released for further processing downstream. This allows us to model a certain type of batching that involves simply the accumulation of parts and their subsequent release—all at a single location. The capacity of the location at which the accumulation takes place must be at least as large as the accumulation limit or the amount of accumulation.

Problem Statement

Visitors arrive at **California Adventure Park** in groups that vary in size from two to four (uniformly distributed). The average time between arrival of two successive groups is five minutes, exponentially distributed. All visitors wait in front of the gate until five visitors have accumulated. At that point the gate opens and allows the visitors to enter the park. On average, a visitor spends 20±10 minutes (uniformly distributed) in the park. Simulate for 1000 hours. Track how many visitors are waiting outside the gate and how many are in the park.

Three locations (Gate_In, Walk_In_Park, and Gate_Out) are defined in this model. Visitor is defined as the entity. The processes and the layout of the adventure park are shown in Figures L7.42 and L7.43.

FIGURE L7.42

Process and routing tables for California Adventure Park.

```
*******************************************************************************
*                               Processing
*******************************************************************************

                        Process                         Routing

Entity    Location      Operation        Blk  Output   Destination    Rule
--------  ------------  ---------------   ---  -------  -------------  --------
Visitor   Gate_In       Accum 5           1    Visitor  Walk_In_Park   FIRST 1
Visitor   Walk_In_Park  Wait U(20,10) min
                                          1    Visitor  Gate_Out       FIRST 1
Visitor   Gate_Out                        1    Visitor  EXIT           FIRST 1
```

FIGURE L7.43

Layout of California Adventure Park.

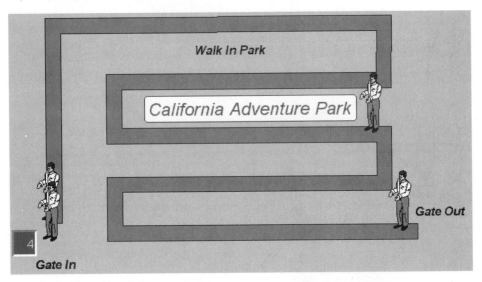

L7.9 Splitting of One Entity into Multiple Entities

When a single incoming entity is divided into multiple entities at a location and processed individually, we can use the SPLIT construct. The SPLIT command splits the entity into a specified number of entities, and optionally assigns them a new name. The resulting entities will have the same attribute values as the original entity. Use SPLIT to divide a piece of raw material into components, such as a silicon wafer into silicon chips or six-pack of cola into individual cans.

Problem Statement

The cafeteria at **San Dimas High School** receives 10 cases of milk from a vendor each day before the lunch recess. On receipt, the cases are split open and individual cartons (10 per case) are stored in the refrigerator for distribution to students during lunchtime. The distribution of milk cartons takes triangular(.1,.15,.2) minute per student. The time to split open the cases takes a minimum of 5 minutes and a maximum of 7 minutes (uniform distribution) per case. Moving the cases from receiving to the refrigerator area takes five minutes per case, and moving the cartons from the refrigerator to the distribution area takes 0.2 minute per carton. Students wait in the lunch line to pick up one milk carton each. There are only 100 students at this high school. Students show up for lunch with a mean interarrival time of 1 minute (exponential). On average, how long does a carton stay in the cafeteria before being distributed and consumed? What are the maximum and the minimum times of stay? Simulate for 10 days.

The layout of the San Dimas High School cafeteria is shown in Figure L7.44. Three entities—Milk_Case, Milk_Carton, and Student—are defined. Ten milk cases arrive with a frequency of 480 minutes. One hundred students show up for lunch each day. The arrival of students and milk cases is shown in Figure L7.45. The processing and routing logic is shown in Figure L7.46.

FIGURE L7.44

Layout of the San Dimas High School cafeteria.

FIGURE L7.45

Arrival of milk and students at the San Dimas High School cafeteria.

Entity...	Location...	Qty each...	First Time	Occurrences	Frequency
Milk_Case	Receiving	10	0	inf	480 min
student	Lunch_Line	100	0	inf	480 min

Figure L7.46

Process and routing logic at the San Dimas High School cafeteria.

		Process			Routing		
Entity	Location	Operation	Blk	Output	Destination	Rule	Move Logic
MIlk_Case	Receiving		1	MIlk_Case	Refrigerator	FIRST 1	Move for 5 min
MIlk_Case	Refrigerator	SPLIT 10 AS Milk_Carton Wait U(6,1) min					
Milk_Carton	Refrigerator		1	Milk_Carton	Cafetaria	JOIN 1	Move for .2 min
student	Lunch_Line		1	student	Cafetaria	FIRST 1	
student	Cafetaria	Join 1 Milk_Carton Wait T(.1, .15, .2) min					
			1	student	EXIT	FIRST 1	

L7.10 Decision Statements

In ProModel you can make use of several general control statements for decision making. ProModel meets the modern standards of structured program design. Although there is no formal definition, computer scientists agree that a structured program should have modular design and use only the following three types of logical structures: sequences, decisions, and loops.

- *Sequences:* Program statements are executed one after another.
- *Decisions:* One of two blocks of program code is executed based on a test for some condition.
- *Loops:* One or more statements are executed repeatedly as long as a specified condition is true.

All the control statements available in ProModel are shown in Figure L7.47. In this section we introduce you to some of these statements with examples.

L7.10.1 *IF-THEN-ELSE* **Statement**

An IF block allows a program to decide on a course of action based on whether a certain condition is true or false. A program block of the form

```
IF condition THEN
        action1
ELSE
        action2
```

causes the program to take action1 if condition is true and action2 if condition is false. Each action consists of one or more ProModel statements. After an action is taken, execution continues with the line after the IF block.

Problem Statement

The **Bombay Restaurant** offers only a drive-in facility. Customers arrive at the rate of six each hour (exponential interarrival time). They place their orders at the first window, drive up to the next window for payment, pick up food from the last window, and then leave. The activity times are given in Table L7.4. The drive-in facility can accommodate 10 cars at most. However, customers typically leave

FIGURE L7.47

Control statements available in ProModel.

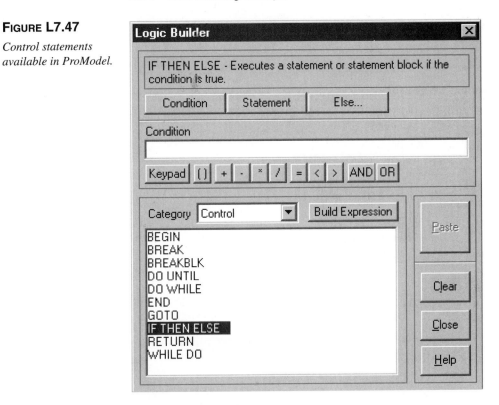

TABLE L7.4   **Process Times at the Bombay Restaurant**

Activity	Process Time (minutes)
Order food	Normal (5,1)
Make payment	Normal (7,2)
Pick up food	Normal (10,2)

and go to the Madras Café across the street if six cars are waiting in line when they arrive. Simulate for 100 days (8 hours each day). Estimate the number of customers served each day. Estimate on average how many customers are lost each day to the competition.

An additional location (Arrive) is added to the model. After the customers arrive, they check if there are fewer than six cars at the restaurant. If yes, they join the line and wait; if not, they leave and go across the street to the Madras Café. An IF-THEN-ELSE statement is added to the logic window in the processing table (Figure L7.48). A variable (Customer_Lost) is added to the model to keep track of the number of customers lost to the competition. The layout of the Bombay

FIGURE L7.48

Process and routing logic at the Bombay Restaurant.

		Process			Routing		
Entity	Location	Operation	Blk	Output	Destination	Rule	Move Logic
Customer	Arrive	If (contents(Q_1)+contents(Q_2)+contents(Q_3)+ contents(order_food)+contents(pay)+contents(pickup_food)) <6 Then Route 1 Else Begin Inc Customer_Lost Route 2 End					
			1	Customer	Q_1	FIRST 1	
			2	Customer	EXIT	FIRST 1	
Customer	Q_1		1	Customer	Order_Food	FIRST 1	
Customer	Order_Food	Wait N(5,1) min	1	Customer	Q_2	FIRST 1	
Customer	Q_2		1	Customer	Pay	FIRST 1	
Customer	Pay	Wait N(7,2)	1	Customer	Q_3	FIRST 1	
Customer	Q_3		1	Customer	Pickup_Food	FIRST 1	
Customer	Pickup_Food	Wait N(10,2) min	1	Customer	EXIT	FIRST 1	

FIGURE L7.49

Layout of the Bombay Restaurant.

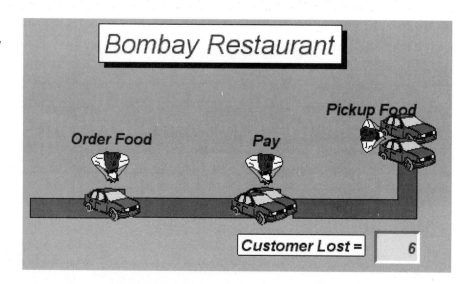

Restaurant is shown in Figure L7.49. Note that Q_1 and Q_2 are each 100 feet long and Q_3 is 200 feet long.

The total number of customers lost is 501 in 100 days. The number of customers served in 100 days is 4791. The average cycle time per customer is 36.6 minutes.

L7.10.2 WHILE-DO Loop

The WHILE-DO block repeats a group of statements continuously while a condition remains true. If the condition is false, the loop is bypassed.

FIGURE L7.50

An example of the WHILE-DO *logic for Shipping Boxes Unlimited.*

		Process			Routing	
Entity	Location	Operation	Blk	Output	Destination	Rule
Monitor	monitor_Q	Inc WIP	1	Monitor	Packing_Table	JOIN 1
Empty_Box	box_Q		1	Empty_Box	Packing_Table	FIRST 1
Empty_Box	Packing_Table	Join 1 Monitor				
		Wait N(5,1) min	1	Full_Box	Inspect_Q	FIRST 1
Full_Box	Inspect_Q	While contents(Inspect_Q) < 5 Do				
		Wait 60 min	1	Full_Box	Inspector	LOAD 1
			1	Empty_Pallet	Inspector	FIRST 1
Empty_Pallet	Pallet_Q					
Empty_Pallet	Inspector	Wait N(4,2) min /*inspection time*/				
		Wait U(3,1) min /*load time*/				
		Load 1				
			1	Full_Pallet	shipping_Q	FIRST 1
			1	Full_Pallet	shipping	FIRST 1
Full_Pallet	shipping_Q					
Full_Pallet	shipping	Wait U(3,1) min /*unload time*/				
		Unload 1				
			1	Empty_Pallet	Pallet_Q	FIRST 1
Full_Box	shipping	Wait U(5,1) min /*loading in truck*/				
		Dec WIP				
		Inc monitor_shipped				
			1	Full_Box	EXIT	FIRST 1

Problem Statement

The inspector in Section L7.7.2 is also the supervisor of the shop. As such, she inspects only when at least five full boxes are waiting for inspection in the Inspect_Q. A WHILE-DO loop is used to check if the queue has five or more boxes waiting for inspection (Figure L7.50). The loop will be executed every hour. Figure L7.51 shows a time-weighted plot of the contents of the inspection queue. Note how the queue builds up to 5 (or more) before the inspector starts inspecting the full boxes.

L7.10.3 DO-WHILE *Loop*

The DO-WHILE block repeats a group of statements continuously while a condition remains true. This loop will be executed at least once; use it for processes that will be executed at least once and possibly more.

Problem Statement

For the **Poly Casting Inc.** example in Section L7.4, we would like to assign an upper limit to the work-in-process inventory in the shop. A DO-WHILE loop will be used to check if the level of WIP is equal to or above five; if so, the incoming castings will wait at the receiving dock (Figure L7.52). The loop will be executed once every hour to check if the level of WIP has fallen below five. Figure L7.53 show a time-weighted plot of the value of WIP at Poly Castings Inc. Note that the level of WIP is kept at or below five.

FIGURE L7.51

A plot of the contents of the inspection queue.

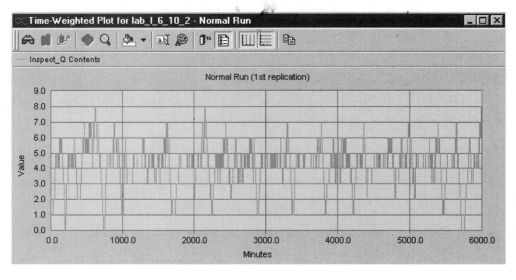

FIGURE L7.52

An example of a DO-WHILE loop.

		Process			Routing		
Entity	Location	Operation	Blk	Output	Destination	Rule	Move Logic
Casting	Receiving_Dock	Do Wait 60 min While WIP >=5 Inc WIP	1	Casting	Mill	FIRST 1	MOVE FOR 1 MIN
Casting	Mill	Wait N(3,1) min	1	Casting	Grinder	FIRST 1	MOVE FOR 1 MIN
Casting	Grinder	Wait U(5,1) min	1	Casting	Finish_Parts_Store	FIRST 1	MOVE FOR 1 MIN
Casting	Finish_Parts_Store	Dec WIP Inc PROD_QTY	1	Casting	EXIT	FIRST 1	

L7.10.4 *GOTO* **Statement**

The GOTO block jumps to the statement identified by the designated label. A label should follow the normal rules for names. A colon follows a label.

Problem Statement

The **Indian Bank** has acquired the Bank of India in Section L7.3. Anthony has been given a golden handshake (that is, fired). As a result of new training provided to the two remaining tellers (Amar and Akbar) there has been a change in their customer service time. Whenever more than three customers are waiting for service, the service time is reduced to an average of five minutes and a standard deviation of two minutes (normally distributed). The layout of the new Indian Bank is shown in Figure L7.54. A GOTO statement is used to jump to the appropriate wait statement depending on the number of customers waiting for service in the Teller_Q, as shown in Figure L7.55.

FIGURE L7.53

A plot of the value of WIP at Poly Castings Inc.

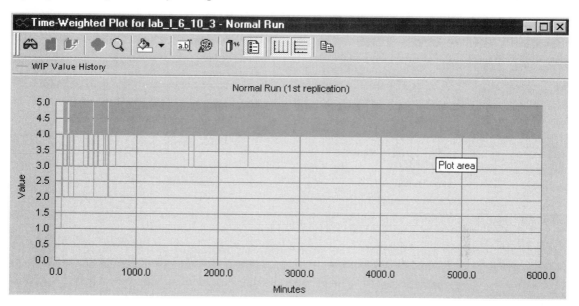

FIGURE L7.54

The layout of the Indian Bank.

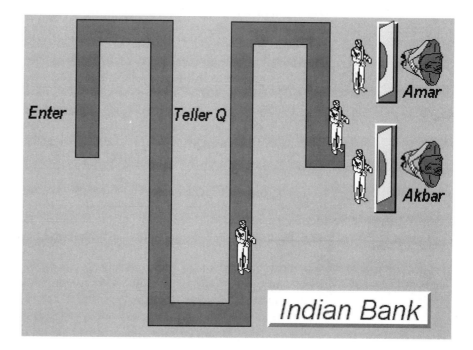

FIGURE L7.55

An example of a GOTO statement.

		Process			**Routing**	
Entity	Location	Operation	Blk	Output	Destination	Rule
Customers	Enter		1	Customers	Teller_Q	FIRST 1
Customers	Teller_Q		1	Customers	Amar	FIRST 1
				Customers	Akbar	FIRST
Customers	Amar	If Contents(Teller_Q) > 3 Then GOTO L1				
		Wait U(10,6) min				
		L1:				
		Wait N(5,2) min	1	Customers	EXIT	FIRST 1
Customers	Akbar	If Contents(Teller_Q) > 3 Then GOTO L1				
		Wait U(10,6) min				
		L1:				
		Wait N(5,2) min	1	Customers	EXIT	FIRST 1

L7.11 Periodic System Shutdown

Some service systems periodically stop admitting new customers for a given duration to catch up on the backlog of work and reduce the congestion. Some automobile service centers, medical clinics, restaurants, amusement parks, and banks use this strategy. The modulus mathematical operator in ProModel is useful for simulating such periodic shutdowns.

Problem Statement

The **Bank of India** in Section L7.3 opens for business each day for 8 hours (480 minutes). Assume that the customers arrive to the bank according to an exponential distribution (mean of 4 minutes). All the tellers service the customers according to a uniform distribution (mean of 10 minutes and half-width of 6 minutes). Customers are routed to the three tellers in turn (selected in rotation).

Each day after 300 minutes (5 hours) of operation, the front door of the bank is locked, and any new customers arriving at the bank are turned away. Customers already inside the bank continue to get served. The bank reopens the front door 90 minutes (1.5 hours) later to new customers. Simulate the system for 480 minutes (8 hours). Make a time-series plot of the Teller_Q to show the effect of locking the front door on the bank.

The logic for locking the front door is shown in Figure L7.56. The simulation clock time clock(min) is a cumulative counter of hours elapsed since the start of the simulation run. The current time of any given day can be determined by modulus dividing the current simulation time, clock(min), by 480 minutes. If the remainder of clock(min) divided by 480 minutes is between 300 minutes (5 hours) and 390 minutes (6.5 hours), the arriving customers are turned away (disposed). Otherwise, they are allowed into the bank. An IF-THEN-ELSE logic block as described in Section L7.10.1 is used here.

A time-series plot of the contents of the Teller_Q is shown in Figure L7.57. This plot clearly shows how the Teller_Q (and the whole bank) builds up during

FIGURE L7.56

Process and routing logic for the Bank of India.

```
**********************************************************************
*                            Processing                             *
**********************************************************************

                    Process                        Routing

Entity     Location  Operation           Blk  Output    Destination Rule      Move Logic
_____   _____  _____           ___  _____  _____ ____      _____

Customers  Enter     If <clock(min) mod 480 >= 300) and <clock(min) mod 480 <= 390) Then
                              //modulus division gives the current hour of the day.
                     {
                     Route 2
                     }
                     else
                     {
                     Route 1
                     }


                                         1    Customers Teller_Q    FIRST 1
                                         2    Customers EXIT         FIRST 1
Customers  Teller_Q                      1    Customers Amar         TURN 1
                                              Customers Akbar        TURN
                                              Customers Anthony      TURN
Customers  Amar      Wait U(10,6) min    1    Customers EXIT         FIRST 1
Customers  Akbar     Wait U(10,6) min    1    Customers EXIT         FIRST 1
Customers  Anthony   Wait U(10,6) min    1    Customers EXIT         FIRST 1
```

FIGURE L7.57

Time-series plot of Teller_Q at the Bank of India.

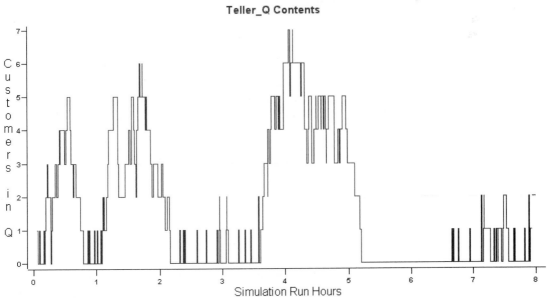

FIGURE L7.58

Histogram of Teller_Q contents at the Bank of India.

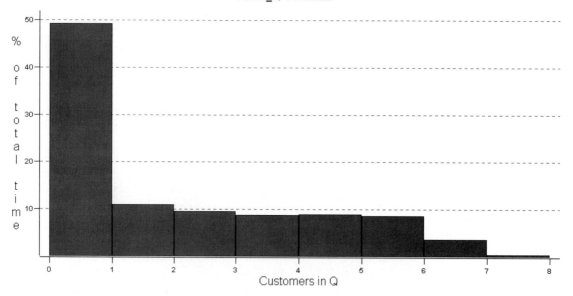

Bank of India
Teller_Q Contents

the day; then, after the front door is locked at the fifth hour (300 minutes) into the simulated day, customers remaining in the queue are processed and the queue length decreases (down to zero in this particular simulation run). The queue length picks back up when the bank reopens the front door at simulation time 6.5 hours (390 minutes).

The histogram of the same queue (Figure L7.58) shows that approximately 49% of the time the queue was empty. About 70% of the time there are 3 or fewer customers waiting in line. What is the average time a customer spends in the bank? Would you recommend that the bank not close the door after 5 hours of operation (customers never liked this practice anyway)? Will the average customer stay longer in the bank?

L7.12 Exercises

1. Visitors arrive at **Kid's World** entertainment park according to an exponential interarrival time distribution with mean 2.5 minutes. The travel time from the entrance to the ticket window is normally distributed with a mean of three minutes and a standard deviation of 0.5 minute. At the ticket window, visitors wait in a single line until one

of four cashiers is available to serve them. The time for the purchase of tickets is normally distributed with a mean of five minutes and a standard deviation of one minute. After purchasing tickets, the visitors go to their respective gates to enter the park. Create a simulation model, with animation, of this system. Run the simulation model for 200 hours to determine

a. The average and maximum length of the ticketing queue.
b. The average number of customers completing ticketing per hour.
c. The average utilization of the cashiers.
d. Whether management should add more cashiers.

2. A consultant for **Kid's World** recommended that four individual queues be formed at the ticket window (one for each cashier) instead of one common queue. Create a simulation model, with animation, of this system. Run the simulation model for 200 hours to determine

a. The average and maximum length of the ticketing queues.
b. The average number of customers completing ticketing per hour.
c. The average utilization of the cashiers.
d. Whether you agree with the consultant's decision. Would you recommend a raise for the consultant?

3. At the **Kid's World** entertainment park in Exercise 1, the operating hours are 8 A.M. till 10 P.M. each day (all week). Simulate for a whole year (365 days) and answer questions *a–d* as given in Exercise 1.

4. At **Southern California Airline**'s traveler check-in facility, three types of customers arrive: passengers with e-tickets (Type E), passengers with paper tickets (Type T), and passengers that need to purchase tickets (Type P). The interarrival distribution and the service times for these passengers are given in Table L7.5. Create a simulation model, with animation, of this system. Run the simulation model for 2000 hours. If separate gate agents serve each type of passenger, determine the following:

a. The average and maximum length of the three queues.
b. The average number of customers of each type completing check-in procedures per hour.
c. The average utilization of the gate agents.

TABLE L7.5 **Interarrival and Service Time Distributions at Southern California Airline**

Type of Traveler	Interarrival Distribution	Service Time Distribution
Type E	Exponential (mean 5.5 min.)	Normal (mean 3 min., std. dev. 1 min.)
Type T	Exponential (mean 10.5 min.)	Normal (mean 8 min., std. dev. 3 min.)
Type P	Exponential (mean 15.5 min.)	Normal (mean 12 min., std. dev. 3 min.)

    d. The percentage of time the number of customers (of each type of customer) waiting in line is ≤ 2.

    e. Would you recommend one single line for check-in for all three types of travelers? Discuss the pros and cons of such a change.

5. **Raja & Rani,** a fancy restaurant in Santa Clara, holds a maximum of 15 diners. Customers arrive according to an exponential distribution with a mean of 5 minutes. Customers stay in the restaurant according to a triangular distribution with a minimum of 45 minutes, a maximum of 60 minutes, and a mode of 75 minutes. Create a simulation model, with animation, of this system. The restaurant operating hours are 3 P.M. till 12 P.M. Run the simulation model for 50 replications of 9 hours each.

    a. Beginning empty, how long (average and standard deviation) does it take for the restaurant to fill?

    b. What is the total number of diners (average and standard deviation) entering the restaurant before it fills?

    c. What is the total number of guests (average and standard deviation) served per night?

    d. What is the average utilization of the restaurant?

6. **Woodland Nursing Home** has six departments—initial exam, X ray, operating room, cast-fitting room, recovery room, and checkout room. The probabilities that a patient will go from one department to another are given in Table L7.6. The time patients spend in each department is given in Table L7.7. Patients arrive at the average rate of 5 per hour (exponential interarrival time). The nursing home remains open 24/7.

TABLE L7.6   Probabilities of Patient Flow

From	To	Probability
Initial exam	X ray	.35
	Operating room	.20
	Recovery room	.15
	Checkout room	.30
X ray	Operating room	.15
	Cast-fitting room	.25
	Recovery room	.40
	Checkout room	.20
Operating room	Cast-fitting room	.30
	Recovery room	.65
	Checkout room	.05
Cast-fitting room	Recovery room	.55
	X ray	.10
	Checkout room	.35
Recovery room	Operating room	.10
	X ray	.20
	Checkout room	.70

TABLE L7.7   Patient Processing Time

Department	Patient Time in Department
Initial exam	Normal (5,2) minutes
X ray	Uniform (6,1) minutes
Operating room	Normal (30,5) minutes
Cast-fitting room	Uniform (10,4) minutes
Recovery room	Normal (8,2) minutes

TABLE L7.8   Process Time Distributions at United Electronics

Assembly	Uniform (13.5±1.5) minutes
Soldering	Normal (36,10) minutes
Painting	Uniform (55±15) minutes
Inspection	Exponential (8) minutes

However, every seven hours (420 minutes) the front door is locked for an hour (60 minutes). No new patients are allowed in the nursing home during this time. Patients already in the system continue to get served. Simulate for one year (365 days, 24 hours per day).

a. Figure out the utilization of each department.

b. What are the average and maximum numbers of patients in each department?

c. Which is the bottleneck department?

d. What is the average time spent by a patient in the nursing home?

7. **United Electronics** manufactures small custom electronic assemblies. Parts must be processed through four stations: assembly, soldering, painting, and inspection. Orders arrive with an exponential interarrival distribution (mean 20 minutes). The process time distributions are shown in Table L7.8.

   The soldering operation can be performed on three jobs at a time. Painting can be done on four jobs at a time. Assembly and inspection are performed on one job at a time. Create a simulation model, with animation, of this system. Simulate this manufacturing system for 100 days, eight hours each day. Collect and print statistics on the utilization of each station, associated queues, and the total number of jobs manufactured during each eight-hour shift (average).

8. In **United Electronics** in Exercise 7, 10 percent of all finished assemblies are sent back to soldering for rework after inspection, five percent are sent back to assembly for rework after inspection, and one

percent of all assemblies fail to pass and are scrapped. Create a simulation model, with animation, of this system. Simulate this manufacturing system for 100 days, eight hours each day. Collect and print statistics on the utilization of each station, associated queues, total number of jobs assembled, number of assemblies sent for rework to assembly and soldering, and the number of assemblies scrapped during each eight-hour shift (average).

9. Small toys are assembled in four stages (Centers 1, 2, and 3 and Inspection) at **Bengal Toy Company.** After each assembly step, the appliance is inspected or tested; if a defect is found, it must be corrected and then checked again. The assemblies arrive at a constant rate of one assembly every two minutes. The times to assemble, test, and correct defects are normally distributed. The means and standard deviations of the times to assemble, inspect, and correct defects, as well as the likelihood of an assembly error, are shown in Table L7.9. If an assembly is found defective, the defect is corrected and it is inspected again. After a defect is corrected, the likelihood of another defect being found is the same as during the first inspection. We assume in this model that an assembly defect is eventually corrected and then it is passed on to the next station. Simulate for one year (2000 hours) and determine the number of good toys shipped in a year.

10. **Salt Lake City Electronics** manufactures small custom communication equipment. Two different job types are to be processed within the following manufacturing cell. The necessary data are given in Table L7.10. Simulate the system to determine the average number of jobs waiting for different operations, number of jobs of each type finished each day, average cycle time for each type of job, and the average cycle time for all jobs.

11. Six dump trucks at the **DumpOnMe** facility in Riverside are used to haul coal from the entrance of a small mine to the railroad. Figure L7.59 provides a schematic of the dump truck operation. Each truck is loaded by one of two loaders. After loading, a truck immediately moves

TABLE L7.9   **Process Times and Probability of Defects at Bengal Toy Company**

Center	Assembly Time		Inspect Time		Correct Time		
	Mean	Standard Deviation	Mean	Standard Deviation	P(error)	Mean	Standard Deviation
1	.7	.2	.2	.05	.1	.2	.05
2	.75	.25	.2	.05	.05	.15	.04
3	.8	.15	.15	.03	.03	.1	.02

FIGURE L7.59

Schematic of dump truck operation for DumpOnMe.

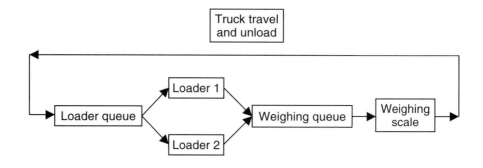

TABLE L7.10   **Data Collected at Salt Lake City Electronics**

Job Type	Number of Batches	Number of Jobs per Batch	Assembly Time	Soldering Time	Painting Time	Inspection Time	Time between Batch Arrivals
1	15	5	Tria (5,7,10)	Normal (36,10)	Uniform (55±15)	Exponential (8)	Exp (14)
2	25	3	Tria (7,10,15)		Uniform (35±5)	Exponential (5)	Exp (10)

Note: All times are given in minutes.

TABLE L7.11   **Various Time Measurement Data at the DumpOnMe Facility**

Loading time	Uniform (7.5±2.5) minutes
Weighing time	Uniform (3.5±1.5) minutes
Travel time	Triangular (10,12,15) minutes

to the scale to be weighed as soon as possible. Both the loaders and the scale have a first-come, first-served waiting line (or queue) for trucks. Travel time from a loader to the scale is considered negligible. After being weighed, a truck begins travel time (during which time the truck unloads), and then afterward returns to the loader queue. The distributions of loading time, weighing time, and travel time are shown in Table L7.11.

a. Create a simulation model, with animation, of this system. Simulate for 200 days, eight hours each day.

b. Collect statistics to estimate the loader and scale utilization (percentage of time busy).

c. About how many trucks are loaded each day on average?

12. At the **Pilot Pen Company,** a molding machine produces pen barrels of two different colors—red and blue—in the ratio of 3:2. The molding time is triangular (3,4,6) minutes per barrel. The barrels go to a filling machine, where ink of appropriate color is filled at the rate of 20 pens per hour (exponentially distributed). Another molding machine makes caps of the same two colors in the ratio of 3:2. The molding time is triangular (3,4,6) minutes per cap. At the next station, caps and filled barrels of matching colors are joined together. The joining time is exponentially distributed with a mean of 1 min. Simulate for 2000 hours. Find the average number of pens produced per hour. Collect statistics on the utilization of the molding machines and the joining equipment.

13. Customers arrive at the **NoWaitBurger** hamburger stand with an interarrival time that is exponentially distributed with a mean of one minute. Out of 10 customers, 5 buy a hamburger and a drink, 3 buy a hamburger, and 2 buy just a drink. One server handles the hamburger while another handles the drink. A person buying both items needs to wait in line for both servers. The time it takes to serve a customer is N(70,10) seconds for each item. Simulate for 100 hours. Collect statistics on the number of customers served per hour, size of the queues, and utilization of the servers. What changes would you suggest to make the system more efficient?

14. Workers who work at the **Detroit ToolNDie** plant must check out tools from a tool crib. Workers arrive according to an exponential distribution with a mean time between arrivals of five minutes. At present, three tool crib clerks staff the tool crib. The time to serve a worker is normally distributed with a mean of 10 minutes and a standard deviation of 2 minutes. Compare the following servicing methods. Simulate for 2000 hours and collect data.
    a. Workers form a single queue, choosing the next available tool crib clerk.
    b. Workers enter the shortest queue (each clerk has his or her own queue).
    c. Workers choose one of three queues at random.

15. At the **ShopNSave,** a small family-owned grocery store, there are only four aisles: aisle 1—fruits/vegetables, aisle 2—packaged goods (cereals and the like), aisle 3—dairy products, and aisle 4—meat/fish. The time between two successive customer arrivals is exponentially distributed with a mean of 5 minutes. After arriving to the store, each customer grabs a shopping cart. Twenty percent of all customers go to aisle 1, 30 percent go to aisle 2, 50 percent go to aisle 3, and 70 percent go to aisle 4. The number of items selected for purchase in each aisle is uniformly distributed between 2 and 8. The time spent to browse and pick up each item is normally distributed: N(5,2) minutes. There are three identical checkout counters; each counter has its own checkout

line. The customer chooses the shortest line. Once a customer joins a line, he or she is not allowed to leave or switch lines. The checkout time is given by the following regression equation:

$$\text{Checkout time} = N(3,0.3) + (\#\text{of items}) * N(0.5,0.15) \text{ minutes}$$

The first term of the checkout time is for receiving cash or a check or credit card from the customer, opening and closing the cash register, and handing over the receipt and cash to the customer. After checking out, a customer leaves the cart at the front of the store and leaves. Build a simulation model for the grocery store. Use the model to simulate a 14-hour day.

a. The percentages of customers visiting each aisle do not add up to 100 percent. Why?

b. What is the average amount of time a customer spends at the grocery store?

c. How many customers check out per cashier per hour?

d. What is the average amount of time a customer spends waiting in the checkout line?

e. What is the average utilization of the cashiers?

f. Assuming there is no limit to the number of shopping carts, determine the average and maximum number of carts in use at any time.

g. On average how many customers are waiting in line to checkout?

h. If the owner adopts a customer service policy that there will never be any more than three customers in any checkout line, how many cashiers are needed?

Embellishments:

I. The store manager is considering designating one of the checkout lines as Express, for customers checking out with 10 or fewer items. Is that a good idea? Why or why not?

II. In reality, there are only 10 shopping carts at the ShopNSave store. If there are no carts available when customers arrive they leave immediately and go to the more expensive ShopNSpend store down the street. Modify the simulation model to reflect this change. How many customers are lost per hour? How many shopping carts should they have so that no more than 5 percent of customers are lost?

16. Planes arrive at the **Netaji Subhash Chandra Bose International Airport, Calcutta** with interarrival times that are exponentially distributed with a mean time of 30 minutes. If there is no room at the airport when a plane arrives, the pilot flies around and comes back to land after a normally distributed time having a mean of 20 minutes and a standard deviation of 5 minutes. There are two runways and three gates at this small airport. The time from touchdown to arrival at a gate is normally distributed having a mean of five minutes and a standard

deviation of one minute. A maximum of three planes can be unloaded and loaded at the airport at any time. The times to unload and load a plane are uniformly distributed between 20 and 30 minutes. The pushoff, taxi, and takeoff times are normally distributed with a mean of six minutes and a standard deviation of one minute. The airport operates 24/7. Build a simulation model and run it for one year.

a. How much will adding a new gate decrease the time that airplanes have to circle before landing?

b. Will adding a new gate affect the turnaround time of airplanes at the airport? How much?

c. Is it better to add another runway instead of another gate? If so, why?

8   MODEL VERIFICATION AND VALIDATION

Dew knot trussed yore spell chequer two fined awl yore mistakes.

—Brendan Hills

In this lab we describe the verification and validation phases in the development and analysis of simulation models. In Section L8.1 we describe an inspection and rework model. In Section L8.2 we show how to verify the model by tracing the events in it. Section L8.3 shows how to debug a model. The ProModel logic, basic, and advanced debugger options are also discussed.

L8.1   Verification of an Inspection and Rework Model

Problem Statement

Bombay Clothing Mill is a textile manufacturer, importer, and reseller. Boxes of garments arrive (three per hour, exponentially distributed) from an overseas supplier to the Bombay Clothing mill warehouse in Bombay for identification and labeling, inspection, and packaging (Figure L8.1). The identification and labeling take U(10,2), inspection takes N(5,1), and packaging takes N(4,0.25) minutes. At times the labeling machine malfunctions. The labeling mistakes are detected during inspection. Sixty percent of the garments are labeled properly and go on to the packaging area; however, 40 percent of the garments are sent back for rework (relabeling). Garments sent back for rework have a higher priority over regular garments at the labeling machine. Movement between all areas takes one minute. Simulate for 100 hours.

Priorities determine which waiting entity or downtime is permitted to access a location or resource when the location or resource becomes available. Priorities may be any value between 0 and 999 with higher values having higher priority. For simple prioritizing, you should use priorities from 0 to 99.

FIGURE L8.1

Layout of the Bombay Clothing Mill.

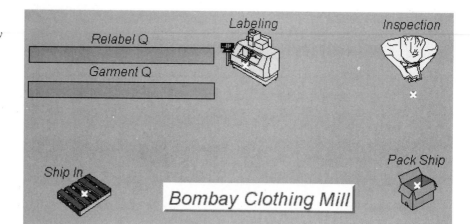

FIGURE L8.2

Locations at the Bombay Clothing Mill warehouse.

```
*                                           Locations
*************************************************************
```

Name	Cap	Units	Stats	Rules
Label_Q	INFINITE	1	Time Series	Oldest, FIFO,
Labeling	1	1	Time Series	Oldest, ,
Inspection	inf	1	Time Series	Oldest, ,
Pack_Ship	inf	1	Time Series	Oldest, ,
Ship_In	1	1	Time Series	Oldest, ,

FIGURE L8.3

Process and routing tables at the Bombay Clothing Mill.

		Process			Routing		
Entity	Location	Operation	Blk	Output	Destination	Rule	Move Logic
Garments	Ship_In		1	Garments	Label_Q	FIRST 1	Move for 1 min
Garments	Label_Q		1	Garments	Labeling,1	FIRST 1	
Garments	Labeling	Wait U(10,2) min	1	Garments	Inspection	FIRST 1	Move for 1 min
Garments	Inspection	Wait N(5,1) min	1	Garments	Pack_Ship	0.600000 1	Move for 1 min
				Relabel	Label_Q	0.400000	Move for 1 min
Garments	Pack_Ship	Wait N(4,0.25) min	1	Garments	EXIT	FIRST 1	
Relabel	Label_Q		1	Relabel	Labeling,2	FIRST 1	
Relabel	Labeling	Wait U(10,2) min	1	Garments	Inspection	FIRST 1	

Five locations (Ship_In, Label_Q, Relabel_Q, Labeling, Inspection, and Pack_Ship) are defined as shown in Figure L8.2. Two entities (Garments and Relabel) are defined next. The processes and routing logic are defined as shown in Figure L8.3. Note that the garments sent back for relabeling have a higher priority of 2 to access the labeling machine location as opposed to a lower priority of 1 for incoming garments.

L8.2 Verification by Tracing the Simulation Model

Now we will run the model developed in the previous section with the Trace feature on. Trace will list events as they happen during a simulation. A *trace* is a list of events that occur during a simulation. For example, a line in the trace could read, "EntA arrives at Loc1, Downtime for Res1 begins." A trace listing also displays assignments, such as variable and array element assignments. A trace listing of a simulation run may be viewed in several ways and is generated in one of two modes, Step or Continuous.

- *Off:* Select this option to discontinue a trace.
- *Step:* Choose this option to step through the trace listing one event at a time. Each time you click the left mouse button, the simulation will advance one event. Clicking and holding the right mouse button while in this mode generates a continuous trace.
- *Continuous:* Choose this option to write the trace continuously to the output device selected from the Trace Output submenu. This is useful when you do not know exactly where to begin or end the trace. Clicking and holding the right mouse button causes the trace to stop until the right mouse button is released.

Run the simulation model. Choose Trace from the Options menu. You can choose either the step mode or the continuous mode for running Trace (Figure L8.4). Choose the step mode for now. The trace output can be sent either to the Window or to a file (Figure L8.5). Let's choose to send the output to the Window.

The trace prompts are displayed at the bottom of the screen as shown in Figure L8.6. Follow the trace prompts and verify that the model is doing what it is supposed to be doing. Make sure through visualization and the trace messages that the rework items are indeed given higher priority than first-time items. Also, from the garment and relabel queue contents shown in Figure L8.7, notice that the relabel queue is quite small, whereas the garment queue grows quite big at times (as a result of its lower priority for the labeling operation).

FIGURE L8.4

Trace modes in ProModel.

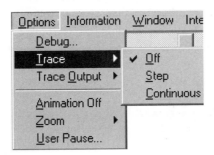

FIGURE L8.5

Trace output options in ProModel.

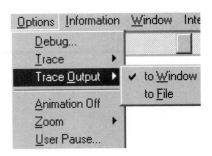

FIGURE L8.6

Tracing the simulation model of the Bombay Clothing Mill warehouse.

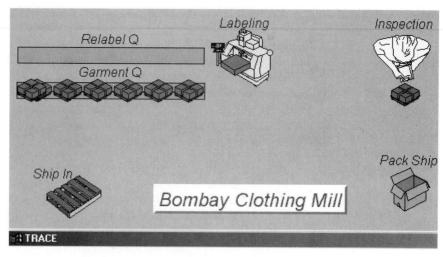

```
26:41.337    Start move to Labeling.
26:41.337 Relabel arrives at Labeling.
26:41.337 For Relabel at Labeling:
26:41.337    Relabel enters Labeling.
26:41.337    Wait 8.950 Min.
26:41.337 For Relabel at Relabel_Q:
26:41.337    Process completed.
26:41.337    Release the captured capacity.
26:42.337 Garments arrives at Inspection.
26:42.337 For Garments at Inspection:
26:42.337    Garments enters Inspection.
26:42.337    Wait 5.657 Min.
```

FIGURE L8.7

Plots of garment and relabel queue contents.

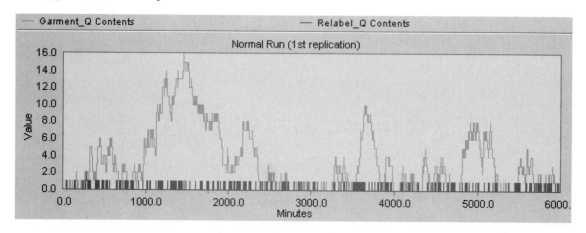

L8.3 Debugging the Simulation Model

The debugger (Figure L8.8) is a convenient, efficient way to test or follow the processing of any logic in your model. The debugger is used to step through the logic one statement at a time and examine variables and attributes while a model is running. All local variables, current entity attributes, and so forth are displayed, and the user has full control over what logic gets tracked, even down to a specific entity at a specific location (Figure L8.9).

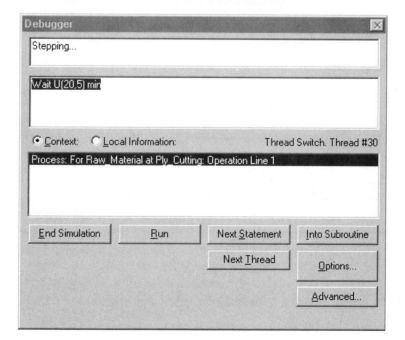

The user can launch the debugger using a DEBUG statement within the model code or from the Options menu during run time. The system state can be monitored to see exactly when and why things occur. Combined with the Trace window, which shows the events that are being scheduled and executed, the debugger enables a modeler to track down logic errors or model bugs.

L8.3.1 Debugging ProModel Logic

Before we discuss the details of the debugger, it is important to understand the following terms:

- *Statement:* A statement in ProModel performs some operation or takes some action—for example, JOIN, WAIT, USE, MOVE, and so on (refer to the ProModel Reference Guide for more information on all ProModel statements).
- *Logic:* Logic is the complete set of statements defined for a particular process record, downtime event, initialization or termination of the simulation, and so forth.
- *Thread:* A thread is a specific execution of any logic. A thread is initiated whenever logic needs to be executed. This can be an entity running through operation logic, the initialization logic, a resource running some node logic, downtime logic, or any other logic. Note that the same logic may be running in several threads at the same time. For example, three entities of the same type being processed simultaneously at the same multicapacity location would constitute three threads.

Even though several threads can execute the same logic at the same time in the simulation, the simulation processor can process them only one at a time. So there is really only one current thread, while all other threads are suspended (either scheduled for some future simulation time or waiting to be executed after the current thread at the same simulation instant).

Every time logic is executed, a thread that is assigned a unique number executes it. THREADNUM returns the number of the thread that called the function. This function can be used with the IF-THEN and DEBUG statements to bring up the debugger at a certain process.

For example, for ElSegundo Composites, from Lab 7, Section L7.6.1, suppose the GROUP quantity has been typed as –5 (typo). A debug statement IF THREADNUM () = 39 THEN DEBUG has been added to the operation logic at the Oven location that causes the error, as shown in Figure L8.10. The simulation will run until the proper process and then bring up the debugger (Figure L8.11). The debugger can then be used to step through the process to find the particular statement causing the error.

L8.3.2 Basic Debugger Options

The debugger can be used in two modes: Basic and Advanced. The Basic Debugger appears initially, with the option of using the Advanced Debugger. The Basic

FIGURE L8.10

An example of a DEBUG *statement in the processing logic.*

		Process			Routing			
Entity	Location	Operation	Blk	Output	Destination	Rule	Move Logic	
Raw_Material	Order_Q	Inc WIP	1	Raw_Material	Ply_Cutting	FIRST 1	Move for 15	
Raw_Material	Ply_Cutting	Wait U(20,5) min	1	Raw_Material	LayUp	FIRST 1	Move for 15	
Raw_Material	LayUp	Wait U(30,10) min						
			1	Raw_Material	Oven	FIRST 1	Move for 15	
Raw_Material	Oven	IF THREADNUM() = 39 THEN DEBUG						
		Group -5 as Batch						
Batch	Oven	Wait U(100,10) min						
		UNGROUP						
Raw_Material	Oven		1	Raw_Material	Ship_Q	FIRST 1	Move for 15	
Raw_Material	Ship_Q		1	Raw_Material	Ship_Clerk	FIRST 1	Move for 15	
Raw_Material	Ship_Clerk	Wait N(20,5) min						
		Dec WIP	1	Raw_Material	EXIT	FIRST 1		

FIGURE L8.11

The Debugger window.

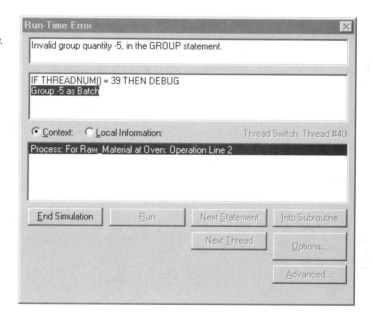

Debugger (Figure L8.11) has the following options:

- *Error display box:* Displays the error message or reason why the debugger is displayed, such as the user condition becoming true.
- *Logic display box:* Displays the statements of the logic being executed.
- *Information box:* Displays either the context of the logic or local information.
- *Context:* Displays the module, operation, and line number in which the debugger stopped.
- *Local information:* Displays local variables and entity attributes with nonzero values for the thread in the information box.
- *End Simulation:* Choose this option to terminate simulation. This will prompt for whether or not you would like to collect statistics.

- *Run:* Continues the simulation, but still checks the debugger options selected in the Debugger Options dialog box.

- *Next Statement:* Jumps to the next statement in the current logic. Note that if the last statement executed suspends the thread (for example, if the entity is waiting to capture a resource), another thread that also meets the debugger conditions may be displayed as the next statement.

- *Next Thread:* Brings up the debugger at the next thread that is initiated or resumed.

- *Into Subroutine:* Steps to the first statement in the next subroutine executed by this thread. Again, if the last statement executed suspends the thread, another thread that also meets debugger conditions may be displayed first. If no subroutine is found in the current thread, a message is displayed in the Error Display box.

- *Options:* Brings up the Debugger Options dialog box. You may also bring up this dialog box from the Simulation menu.

- *Advanced:* Changes the debugger to Advanced mode.

L8.3.3 Advanced Debugger Options

The Advanced Debugger (Figure L8.12) contains all the options in the Basic Debugger plus a few advanced features:

- *Next (Thread):* Jumps to the next thread that is initiated or resumed. This button has the same functionality as the Next Thread button in the Basic Debugger.

- *New (Thread):* Jumps to the next thread that is initiated.

- *Disable (Thread):* Temporarily disables the debugger for the current thread.

- *Exclusive (Thread):* The Debugger displays the statements executed within the current thread only. When the thread terminates, the Exclusive setting is removed.

- *Next (Logic):* Jumps to the next initiated or resumed thread that is not executing the same logic as the current thread.

- *New (Logic):* Jumps over any resumed threads to the next initiated thread that is not executing the same logic as the current thread. This will automatically jump to a new thread.

- *Disable (Logic):* Temporarily disables the debugger for all threads executing the current logic.

- *Exclusive (Logic):* The debugger displays only the statements executed in any thread that are an instance of the current logic.

- *Enable disabled threads and logics:* Enables the threads and logics that were disabled previously.

For more information and detailed examples of how to use the various debugger options, consult the ProModel Users Guide.

L8.4 Exercises

1. For the example in Section L8.1, insert a DEBUG statement when a garment is sent back for rework. Verify that the simulation model is actually sending back garments for rework to the location named Label_Q.

2. For the example in Section L7.1 (Pomona Electronics), trace the model to verify that the circuit boards of type B are following the routing given in Table L7.1.

3. For the example in Section L7.5 (Poly Casting Inc.), run the simulation model and launch the debugger from the Options menu. Turn on the Local Information in the Basic Debugger. Verify the values of the variables WIP and PROD_QTY.

4. For the example in Section L7.3 (Bank of India), trace the model to verify that successive customers are in fact being served by the three tellers in turn.

5. For the example in Section L7.7.2 (Shipping Boxes Unlimited), trace the model to verify that the full boxes are in fact being loaded on empty pallets at the Inspector location and are being unloaded at the Shipping location.

9   SIMULATION OUTPUT ANALYSIS

Nothing has such power to broaden the mind as the ability to investigate systematically and truly all that comes under thy observation in life.

—Marcus Aurelius

In this lab we review the differences between terminating and nonterminating simulations. You will learn how to collect observations of a performance metric from both terminating and nonterminating simulations using ProModel. Key problems of identifying the warm-up period and selecting the run length of a nonterminating simulation are addressed in the lab. We also illustrate how supporting software can be used to help determine if the observations reported by ProModel for a simulation are independent and normally distributed.

L9.1 Terminating versus Nonterminating Simulations

A terminating simulation has a fixed starting condition and a fixed ending condition (terminating point). For example, a simulation model of a store that opens at 9:00 A.M. begins with a starting condition of no customers in the system and terminates after the store closes at 7:00 P.M. and the last customer is processed. Many service-oriented systems (banks, doctors' offices, post offices, schools, amusement parks, restaurants, and department stores) are analyzed with a terminating simulation. The system returns to a fixed starting condition after reaching a natural terminating event.

A nonterminating simulation has neither a fixed starting condition nor a natural ending condition. The simulation cannot run indefinitely, so we end it after its output reflects the long-term average behavior of the system. A hospital or a 24-hour gas station could be analyzed with a nonterminating simulation depending on the objective of the study. Many systems appear to be terminating but in reality are nonterminating. Systems that close/terminate but do not go back to a fixed starting condition (empty for example) are usually analyzed with nonterminating simulations. For example, a production system that uses the current day's unfinished work as the next day's starting work is really a nonterminating system. If we compress time by removing the intervals during which the system is closed, it behaves as a continuously operating system with no point at which the system starts anew. In general, most manufacturing systems are nonterminating. The work-in-process inventory from one day is typically carried forward to the next day.

When we report the simulation model's response based on the output from a terminating simulation or a nonterminating simulation, it is important to use interval estimates rather than simple point estimates of the parameter being measured. The advantage of the interval estimate is that it provides a measure of the variability in output produced by the simulation. Interval estimates or confidence intervals provide a range of values around the point estimate that have a specified likelihood of containing the parameter's true but unknown value. For a given confidence level, a smaller confidence interval is considered better than a larger one. Similarly, for a given confidence interval, a higher confidence level indicates a better estimate. To construct the confidence interval, the observations produced by replications or batch intervals need to be independent and identically distributed. Furthermore, the observations should be normally distributed when using the Student's t distribution to calculate the confidence interval. We can fudge a little on the normality assumption but not on the independence and identically distribution requirements. These topics are elaborated on in Chapter 9.

L9.2 Terminating Simulation

Problem Statement

Spuds-n-More by the Seashore is a favorite among vacationers at the seaside resort. Spuds-n-More, a takeout fast-food restaurant, specializes in potato foods, serving many different varieties of french fries, waffle fries, and potato logs along with hamburgers and drinks. Customers walking along the boardwalk access the restaurant via a single ordering window, where they place their orders, pay, and receive their food. There is no seating in the restaurant. The interarrival time of customers to Spuds-n-More is exponentially distributed with a mean of six minutes. The activity times are given in Table L9.1. The waiting area in front of the walkup window can accommodate only five customers. If the waiting area is filled

TABLE L9.1   Process Times at Spuds-n-More by the Seashore

Activity	Process Time (minutes)
Order food	E(2)
Make payment	E(1.5)
Pick up food	E(5)

to capacity when a customer arrives, the customer bypasses Spuds-n-More. Mr. Taylor closes the entryway to the waiting area in front of his restaurant after eight hours of operation. However, any customers already in the waiting area by the closing time are served before the kitchen shuts down for the day.

Mr. Taylor is thinking about expanding into the small newspaper stand next to Spuds-n-More that has been abandoned. This would allow him to add a second window to serve his customers. The first window would be used for taking orders and collecting payments, and the second window would be used for filling orders. Before investing in the new space, however, Mr. Taylor needs to convince the major stockholder of Spuds-n-More, Pritchard Enterprises, that the investment would allow the restaurant to serve more customers per day. The first step is to build a baseline (as-is) model of the current restaurant configuration and then to embellish the model with the new window and measure the effect it has on the number of customers processed per day. Mr. Taylor is interested in seeing a 90 percent confidence interval for the number of customers processed per day in the baseline simulation.

L9.2.1 Starting and Terminating Conditions (Run Length)

Build the baseline model of the as-is restaurant and conduct a terminating simulation to estimate the expected number of customers served each day. The starting condition for the simulation is that the system begins in the morning empty of customers with the first customer arriving at time zero. Customers are admitted into the restaurant's queue over the eight hours of operation. The simulation terminates after reaching eight hours and having processed the last customer admitted into the queue. The layout of the simulation model is shown in Figure L9.1, and a printout of the model is given in Figure L9.2. Note that the length of the Order_Q location is 25 feet and the Customer entity travel speed is 150 feet per minute. These values affect the entity's travel time in the queue (time for a customer to walk to the end of the queue) and are required to match the results presented in this lab. The model is included on the CD accompanying the book under file name Lab 9_2_1 Spuds-n-More1.MOD.

Let's take a look at the model listing in Figure L9.2 to see how the termination criterion was implemented. The Entry location serves as a gate for allowing customers into the queue until the simulation clock time reaches eight hours. Entities arriving after the simulation clock reaches eight hours are turned away (routed to Exit) by the routing rule User Condition (IF clock(hr) > 8) at the Entry location. Entities arriving within eight hours are routed to the Order_Q location (see routing rule User Condition IF clock(hr) <= 8). Two global variables are declared to count the number of entities arriving at the Order_Q location and the number of entities processed at the Order_Clerk location. See the statement Inc Arrivals at the Order_Q location and the statement Inc Processed at the Order_Clerk location. The Inc statement increments the variable's value by one each time it is executed. As the entities that arrive after eight hours of simulation time are turned away to the Exit, the number of entities processed by the Order_Clerk is compared to the number of entities allowed into the Order_Q (see Move Logic). If Processed = Arrivals, then all entities that were allowed into the

FIGURE L9.1

Layout of the Spuds-n-More simulation model.

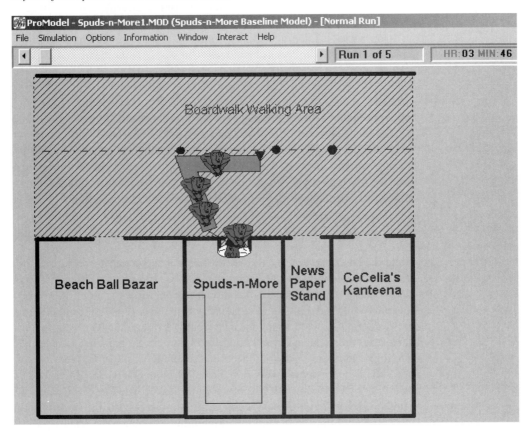

Order_Q have been processed and the simulation is terminated by the Stop statement. Notice that the combined capacity of the Entry and Order_Q locations is five in order to satisfy the requirement that the waiting area in front of the restaurant accommodates up to five customers.

We have been viewing the output from our simulations with ProModel's new Output Viewer 3DR. Let's conduct this lab using ProModel's traditional Output Viewer, which serves up the same information as the 3DR viewer but does so a little faster. To switch viewers, select Tools from the ProModel main menu bar and then select Options. Select the Output Viewer as shown in Figure L9.3.

L9.2.2 Replications

Run the simulation for five replications to record the number of customers served each day for five successive days. To run the five replications, select Options from under the Simulation main menu. Figure L9.4 illustrates the simulation options set to run five replications of the simulation. Notice that no run hours are specified.

Figure L9.2

The simulation model of Spuds-n-More.

```
Time Units:  Minutes
Distance Units:  Feet
*****************************************************************************
*                              Locations                              *
*****************************************************************************
  Name        Cap Units Stats      Rules          Cost
  ----------- --- ----- ---------- -------------- ------------
  Entry         1   1   None       Oldest, ,
  Order_Q       4   1   None       Oldest, FIFO,
  Order_Clerk   1   1   None       Oldest, ,

*****************************************************************************
*                               Entities                              *
*****************************************************************************
  Name        Speed (fpm)  Stats      Cost
  ----------  -----------  ---------- ------------
  Customer    150          None

*****************************************************************************
*                              Processing                             *
*****************************************************************************
                Process                            Routing

Entity    Location    Operation          Blk  Output    Destination  Rule                 Move Logic
--------  ----------  ------------------  ---  --------  -----------  -------------------  ------------
Customer  Entry                           1   Customer  Order_Q      IF clock(hr) <= 8, 1
                                              Customer  EXIT         IF clock(hr) > 8     If Processed = Arrivals
                                                                                          Then Stop

Customer  Order_Q     Inc Arrivals        1   Customer  Order_Clerk  FIRST 1
Customer  Order_Clerk // Order
                      Wait e(2) min
                      // Pay
                      Wait e(1.5) min
                      // Pickup
                      Wait e(5) min
                      Inc Processed       1   Customer  EXIT         FIRST 1

*****************************************************************************
*                               Arrivals                              *
*****************************************************************************
  Entity    Location Qty Each  First Time Occurrences Frequency  Logic
  --------  -------- ---------- ---------- ----------- ---------- ------------
  Customer  Entry    1          0          inf         E(6) min

*****************************************************************************
*                           Variables (global)                       *
*****************************************************************************
  ID         Type         Initial value Stats
  ---------  -----------  ------------- -----------
  Arrivals   Integer      0             None          Note that the observation-based
  Processed  Integer      0             Time Series ◄── suboption of Time Series was selected.
```

After ProModel runs the five replications, it displays a message asking if you wish to see the results. Answer yes.

Next ProModel displays the General Report Type window (Figure L9.5). Here you specify your desire to see the output results for "<All>" replications and then click on the Options . . . button to specify that you wish the output report to also include the sample mean (average), sample standard deviation, and 90 percent confidence interval of the five observations collected via the five replications.

The ProModel output report is shown in Figure L9.6. The number of customers processed each day can be found under the Current Value column of the output report in the VARIABLES section. The simulation results indicate that the number of customers processed each day fluctuates randomly. The fluctuation may be

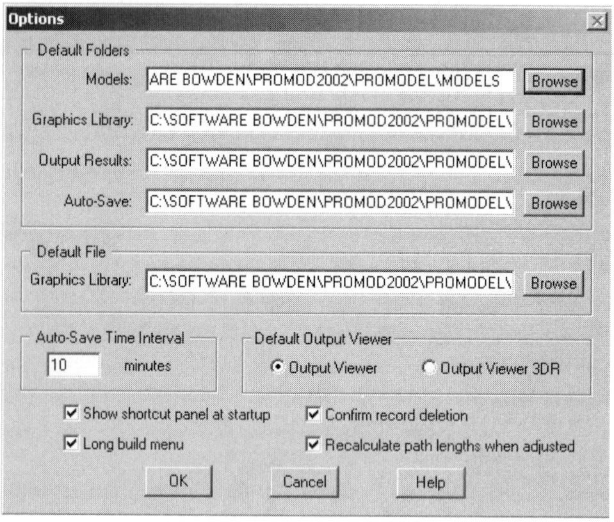

FIGURE L9.3

*ProModel's Default
Output Viewer set to
Output Viewer.*

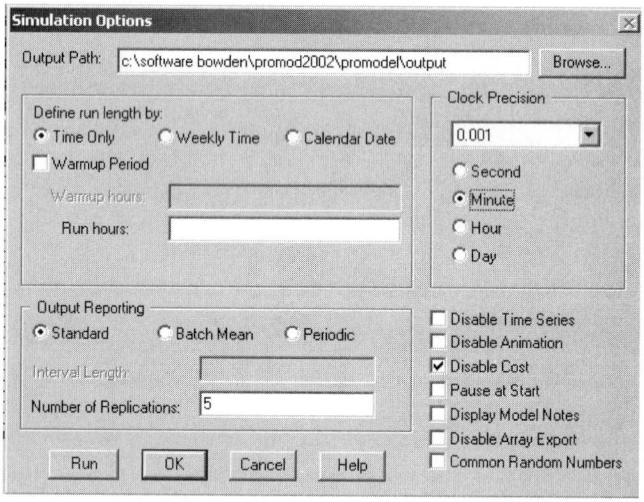

FIGURE L9.4

*ProModel's Simulation
Options window set to
run five replications of
the simulation.*

attributed to variations in processing times or customer arrivals. It is obvious that
we cannot reliably use the observation from a single day's simulation to estimate
the expected number of customers served each day. The results from a single day
could produce an estimate as small as 59 or as large as 67. Either value could be a
very poor estimate of the true expected number of customers served each day.

The average number of customers processed over the five days is computed
as $\bar{x} = 62.20$ (Figure L9.6). This provides a much better point estimate of the expected daily service rate (number of customers processed each day), but it fails to
capture any information on the variability of our estimate. The sample deviation
of $s = 3.27$ customers per day provides that information. The most valuable piece
of information is the 90 percent confidence interval computed by ProModel using

FIGURE L9.5

ProModel General Report Options set to display the results from all replications as well as the average, standard deviation, and 90 percent confidence interval.

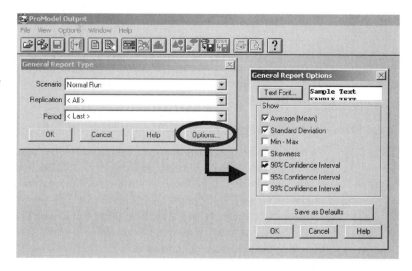

FIGURE L9.6

ProModel output report for five replications of the Spuds-n-More simulation.

FAILED ARRIVALS

Entity Name	Location Name	Total Failed	
Customer	Entry	21	(Rep 1)
Customer	Entry	18	(Rep 2)
Customer	Entry	17	(Rep 3)
Customer	Entry	14	(Rep 4)
Customer	Entry	30	(Rep 5)
Customer	Entry	20	(Average)
Customer	Entry	6.123	(Std. Dev.)
Customer	Entry	14.161	(90% C.I. Low)
Customer	Entry	25.838	(90% C.I. High)

VARIABLES (* indicates observation based variables)

Variable Name	Total Changes	Average Minutes Per Change	Minimum Value	Maximum Value	Current Value	Average Value	
Processed*	64	8.136	1	64	64	32.5	(Rep 1)
Processed*	67	8.000	1	67	67	34	(Rep 2)
Processed*	61	8.549	1	61	61	31	(Rep 3)
Processed*	59	9.002	1	59	59	30	(Rep 4)
Processed*	60	8.753	1	60	60	30.5	(Rep 5)
Processed*	62.2	8.488	1	62.2	62.2	31.6	(Average)
Processed*	3.271	0.418	0	3.271	3.271	1.635	(Std. Dev.)
Processed*	59.081	8.089	1	59.081	59.081	30.040	(90% C.I. Low)
Processed*	65.318	8.887	1	65.318	65.318	33.159	(90% C.I. High)

the sample mean and sample standard deviation. With approximately 90 percent confidence, the true but unknown mean number of customers processed per day falls between 59.08 and 65.32 customers. These results convince Mr. Taylor that the model is a valid representation of the actual restaurant, but he wants to get a better estimate of the number of customers served per day.

L9.2.3 Required Number of Replications

The advantage of the confidence interval estimate is that it quantifies the error in the point estimate. A wide confidence interval implies that the point estimate is not very accurate and should be given little credence. The width of the confidence interval is a function of both the variability of the system and the amount of data collected for estimation. We can make the confidence interval smaller by collecting more data (running more replications). However, for highly variable systems, one may need a lot of data to make extremely precise statements (smaller confidence intervals). Collecting the data may be very time consuming if the simulation takes long to run. Hence, we sometimes have to make a judicious choice about the amount of precision required.

The half-width of the 90 percent confidence interval based on the five replications is 3.12 customers (65.32 C.I. High − 62.20 Average). How many additional replications are needed if Mr. Taylor desires a confidence interval half-width of 2.0 customers? Using the technique described in Section 9.2.3 of Chapter 9 we have

$e = hw = 2.0$ customers

$\alpha =$ significance level $= 0.10$ for a 0.90 confidence level

$$n' = \left[\frac{(Z_{\alpha/2})s}{e} \right]^2 = \left[\frac{(t_{\infty,\alpha/2})s}{e} \right]^2 = \left[\frac{(1.645)3.27}{2.0} \right]^2 = 7.23 \text{ replications}$$

The value of 7.23 is appropriately rounded up to 8 replications. The $n' = 8$ provides a rough approximation of the number of replications that are needed to achieve the requested error amount $e = 2.0$ customers at the 0.10 significance level. Running an additional three replications of the Spuds-n-More baseline simulation and combining the output with the output from the original five replications produces a sample mean of 61.50 customers and an approximate 90 percent confidence interval of 58.99 to 64.00 customers. You should verify this. Although the confidence interval half-width of 2.51 customers is larger than the target value of 2.0 customers, it is smaller than the half-width of the confidence interval based on five replications. We could keep running additional replications to get closer to the target half-width of 2.0 customers if we like.

L9.2.4 Simulation Output Assumptions

To compute the confidence interval, we assumed that the observations of the number of customers processed each day were independent and normally distributed. The observations are independent because ProModel used a different stream of random numbers to simulate each replication. We do not know that the observations are normally distributed, however. Therefore, we should specify that the confidence level of the interval is approximately 0.90. Nevertheless, we followed an acceptable procedure to get our estimate for the service rate, and we can stand by our approximate confidence interval. To illustrate a point, however, let's take a moment to use the Stat::Fit software provided with ProModel to test if the simulation output is normally distributed. To do so, we need a lot more than eight observations. Let's pick 100 just because it is a round number and we can quickly generate them by configuring ProModel to run 100 replications of the simulation. Try this now. Notice that you get a very small confidence interval with this many observations.

It would take a while to enter the 100 observations into Stat::Fit one at a time. A faster method is to save the ProModel output report with the results from the 100 replications into a Microsoft Excel file format that can then be copied into the Stat::Fit data table. To do this from the ProModel output report viewer, select File from the main menu and click on the Save As option, then enter a name for the file, and save the file as type MS Excel (Figure L9.7). Open the Excel file that you just

FIGURE L9.7

Saving a ProModel output report into a Microsoft Excel file format.

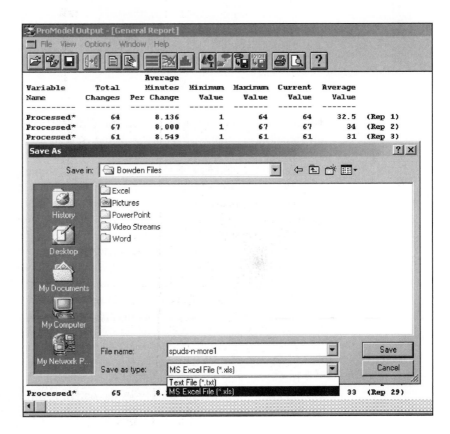

FIGURE L9.8

The ProModel output report displayed within a Microsoft Excel spreadsheet.

	A	B	C	D	E	F	G	H	I	J
1	Variable_Name	Total_Changes	Average_Min	Minimum	Maximum	Current_Value	Average_Value			
2	Processed*	64	8.136	1	64	64	32.5			
3	Processed*	67	8	1	67	67	34			
4	Processed*	61	8.549	1	61	61	31			
5	Processed*	59	9.002	1	59	59	30			
6	Processed*	60	8.753	1	60	60	30.5			
7	Processed*	66	8.164	1	66	66	33.5			
8	Processed*	57	9.244	1	57	57	29			
9	Processed*	58	8.904	1	58	58	29.5			
10	Processed*	56	9.054	1	56	56	28.5			
11	Processed*	55	9.406	1	55	55	28			
12	Processed*	61	8.777	1	61	61	31			
13	Processed*	59	8.766	1	59	59	30			
14	Processed*	57	9.011	1	57	57	29			
15	Processed*	63	8.171	1	63	63	32			
16	Processed*	65	7.858	1	65	65	33			
17	Processed*	57	8.994	1	57	57	29			
18	Processed*	53	9.366	1	53	53	27			
19	Processed*	58	8.709	1	58	58	29.5			
20	Processed*	65	7.874	1	65	65	33			
21	Processed*	58	8.919	1	58	58	29.5			
22	Processed*	55	9.331	1	55	55	28			
23	Processed*	66	7.901	1	66	66	33.5			
24	Processed*	63	8.693	1	63	63	32			
25	Processed*	53	9.52	1	53	53	27			

saved the ProModel output to, click on the VARIABLES sheet tab at the bottom of the spreadsheet (Figure L9.8), highlight the 100 Current Value observations of the Processed variable—making sure not to include the average, standard deviation, 90% C.I. Low, and 90% C.I. High values (the last four values in the column)—and paste the 100 observations into the Stat::Fit data table (Figure L9.9).

After the 100 observations are pasted into the Stat::Fit Data Table, display a histogram of the data. Based on the histogram, the observations appear somewhat normally distributed (Figure L9.9). Furthermore, Stat::Fit estimates that the normal distribution with mean $\mu = 59.30$ and standard deviation $\sigma = 4.08$ provides an acceptable fit to the data. Therefore, we probably could have dropped the word "approximate" when we presented our confidence interval to Mr. Taylor. Be sure to verify all of this for yourself using the software.

Note that if you were to repeat this procedure in practice because the problem requires you to be as precise and as sure as possible in your conclusions, then you would report your confidence interval based on the larger number of observations. Never discard precious data.

Before leaving this section, look back at Figure L9.6. The Total Failed column in the Failed Arrivals section of the output report indicates the number of customers that arrived to eat at Spuds-n-More but left because the waiting line

FIGURE L9.9

Stat::Fit analysis of the 100 observations of the number of customers processed per day by Spuds-n-More.

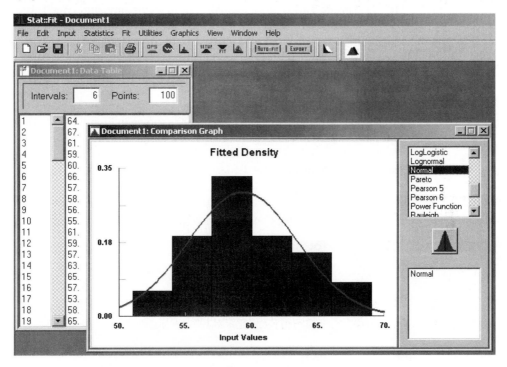

was full. Mr. Taylor thinks that his proposed expansion plan will allow him to capture some of the customers he is currently losing. In Lab Chapter 10, we will add embellishments to the as-is simulation model to reflect Mr. Taylor's expansion plan to see if he is right.

L9.3 Nonterminating Simulation

We run a nonterminating simulation of a system if we are interested in understanding the steady-state behavior of the system—that is, its long term average behavior. Because a nonterminating system does not return to a fixed initial condition and theoretically never shuts down, it is not easy to select either the starting condition for the simulation or the amount of time to run the simulation.

When a simulation has just begun running, its state (number of customers in the system for example) will be greatly affected by the initial state of the simulation and the elapsed time of the run. In this case, the simulation is said to be in a transient condition. However, after sufficient time has elapsed, the state of the simulation becomes essentially independent of the initial state and the elapsed time. When this happens, the simulation has reached a steady-state condition. Although the simulation's output will continue to be random, the statistical distribution of the simulation's output does not change after the simulation reaches steady state.

The period during which the simulation is in the transient phase is known as the warm-up period. This is the amount of time to let the simulation run before gathering statistics. Data collection for statistics begins at the end of the warm-up and continues until the simulation has run long enough to allow all simulation events (even rare ones) to occur many times (hundred to thousands of times if practical).

Problem Statement

A simulation model of the **Green Machine Manufacturing Company (GMMC)** owned and operated by Mr. Robert Vaughn is shown in Figure L9.10. The interarrival time of jobs to the GMMC is constant at 1.175 minutes. Jobs require processing by each of the four machines. The processing time for a job at each green machine is given in Table L9.2.

FIGURE L9.10

The layout of the Green Machine Manufacturing Company simulation model.

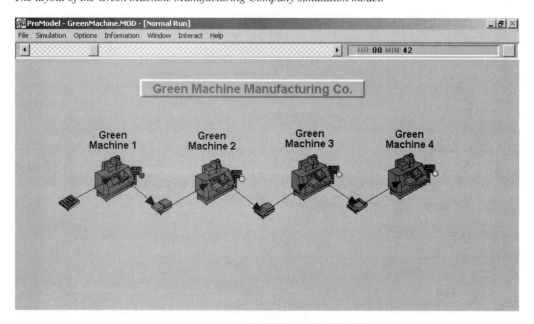

TABLE L9.2   Process Times at Green Machine Manufacturing Company

Activity	Process Time (minutes)	ProModel Format
Machine 1	Exponential with mean = 1.0	Wait E(1.0) min
Machine 2	Weibull with minimum = 0.10, $\alpha = 5$, $\beta = 1$	Wait (0.10 + W(5,1)) min
Machine 3	Weibull with minimum = 0.25, $\alpha = 5$, $\beta = 1$	Wait (0.25 + W(5,1)) min
Machine 4	Weibull with minimum = 0.25, $\alpha = 5$, $\beta = 1$	Wait (0.25 + W(5,1)) min

A plentiful amount of storage is provided for waiting jobs in front of each of the green machines. Our task is to conduct a nonterminating simulation to estimate the time-average amount of work-in-process (WIP) inventory (the steady-state expected number of jobs in the system). A printout of the model is given in Figure L9.11. The model is included on the CD accompanying the book under the file name Lab 9_3 GreenMachine.MOD.

L9.3.1 Warm-up Time and Run Length

In this section we estimate the amount of time needed for the GMMC model to warm up, and the amount of time to run the model past its warm-up period. We first declare a model variable called WIP to keep track of the work-in-process inventory in the simulated manufacturing system. Notice we are collecting statistics only for the WIP variable and the entity. You should configure your model the same and run the simulation 250 hours to produce a time-series plot of the WIP inventory levels (Figure L9.12(a)).

The simplest method for selecting the end of the warm-up phase is visual determination. In this method, we plot the simulation model's output over time, as illustrated in Figure L9.12(a), and then visually determine the point at which the initial transient phase is over. In this example, it seems that the transient phase created by the initial condition of zero WIP inventory is over at around 20 hours. Usually you will want to select a value beyond this to guard against underestimating the end of the warm-up phase. So we will pick 100 hours as the end of the warm-up. Actually, you should not make this decision based on the output from one replication of the simulation. We are doing this here to quickly illustrate the need for allowing the model to warm up. We will use the preferred method to estimate the end of the warm-up phase in a moment.

In most nonterminating simulation models, the quality of the results substantially improves if we throw away the output generated during the warm-up phase of the simulation. In Figure L9.12(a), the observations during the transient phase of the first 100 hours of the simulation are alleged to be artificially low. As such, we do not want them to bias our estimate of the time-average WIP inventory. Therefore, in our next run of the GMMC simulation, we specify a warm-up time of 100 hours (Figure L9.13). Notice how the simulation output during the 100-hour warm-up has been discarded in Figure L9.12(b). The discarded output will no longer affect the observations reported in the simulation output.

The simplified process we just used to identify the end of the warm-up phase should be avoided. The process can be substantially improved by running multiple replications of the simulation and then combining the time-series output from each replication into an average time-series plot. Averaging the time-series data from multiple replications together will help filter out the large fluctuations (high values from one replication are offset by small values from another replication) and *will give you a better estimate of the true but unknown time-series output*. In many cases, you may need to add an additional layer of filtering to smooth the "averaged" time-series plot. This can be done by computing a Welch moving average of the time-series data. Section 9.6.1 of Chapter 9 provides additional

FIGURE L9.11

The simulation model of the Green Machine Manufacturing Company.

```
Time Units:  Minutes
Distance Units:   Feet
**********************************************************************
*                               Locations                           *
**********************************************************************
  Name          Cap Units Stats        Rules      Cost
  ----------    --- ----- -----------  ---------- ------------
  Pallet1       inf 1     None         Oldest, ,
  Machining1    1   1     None         Oldest, ,
  Pallet2       inf 1     None         Oldest, ,
  Machining2    1   1     None         Oldest, ,
  Pallet3       inf 1     None         Oldest, ,
  Machining3    1   1     None         Oldest, ,
  Pallet4       inf 1     None         Oldest, ,
  Machining4    1   1     None         Oldest, ,


**********************************************************************
*                               Entities                            *
**********************************************************************
  Name          Speed (fpm)  Stats        Cost
  ----------    ------------ -----------  ------------
  Job           150          Basic


**********************************************************************
*                               Processing                          *
**********************************************************************
                        Process                    Routing

  Entity   Location    Operation              Blk  Output    Destination Rule     Move Logic
  -------- ----------  -----------------      ---- --------   ----------- -------  ------------
  Job      Pallet1     Inc WIP                 1   Job        Machining1  FIRST 1
  Job      Machining1  Wait E(1) min           1   Job        Pallet2     FIRST 1
  Job      Pallet2                             1   Job        Machining2  FIRST 1
  Job      Machining2  Wait (0.10 + W(5,1)) min
                                               1   Job        Pallet3     FIRST 1
  Job      Pallet3                             1   Job        Machining3  FIRST 1
  Job      Machining3  Wait (0.25 + W(5,1)) min
                                               1   Job        Pallet4     FIRST 1
  Job      Pallet4                             1   Job        Machining4  FIRST 1
  Job      Machining4  Wait (0.25 + W(5,1)) min
                        Dec WIP                1   Job        EXIT        FIRST 1


**********************************************************************
*                               Arrivals                            *
**********************************************************************
  Entity   Location Qty Each   First Time Occurrences Frequency  Logic
  -------- -------- --------   ---------- ----------- ---------- -----------
  Job      Pallet1  1          0          inf         1.175 min


**********************************************************************
*                         Variables (global)                        *
**********************************************************************
  ID      Type          Initial value Stats
  ----    ------------  ------------- ------------
  WIP     Integer       0             Time Series  <--- Note that the Time-weighted suboption
                                                        of Time Series was selected.


**********************************************************************
*                               Macros                              *
**********************************************************************
  ID              Text
  --------------  ------------
  dummy           1
```

FIGURE L9.12

Work-in-process (WIP) inventory value history for one replication of the GMMC simulation. (a) WIP value history without warm-up phase removed. Statistics will be biased low by the transient WIP values. (b) WIP value history with 100-hour warm-up phase removed. Statistics will not be biased low.

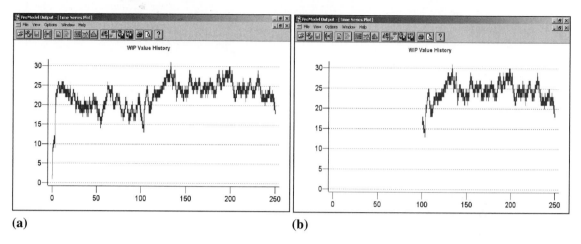

(a) **(b)**

FIGURE L9.13

ProModel's Simulation Options window set for a single replication with a 100-hour warm-up period followed by a 150-hour run length.

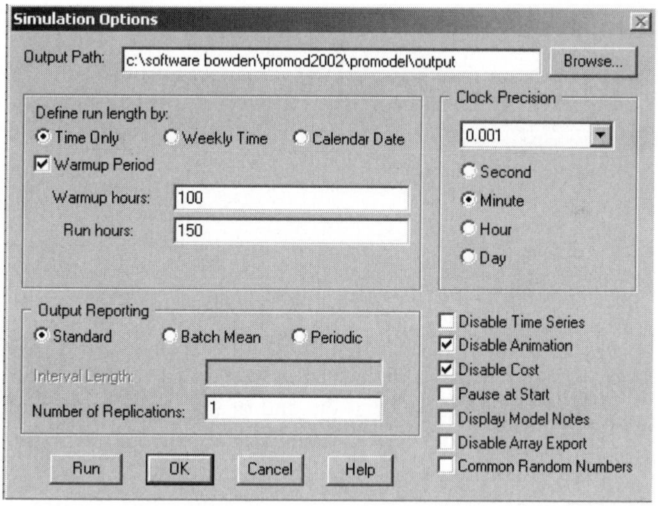

details on this subject, and exercise 4 in Lab Chapter 11 illustrates how to use the Welch method implemented in SimRunner.

Figure L9.14 was produced by SimRunner by recording the time-average WIP levels of the GMMC simulation over successive one-hour time periods. The results from each period were averaged across five replications to produce the "raw data" plot, which is the more erratic line that appears red on the computer screen. A 54-period moving average (the smooth line that appears green on the

FIGURE L9.14

SimRunner screen for estimating the end of the warm-up time.

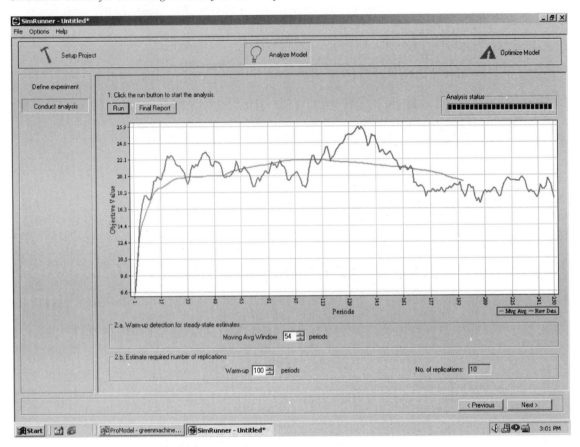

computer) indicates that the end of the warm-up phase occurs between the 33rd and 100th periods. Given that we need to avoid underestimating the warm-up, let's declare 100 hours as the end of the warm-up time. We feel much more comfortable basing our estimate of the warm-up time on five replications. Notice that SimRunner indicates that at least 10 replications are needed to estimate the average WIP to within a 7 percent error and a confidence level of 90 percent using a warm-up of 100 periods (hours). You will see how this was done with SimRunner in Exercise 4 of Section L11.4 in Lab Chapter 11.

Why did we choose an initial run length of 250 hours to produce the time-series plot in Figure L9.14? Well, you have to start with something, and we picked 250 hours. You can rerun the experiment with a longer run length if the time-series plot, produced by averaging the output from several replications of the simulation, does not indicate a steady-state condition. Long runs will help prevent you from wondering what the time-series plot looks like beyond the point at which you stopped it. Do you wonder what our plot does beyond 250 hours?

Let's now direct our attention to answering the question of how long to run the model past its warm-up time to estimate our steady-state statistic, mean WIP inventory. We will somewhat arbitrarily pick 100 hours not because that is equal to the warm-up time but because it will allow ample time for the simulation events to happen thousands of times. In fact, the 100-hour duration will allow approximately 5,100 jobs to be processed per replication, which should give us decently accurate results. How did we derive the estimate of 5,100 jobs processed per replication? With a 1.175-minute interarrival time of jobs to the system, 51 jobs arrive per hour (60 minutes/1.175 minutes) to the system. Running the simulation for 100 hours should result in about 5,100 jobs (100 hours × 51 jobs) exiting the system. You will want to check the number of Total Exits in the Entity Activity section of the ProModel output report that you just produced to verify this.

L9.3.2 Replications or Batch Intervals

We have a choice between running replications or batch intervals to estimate the time-average WIP inventory. Replications are our favorite choice because we get independent observations, guaranteed. Figure L9.15 illustrates how this is done in ProModel as well as the specification of the 100 hours of run time beyond the warm-up. Running the simulation for 10 replications produces the results in Figure L9.16. We are approximately 90 percent confident that the true but unknown mean WIP inventory is between 17.64 and 21.09 jobs.

FIGURE L9.15

Specification of warm-up hours and run hours in the Simulation Options menu for running replications.

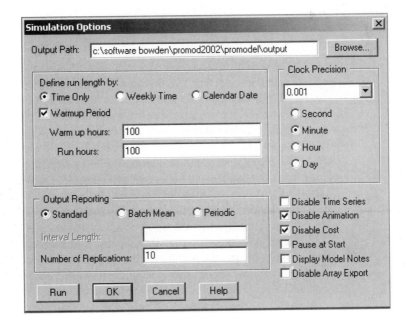

FIGURE L9.16

Ten replications of the Green Machine Manufacturing Company simulation using a 100-hour warm-up and a 100-hour run length.

```
----------------------------------------------------------------------------------
General Report
Output from C:\Bowden Files\Word\McGraw 2nd Edition\GreenMachine.MOD
Date: Jul/16/2002   Time: 01:04:42 AM
----------------------------------------------------------------------------------
Scenario          : Normal Run
Replication       : All
Period            : Final Report (100 hr to 200 hr Elapsed: 100 hr)
Warmup Time       : 100
Simulation Time   : 200 hr
----------------------------------------------------------------------------------
VARIABLES
```

Variable Name	Total Changes	Average Minutes Per Change	Minimum Value	Maximum Value	Current Value	Average Value	
WIP	10202	0.588	13	31	27	24.369	(Rep 1)
WIP	10198	0.588	9	32	29	18.928	(Rep 2)
WIP	10207	0.587	14	33	22	24.765	(Rep 3)
WIP	10214	0.587	10	26	22	18.551	(Rep 4)
WIP	10208	0.587	8	27	25	16.929	(Rep 5)
WIP	10216	0.587	13	27	17	19.470	(Rep 6)
WIP	10205	0.587	8	30	24	19.072	(Rep 7)
WIP	10215	0.587	8	25	14	15.636	(Rep 8)
WIP	10209	0.587	9	26	20	17.412	(Rep 9)
WIP	10209	0.587	11	27	24	18.504	(Rep 10)
WIP	10208.3	0.587	10.3	28.4	22.4	19.364	(Average)
WIP	5.735	0.0	2.311	2.836	4.501	2.973	(Std. Dev.)
WIP	10205	0.587	8.959	26.756	19.790	17.640	(90% C.I. Low)
WIP	10211.6	0.587	11.64	30.044	25.009	21.087	(90% C.I. High)

If we decide to make one single long run and divide it into batch intervals to estimate the expected WIP inventory, we would select the Batch Mean option in the Output Reporting section of the Simulation Options menu (Figure L9.17). A guideline in Chapter 9 for making an initial assignment to the length of time for each batch interval was to set the batch interval length to the simulation run time you would use for replications, in this case 100 hours. Based on the guideline, the Simulation Options menu would be configured as shown in Figure L9.17 and would produce the output in Figure L9.18. To get the output report to display the results of all 10 batch intervals, you specify "<All>" Periods in the General Report Type settings window (Figure L9.19). We are approximately 90 percent confident that the true but unknown mean WIP inventory is between 18.88 and 23.19 jobs. If we desire a smaller confidence interval, we can increase the sample size by extending the run length of the simulation in increments of the batch interval length. For example, we would specify run hours of 1500 to collect 15 observations of average WIP inventory. Try this and see if the confidence interval becomes smaller. Note that increasing the batch interval length is also helpful.

Figure L9.17

Warm-up hours, run hours, and batch interval length in the Simulation Options menu for running batch intervals. Note that the time units are specified on the batch interval length.

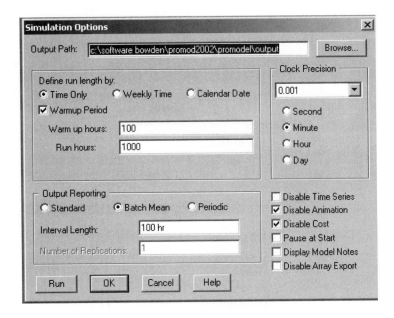

Figure L9.18

Ten batch intervals of the Green Machine Manufacturing Company simulation using a 100-hour warm-up and a 100-hour batch interval length.

```
-------------------------------------------------------------------------------
General Report
Output from C:\Bowden Files\Word\McGraw 2nd Edition\GreenMachine.MOD
Date: Jul/16/2002   Time: 01:24:24 AM
-------------------------------------------------------------------------------
Scenario         : Normal Run
Replication      : 1 of 1
Period           : All
Warmup Time      : 100 hr
Simulation Time  : 1100 hr
-------------------------------------------------------------------------------
VARIABLES
                      Average
Variable    Total     Minutes    Minimum   Maximum   Current   Average
Name        Changes   Per Change  Value     Value     Value     Value
--------    -------   ----------  -------   -------   -------   -------
WIP          10202      0.588       13        31        27      24.369   (Batch 1)
WIP          10215      0.587       16        29        26      23.082   (Batch 2)
WIP          10216      0.587        9        30        22      17.595   (Batch 3)
WIP          10218      0.587       13        30        16      22.553   (Batch 4)
WIP          10202      0.588       13        33        28      23.881   (Batch 5)
WIP          10215      0.587       17        37        25      27.335   (Batch 6)
WIP          10221      0.587        8        29        18      20.138   (Batch 7)
WIP          10208      0.587       10        26        22      17.786   (Batch 8)
WIP          10214      0.587        9        29        20      16.824   (Batch 9)
WIP          10222      0.586        9        23        12      16.810   (Batch 10)
WIP        10213.3      0.587     11.7      29.7      21.6      21.037   (Average)
WIP          7.103      0.0       3.164     3.743     5.168      3.714   (Std. Dev.)
WIP        10209.2      0.587     9.865    27.530    18.604     18.884   (90% C.I. Low)
WIP        10217.4      0.587    13.534    31.869    24.595     23.190   (90% C.I. High)
```

FIGURE L9.19

*ProModel General
Report Options set to
display the results
from all batch
intervals.*

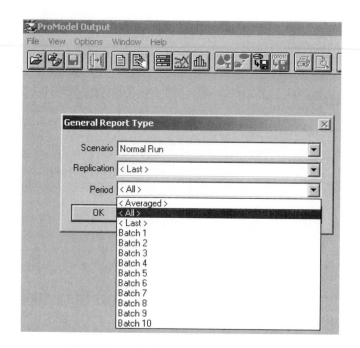

L9.3.3 Required Batch Interval Length

The worrisome part of the batch interval method is whether the batch interval is long enough to acquire "approximately" independent observations. The word "approximately" is emphasized because we know that the observations within a simulation run are often autocorrelated (not independent). In fact, they are often positively correlated, which would bias the sample standard deviation of the batch means to the low side. And this would bias our estimate of the confidence interval's half-width to the low side. We do not know if this is the case with our WIP inventory observations. However, if it was the case and we were not aware of it, we would think that our estimate of the mean WIP inventory was better than it really is. Therefore, we should do ourselves a favor and use Stat::Fit to check the lag-1 autocorrelation of the observations before we interpret the results. To get a decent estimate of the lag-1 autocorrelation, however, we need more than 10 observations. Section 9.6.2 of Chapter 9 stated that at least 100 observations are necessary to get a reliable estimate of the lag-1 autocorrelation. So let's increase the simulation run length to 10,000 hours and produce an output report with 100 observations of the time-average WIP inventory.

It would take a while to enter the 100 observations into Stat::Fit one at a time. Therefore, save the ProModel output report with the results from the 100 batch intervals into a Microsoft Excel file format that can then be copied into the Stat::Fit data table. The procedure is the same as used in Section L9.2.4 to enter the observations under the Current Value column from the VARIABLES sheet tab of the Excel spreadsheet into Stat::Fit—except that you are now entering the

FIGURE L9.20

Stat::Fit Autocorrelation plot of the observations collected over 100 batch intervals. Lag-1 autocorrelation is within the −0.20 to +0.20 range.

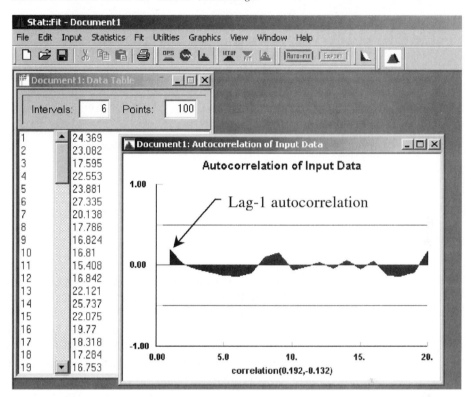

observations under the Average Value column from the VARIABLES sheet tab of the Excel spreadsheet into Stat::Fit.

Figure L9.20 illustrates the Stat::Fit results that you will want to verify. The lag-1 autocorrelation value is the first value plotted in Stat::Fit's Autocorrelation of Input Data plot. Note that the plot begins at lag-1 and continues to lag-20. For this range of lag values, the highest autocorrelation is 0.192 and the lowest value is −0.132 (see correlation(0.192, −0.132) at the bottom of the plot). Therefore, we know that the lag-1 autocorrelation is within the −0.20 to +0.20 range recommended in Section 9.6.2 of Chapter 9, which is required before proceeding to the final step of "rebatching" the data into between 10 and 30 larger batches. We have enough data to form 10 batches with a length of 1000 hours (10 batches at 1000 hours each equals 10,000 hours of simulation time, which we just did). The results of "rebatching" the data in the Microsoft Excel spreadsheet are shown in Table L9.3. Note that the first batch mean of 21.037 in Table L9.3 is the average of the first 10 batch means that we originally collected. The second batch mean of 19.401 is the average of the next 10 batch means and so on. The confidence

---

TABLE L9.3   **Final Results of the 100 Batch Intervals Combined into 10 Batch Intervals**

WIP Average Value	
21.037	(Batch 1)
19.401	(Batch 2)
21.119	(Batch 3)
20.288	(Batch 4)
20.330	(Batch 5)
20.410	(Batch 6)
22.379	(Batch 7)
22.024	(Batch 8)
19.930	(Batch 9)
22.233	(Batch 10)
20.915	(Average)
1.023	(Std. Dev.)
20.322	(90% C.I. Low)
21.508	(90% C.I. High)

---

interval is very narrow due to the long simulation time (1000 hours) of each batch. The owner of GMMC, Mr. Robert Vaughn, should be very pleased with the precision of the time-average WIP inventory estimate.

The batch interval method requires a lot of work to get it right. Therefore, if the time to simulate through the warm-up period is relatively short, the replications method should be used. Reserve the batch interval method for simulations that require a very long time to reach steady state.

L9.4 Exercises

1. An average of 100 customers per hour arrive to the Picayune Mutual Bank. It takes a teller an average of two minutes to serve a customer. Interarrival and service times are exponentially distributed. The bank currently has four tellers working. Bank manager Rich Gold wants to compare the following two systems with regard to the average time customers spend in the bank.

System #1
A separate queue is provided for each teller. Assume that customers choose the shortest queue when entering the bank, and that customers cannot jockey between queues (jump to another queue).

System #2
A single queue is provided for customers to wait for the first available teller.

Assume that there is no move time within the queues. Run 15 replications of an eight-hour simulation to complete the following:

a. For each system, record the 90 percent confidence interval using a 0.10 level of significance for the average time customers spend in the bank.

b. Estimate the number of replications needed to reduce the half-width of each confidence interval by 25 percent. Run the additional replications for each system and compute new confidence intervals using a 0.10 level of significance.

c. Based on the results from part *b*, would you recommend one system over the other? Explain.

2. Rerun the simulation of the Green Machine Manufacturing Company of Section L9.3 to produce 100 batch means with a 50-hour interval length. Use a 100-hour warm-up. Check the lag-1 autocorrelation of the 100 time-average WIP observations to see if it falls between −0.20 and +0.20. If it does not, increase the batch interval length by 50 percent and continue the process until an acceptable lag-1 autocorrelation is achieved. Once the lag-1 autocorrelation is acceptable, check to see if the 100 observations are normally distributed. After that, rebatch the 100 observations into 20 batch means and compute a 95 percent confidence interval. Are you comfortable with the resulting confidence interval? Explain.

3. Use the simulation model created for the DumpOnMe facility presented in Exercise 11 of Lab Section L7.12 to complete the following:

a. For a simulation run length of 60 minutes, record the 90 percent confidence interval for the scale's average utilization based on 30 replications.

b. For a 60-minute warm-up and a 60-minute simulation run length beyond the warm-up time, record the 90 percent confidence interval for the scale's average utilization based on 30 replications.

c. Based only on the two confidence intervals, do you think that a warm-up period is necessary for the DumpOnMe model? Explain.

4. See Lab Chapter 11 for exercises involving the use of SimRunner's implementation of the Welch moving average technique.

10   COMPARING ALTERNATIVE SYSTEMS

The great tragedy of Science—the slaying of a beautiful hypothesis by an ugly fact.
—Thomas H. Huxley

In this lab we see how ProModel is used with some of the statistical methods presented in Chapter 10 to compare alternative designs of a system with the goal of identifying the superior system relative to some performance measure. We will also learn how to program ProModel to use common random numbers (CRN) to drive the simulation models that represent alternative designs for the systems being compared. The use of CRN allows us to run the opposing simulations under identical experimental conditions to facilitate an objective evaluation of the systems.

L10.1 Overview of Statistical Methods

Often we want to compare alternative designs of a system. For example, we may wish to compare two different material handling systems, production scheduling rules, plant layouts, or staffing schedules to determine which yields better results. We compare the two alternative system designs based on a certain measure of performance—for example, production rate, number of customers served, time in system, or waiting time for service. The comparison is based on the estimates of the expected value of the performance measure that we derive from simulating the alternative system designs. If the estimates provide evidence that the performances of the two alternative simulation models are not equal, we can justify picking the alternative that produced the better estimate as the best alternative.

Chapter 10 covers the paired-*t* confidence interval method and the Welch confidence interval method for testing hypotheses about two alternatives. The

Bonferroni approach is used for comparing from three to about five alternatives. Either the Welch method or paired-t method is used with the Bonferroni approach to make the comparisons. The Bonferroni approach does not identify the best alternative but rather identifies which pairs of alternatives perform differently. This information is helpful for identifying the better alternatives, if in fact there is a significant difference in the performance of the competing alternatives.

Chapter 10 recommends that the technique of analysis of variance (ANOVA) be used when more than about five alternative system designs are being compared. The first step in ANOVA is to evaluate the hypothesis that the mean performances of the systems are equal against the alternative hypothesis that at least one pair of the means is different. To figure out which means differ from which other ones, a multiple comparison test is used. There are several multiple comparison tests available, and Chapter 10 presents the Fisher's protected least significant difference (LSD) test.

The decision to use paired-t confidence intervals, Welch confidence intervals, the Bonferroni approach, or ANOVA with protected LSD depends on the statistical requirements placed on the data (observations from the simulations) by the procedure. In fact, simulation observations produced using common random numbers (CRN) cannot be used with the Welch confidence interval method or the technique of ANOVA. Please see Chapter 10 for details.

In this lab we will use the paired-t confidence interval method because it places the fewest statistical requirements on the observations collected from the simulation model. Therefore, we are less likely to get into trouble with faulty assumptions about our data (observations) when making decision based on paired-t confidence intervals. Furthermore, one of the main purposes of this lab is to demonstrate how the CRN technique is implemented using ProModel. Out of the comparison procedures of Chapter 10, only the ones based on paired-t confidence intervals can be used in conjunction with the CRN technique.

L10.2 Three Alternative Systems

Problem Statement

The analysis of the Spuds-n-More by the Seashore restaurant started in Section L9.2 of Lab Chapter 9 will be completed in this lab. The details of the system and objective for the study are reviewed here. Spuds-n-More, a takeout fast-food restaurant, is a favorite among vacationers at a seaside resort. Customers walking along the boardwalk access the restaurant via a single ordering window, where they place their orders, pay, and receive their food. There is no seating in the restaurant. The interarrival time of customers to Spuds-n-More is exponentially distributed with a mean of six minutes. The activity times are given in Table L9.1. The waiting area in front of the walkup window can accommodate only five customers. If the waiting area is filled to capacity when a customer arrives, the customer bypasses Spuds-n-More. Mr. Taylor closes the entry to the waiting area in front of his restaurant after eight hours of operation. However, any customers already in the waiting area by closing time are served before the kitchen shuts down for the day.

Mr. Taylor is thinking about expanding into the small newspaper stand next to Spuds-n-More, which has been abandoned. This would allow him to add a second window to serve his customers. The first window would be used for taking orders and collecting payments, and the second window would be used for filling orders. Before investing in the new space, however, Mr. Taylor needs to convince the major stockholder of Spuds-n-More, Pritchard Enterprises, that the investment would allow the restaurant to serve more customers per day.

A baseline (as-is) model of the current restaurant configuration was developed and validated in Section L9.2 of Lab Chapter 9. See Figures L9.1 and L9.2 for the layout of the baseline model and the printout of the ProModel model. Our task is to build a model of the proposed restaurant with a second window to determine if the proposed design would serve more customers per day than does the current restaurant. Let's call the baseline model Spuds-n-More1 and the proposed model Spuds-n-More2. The model is included on the CD accompanying the book under file name Lab 10_2 Spuds-n-More2.MOD.

The ProModel simulation layout of Spuds-n-More2 is shown in Figure L10.1, and the ProModel printout is given in Figure L10.2. After customers wait in the order queue (location Order_Q) for their turn to order, they move to the first window to place their orders and pay the order clerk (location Order_Clerk). Next customers proceed to the pickup queue (location Pickup_Q) to wait for their turn to be served by the pickup clerk (Pickup_Clerk) at the second window. There

FIGURE L10.1

Layout of Spuds-n-More2 simulation model.

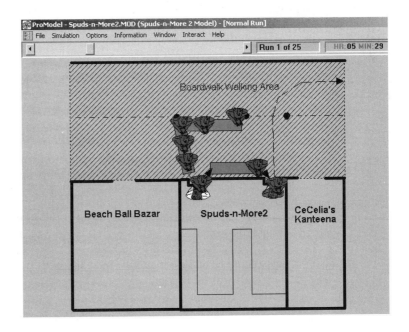

FIGURE L10.2

The Spuds-n-More2 simulation model.

```
Time Units:  Minutes
Distance Units:  Feet
***************************************************************************
*                              Locations                                  *
***************************************************************************

  Name         Cap Units Stats      Rules           Cost
  -----------  --- ----- ---------- --------------- ------------
  Entry         1   1    None       Oldest, ,
  Order_Q       4   1    None       Oldest, FIFO,
  Order_Clerk   1   1    None       Oldest, ,
  Pickup_Q      1   1    None       Oldest, FIFO,
  Pickup_Clerk  1   1    None       Oldest, ,

***************************************************************************
*                              Entities                                   *
***************************************************************************

  Name        Speed (fpm)  Stats      Cost
  ----------  -----------  ---------- ------------
  Customer    150          None

***************************************************************************
*                              Processing                                 *
***************************************************************************
                Process                           Routing

Entity   Location     Operation       Blk Output    Destination   Rule                  Move Logic
-------- -----------  --------------- --- --------   -----------   -------------------   ------------
Customer Entry                         1  Customer   Order_Q       IF clock(hr) <= 8,1
                                          Customer   EXIT          IF clock(hr) > 8      If Processed = Arrivals
                                                                                         Then Stop
Customer Order_Q      Inc Arrivals     1  Customer   Order_Clerk   FIRST 1
Customer Order_Clerk  // Order
                      Wait e(2) min
                      // Pay
                      Wait e(1.5) min  1  Customer   Pickup_Q      FIRST 1
Customer Pickup_Q                      1  Customer   Pickup_Clerk  FIRST 1
Customer Pickup_Clerk // Pickup
                      Wait e(5) min
                      Inc Processed    1  Customer   EXIT          FIRST 1

***************************************************************************
*                              Arrivals                                   *
***************************************************************************

  Entity   Location Qty Each  First Time Occurrences Frequency  Logic
  -------- -------- --------  ---------- ----------- ---------- ------------
  Customer Entry    1          0          inf         E(6) min

***************************************************************************
*                              Variables (global)                         *
***************************************************************************
  ID         Type         Initial value Stats
  ---------- -----------  ------------- -----------
  Arrivals   Integer       0             None
  Processed  Integer       0             Basic
```

is enough space for one customer in the pickup queue. The pickup clerk processes one order at a time.

As the Spuds-n-More2 model was being built, Mr. Taylor determined that for additional expense he could have a carpenter space the two customer service windows far enough apart to accommodate up to three customers in the order pickup queue. Therefore, he requested that a third alternative design with an order pickup queue capacity of three be simulated. To do this, we only need to change the capacity of the pickup queue from one to three in our Spuds-n-More2 model. We shall call this model Spuds-n-More3. Note that for our Spuds-n-More2 and Spuds-n-More3 models, we assigned a length of 25 feet to the Order_Q, a length of 12 feet to the Pickup_Q, and a Customer entity travel speed of 150 feet per minute. These values affect the entity's travel time in the queues (time for a customer to walk to the end of the queues) and are required to match the results presented in this lab.

L10.3 Common Random Numbers

Let's conduct the analysis for Mr. Taylor using Common Random Numbers (CRN). The CRN technique provides a way of comparing alternative system designs under more equal experimental conditions. This is helpful in ensuring that the observed differences in the performance of system designs are due to the differences in the designs and not to differences in experimental conditions. The goal is to evaluate each system under the exact same set of circumstances to ensure that a fair comparison is made. Therefore, the simulation of each alternative system design is driven by the same stream of random numbers to ensure that the differences in key performance measures are due only to differences in the alternative designs and not because some designs were simulated with a stream of random numbers that produced more extreme conditions. Thus we need to use the exact same random number from the random number stream for the exact same purpose in each simulated system. A common practice that helps to keep random numbers synchronized across systems is to assign a different random number stream to each stochastic element in the model. For this reason, ProModel provides 100 unique streams (Stream 1 to Stream 100) of random numbers for generating random variates from the statistical distributions available in the software.

The Spuds-n-More models use observations (random variates) from an exponential distribution for the customer interarrival time, order time, order payment time, and order pickup time. For CRN, in this example we can use Stream 1 to generate interarrival times, Stream 2 to generate order times, Stream 3 to generate payment times, and Stream 4 to generate pickup times. The ProModel function for generating random variates from the exponential distribution is E(a, *stream*), where the parameter a is the mean of the random variable and the parameter *stream* is the ProModel stream number between 1 and 100. All ProModel distribution functions include the *stream* parameter as the last parameter of the function. The *stream* parameter is optional (you do not have to specify a value). If you do not provide a value for the *stream* parameter, ProModel uses *stream* = 1 to generate observations (random variates) from the specified distribution. Figure L10.3 illustrates the desired assignment of random number streams in the processing and arrivals section of a Spuds-n-More model. Configure all three of your Spuds-n-More models accordingly.

FIGURE L10.3

Unique random number streams assigned to the Spuds-n-More model.

```
***********************************************************************
*                            Processing                               *
***********************************************************************
              Process                        Routing

Entity    Location    Operation    Blk Output   Destination  Rule                 Move Logic
--------  ----------  -----------  --- --------  -----------  -------------------  ----------
Customer  Entry                     1  Customer  Order_Q      IF clock(hr) <= 8,1
                                       Customer  EXIT         IF clock(hr) > 8     If Processed = Arrivals
                                                                                   Then Stop
Customer  Order_Q      Inc Arrivals 1  Customer  Order_Clerk  FIRST 1
Customer  Order_Clerk  // Order
                       Wait e(2,2) min
                       // Pay
                       Wait e(1.5,3) min 1 Customer Pickup_Q      FIRST 1
Customer  Pickup_Q                   1  Customer  Pickup_Clerk FIRST 1
Customer  Pickup_Clerk // Pickup
                       Wait e(5,4) min
                       Inc Processed 1  Customer  EXIT         FIRST 1

***********************************************************************
*                            Arrivals                                 *
***********************************************************************
 Entity   Location Qty Each  First Time Occurrences Frequency  Logic
 -------  -------- --------  ---------- ----------- ---------  --------
 Customer Entry      1           0         inf       E(6,1) min
```

L10.4 Bonferroni Approach with Paired-*t* Confidence Intervals

Our objective is to compare the performance of the three alternative Spuds-n-More restaurant designs with respect to the number of customers processed per day. Let μ_i represent the true number of customers processed per day by restaurant design i, where $i = 1$ denotes Spuds-n-More1, $i = 2$ denotes Spuds-n-More2, and $i = 3$ denotes Spuds-n-More3. Specifically, we wish to evaluate the hypotheses

$$H_0: \mu_1 = \mu_2 = \mu_3$$
$$H_1: \mu_1 \neq \mu_2 \text{ or } \mu_1 \neq \mu_3 \text{ or } \mu_2 \neq \mu_3$$

at the $\alpha = 0.06$ significance level. To evaluate these hypotheses, we estimate the number of customers processed by each restaurant design by simulating each design for 25 days (25 replications). The simulations are to be driven by Common Random Numbers (CRN). Therefore, select the CRN option in the bottom right corner of the Simulation Options window before running the simulations (Figure L10.4). This option helps to synchronize the use of random numbers across the three simulated systems. When the option is selected, ProModel starts each replication at predetermined locations on the random number streams. At the end of a replication, ProModel advances to the next set of predetermined locations on the streams before starting the next replication. Even if one model uses more random numbers than another during a replication, both models will begin the next replication from the same positions on the random number streams. The numbers of customers processed by each restaurant design for 25 replications using the CRN option are shown in Table L10.1.

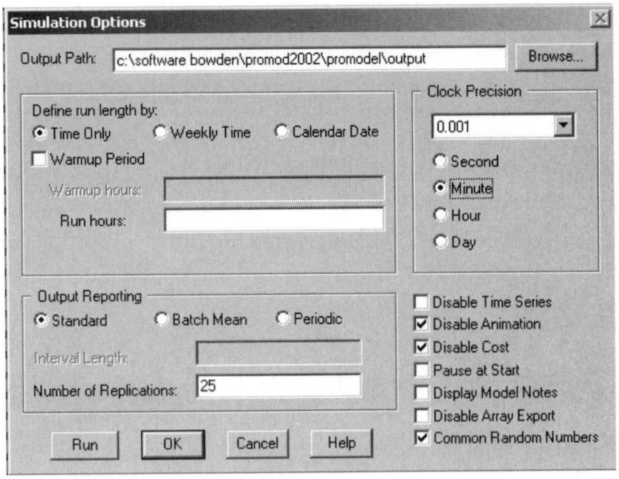

FIGURE L10.4

ProModel's Simulation Options set to run 25 replications using Common Random Numbers.

TABLE L10.1   Comparison of the Three Restaurant Designs Based on Paired Differences

(A) Rep. (j)	(B) Spuds-n-More1 Customers Processed x_{1j}	(C) Spuds-n-More2 Customers Processed x_{2j}	(D) Spuds-n-More3 Customers Processed x_{3j}	(E) Difference (B − C) $x_{(1-2)j}$	(F) Difference (B − D) $x_{(1-3)j}$	(G) Difference (C − D) $x_{(2-3)j}$
1	60	72	75	−12	−15	−3
2	57	79	81	−22	−24	−2
3	58	84	88	−26	−30	−4
4	53	69	72	−16	−19	−3
5	54	67	69	−13	−15	−2
6	56	72	74	−16	−18	−2
7	57	70	71	−13	−14	−1
8	55	65	65	−10	−10	0
9	61	84	84	−23	−23	0
10	60	76	76	−16	−16	0
11	56	66	70	−10	−14	−4
12	66	87	90	−21	−24	−3
13	64	77	79	−13	−15	−2
14	58	66	66	−8	−8	0
15	65	85	89	−20	−24	−4
16	56	78	83	−22	−27	−5
17	60	72	72	−12	−12	0
18	55	75	77	−20	−22	−2
19	57	78	78	−21	−21	0
20	50	69	70	−19	−20	−1
21	59	74	77	−15	−18	−3
22	58	73	76	−15	−18	−3
23	58	71	76	−13	−18	−5
24	56	70	72	−14	−16	−2
25	53	68	69	−15	−16	−1
Sample mean $\bar{x}_{(i-i')}$, for all i and i' between 1 and 3, with $i < i'$				−16.20	−18.28	−2.08
Sample standard dev $s_{(i-i')}$, for all i and i' between 1 and 3, with $i < i'$				4.66	5.23	1.61

The Bonferroni approach with paired-t confidence intervals is used to evaluate our hypotheses for the three alternative designs for the restaurant. The evaluation of the three restaurant designs requires that three pairwise comparisons be made:

Spuds-n-More1 vs. Spuds-n-More2
Spuds-n-More1 vs. Spuds-n-More3
Spuds-n-More2 vs. Spuds-n-More3

Following the procedure described in Section 10.4.1 of Chapter 10 for the Bonferroni approach, we begin constructing the required three paired-t confidence intervals by letting $\alpha_1 = \alpha_2 = \alpha_3 = \alpha/3 = 0.06/3 = 0.02$.

The computation of the three paired-t confidence intervals follows:

Spuds-n-More1 vs. Spuds-n-More2 ($\mu_{(1-2)}$)

$$\alpha_1 = 0.02$$

$$t_{n-1,\alpha_1/2} = t_{24,0.01} = 2.485 \text{ from Appendix B}$$

$$hw = \frac{(t_{24,0.01})s_{(1-2)}}{\sqrt{n}} = \frac{(2.485)4.66}{\sqrt{25}} = 2.32 \text{ customers}$$

The approximate 98 percent confidence interval is

$$\bar{x}_{(1-2)} - hw \leq \mu_{(1-2)} \leq \bar{x}_{(1-2)} + hw$$
$$-16.20 - 2.32 \leq \mu_{(1-2)} \leq -16.20 + 2.32$$
$$-18.52 \leq \mu_{(1-2)} \leq -13.88$$

Spuds-n-More1 vs. Spuds-n-More3 ($\mu_{(1-3)}$)

$$\alpha_2 = 0.02$$

$$t_{n-1,\alpha_2/2} = t_{24,0.01} = 2.485 \text{ from Appendix B}$$

$$hw = \frac{(t_{24,0.01})s_{(1-3)}}{\sqrt{n}} = \frac{(2.485)5.23}{\sqrt{25}} = 2.60 \text{ customers}$$

The approximate 98 percent confidence interval is

$$\bar{x}_{(1-3)} - hw \leq \mu_{(1-3)} \leq \bar{x}_{(1-3)} + hw$$
$$-18.28 - 2.60 \leq \mu_{(1-3)} \leq -18.28 + 2.60$$
$$-20.88 \leq \mu_{(1-3)} \leq -15.68$$

Spuds-n-More2 vs. Spuds-n-More3 ($\mu_{(2-3)}$)

The approximate 98 percent confidence interval is

$$-2.88 \leq \mu_{(2-3)} \leq -1.28$$

Given that the confidence interval for $\mu_{(1-2)}$ excludes zero, we conclude that there is a significant difference in the mean number of customers processed by the current "as-is" restaurant Spuds-n-More1 and the proposed Spuds-n-More2 design. The confidence interval further suggests that Spuds-n-More2 processes an estimated 13.88 to 18.52 more customers per day on average than does Spuds-n-More1. Likewise, there is a significant difference in the mean number of customers processed by Spuds-n-More1 and Spuds-n-More3. The proposed Spuds-n-More3 processes an estimated 15.68 to 20.88 more customers per day on average than does Spuds-n-More1. The third confidence interval suggests that the difference between Spuds-n-More2 and Spuds-n-More3, although statistically significant, is fairly small with Spuds-n-More3 processing an estimated 1.28 to 2.88 more customers per day on average than does Spuds-n-More2.

With $\alpha = 0.06$ the *overall confidence* for our conclusions is approximately 94 percent. Based on these results we are justified in concluding that both of the new designs (Spuds-n-More2 and Spuds-n-More3) serve more customers per day than does the current "as-is" restaurant. It appears that Spuds-n-More3 is not that much better than Spuds-n-More2 from a customer service rate perspective. Perhaps we will advise Mr. Taylor to request funding from the restaurant's stockholders to implement the Spuds-n-More2 design because the confidence interval suggests that it processes an estimated 14 to 19 more customers per day than does the current restaurant. The extra expense of Spuds-n-More3 above the implementation cost of Spuds-n-More2 for spacing the order and pickup windows farther apart will probably not be offset by the extra income from processing only an estimated 1 to 3 more customers per day.

In the first exercise problem in the next section, we will find out how much of a reduction, if any, we achieved in the half-widths of the confidence intervals by using the CRN technique. Theoretically, the CRN technique produces a positive correlation between the observations in columns B, C, and D of Table L10.1. The positive correlation reduces the value of the sample standard deviation of the observations in the difference columns (columns E, F, and G of Table L10.1). The smaller standard deviation results in a smaller confidence interval half-width (a more precise statement). This is important because, for example, if the third confidence interval was $-10.93 \leq \mu_{(2-3)} \leq -1.79$, which suggests that Spuds-n-More3 processes an estimated 1.79 to 10.93 more customers per day on average than does Spuds-n-More2, then we may not be so quick to rule out Spuds-n-More3 as a serious competitor to Spuds-n-More2 due to the greater range of the confidence interval.

L10.5 Exercises

1. Without using the Common Random Numbers (CRN) technique, compare the Spuds-n-More1, Spuds-n-More2, and Spuds-n-More3 models with respect to the average number of customers served per day. To avoid the use of CRN, assign ProModel Stream 1 to each stochastic element in the model and do not select the CRN option from the

Simulation Options menu. Simulate each of the three restaurants for 25 replications.

  a. Use the Bonferroni approach with an overall significance level of 0.06, and construct the confidence intervals using the paired-t method. How do the half-widths of the confidence intervals compare to the ones computed in Section L10.4? Are the differences in half-widths what you expected? Explain.

  b. Use the Analysis of Variance (ANOVA) technique with the Fisher's protected Least Significant Difference (LSD) test to compare the three systems. Use a 0.05 significance level. Explain any differences between your conclusions here and those made in Section L10.4 about the three systems.

2. Increase the capacity of the Pickup_Q location in the Spuds-n-More3 model from three to six. Call this model Spuds-n-More4. Using the Common Random Numbers (CRN) technique with ProModel Streams 1, 2, 3, and 4 assigned as in Figure L10.3, see if Spuds-n-More4 processes more customers per day on average than does Spuds-n-More3. Use a paired-t confidence interval with a 0.02 significance level. Run each simulation model for 25 replications.

3. For the Picayune Mutual Bank (Exercise 1 of Section L9.4), run 15 eight-hour replications of each alternative system design with the objective of identifying which design minimizes the average time customers spend in the bank. Use the Welch confidence interval method with a 0.10 level of significance to make the comparison. Contrast this comparison with the ballpark guess you made in Section L9.4.

4. The DumpOnMe facility presented in Exercise 11 of Lab Section L7.12 operates with six dump trucks. The owner of the facility wants to increase the amount of coal hauled to the railroad. The loader operators recommend adding two more trucks for a total of eight trucks. Not to be outdone, the scale operator recommends running the operation with 10 trucks. Trucks are expensive, so the owner hires you to simulate the system as is and then with the additional trucks before making her decision. Since each truck hauls approximately the same amount of coal, you will determine which alternative delivers the most coal to the railroad by evaluating the number of trucks processed by the scale. To compare the performance of the 6-, 8-, and 10-truck systems, run five replications of each alternative system using a 100-minute warm-up time followed by a 480-minute simulation run length beyond the warm-up time. Use the Welch confidence interval method with a 0.06 level of significance to make the comparison. Who is correct, the loader operators or the scale operator? Are you a little surprised by the answer? Look over the output reports from the simulations to find an explanation for the answer.

11 SIMULATION OPTIMIZATION WITH SIMRUNNER

Climb mountains to see lowlands.
—Chinese Proverb

The purpose of this lab is to demonstrate how to solve simulation-based optimization problems using SimRunner. The lab introduces the five major steps for formulating and solving optimization problems with SimRunner. After stepping through an example application of SimRunner, we provide additional application scenarios to help you gain experience using the software.

L11.1 Introduction to SimRunner

When you conduct an analysis using SimRunner, you build and run projects. With each project, SimRunner applies its evolutionary algorithms to your simulation model to seek optimal values for multiple decision variables. In SimRunner, decision variables are called *input factors* (Figure L11.1). For each project, you will need to give SimRunner a model to optimize, identify which input factors to change, and define how to measure system performance using an objective function. The following describes the terminology and procedure used to conduct experiments using SimRunner.

Step 1. Create, verify, and validate a simulation model using ProModel, Med-Model, or ServiceModel. Next, create a macro and include it in the run-time interface for each input factor that is believed to influence the output of the simulation model. The input factors are the variables for which you are seeking optimal values, such as the number of nurses assigned to a shift or the number of machines to be placed in a work cell. Note that SimRunner can test only those factors identified as macros in ProModel, MedModel, or ServiceModel.

Figure L11.1

Relationship between SimRunner's optimization algorithms and ProModel simulation model.

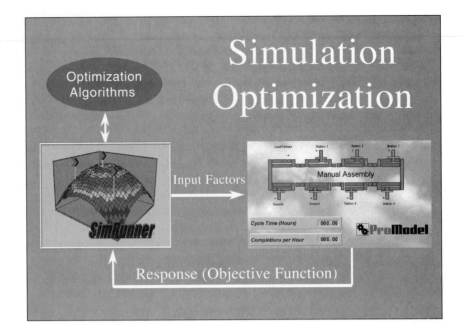

Step 2. Create a new SimRunner project and select the input factors you wish to test. For each input factor, define its numeric data type (integer or real) and its lower bound (lowest possible value) and upper bound (highest possible value). SimRunner will generate solutions by varying the values of the input factors according to their data type, lower bounds, and upper bounds. Care should be taken when defining the lower and upper bounds of the input factors to ensure that a combination of values will not be created that leads to a solution that was not envisioned when the model was built.

Step 3. After selecting the input factors, define an objective function to measure the utility of the solutions tested by SimRunner. The objective function is built using terms taken from the output report generated at the end of the simulation run. For example, the objective function could be based on entity statistics, location statistics, resource statistics, variable statistics, and so forth. In designing the objective function, the user specifies whether a term is to be minimized or maximized as well as the overall weighting of that term in the objective function. Some terms may be more important than other terms to the decision maker. SimRunner also allows you to seek a target value for an objective function term.

Step 4. Select the optimization profile and begin the search by starting the optimization algorithms. The optimization profile sets the size of the evolutionary algorithm's population. The population size defines the number of solutions evaluated by the algorithm during each generation of its search. SimRunner provides

FIGURE L11.2

Generally, the larger the size of the population the better the result.

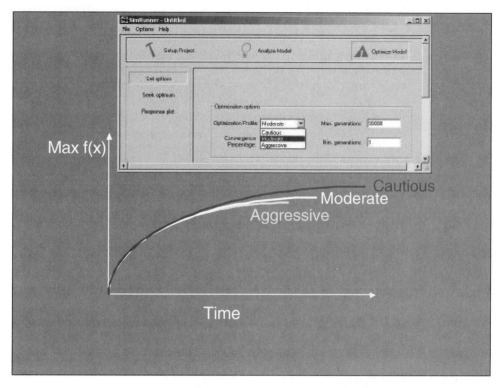

three population sizes: small, medium, and large. The small population size corresponds to the aggressive optimization profile, the medium population size corresponds to the moderate optimization profile, and the large population size corresponds to the cautious profile. In general, as the population size is increased, the likelihood that SimRunner will find the optimal solution increases, as does the time required to conduct the search (Figure L11.2).

Step 5. Study the top solutions found by SimRunner and pick the best. SimRunner will show the user the data from all experiments conducted and will rank each solution based on its utility, as measured by the objective function. Remember that the value of an objective function is a random variable because it is produced from the output of a stochastic simulation model. Therefore, be sure that each experiment is replicated an appropriate number of times during the optimization.

Another point to keep in mind is that the list of solutions presented by SimRunner represents a rich source of information about the behavior, or response surface, of the simulation model. SimRunner can sort and graph the solutions many different ways to help you interpret the "meaning" of the data.

L11.2 SimRunner Projects

Problem Statement

Prosperity Company has selected what it thinks is the ideal product to manufacture and has also designed the "ideal production system" (see Figure L11.3). Plates of raw material arrive to the ideal production system and are transformed into gears by a milling operation. The time between arrivals of plates is exponentially distributed, as is the processing time at the milling machine. Plates are processed in a first-in, first-out (FIFO) fashion. The time to move material between the pallets and the milling machine is negligible. The input and output queues have infinite capacities and the milling machine never fails or requires maintenance. It is the ideal production system. However, we have been asked to look for optimal operational parameters under three different scenarios.

We begin by building a simulation model of the ideal production system. Set the default time units to minutes in the General Information dialog box. The model consists of three locations: InputPalletQueue, MillingMachine, and OutputPalletQueue. Set the capacity of the InputPalletQueue to infinity (Inf) and the capacity of the milling machine and OutputPalletQueue to one. For simplicity, use a single entity type to represent both plates and gears. Assign Gear as the name of the entity. The parameters for the exponentially distributed time between arrivals and processing time will be given later. For now, set the model's run hours to 250 and warm-up hours to 50. The complete model is shown in Figure L11.4. The model is included on the CD accompanying the book under file name Lab 11_2 ProsperityCo.Mod.

Before continuing, we would like to point out that this fictitious production system was chosen for its simplicity. The system is not complex, nor are the example application scenarios that follow. This deliberate choice will allow us to

FIGURE L11.3

The ideal production system for Prosperity Company.

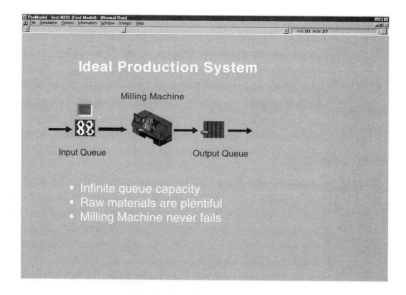

FIGURE L11.4

ProModel model of Prosperity Company.

```
***********************************************************************

    Time Units:                        Minutes
    Distance Units:                    Feet


***********************************************************************
*                            Locations                               *
***********************************************************************

    Name              Cap Units Stats        Rules           Cost
    ---------------   --- ----- -----------  -------------- ------------
    InputPalletQue    inf 1     Time Series  Oldest, FIFO,
    MillingMachine    1   1     None         Oldest, FIFO,
    OutputPalletQue   1   1     None         Oldest, FIFO,


***********************************************************************
*                            Entities                                *
***********************************************************************

    Name         Speed (fpm)  Stats        Cost
    ----------   -----------  -----------  ------------
    Gear         150          Time Series


***********************************************************************
*                            Processing                              *
***********************************************************************

                          Process                        Routing

    Entity    Location         Operation          Blk  Output  Destination       Rule
    Move Logic
    --------  ---------------  -----------------  ----  ------- ---------------- ------- ---
    ---------
    Gear      InputPalletQue                       1    Gear    MillingMachine   FIRST 1
    Gear      MillingMachine   wait E(ProcessTime)
                                                   1    Gear    OutputPalletQue  FIRST 1
    Gear      OutputPalletQue                      1    Gear    EXIT             FIRST 1


***********************************************************************
*                            Arrivals                                *
***********************************************************************

    Entity   Location        Qty each   First Time Occurrences Frequency  Logic
    -------- --------------- ---------- ---------- ----------- ---------- ------------
    Gear     InputPalletQue  1          0          INF         E(TBA)


***********************************************************************
*                            Macros                                  *
***********************************************************************

    ID               Text
    ---------------- ------------
    ProcessTime      2
    TBA              3
```

focus on learning about the SimRunner software as opposed to getting bogged down in modeling details. Additionally, the problems that are presented are easily solved using queuing theory. Therefore, if so inclined, you may want to compare the estimates obtained by SimRunner and ProModel with the actual values obtained using queuing theory. In the real world, we seldom have such opportunities. We can take refuge in knowing that ProModel and SimRunner are robust tools that can be effectively applied to both trivial and complex real-world problems, as was demonstrated in Chapter 11.

L11.2.1 Single Term Objective Functions

In the first scenario, the average time between arrivals of a plate of raw material to the input pallet queue is E(3.0) minutes. For this scenario, our objective is to find a value for the mean processing time at the milling machine that minimizes the average number of plates waiting at the input pallet queue. (We have a great deal of latitude when it comes to adjusting processing times for the ideal production system.) To do this, we will create a macro in ProModel that represents the mean processing time and give it an identification of ProcessTime. Thus, the processing time for entities at the milling machine is modeled as E(ProcessTime) minutes. See Operation field for the MillingMachine location in Figure L11.4.

Before continuing, let's cheat by taking a moment to look at how varying the mean processing time affects the mean number of plates waiting in the input pallet queue. The plot would resemble the one appearing in Figure L11.5. Therefore, we can minimize the average number of plates waiting in the queue by setting the

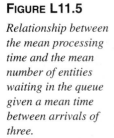

FIGURE L11.5

Relationship between the mean processing time and the mean number of entities waiting in the queue given a mean time between arrivals of three.

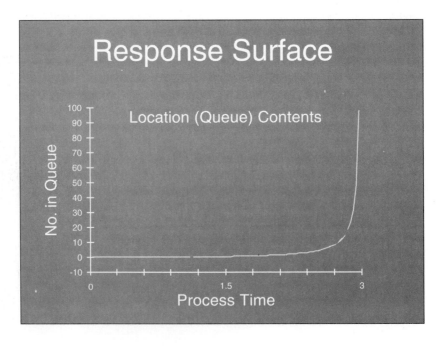

FIGURE L11.6

ProModel macro editor.

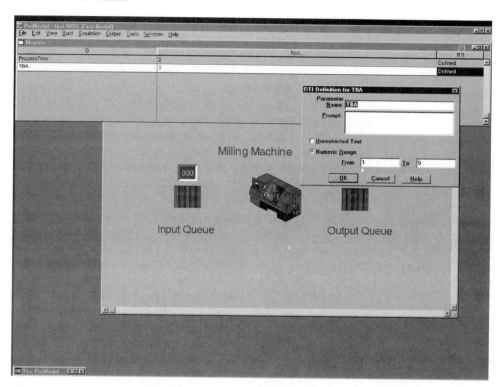

mean processing time of the milling machine to zero minutes, which of course is a theoretical value. For complex systems, you would not normally know the answer in advance, but it will be fun to see how SimRunner moves through this known response surface as it seeks the optimal solution.

The first step in the five-step process for setting up a SimRunner project is to define the macros and their Run-Time Interface (RTI) in the simulation model. In addition to defining ProcessTime as a macro (Figure L11.6), we shall also define the time between arrivals (TBA) of plates to the system as a macro to be used later in the second scenario that management has asked us to look into. The identification for this macro is entered as TBA. Be sure to set each macro's "Text . . ." value as shown in Figure L11.6. The Text value is the default value of the macro. In this case, the default value for ProcessTime is 2 and the default value of TBA is 3. If you have difficulty creating the macros or their RTI, please see Lab Chapter 14.

Next we activate SimRunner from ProModel's Simulation menu. SimRunner opens in the Setup Project mode (Figure L11.7). The first step in the Setup Project module is to select a model to optimize or to select an existing optimization project (the results of a prior SimRunner session). For this scenario, we are optimizing a

FIGURE L11.7

The opening SimRunner screen.

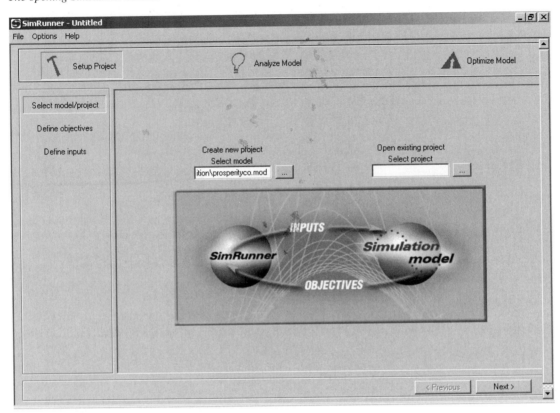

model for the first time. Launching SimRunner from the ProModel Simulation menu will automatically load the model you are working on into SimRunner. See the model file name loaded in the box under "Create new project—Select model" in Figure 11.7. Note that the authors named their model ProsperityCo.Mod.

With the model loaded, the input factors and objective function are defined to complete the Setup Project module. Before doing so, however, let's take a moment to review SimRunner's features and user interface. After completing the Setup Project module, you would next run either the Analyze Model module or the Optimize Model module. The Analyze Model module helps you determine the number of replications to run to estimate the expected value of performance measures and/or to determine the end of a model's warm-up period using the techniques described in Chapter 9. The Optimize Model module automatically seeks the values for the input factors that optimize the objective function using the techniques described in Chapter 11. You can navigate through SimRunner by selecting items from the menus across the top of the window and along the left

FIGURE L11.8

Single term objective function setup for scenario one.

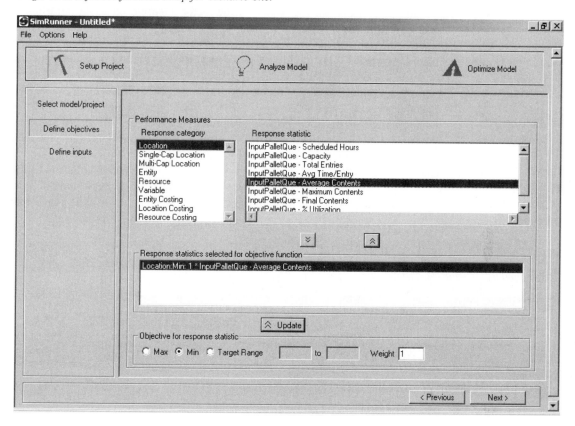

side of the window or by clicking the <Previous or Next> buttons near the bottom right corner of the window.

Clicking the Next> button takes you to the section for defining the objective function. The objective function, illustrated in Figure L11.8, indicates the desire to minimize the average contents (in this case, plates) that wait in the location called InputPalletQue. The InputPalletQue is a location category. Therefore, to enter this objective, we select Location from the Response Category list under Performance Measures by clicking on Location. This will cause SimRunner to display the list of location statistics in the Response Statistic area. Click on the response statistic InputPalletQue—AverageContents and then press the button below with the down arrows. This adds the statistic to the list of response statistics selected for the objective function. The default objective for each response statistic is maximize. In this example, however, we wish to minimize the average contents of the input pallet queue. Therefore, click on Location:Max:1*Input-PalletQue—AverageContents, which appears under the area labeled Response

Statistics Selected for the Objective Function; change the objective for the response statistic to Min; and click the Update button. Note that we accepted the default value of one for the weight of the factor. Please refer to the SimRunner Users Guide if you have difficulty performing this step.

Clicking the Next> button takes you to the section for defining the input factors. The list of possible input factors (macros) to optimize is displayed at the top of this section under Macros Available for Input (Figure L11.9). The input factor to be optimized in this scenario is the mean processing time of the milling machine, ProcessTime. Select this macro by clicking on it and then clicking the button below with the down arrows. This moves the ProcessTime macro to the list of Macros Selected as Input Factors (Figure L11.9). Next, indicate that you want to consider integer values between one and five for the ProcessTime macro. Ignore the default value of 2.00. If you wish to change the data type or lower and upper bounds, click on the input factor, make the desired changes, and click the Update button. Please note that an input factor is designated as an integer when

FIGURE L11.9

Single input factor setup for scenario one.

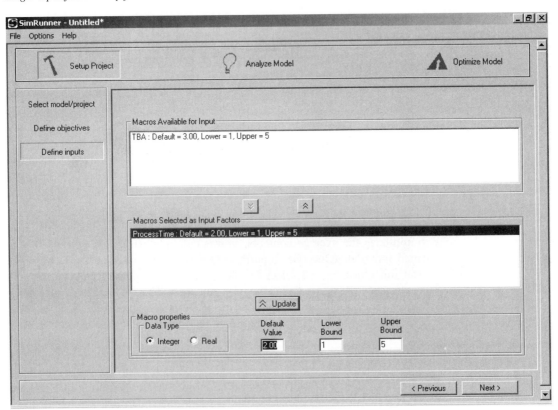

the lower and upper bounds appear without a decimal point in the Macros properties section. When complete, SimRunner should look like Figure L11.9.

From here, you click the Next> button until you enter the Optimize Model module, or click on the Optimize Model module button near the top right corner of the window to go directly to it. The first step here is to specify Optimization options (Figure L11.10). Select the Aggressive Optimization Profile. Accept the default value of 0.01 for Convergence Percentage, the default of one for Min Generations, and the default of 99999 for Max Generations.

The convergence percentage, minimum number of generations, and maximum number of generations control how long SimRunner's optimization algorithms will run experiments before stopping. With each experiment, SimRunner records the objective function's value for a solution in the population. The evaluation of all solutions in the population marks the completion of a generation. The maximum number of generations specifies the most generations SimRunner will

FIGURE L11.10

Optimization and simulation options.

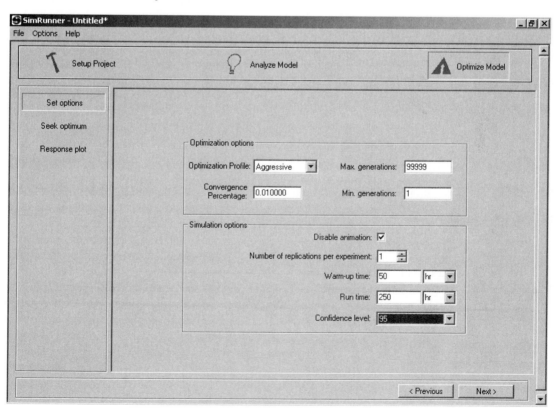

use to conduct its search for the optimal solution. The minimum number of generations specifies the fewest generations SimRunner will use to conduct its search for the optimal solution. At the end of a generation, SimRunner computes the population's average objective function value and compares it with the population's best (highest) objective function value. When the best and the average are at or near the same value at the end of a generation, all the solutions in the population are beginning to look alike (their input factors are converging to the same setting). It is difficult for the algorithms to locate a better solution to the problem once the population of solutions has converged. Therefore, the optimization algorithm's search is usually terminated at this point.

The convergence percentage controls how close the best and the average must be to each other before the optimization stops. A convergence percentage near zero means that the average and the best must be nearly equal before the optimization stops. A high percentage value will stop the search early, while a very small percentage value will run the optimization longer. High values for the maximum number of generations allow SimRunner to run until it satisfies the convergence percentage. If you want to force SimRunner to continue searching after the convergence percentage is satisfied, specify very high values for both the minimum number of generations and maximum number of generations. Generally, the best approach is to accept the default values shown in Figure L11.10 for the convergence percentage, maximum generations, and minimum generations.

After you specify the optimization options, set the simulation options. Typically you will want to disable the animation as shown in Figure L11.10 to make the simulation run faster. Usually you will want to run more than one replication to estimate the expected value of the objective function for a solution in the population. When more than one replication is specified, SimRunner will display the objective function's confidence interval for each solution it evaluates. Note that the confidence level for the confidence interval is specified here. Confidence intervals can help you to make better decisions at the end of an optimization as discussed in Section 11.6.2 of Chapter 11. In this case, however, use one replication to speed things along so that you can continue learning other features of the SimRunner software. As an exercise, you should revisit the problem and determine an acceptable number of replications to run per experiment. As indicated in Figure L11.10, set the simulation warm-up time to 50 hours and the simulation run time to 250 hours. You are now ready to have SimRunner seek the optimal solution to the problem.

With these operations completed, click the Next> button (Figure L11.10) and then click the Run button on the Optimize Model module (Figure L11.11) to start the optimization. For this scenario, SimRunner runs all possible experiments, locating the optimum processing time of one minute on its third experiment. The Experimental Results table shown in Figure L11.11 records the history of SimRunner's search. The first solution SimRunner evaluated called for a mean processing time at the milling machine of three minutes. The second solution evaluated assigned a processing time of two minutes. These sequence numbers are recorded in the

Figure L11.11

Experimental results table for scenario one.

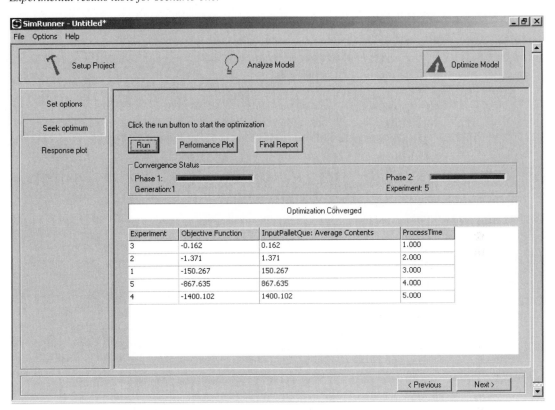

Experiment column, and the values for the processing time (ProcessTime) input factor are recorded in the ProcessTime column. The value of the term used to define the objective function (minimize the mean number of plates waiting in the input pallet queue) is recorded in the InputPalletQue:AverageContents column. This value is taken from the output report generated at the end of a simulation. Therefore, for the third experiment, we can see that setting the ProcessTime macro equal to one results in an average of 0.162 plates waiting in the input pallet queue. If you were to conduct this experiment manually with ProModel, you would set the ProcessTime macro to one, run the simulation, display output results at the end of the run, and read the average contents for the InputPalletQue location from the report. You may want to verify this as an exercise.

Because the objective function was to minimize the mean number of plates waiting in the input pallet queue, the same values from the InputPalletQue:Average-Contents column also appear in the Objective Function column. However, notice that the values in the Objective Function column are preceded by a negative sign

FIGURE L11.12

SimRunner's process for converting minimization problems to maximization problems.

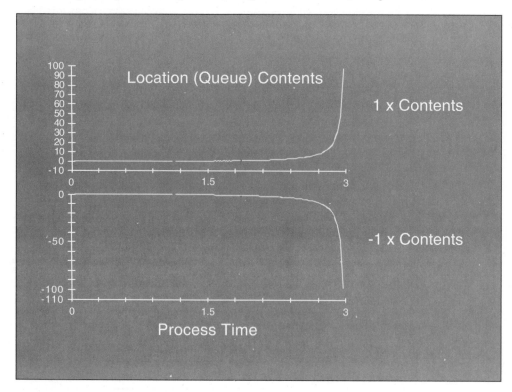

(Figure L11.11). This has to do with the way SimRunner treats a minimization objective. SimRunner's optimization algorithms view all problems as maximization problems. Therefore, if we want to minimize a term called Contents in an objective function, SimRunner multiplies the term by a negative one $\{(-1)Contents\}$. Thus SimRunner seeks the minimal value by seeking the maximum negative value. Figure L11.12 illustrates this for the ideal production system's response surface.

Figure L11.13 illustrates SimRunner's Performance Measures Plot for this optimization project. The darker colored line (which appears red on the computer screen) at the top of the Performance Measures Plot represents the best value of the objective function found by SimRunner as it seeks the optimum. The lighter colored line (which appears green on the computer screen) represents the value of the objective function for all of the solutions that SimRunner tried.

The last menu item of the Optimize Model module is the Response Plot (Figure L11.11), which is a plot of the model's output response surface based on the solutions evaluated during the search. We will skip this feature for now and cover it at the end of the lab chapter.

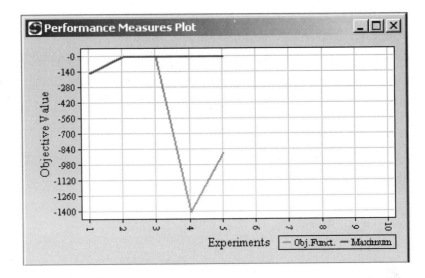

L11.2.2 Multiterm Objective Functions

Objective functions may be composed of any number of terms taken from the output of a simulation model. This application scenario demonstrates the construction of an objective function with two terms.

The managers of the ideal production system presented in Section L11.2.1 have a not-so-ideal objective for the system. They have conflicting objectives of (1) maximizing the number of gears produced and (2) minimizing the amount of space allocated for storing work-in-process at the input pallet queue area. Because space has to be available for the maximum number of plates that could wait in the queue, the second objective can be restated as minimizing the maximum number of plates waiting at the input pallet queue. In attempting to satisfy the managers' objectives, both the mean processing time at the milling machine (ProcessTime) and the mean time between arrivals (TBA) of plates to the input pallet queue can be varied. Each of these input factors can be assigned integer values between one and five minutes. To allow SimRunner to change the mean time between arrivals (TBA) in the simulation model, enter E(TBA) in the Frequency column of the model's Arrivals table (Figure L11.4). Remember that you previously defined TBA as a macro.

The SimRunner objective function for this scenario is shown in Figure L11.14. The first segment of the objective function can be implemented using the output response statistic that records the total number of gear entities that exit the system (Gear:TotalExits), which is an Entity response category. The second segment is implemented using the output response statistic that records the maximum number of plate entities that occupied the input pallet queue location during the simulation (InputPalletQue:MaximumContents), which is a Location response

FIGURE L11.14

Multiterm objective function for scenario two.

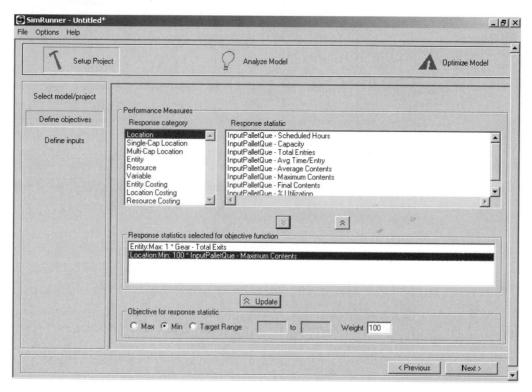

category. Management has indicated that a fairly high priority should be assigned to minimizing the space required for the input pallet queue area. Therefore, a weight of 100 is assigned to the second term in the objective function and a weight of one is assigned to the first term. Thus the objective function consists of the following two terms:

$$\text{Maximize } [(1)(\text{Gear:TotalExits})]$$

$$\text{and} \quad \text{Minimize } [(100)(\text{InputPalletQue:Maximum Contents})]$$

SimRunner minimizes terms by first multiplying each minimization term appearing in the objective function by a negative one, as explained in Section L11.2.1. Similarly, SimRunner multiplies each maximization term by a positive one. Next the terms are arranged into a linear combination as follows:

$$(+1)[(1)(\text{Gear:TotalExits})] + (-1)[(100)(\text{InputPalletQue:MaximumContents})]$$

which reduces to

$$[(1)(\text{Gear:TotalExits})] + [(-100)(\text{InputPalletQue:MaximumContents})]$$

Given that SimRunner's optimization algorithms view all problems as maximization problems, the objective function F becomes

$$F = \text{Maximize } \{[(1)(\text{Gear:TotalExits})]$$
$$+ [(-100)(\text{InputPalletQue:MaximumContents})]\}$$

The highest reward is given to solutions that produce the largest number of gears without allowing many plates to accumulate at the input pallet queue. In fact, a solution is penalized by 100 points for each unit increase in the maximum number of plates waiting in the input pallet queue. This is one way to handle competing objectives with SimRunner.

Develop a SimRunner project using this objective function to seek the optimal values for the input factors (macros) TBA and ProcessTime, which are integers between one and five (Figure L11.15). Use the aggressive optimization profile with the convergence percentage set to 0.01, max generations equal to 99999, and min generations equal to one. To save time, specify one replication

FIGURE L11.15

Multiple input factors setup for scenario two.

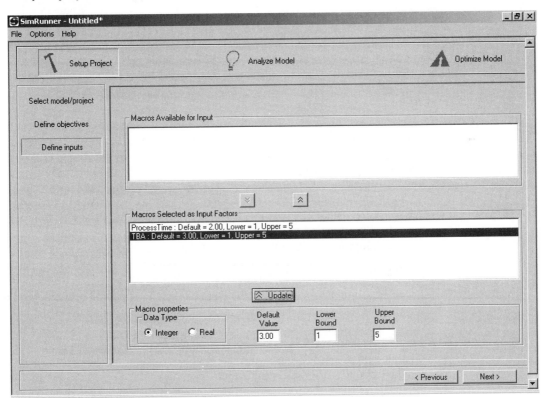

FIGURE L11.16

Experimental results table for scenario two.

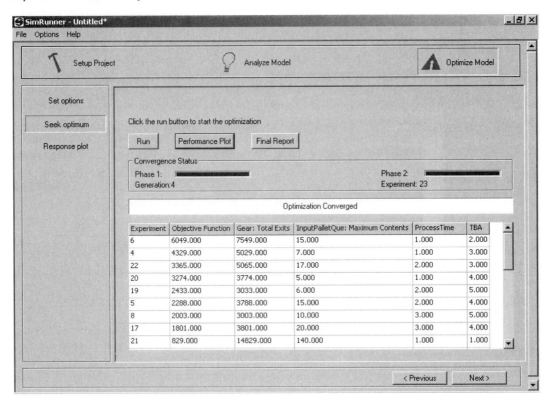

per experiment. (Remember, you will want to run multiple replications on real applications.) Also, set the simulation run hours to 250 and the warm-up hours to 50 for now.

At the conclusion of the optimization, SimRunner will have run four generations as it conducted 23 experiments (Figure L11.16). What values do you recommend to management for TBA and ProcessTime?

Explore how sensitive the solutions listed in the Experimental Results table for this project are to changes in the weight assigned to the maximum contents statistic. Change the weight of this second term in the objective function from 100 to 50. To do this, go back to the Define Objectives section of the Setup Project module and update the weight assigned to the InputPalletQue—Maximum-Contents response statistic from 100 to 50. Upon doing this, SimRunner warns you that the action will clear the optimization data that you just created. You can save the optimization project with the File Save option if you wish to keep the results from the original optimization. For now, do not worry about saving the data and click the Yes button below the warning message. Now rerun the

Figure L11.17

Experimental results table for scenario two with modified objective function.

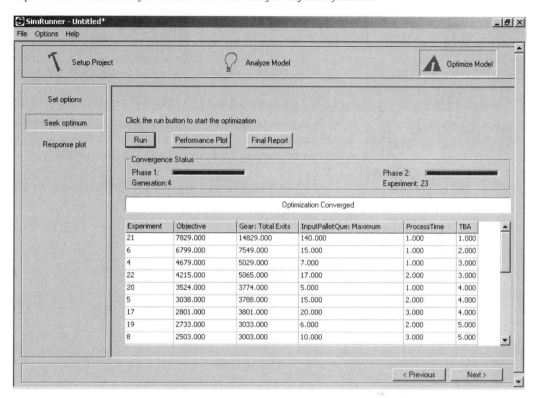

optimization and study the result (Figure L11.17). Notice that a different solution is reported as optimum for the new objective function. Running a set of preliminary experiments with SimRunner is a good way to help fine-tune the weights assigned to terms in an objective function. Additionally, you may decide to delete terms or add additional ones to better express your desires. Once the objective function takes its final form, rerun the optimization with the proper number of replications.

L11.2.3 Target Range Objective Functions

The target range objective function option directs SimRunner to seek a "target" value for the objective function term instead of a maximum or minimum value. For example, you may wish to find an arrangement for the ideal production system that produces from 100 to 125 gears per day. Like the maximization and minimization objective options, the target range objective option can be used alone or in combination with the maximization and minimization options.

For this application scenario, the managers of the ideal production system have specified that the mean time to process gears through the system should range between four and seven minutes. This time includes the time a plate waits in the input pallet queue plus the machining time at the mill. Recall that we built the model with a single entity type, named Gear, to represent both plates and gears. Therefore, the statistic of interest is the average time that the gear entity is in the system. Our task is to determine values for the input factors ProcessTime and TBA that satisfy management's objective.

The target range objective function is represented in SimRunner as shown in Figure L11.18. Develop a SimRunner project using this objective function to seek the optimal values for the input factors (macros) TBA and ProcessTime. Specify that the input factors are integers between one and five, and use the aggressive optimization profile with the convergence percentage set to 0.01, maximum generations equal to 99999, and minimum generations equal to one. To save time, set

FIGURE L11.18

Target range objective function setup for scenario three.

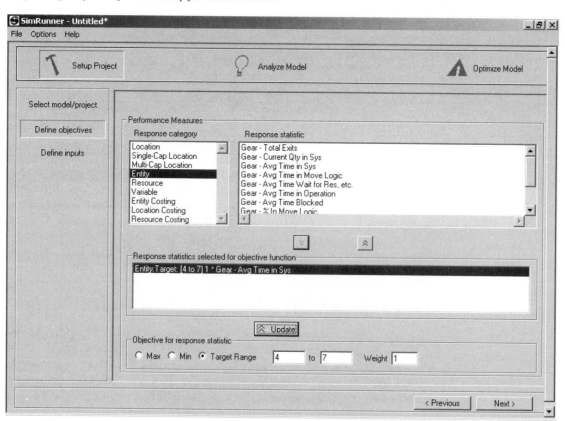

FIGURE 11.19

Experimental results table with Performance Measures Plot for scenario three.

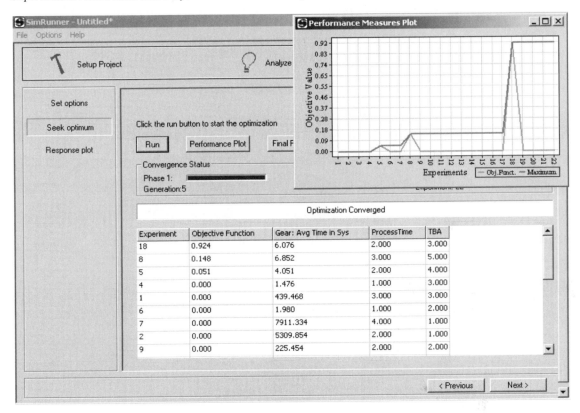

the number of replications per experiment to one. (Remember, you will want to run multiple replications on real applications.) Also, set the simulation run hours to 250 and the warm-up hours to 50 and run the optimization. Notice that only the solutions producing a mean time in the system of between four and seven minutes for the gear received a nonzero value for the objective function (Figure L11.19). What values do you recommend to management for TBA and ProcessTime?

Now plot the solutions SimRunner presented in the Experimental Results table by selecting the Response Plot button on the Optimize Model module (Figure L11.19). Select the independent variables as shown in Figure L11.20 and click the Update Chart button. The graph should appear similar to the one in Figure L11.20. The plot gives you an idea of the response surface for this objective function based on the solutions that were evaluated by SimRunner. Click the Edit Chart button to access the 3D graph controls to format the plot and to reposition it for different views of the response surface.

FIGURE L11.20

Surface response plot for scenario three.

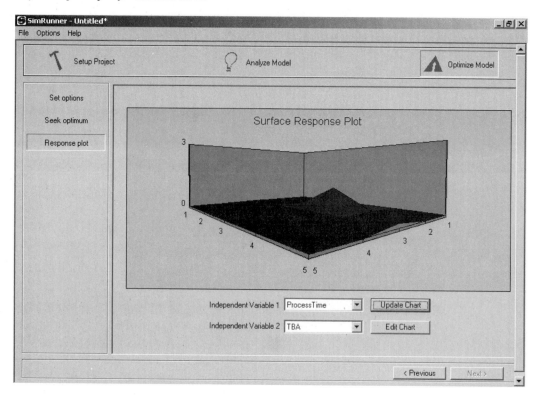

L11.3 Conclusions

Sometimes it is useful to conduct a preliminary optimization project using only one replication to help you set up the project. However, you should rarely, if ever, make decisions based on an optimization project that used only one replication per experiment. Therefore, you will generally conduct your final project using multiple replications. In fact, SimRunner displays a confidence interval about the objective function when experiments are replicated more than once. Confidence intervals indicate how accurate the estimate of the expected value of the objective function is and can help you make better decisions, as noted in Section 11.6.2 of Chapter 11.

Even though it is easy to use SimRunner, do not fall into the trap of letting SimRunner, or any other optimizer, become the decision maker. Study the top solutions found by SimRunner as you might study the performance records of different cars for a possible purchase. Kick their tires, look under their hoods, and drive them around the block before buying. Always remember that the optimizer is not the decision maker. SimRunner can only suggest a possible course of action. It is your responsibility to make the final decision.

L11.4 Exercises

Simulation Optimization Exercises

1. Rerun the optimization project presented in Section L11.2.1, setting the number of replications to five. How do the results differ from the original results?

2. Conduct an optimization project on the buffer allocation problem presented in Section 11.6 of Chapter 11. The model's file name is Lab 11_4 BufferOpt Ch11.Mod and is included on the CD accompanying the textbook. To get your results to appear as shown in Figure 11.5 of Chapter 11, enter Buffer3Cap as the first input factor, Buffer2Cap as the second input factor, and Buffer1Cap as the third input factor. For each input factor, the lower bound is one and the upper bound is nine. The objective is to maximize profit. Profit is computed in the model's termination logic by

$$\text{Profit} = (10*\text{Throughput})$$
$$- (1000*(\text{Buffer1Cap} + \text{Buffer2Cap} + \text{Buffer3Cap}))$$

Figure L11.21 is a printout of the model. See Section 11.6.2 of Chapter 11 for additional details. Use the Aggressive optimization profile and set the number of replications per experiment to 10. Specify a warm-up time of 240 hours, a run time of 720 hours, and a confidence level of 95 percent. Note that the student version of SimRunner will halt at 25 experiments, which will be before the search is completed. However, it will provide the data necessary for answering these questions:

   a. How do the results differ from those presented in Chapter 11 when only five replications were run per experiment?
   b. Are the half-widths of the confidence intervals narrower?
   c. Do you think that the better estimates obtained by using 10 replications will make it more likely that SimRunner will find the true optimal solution?

3. In Exercise 4 of Lab Section L10.5, you increased the amount of coal delivered to the railroad by the DumpOnMe facility by adding more dump trucks to the system. Your solution received high praise from everyone but the lead engineer at the facility. He is concerned about the maintenance needs for the scale because it is now consistently operated in excess of 90 percent. A breakdown of the scale will incur substantial repair costs and loss of profit due to reduced coal deliveries. He wants to know the number of trucks needed at the facility to achieve a target scale utilization of between 70 percent and 75 percent. This will allow time for proper preventive maintenance on the scale. Add a macro to the simulation model to control the number of dump trucks circulating in the system. Use the macro in the Arrivals Table to specify the number of dump trucks that are placed into the system at the start of each simulation. In SimRunner, select the macro as an input factor and assign

FIGURE L11.21

Buffer allocation model from Chapter 11 (Section 11.6.2).

```
Time Units: Minutes
Distance Units: Feet
Termination Logic: Profit=(10*Throughput)-(1000*(Buffer1Cap+Buffer2Cap+Buffer3Cap))
```

```
************************************************************************
*                            Locations                                *
************************************************************************
  Name         Cap         Units Stats      Rules      Cost
  ----------   ----------  ----- ---------- ---------- ------------
  Machine1     1           1     None       Oldest, ,
  Machine2     1           1     None       Oldest, ,
  Machine3     1           1     None       Oldest, ,
  Machine4     1           1     None       Oldest, ,
  Buffer1      Buffer1Cap  1     None       Oldest, ,
  Buffer2      Buffer2Cap  1     None       Oldest, ,
  Buffer3      Buffer3Cap  1     None       Oldest, ,
  Loc1         1           1     None       Oldest, ,
```

```
************************************************************************
*                            Entities                                 *
************************************************************************
  Name         Speed (fpm)  Stats        Cost
  ----------   ------------ ----------- ------------
  Part         150          None
```

```
************************************************************************
*                            Processing                               *
************************************************************************
                    Process                          Routing

  Entity   Location Operation          Blk  Output  Destination Rule     Move Logic
  -------- -------- ------------------  ---- ------- ----------- -------  -----------
  Part     Loc1                        1    Part    Machine1    FIRST 1  move for MoveTime
  Part     Machine1 Wait E(1.0)        1    Part    Buffer1     FIRST 1  move for MoveTime
                                       2*   Part    Loc1        FIRST 1  move for MoveTime
  Part     Buffer1                     1    Part    Machine2    FIRST 1  move for MoveTime
  Part     Machine2 Wait E(1.3)        1    Part    Buffer2     FIRST 1  move for MoveTime
  Part     Buffer2                     1    Part    Machine3    FIRST 1  move for MoveTime
  Part     Machine3 Wait E(0.70)       1    Part    Buffer3     FIRST 1  move for MoveTime
  Part     Buffer3                     1    Part    Machine4    FIRST 1  move for MoveTime
  Part     Machine4 Wait E(1.0)
                    If clock (HR) >= 240 Then Throughput = Throughput + 1
                                       1    Part    EXIT        FIRST 1
```

```
************************************************************************
*                            Arrivals                                 *
************************************************************************
  Entity   Location Qty Each  First Time Occurrences Frequency Logic
  -------- -------- ---------- ---------- ----------- ---------- ------------
  Part     Loc1     1          0          1
```

```
************************************************************************
*                            Attributes                               *
************************************************************************
  ID           Type         Classification
  ------------ ------------ ---------------
```

```
************************************************************************
*                            Variables (global)                       *
************************************************************************
  ID                Type         Initial value Stats
  ----------------- ------------ ------------- -----------
  MoveTime          Real         0             None
  Throughput        Integer      0             None
  Profit            Real         0             Basic
```

```
************************************************************************
*                            Macros                                   *
************************************************************************
  ID              Text
  --------------- ------------
  Buffer3Cap      2
  Buffer2Cap      2
  Buffer1Cap      2
```

it a lower bound of one and an upper bound of 15. Conduct an optimization project to seek the number of trucks that will achieve the target scale utilization using the Aggressive optimization profile, five replications per experiment, a warm-up time of 100 minutes, and a run time of 480 minutes.

SimRunner Warm-up Detection Exercises

SimRunner's Analyze Model module implements the Welch moving average technique presented in Section 9.6.1 of Chapter 9 to help you determine the end of a nonterminating simulation's warm-up phase before beginning an optimization project. It also helps you evaluate the number of replications needed to obtain a point estimate to within a specified percentage error and confidence level. Although the module was created to help you set up an optimization project, it is also useful for nonoptimization projects. To use the module without optimization, you declare a dummy macro in your simulation model and select it as an input factor in SimRunner. Then you select the output statistic that you wish to use in order to evaluate the end of the simulation's warm-up. The output statistic is entered as an objective function term. Exercises 4 and 5 here involve this feature of SimRunner.

4. This exercise will help you duplicate the SimRunner result presented in Figure L9.14 of Lab Chapter 9, which was used to estimate the end of the warm-up phase for the Green Machine Manufacturing Company (GMMC) simulation model. Load the GMMC model into ProModel, declare a dummy macro, and define its Run-Time Interface (RTI). Start SimRunner. The purpose of the GMMC model was to estimate the steady-state value of the time-average amount of work-in-process (WIP) inventory in the system. Therefore, select the WIP—Average Value response statistic from the Variable response category as a SimRunner maximization objective. Select the dummy macro as an input factor. Click the Next> button to move into the Analyze Model module, and fill in the parameters as shown in Figure L11.22.

   Percentage error in the objective function estimate is the desired amount of relative error in our average WIP inventory estimate expressed as a percentage (see Section 9.2.3 of Chapter 9 for additional details on relative error). In this case, we are seeking to approximate the number of replications needed to estimate the average WIP with a percentage error of 7 percent and a confidence level of 90 percent as we also estimate the end of the warm-up phase. Click the Next> button and then the Run button on the Conduct Analysis window to start the analysis. After SimRunner runs the simulation for five replications, your screen should appear similar to Figure L9.14. Here you adjust the number of periods for the moving average window to help you identify the end of simulation's warm-up, which seems to occur between periods 33 and 100. SimRunner computes that at least 10 replications are needed

FIGURE L11.22

SimRunner parameters for the GMMC warm-up example.

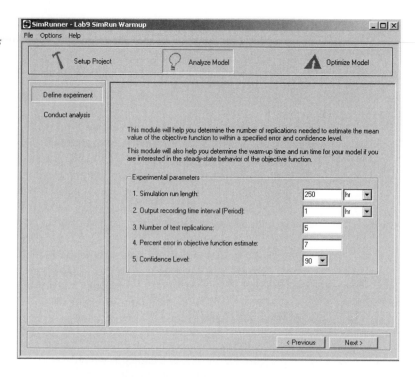

to estimate the average WIP inventory with a 7 percent error and a confidence level of 90 percent assuming a 100-period (hour) warm-up.

5. Use SimRunner's Analyze Model module to determine the end of the warm-up phase of the DumpOnMe simulation model with six dump trucks as presented in Exercise 11 of Lab Section L7.12. Base your assessment of the warm-up phase on five replications of the simulation with a run time of 300 minutes, and an output recording time interval of one minute.

12   INTERMEDIATE MODELING CONCEPTS

All truths are easy to understand once they are discovered; the point is to discover them.
—Galileo Galilei

In this lab we expand on the ProModel concepts discussed in Chapters 3 and 6. Section L12.1 introduces the concept of attributes; Section L12.2 shows its application in the calculation of cycle times; and Section L12.3 shows the process of sortation, sampling inspection, and rework. In Section L12.4 we show how to merge two different submodules. Section L12.5 discusses various aspects of machine breakdowns and maintenance; Section L12.6 shows how ProModel can conveniently model shift-working patterns; Section L12.7 shows the application of ProModel in a job shop; Section L12.8 introduces the modeling of priorities; and Section L12.9 has a couple of examples of pull system applications in manufacturing. Costs are modeled and tracked in Section L12.10. Section L12.11 shows how to import background graphics into a model, and Section L12.12 shows how to define and display various views of a model.

L12.1 Attributes

Attributes can be defined for entities or for locations. Attributes are placeholders similar to variables but are attached to specific entities or locations and usually contain information about that entity or location. Attributes are changed and assigned when an entity executes the line of logic that contains an operator, much like the way variables work. Some examples of attributes are part type, customer number, and time of arrival of an entity, as well as length, weight, volume, or some other characteristic of an entity.

To define an attribute use the attribute editor, as follows:

1. Go to the Build/More Elements/Attributes menu and create a name (ID) for the attribute.
2. Select the type of attribute—integer or real.
3. Select the class of attribute—entity or location.

L12.1.1 Using Attributes to Track Customer Types

Problem Statement

Customers visit the neighborhood barbershop **Fantastic Dan** for a haircut. Among the customers there are 20 percent children, 50 percent women, and 30 percent men. The customer interarrival time is triangularly distributed with a minimum, mode, and maximum of seven, eight, and nine minutes respectively. The haircut time (in minutes) depends on the type of customer and is given in Table L12.1. This time also includes the initial greetings and the transaction of money at the end of the haircut. Run the simulation model for one day (480 minutes).

a. About how many customers of each type does Dan process per day?

b. What is the average number of customers of each type waiting to get a haircut? What is the maximum?

c. What is the average time spent by a customer of each type in the salon? What is the maximum?

Two locations (Barber Dan and Waiting for Dan) and an entity (Customer) are defined. Customer_Type is defined as an attribute (type = integer and classification = entity) as shown in Figure L12.1. The process/routing and customer arrivals are defined as shown in Figures L12.2 and L12.3. A snapshot of the simulation model is shown in Figure L12.4.

FIGURE L12.1

Customer_Type declared as an attribute.

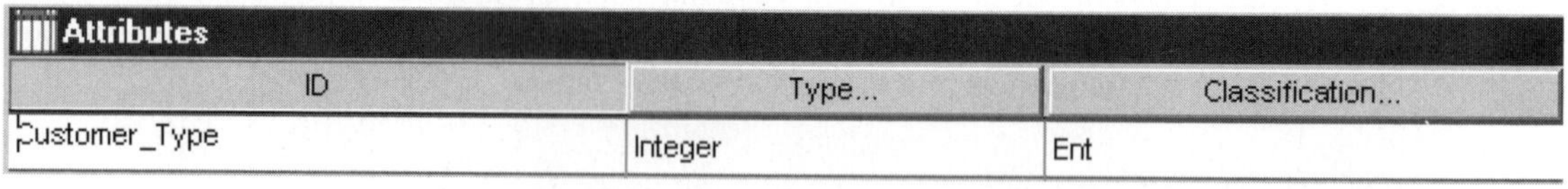

Attributes		
ID	Type...	Classification...
Customer_Type	Integer	Ent

TABLE L12.1   The Haircut Time for All Customers

	Haircut Time (minutes)	
Customers	*Mean*	*Half-Width*
Children	8	2
Women	12	3
Men	10	2

FIGURE L12.2

Process and routing tables for Fantastic Dan.

		Process				Routing		
Entity	Location	Operation	Blk	Output	Destination	Rule	Move Logic	
Customer	Waiting_for_Barber		1	Customer	Waiting_for_Barber	0.300000 1	customer_Type = 1 graphic 1	
				Customer	Waiting_for_Barber	0.500000	customer_Type = 2 graphic 2	
				Customer	Waiting_for_Barber	0.200000	customer_Type = 3 graphic 3	
Customer	Waiting_for_Barber		1	Customer	Barber_Dan	FIRST 1		
Customer	Barber_Dan	if customer_type = 1 then wait u(10,2) min if customer_type = 2 then wait u(12,3) min if customer_type = 3 then wait u(8,2) min						
			1	Customer	EXIT	FIRST 1		

FIGURE L12.3

Arrival of customers at Fantastic Dan.

```
*********************************************************************************
*                                  Arrivals                                     *
*********************************************************************************
```

Entity	Location	Qty each	First Time	Occurrences	Frequency	Logic
Customer	Waiting_for_Barber	1	0	inf	t(7,8,9) min	

FIGURE L12.4

Simulation model for Fantastic Dan.

L12.2 Cycle Time

The Clock and Log are functions built into ProModel to allow us to keep track of system events such as cycle time, lead time, or flow time within the system. The Clock function returns the current simulation clock time in hours, minutes, or seconds. The value returned is real.

The Log function is used to subtract an expression from the current simulation clock time and stores the result with a text string header.

```
Time_In = Clock()
Log "Cycle Time =", Time_In
```

For the example in Section L12.1, find

a. The cycle time for each type of customer of Barber Dan.

b. The average cycle time for all customers.

Define Time_In as an attribute to track the time of arrival for all customer types. Figure L12.5 shows the setting of the attribute Time_In and also the logging of the average cycle time. The cycle times for children, women, and men are reported in Figure L12.6.

FIGURE L12.5

Setting the attribute Time_In and logging the cycle time.

		Process			Routing	
Entity	Location	Operation	Blk	Output	Destination	Rule
Customer	Waiting_for_Barber	Time_In=Clock()				
			1	Child	Waiting_for_Barber	0.200000 1
				Woman	Waiting_for_Barber	0.500000
				Man	Waiting_for_Barber	0.300000
Child	Waiting_for_Barber		1	Child	Barber_Dan	FIRST 1
Child	Barber_Dan	wait u(8,2) min				
		Log "Child Cycle Time=",time_in				
			1	Child	EXIT	FIRST 1
Woman	Waiting_for_Barber		1	Woman	Barber_Dan	FIRST 1
Woman	Barber_Dan	wait u(12,3) min				
		Log "Woman Cycle Time=",time_in				
			1	Woman	EXIT	FIRST 1
Man	Waiting_for_Barber		1	Man	Barber_Dan	FIRST 1
Man	Barber_Dan	wait u(10,2) min				
		Log "Man Cycle Time=",time_in				
			1	Man	EXIT	FIRST 1

FIGURE L12.6

The minimum, maximum, and average cycle times for various customers.

Report for lab_I_11_2 - Normal Run

| Entity States | Variables | Location Costing | Resource Costing | Entity Costing | Logs |

Logs for lab_I_11_2, Normal Run				
Name	Number Observations	Minimum Value	Maximum Value	Avg Value
Child Cycle Ti...	8	27	124	74
Man Cycle Ti...	14	19	115	62
Woman Cycle...	23	12	128	69

L12.3 Sorting, Inspecting a Sample, and Rework

Problem Statement

Orders for two types of widgets (widget A and widget B) are received by **Widgets-R-Us Manufacturing Inc.** Widget A orders arrive on average every five minutes (exponentially distributed), while widget B orders arrive on average every ten minutes (exponentially distributed). Both widgets arrive at the input queue. An attribute Part_Type is defined to differentiate between the two types of widgets.

Widget A goes to the lathe for turning operations that take Normal(5,1) minutes. Widget B goes on to the mill for processing that takes Uniform(6,2) minutes. Both widgets go on to an inspection queue, where every fifth part is inspected. Inspection takes Normal(6,2) minutes. After inspection, 70 percent of the widgets pass and leave the system; 30 percent of the widgets fail and are sent back to the input queue for rework. Determine the following:

a. How many widgets of each type are shipped each week (40-hour week)?

b. What is the cycle time for each type of widget?

c. What are the maximum and minimum cycle times?

d. What is the number of widgets reworked each week?

e. What is the average number of widgets waiting in the inspection queue?

Five locations (Mill, Input_Queue, Lathe, Inspect, and Inspect_Q) are defined for this model. Three variables are defined, as in Figure L12.7. Figure L12.8 shows how we keep track of the machined quantity as well as the probabilistic routings

FIGURE L12.7

Variables for Widgets-R-Us.

Icon	ID	Type...	Initial value	Stats...
No	qty	Integer	0	Time Series,
Yes	machined_qty	Integer	0	Time Series,
Yes	inspect_qty	Integer	0	Time Series,

FIGURE L12.8

Keeping track of machined_qty and probabilistic routings at the Inspect location.

```
                        Process                      Routing

Entity     Location    Operation        Blk  Output    Destination  Rule        Move Logic
-------    --------    ---------        ---  ------    -----------  ----        ----------
Widget_A   Input_Queue wait 0           1    Widget_A  Lathe        FIRST 1     move for 1
Widget_A   Lathe       Wait N(5,1) min  1    Widget_A  Inspect_Q    FIRST 1     move for 1
Widget_B   Input_Queue wait 0           1    Widget_B  Mill         FIRST 1     move for 1
Widget_B   Mill        wait U(6,2) min  1    Widget_B  Inspect_Q    FIRST 1     move for 1
ALL        Inspect_Q   INC machined_qty
                       INC QTY          1    ALL       EXIT         IF qty < 5, 1  move for 1
                                             ALL       Inspect      IF qty = 5     INC inspect_qty
                                                                                   qty = 0
ALL        Inspect     Wait N(6,2) min  1    ALL       Input_Queue  0.300000 1  move for 1
                                             ALL       EXIT         0.700000
```

FIGURE L12.9

Simulation model for Widgets-R-Us.

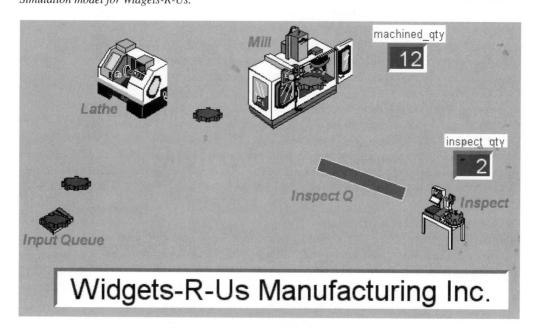

after inspection. Figure L12.9 shows the complete simulation model with counters added for keeping track of the number of widgets reworked and the number of widgets shipped.

L12.4 Merging a Submodel

Sometimes a large model is built in smaller segments. A model segment can be a manufacturing cell or a department. Different analysts can build each segment. After all segments are complete, they can be merged together to form a single unified model.

The Merge Model option in the File menu allows two or more independent (complete or incomplete) models to be merged together into a single model. Entity and attribute names common to both models are considered common elements in the merged model. Duplicate locations, resources, or path networks must first be renamed or deleted from the original merging model. If the graphic libraries are different, the user has an option to append the merging model's graphic library to the base model's graphic library. All other duplicate elements cause a prompt to appear with a choice to delete the duplicate element.

Problem Statement

Poly Casting Inc. (Lab 7, Section L7.4) decides to merge with **El Segundo Composites** (Lab 7, Section L7.6.1). The new company is named **El Segundo**

Castings N' Composites. Merge the model for Section L7.4 with the model for Section L7.6.1. All the finished products—castings as well as composites—are now sent to the shipping queue and shipping clerk. The model in Section L7.4 is shown in Figure L12.10. The complete simulation model, after the model for Section L7.6.1 is merged, is shown in Figure L12.11. After merging, make the necessary modifications in the process and routing tables.

We will make suitable modifications in the Processing module to reflect these changes (Figure L12.12). Also, the two original variables in Section L7.4 (WIP and PROD_QTY) are deleted and four variables are added: WIPCasting, WIP-Composite, PROD_QTY_Casting, and PROD_QTY_Composite.

FIGURE L12.10

The layout of the simulation model for Section L7.4.

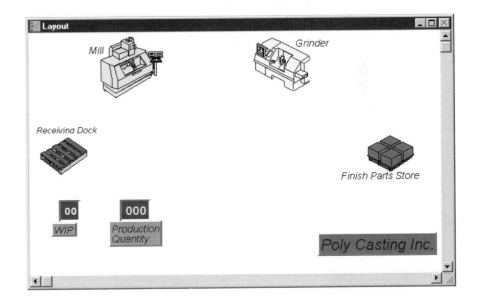

FIGURE L12.11

Merging the models from Section L7.4 and Section L7.6.1.

FIGURE L12.12

Changes made to the process table after merging.

Entity	Location	Operation	Blk	Output	Destination	Rule	Move Logic
Casting	Receiving_Dock	WIPCasting = WIPCasting +1					
			1	Casting	Mill	FIRST 1	MOVE FOR 1 MIN
Casting	Mill	Wait N(3,1) min	1	Casting	Grinder	FIRST 1	MOVE FOR 1 MIN
Casting	Grinder	Wait U(5,1) min	1	Casting	Finish_Parts_Store	FIRST 1	MOVE FOR 1 MIN
Casting	Finish_Parts_Store		1	Casting	Ship_Q	FIRST 1	move for 1
Casting	Ship_Q		1	Casting	Ship_Clerk	FIRST 1	
Casting	Ship_Clerk	wait N(40,5) min					
		WIPCasting = WIPCasting -1					
		PROD_QTY_Casting = PROD_QTY_Casting + 1					
			1	Casting	EXIT	FIRST 1	
Composite	Order_Q	Inc WIPComposite	1	Composite	Ply_Cutting	FIRST 1	Move for 15 min
Composite	Ply_Cutting	Wait U(20,5) min	1	Composite	LayUp	FIRST 1	Move for 15 min
Composite	LayUp	Wait U(30,10) min					
			1	Composite	Oven	FIRST 1	Move for 15 min
Composite	Oven	Group 5 as Batch					
Batch	Oven	Wait U(100,10) min					
		UNGROUP					
Composite	Oven		1	Composite	Ship_Q	FIRST 1	Move for 15 min
Composite	Ship_Q		1	Composite	Ship_Clerk	FIRST 1	Move for 15 min
Composite	Ship_Clerk	Wait N(20,5) min					
		Dec WIPComposite					
		PROD_QTY_Composite=PROD_QTY_Composite+1					
			1	Composite	EXIT	FIRST 1	

L12.5   Preventive Maintenance and Machine Breakdowns

Downtime stops a location or resource from operating. Downtime can occur in one of two ways: preventive maintenance or breakdown. A down resource (or location) no longer functions and is not available for use. Downtimes may represent scheduled interruptions such as shifts, breaks, or scheduled maintenance. They may also represent unscheduled and random interruptions such as equipment failures.

The mean time between failures (MTBF) can be calculated as the reciprocal of the failure rate (distribution of failure over time). Often MTBF will follow a negative exponential distribution. In particular, the probability of failure before time T is given by $1 - e^{-T/\text{MTBF}}$. The reliability is viewed as performance over time and is the probability that a given location or resource will perform its intended function for a specified length of time T under normal conditions of use:

$$\text{Reliability} = P(\text{no failure before time } T) = e^{-T/\text{MTBF}}$$

The mean time to repair (MTTR) refers to the time the location or resource remains down for repair. The MTTR depends on the ease and/or cost with which a location or resource can be maintained or repaired.

For single-capacity locations, downtime of a location or resource can be scheduled at regular intervals based on the clock time that has expired, the number of entities processed at a location, the usage of the machine (resource) in time, or a change in entity type. Unscheduled breakdown of a machine can also be modeled in a similar fashion.

To model preventive maintenance or breakdowns, use the following procedure:

1. Go to the Build/Locations menu.
2. Click on the DT button.

3. Enter the frequency of downtime:
   a. Clock based.
   b. Number of entries based.
   c. Usage based.

4. Enter the first time downtime occurs.

5. Enter the priority of the downtime.

6. In the logic field, enter any logic associated with downtime.

Example:

```
DISPLAY "The Lathe is down for preventive maintenance"
WAIT N(20,5) min
```

7. To disable the downtime feature, click Yes in the disable field. Otherwise, leave this field as No.

L12.5.1 Downtime Using MTBF and MTTR Data

The lathe and the mill at **Widgets-R-Us Manufacturing Inc.** (Section L12.3) have the maintenance schedule shown in Table L12.2. The logic for the preventive maintenance operation is created in Figure L12.13. The processes and

FIGURE L12.13

The logic for preventive maintenance at Widgets-R-Us Manufacturing Inc.

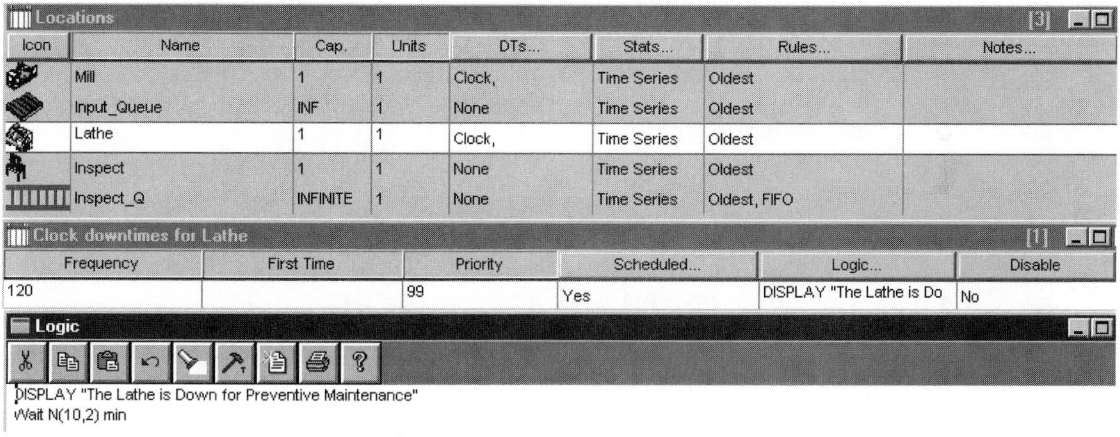

TABLE L12.2   Maintenance Schedule for Machines at Widgets-R-Us

Machine	Time between Repairs	Time to Repair
Lathe	120 minutes	N(10,2) minutes
Mill	200 minutes	T(10,15,20) minutes

FIGURE L12.14

Processes and routings at Widgets-R-Us Manufacturing Inc.

Entity	Location	Operation	Blk	Output	Destination	Rule	Move Logic
		Process			**Routing**		
Widget_A	Input_Queue	wait 0	1	Widget_A	Lathe	FIRST 1	move for 1
Widget_A	Lathe	Wait N(5,1) min	1	Widget_A	Inspect_Q	FIRST 1	move for 1
Widget_B	Input_Queue	wait 0	1	Widget_B	Mill	FIRST 1	move for 1
Widget_B	Mill	wait U(6,2) min	1	Widget_B	Inspect_Q	FIRST 1	move for 1
ALL	Inspect_Q	INC machined_qty					
		INC QTY	1	ALL	EXIT	IF qty < 5, 1	
				ALL	Inspect	IF qty = 5	INC inspect_qty
							qty = 0
ALL	Inspect	Wait N(6,2) min	1	ALL	Input_Queue	0.300000 1	move for 1
				ALL	EXIT	0.700000	

FIGURE L12.15

Complete simulation model for Widgets-R-Us Manufacturing Inc.

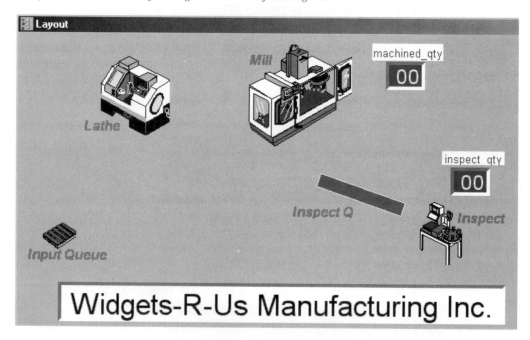

routings are shown in Figure L12.14. Figure L12.15 shows the complete simulation model.

L12.5.2 Downtime Using MTTF and MTTR Data

Reliability of equipment is frequently calibrated in the field using the mean time to failure (MTTF) data instead of the MTBF data. MTTR is the expected time between the end of repair of a machine and the time the machine fails. Data collected in the field are often based on TTF for clock-based failures.

Problem Statement

The turning center in this machine shop (Figure L12.16) has a time to failure (TTF) distribution that is exponential with a mean of 10 minutes. The repair time (TTR) is also distributed exponentially with a mean of 10 minutes.

This model shows how to get ProModel to implement downtimes that use time to failure (TTF) rather than time between failures (TBF). In practice, you most likely will want to use TTF because that is how data will likely be available to you, assuming you have unexpected failures. If you have regularly scheduled downtimes, it may make more sense to use TBF. In this example, the theoretical percentage of uptime is MTTF/(MTTF + MTTR), where M indicates a mean value. The first time to failure and time to repair are set in the variable initialization section (Figure L12.17). Others are set in the downtime logic (Figure L12.18).

The processing and routing tables are shown in Figure L12.19. Run this model for about 1000 hours, then view the batch mean statistics for downtime by picking "averaged" for the period (Figure L12.20) when the output analyzer (classical) comes up. The batch mean statistics for downtime for the turning center are shown in Figure L12.21. (This problem was contributed by Dr. Stephen Chick, University of Michigan, Ann Arbor.)

FIGURE L12.16

Layout of the machine shop—modeling breakdown with TTF and TTR.

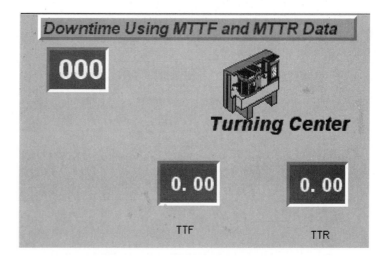

FIGURE L12.17

Variable initializations.

```
************************************************************
*                                       Variables (global)
************************************************************

     ID            Type           Initial value  Stats
     ---------     ------------   -------------  -----------
     ttf           Real           e(MTTF)        Time Series
     ttr           Real           e(MTTR)        Time Series
```

FIGURE L12.18

Clock downtime logic.

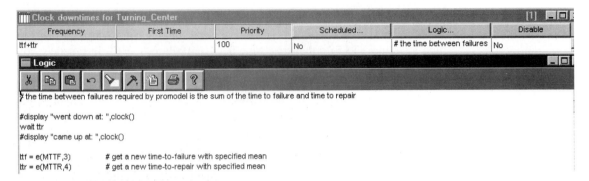

FIGURE L12.19

Process and routing tables.

		Process			Routing		
Entity	Location	Operation	Blk	Output	Destination	Rule	Move Logic
Machinist	Loc1		1	Machinist	Turning_Center	FIRST 1	
Machinist	Turning_Center	wait e(.5,5)	1	Machinist	EXIT	FIRST 1	

FIGURE L12.20

Average of all the batches in the Classical ProModel Output Viewer.

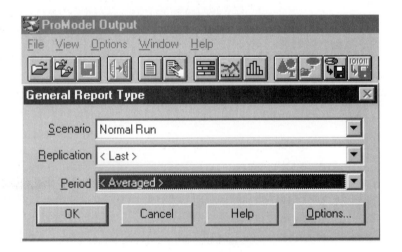

FIGURE L12.21

Batch mean statistics for downtime.

BATCH MEAN ANALYSIS (Sample size 50)

Statistic	Avg	Median	Min	Max	Std Dev	Low 90% CI	High 90% CI
Turning_Center - % Down	49.18	49.59	38.41	55.88	4.56	48.08	50.28

L12.6 Operator Shifts

ProModel offers an excellent interface to include the operator (location or resource) shift working schedules in the simulation model. The shift work for each operator in the model can be defined separately. Work cycles that repeat either daily or weekly can be defined.

Problem Statement

Orders for two types of widgets (widget A and widget B) are received by **Widgets-R-Us Manufacturing Inc.** Widget A orders arrive on average every 5 minutes (exponentially distributed), while widget B orders arrive on average every 10 minutes (exponentially distributed). Both widgets arrive at the input queue. An attribute Part_Type is defined to differentiate between the two types of widgets.

Widget A goes on to the lathe for turning operations that take Normal(5,1) minutes. Widget B goes on to the mill for processing that takes Uniform(4,8) minutes. Both widgets go on to an inspection queue, where every fifth part is inspected. Inspection takes Normal(6,2) minutes. After inspection, 70 percent of the widgets pass and leave the system, while 30 percent of the widgets fail and are sent back to the input queue for rework.

An operator is used to process the parts at both the lathe and the mill. The operator is also used to inspect the part. The operator moves the parts from the input queue to the machines as well as to the inspection station. The operator is on a shift from 8 A.M. until 5 P.M. with breaks as shown in Table L12.3 and Figures L12.22 and L12.23.

Use the DISPLAY statement to notify the user when the operator is on a break. Set up a break area (Figure L12.24) for the operator by extending the path network to a break room and indicating the node as the break node in Resource specs. The processes and the routings at Widgets-R-Us are shown in Figure L12.25. Determine the following:

a. How many widgets of each type are shipped each week (40-hour week).

b. The cycle time for each type of widgets.

c. The maximum and minimum cycle times.

TABLE L12.3 **The Operator Break Schedule at Widgets-R-Us Manufacturing Inc.**

Breaks	*From*	*To*
Coffee break	10 A.M.	10:15 A.M.
Lunch break	12 noon	12:45 P.M.
Coffee break	3 P.M.	3:15 P.M.

FIGURE L12.22

The operator Joe's weekly work and break times at Widgets-R-Us.

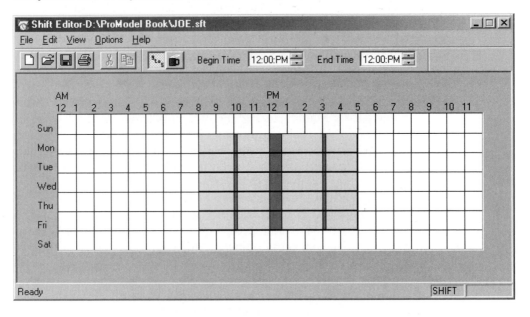

FIGURE L12.23

Assigning the Shift File to the operator Joe.

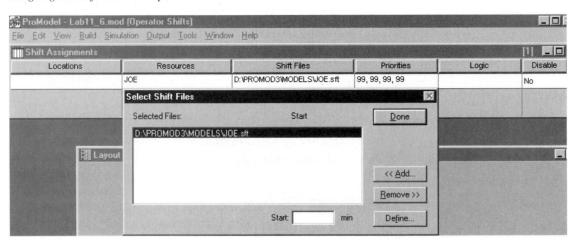

    d. The number of widgets reworked each week.

    e. The average number of widgets waiting in the inspection queue.

       Run the model with and without shift breaks (Figure L12.26). What difference do you notice in these statistics? How can you improve the system?

FIGURE L12.24

The layout and path network at Widgets-R-Us.

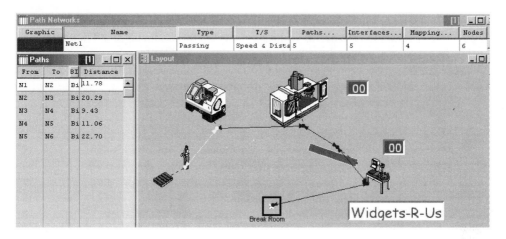

FIGURE L12.25

Processes and routings at Widgets-R-Us.

```
                        Process                        Routing

Entity    Location    Operation        Blk  Output    Destination Rule        Move Logic
-------   --------    ---------        ---  ------    ----------- ----        ----------
Widget_A  Input_Queue wait 0           1    Widget_A  Lathe       FIRST 1     move  WITH JOE THEN FREE
Widget_A  Lathe       GET JOE
                      Wait N(5,1)
                      FREE JOE         1    Widget_A  iNSPECT_Q   FIRST 1     Move  WITH JOE THEN FREE
Widget_B  Input_Queue wait 0           1    Widget_B  Mill        FIRST 1     Move  WITH JOE THEN FREE
Widget_B  Mill        GET JOE
                      wait U(4,8)
                      FREE JOE         1    Widget_B  iNSPECT_Q   FIRST 1     Move  WITH JOE THEN FREE
ALL       iNSPECT_Q   inc machined_qty
                      INC QTY          1    ALL       EXIT        IF qty < 5, 1
                                            ALL       Inspect     IF qty = 5  inc inspect_qty
                                                                              qty = 0
ALL       Inspect     GET JOE
                      Wait N(6,2)
                      FREE JOE         1    ALL       Input_Queue 0.300000 1  Move  WITH JOE THEN FREE
                                            ALL       EXIT        0.700000
```

FIGURE L12.26

A snapshot during the simulation model run for Widgets-R-Us.

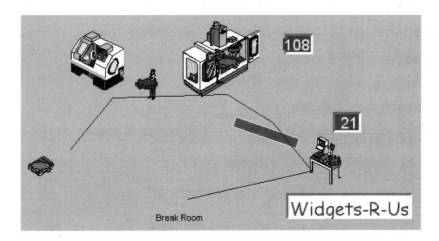

L12.7 Job Shop

A job shop is a collection of processing centers through which jobs are routed in different sequences depending on the characteristics of the particular job. It is a common method of grouping resources for producing small lots with widely varying processing requirements.

Each job in a job shop usually has a unique routing sequence. Job shops usually have a process layout. In other words, similar processes are geographically grouped together.

Problem Statement

In **Joe's Jobshop** there are three machines through which three types of jobs are routed. All jobs go to all machines, but with different routings. The data for job routings and processing times (minutes) are given in Table L12.4. The process times are exponentially distributed with the given average values. Jobs arrive at the rate of Exponential(30) minutes. Simulate for 10 days (80 hours). How many jobs of each type are processed in 10 days?

Figures L12.27, L12.28, L12.29, L12.30, and L12.31 show the locations, entities, variables, processes, and layout of Joe's Jobshop.

FIGURE L12.27

Locations at Joe's Jobshop.

Icon	Name	Cap.	Units	DTs...	Stats...	Rules...
	Machining_Center1	1	1	None	Time Series	Oldest
	Machining_Center2	1	1	None	Time Series	Oldest
	Machining_Center3	1	1	None	Time Series	Oldest
	Incoming	INFINITE	1	None	Time Series	Oldest, FIFO
	MC1_Q	INFINITE	1	None	Time Series	Oldest, FIFO
	MC2_Q	INFINITE	1	None	Time Series	Oldest, FIFO
	MC3_Q	INFINITE	1	None	Time Series	Oldest, FIFO

TABLE L12.4   Summary of Process Plan for Joe's Jobshop

Jobs	Machines	Process Times (minutes)	Job Mix
A	2–3–1	45–60–75	25%
B	1–2–3	70–70–50	35%
C	3–1–2	50–60–60	40%

FIGURE L12.28

Entities at Joe's Jobshop.

```
*************************************************************
                                           Entities
*************************************************************

Name            Speed (fpm)   Stats          Cost
----------      -----------   ----------     -----------
Job             150           Time Series
Job_A           150           Time Series
Job_B           150           Time Series
Job_C           150           Time Series
```

FIGURE L12.29

Variables to track jobs processed at Joe's Jobshop.

```
*************************************************************
                                      Variables (global)
*************************************************************

ID              Type          Initial value  Stats
----------      -----------   -------------- ----------
qty_A           Integer       0              Time Series
qty_B           Integer       0              Time Series
qty_C           Integer       0              Time Series
```

FIGURE L12.30

Processes and routings at Joe's Jobshop.

```
                              Process                                      Routing
Entity   Location          Operation              Blk  Output    Destination          Rule
------   --------------    -----------------      ---  --------  -----------------    ----------
Job      Incoming                                 1    Job_A     MC2_Q                0.250000 1
                                                       Job_B     MC1_Q                0.350000
                                                       Job_C     MC3_Q                0.400000
Job_A    MC2_Q                                     1    Job_A     Machining_Center2    FIRST 1
Job_A    Machining_Center2 wait e(45) min         1    Job_A     MC3_Q                FIRST 1
Job_A    MC3_Q                                     1    Job_A     Machining_Center3    FIRST 1
Job_A    Machining_Center3 wait e(60) min         1    Job_A     MC1_Q                FIRST 1
Job_A    MC1_Q                                     1    Job_A     Machining_Center1    FIRST 1
Job_A    Machining_Center1 wait e(75) min
                           qty_A = qty_A + 1
                                                  1    Job_A     EXIT                 FIRST 1
Job_B    MC1_Q                                     1    Job_B     Machining_Center1    FIRST 1
Job_B    Machining_Center1 wait e(70) min         1    Job_B     MC2_Q                FIRST 1
Job_B    MC2_Q                                     1    Job_B     Machining_Center2    FIRST 1
Job_B    Machining_Center2 wait e(70) min         1    Job_B     MC3_Q                FIRST 1
Job_B    MC3_Q                                     1    Job_B     Machining_Center3    FIRST 1
Job_B    Machining_Center3 wait e(50) min
                           qty_B = qty_B + 1
                                                  1    Job_B     EXIT                 FIRST 1
Job_C    MC3_Q                                     1    Job_C     Machining_Center3    FIRST 1
Job_C    Machining_Center3 wait e(50) min         1    Job_C     MC1_Q                FIRST 1
Job_C    MC1_Q                                     1    Job_C     Machining_Center1    FIRST 1
Job_C    Machining_Center1 wait e(60) min         1    Job_C     MC2_Q                FIRST 1
Job_C    MC2_Q                                     1    Job_C     Machining_Center2    FIRST 1
Job_C    Machining_Center2 wait e(60) min
                           qty_C = qty_C + 1
                                                  1    Job_C     EXIT                 FIRST 1
```

Figure L12.31

Layout of Joe's Jobshop.

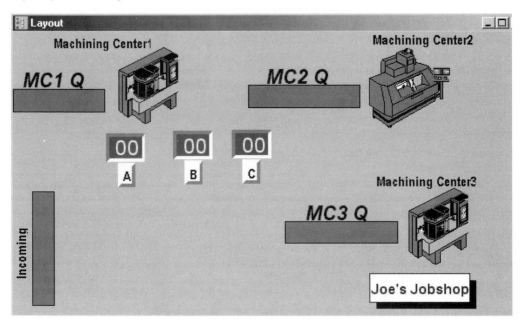

L12.8 Modeling Priorities

Priorities allow us to determine the order in which events occur in the simulation. The most common uses of priorities are

1. Selecting among upstream processes.
2. Selecting among downstream processes—already discussed in Lab 7, Section L7.3.
3. Selecting resources.
4. Prioritizing downtimes—discussed in Section L12.5.

L12.8.1 Selecting among Upstream Processes

This is the process of choosing among a set of upstream processes or queues for entity removal and routing to a downstream process. When one downstream destination exists and two or more upstream entities are competing to get there, priorities can be used. In the example in Figure L12.32 two entities at different locations (Process A and Process B) are both trying to get to Process C.

Problem Statement
At **Wang's Export Machine Shop** in the suburbs of Chicago, two types of jobs are processed: domestic and export. The rate of arrival of both types of jobs is

FIGURE L12.32

Choosing among upstream processes.

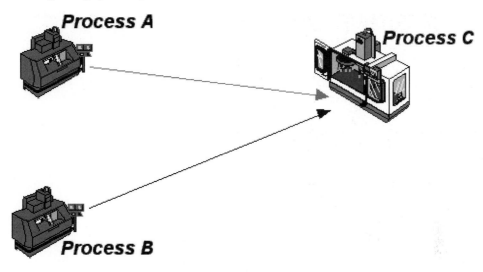

TABLE L12.5   Process Times at Wang's Export Machine Shop

Machine	Process Time
Machining center	Triangular(10,12,18) minutes
Lathe	Triangular(12,15,20) minutes

TABLE L12.6   Distances between Locations

From	To	Distance (feet)
Incoming	Machining center	200
Machining center	Lathe	400
Lathe	Outgoing	200
Outgoing	Incoming	400

Exponential(60) minutes. All jobs are processed through a machining center and a lathe. The processing times (Table L12.5) for all jobs are triangularly distributed. A forklift truck that travels at the rate of 50 feet/minute handles all the material. Export jobs are given priority over domestic jobs for shop release—that is, in moving from the input queue to the machining center. The distance between the stations is given in Table L12.6. Simulate for 16 hours.

FIGURE L12.33

Forklift resource specified for Wang's Export Machine Shop.

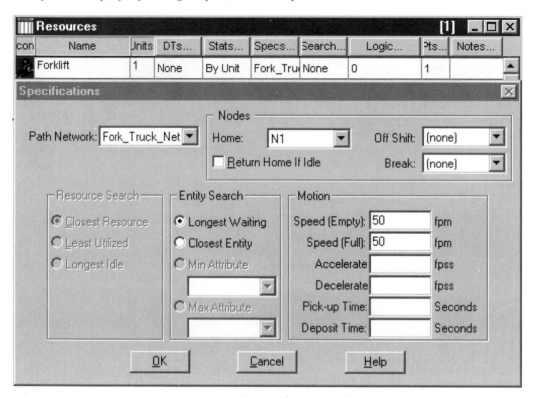

Five locations are defined: Lathe, Machining_Center, In_Q_Domestic, In_Q_Export, and Outgoing_Q. Two entities (domestic and export) are defined. Both these entities arrive with an interarrival time that is exponentially distributed with a mean of 60 minutes. The resource (forklift) and its path network are shown in Figures L12.33 and L12.34, respectively. The processes and routings are shown in Figure L12.35. Note that the priority of domestic jobs is set at a value of 1 (Figure L12.33), while that of export jobs is set at 10. Higher priority numbers signify higher priority in terms of selecting the upstream process. The priorities can range from 0 to 999. The default priority is 0.

L12.8.2 Selecting Resources

Priorities can also be used to decide which process, among two or more competing processes requesting the same resource, will have priority in capturing that resource.

FIGURE L12.34

Definition of path network for forklift for Wang's Export Machine Shop.

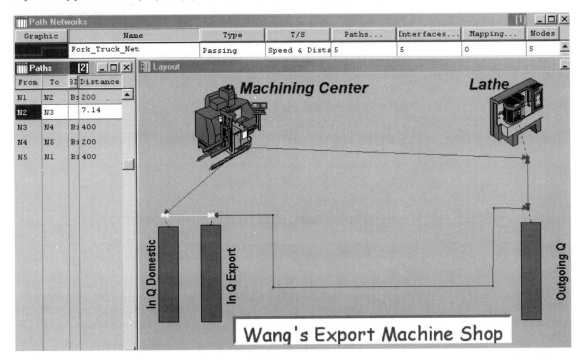

FIGURE L12.35

Processes and routings defined for Wang's Export Machine Shop.

```
********************************************************************************
*                              Processing                                     *
********************************************************************************

                         Process                        Routing

Entity    Location      Operation          Blk  Output    Destination         Rule     Move Logic
-------   -----------   --------------     ---  -------   -------------------  ------   -----------
Domestic  In_Q_Domestic
                                            1   Domestic  Machining_Center.1   FIRST 1  Move with Forklift Then Free


Domestic  Machining_Center  wait t(10,12,18) min

                                            1   Domestic  Lathe                FIRST 1  Move with Forklift Then Free
Domestic  Lathe         wait t(12,15,20) min

                                            1   Domestic  Outgoing_Q           FIRST 1  Move with Forklift Then Free
Domestic  Outgoing_Q                        1   Domestic  EXIT                 FIRST 1
Export    In_Q_Export                       1   Export    Machining_Center.10  FIRST 1  Move with Forklift Then Free
Export    Machining_Center  wait t(10,12,18) min

                                            1   Export    Lathe                FIRST 1  Move with Forklift Then Free
Export    Lathe         wait t(12,15,20) min
                                            1   Export    Outgoing_Q           FIRST 1  Move with Forklift Then Free
Export    Outgoing_Q                        1   Export    EXIT                 FIRST 1
```

Problem Statement

At **Wang's Export Machine Shop,** two types of jobs are processed: domestic and export. Mr. Wang is both the owner and the operator. The rate of arrival of both types of jobs is Exponential(60) minutes. Export jobs are processed on machining center E, and the domestic jobs are processed on machining center D. The processing times for all jobs are triangularly distributed (10, 12, 18) minutes. Mr. Wang gives priority to export jobs over domestic jobs. The distance between the stations is given in Table L12.7.

Five locations (Machining_Center_D, Machining_Center_E, In_Q_Domestic, In_Q_Export, and Outgoing_Q) are defined. Two types of jobs (domestic and export) arrive with an exponential interarrival frequency distribution of 60 minutes. Mr. Wang is defined as a resource in Figure L12.36. The path network and the processes are shown in Figures L12.37 and L12.38 respectively. Mr. Wang is getting old and can walk only 20 feet/minute with a load and 30 feet/minute without a load. Simulate for 100 hours.

Priorities of resource requests can be assigned through a GET, JOINTLY GET, or USE statement in operation logic, downtime logic, or move logic or the subroutines called from these logics. Priorities for resource downtimes are assigned in the Priority field of the Clock and Usage downtime edit tables.

Note that the priority of the resource (Mr_Wang) is assigned through the GET statement in the operation logic (Figure L12.38). The domestic orders have a resource request priority of 1, while that of the export orders is 10.

FIGURE 12.36

Resource defined for Wang's Export Machine Shop.

```
*******************************************************************
*                          Resources
*******************************************************************

                        Res       Ent
Name       Units Stats  Search    Search  Path        Motion           Cost
--------   ----- -----  --------  ------  ----------  --------------   -----
Mr_Wang    1     By Unit Closest  Oldest  Net1        Empty: 30 fpm
                                          Home: N1    Full: 20 fpm
```

TABLE L12.7   **Distances between Stations at Wang's Export Machine Shop**

From	To	Distance (feet)
Machining_Center_E	Machining_Center_D	200
Machining_Center_D	Outgoing	200
Outgoing	Machining_Center_E	200

FIGURE L12.37

Path network defined at Wang's Export Machine Shop.

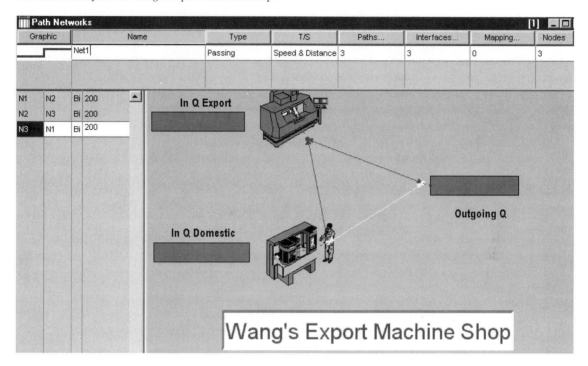

FIGURE L12.38

Processes and routings defined at Wang's Export Machine Shop.

		Process				Routing		
Entity	Location	Operation	Blk	Output	Destination	Rule	Move Logic	
Domestic	In_Q_Domestic		1	Domestic	Machining_Center_D	FIRST 1		
Domestic	Machining_Center_D	Get Mr_Wang, 1 wait t(10,12,18) min Free Mr_Wang						
			1	Domestic	Outgoing_Q	FIRST 1	Move with Mr_Wang Then Free	
Domestic	Outgoing_Q		1	Domestic	EXIT	FIRST 1		
Export	In_Q_Export		1	Export	Machining_Center_E	FIRST 1		
Export	Machining_Center_E	Get Mr_Wang, 10 wait t(10,12,18) min Free Mr_Wang						
			1	Export	Outgoing_Q	FIRST 1	Move with Mr_Wang Then Free	
Export	Outgoing_Q		1	Export	EXIT	FIRST 1		

The results (partial) are shown in Figure L12.39. Note that the average time waiting for the resource (Mr_Wang) is about 50 percent more for the domestic jobs (with lower priority) than for the export jobs. The average time in the system for domestic jobs is also considerably more than for the export jobs.

Part of the results showing the entity activities.

| Resources | Resource States | Node Entries | Failed Arrivals | Entity Activity | Entity States | Variables | Location Costing | |

Entity Activity for lab_L_11_8_2, Normal Run							
Name	**Total Exits**	**Current Qty In System**	**Avg Time In System (MIN)**	**Avg Time In Move Logic (MIN)**	**Avg Time Wait For Res (MIN)**	**Avg Time In Operation (MIN)**	**Avg Time Blocked (MIN)**
Domestic	81	35	894	53	755	14	72
Export	81	21	643	53	504	14	72

L12.9 Modeling a Pull System

A pull system is a system in which locations produce parts only on downstream demand. There are two types of pull systems:

1. Those based on limited buffer or queue sizes.
2. Those based on more distant "downstream" demand.

The first type of pull system is modeled in ProModel by defining locations with limited capacity. In this type of system, upstream locations will be able to send parts to downstream locations only when there is capacity available.

L12.9.1 Pull Based on Downstream Demand

The second type of pull system requires the use of a SEND statement from a downstream location to trigger part movement from an upstream location.

Problem Statement

In the **Milwaukee Machine Shop,** two types of jobs are processed within a machine cell. The cell consists of one lathe and one mill. Type 1 jobs must be processed first on the lathe and then on the mill. Type 2 jobs are processed only on the mill (Table L12.8). All jobs are processed on a first-in, first-out basis.

Brookfield Forgings is a vendor for the Milwaukee Machine Shop and produces all the raw material for them. Forgings are produced in batches of five every day (exponential with a mean of 24 hours). However, the customer supplies them only on demand. In other words, when orders arrive at the Milwaukee Machine Shop, the raw forgings are supplied by the vendor (a pull system of shop loading). Simulate for 100 days (2400 hours). Track the work-in-process inventories and the production quantities for both the job types.

Six locations (Mill, Lathe, Brookfield_Forgings, Order_Arrival, Lathe_Q, and Mill_Q) and four entities (Gear_1, Gear_2, Orders_1, and Orders_2) are defined. The arrivals of various entities are defined as in Figure L12.40. Four variables are defined as shown in Figure L12.41. The processes and routings are

FIGURE L12.40

Arrival of orders at the Milwaukee Machine Shop.

```
************************************************************************
                              Arrivals                               *
************************************************************************

Entity    Location            Qty each    First Time  Occurrences  Frequency  Logic
_____  _____  _____    _____  _____  _____  _____

Gear_1    Brookfield_Forgings 5           0           inf          e(24)
Gear_2    Brookfield_Forgings 5           0           INF          e(24)
Orders_1  Orders_Arrival      1           0           inf          e(8)
Orders_2  Orders_Arrival      1           0           inf          e(10)
```

FIGURE L12.41

Variables defined for the Milwaukee Machine Shop.

```
**********************************************************
                                    Variables (global)
**********************************************************

ID          Type            Initial value  Stats
_____   _____     _____  _____

wip1        Integer         0              Time Series
wip2        Integer         0              Time Series
prod1       Integer         0              Time Series
prod2       Integer         0              Time Series
```

TABLE L12.8   Order Arrival and Process Time Data for the Milwaukee Machine Shop

Job Type	Number of Orders	Time between Order Arrival	Processing Time on Lathe	Processing Time on Mill
1	120	E(8) hrs	E(3) hrs	U(3,1) hrs
2	100	E(10) hrs	—	U(4,1) hrs

shown in Figure L12.42. Note that as soon as a customer order is received at the Milwaukee Machine Shop, a signal is sent to Brookfield Forgings to ship a gear forging (of the appropriate type). Thus the arrival of customer orders pulls the raw material from the vendor. When the gears are fully machined, they are united (JOINed) with the appropriate customer order at the orders arrival location. Figure L12.43 shows a layout of the Milwaukee Machine Shop and a snapshot of the simulation model.

L12.9.2 Kanban System

The kanban system is one of the methods of control utilized within the Toyota production system (TPS). The basic philosophy in TPS is total elimination of

Figure L12.42

Processes and routings defined for the Milwaukee Machine Shop.

Entity	Location	Operation	Blk	Output	Destination	Rule	Move Logic
						Process	Routing
Gear_1	Brookfield_Forgings		1	Gear_1	Lathe_Q	SEND 1	WIP1=WIP1+1
Gear_1	Lathe_Q		1	Gear_1	Lathe	FIRST 1	
Gear_1	Lathe	WAIT E(3)	1	Gear_1	Mill_Q	FIRST 1	
Gear_1	Mill_Q		1	Gear_1	Mill	FIRST 1	
Gear_1	Mill	WAIT U(3,1)	1	Gear_1	Orders_Arrival	JOIN 1	
Gear_2	Brookfield_Forgings		1	Gear_2	Mill_Q	SEND 1	wip2=wip2+1
Gear_2	Mill_Q		1	Gear_2	Mill	FIRST 1	
Gear_2	Mill	WAIT U(4,1)	1	Gear_2	Orders_Arrival	JOIN 1	
Orders_1	Orders_Arrival	SEND 1 GEAR_1 TO LATHE_Q JOIN 1 GEAR_1 wip1=wip1-1 prod1=prod1+1	1	Orders_1	EXIT	FIRST 1	
Orders_2	Orders_Arrival	SEND 1 GEAR_2 TO MILL_Q JOIN 1 GEAR_2 wip2=wip2-1 prod2=prod2+1	1	Orders_2	EXIT	FIRST 1	

Figure L12.43

Simulation model for the Milwaukee Machine Shop.

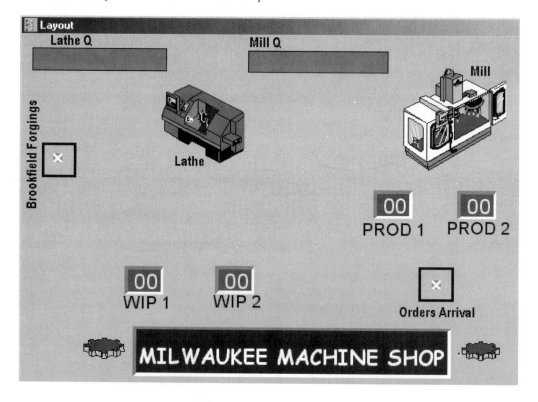

waste in machines, equipment, and personnel. To make the flow of things as close as possible to this ideal condition, a system of just-in-time procurement of material is used—that is, obtain material when needed and in the quantity needed.

"Kanban" literally means "visual record." The word *kanban* refers to the signboard of a store or shop, but at Toyota it simply means any small sign displayed in front of a worker. The kanban contains information that serves as a work order. It gives information concerning what to produce, when to produce it, in what quantity, by what means, and how to transport it.

Problem Statement

A consultant recommends implementing a production kanban system for Section 12.9.1's **Milwaukee Machine Shop.** Simulation is used to find out how many kanbans should be used. Model the shop with a total of five kanbans.

The kanban procedure operates in the following manner:

1. As soon as an order is received by the Milwaukee Machine Shop, they communicate it to Brookfield Forgings.
2. Brookfield Forgings holds the raw material in their own facility in the forging queue in the sequence in which the orders were received.
3. The production of jobs at the Milwaukee Machine Shop begins only when a production kanban is available and attached to the production order.
4. As soon as the production of any job type is finished, the kanban is detached and sent to the kanban square, from where it is pulled by Brookfield Forgings and attached to a forging waiting in the forging queue to be released for production.

The locations at the Milwaukee Machine Shop are defined as shown in Figure L12.44. Kanbans are defined as virtual entities in the entity table (Figure L12.45).

The arrival of two types of gears at Brookfield Forgings is shown in the arrivals table (Figure L12.46). This table also shows the arrival of two types of customer orders at the Milwaukee Machine Shop. A total of five kanbans are generated at the beginning of the simulation run. These are recirculated through the system.

FIGURE L12.44

Locations at the Milwaukee Machine Shop.

```
(*****************************************************************************
                               Locations
(*****************************************************************************

Name                    Cap        Units  Stats         Rules            Cost
-----------------       --------   -----  -----------   --------------   -----
Mill                    1          1      Time Series   Oldest, ,
Lathe                   1          1      Time Series   Oldest, ,
Brookfield_Forgings     INF        1      Time Series   Oldest, ,
Orders_Arrival          INF        1      Time Series   Oldest, ,
Lathe_Q                 INFINITE   1      Time Series   Oldest, FIFO,
Mill_Q                  INFINITE   1      Time Series   Oldest, FIFO,
kanban_square           5          1      Time Series   Oldest, ,
Order_Q                 INFINITE   1      Time Series   Oldest, FIFO,
```

FIGURE L12.45

Entities defined for Milwaukee Machine Shop.

```
**************************************************
                                          Entities
**************************************************

Name          Speed (fpm)   Stats          Cost
----------    -----------   -----------    -------
Gear_1        150           Time Series
Gear_2        150           Time Series
Orders_1      150           Time Series
Orders_2      150           Time Series
kanban        150           Time Series
```

FIGURE L12.46

Arrival of orders at Milwaukee Machine Shop.

```
************************************************************************
                              Arrivals                               *
************************************************************************

Entity    Location              Qty each   First Time   Occurrences   Frequency   Logic
--------  ------------------    --------   ----------   -----------   ---------   -----
Gear_1    Brookfield_Forgings   5          0            inf           e(24)
Gear_2    Brookfield_Forgings   5          0            INF           e(24)
Orders_1  Orders_Arrival        1          0            inf           e(8)
Orders_2  Orders_Arrival        1          0            inf           e(10)
kanban    kanban_square         5          0            1             0
```

FIGURE L12.47

Simulation model of a kanban system for the Milwaukee Machine Shop.

FIGURE L12.48

Process and routing tables for the Milwaukee Machine Shop.

		Process			Routing		
Entity	Location	Operation	Blk	Output	Destination	Rule	
kanban	kanban_square		1	kanban	Order_Q	LOAD 1	
Gear_1	Brookfield_Forgings		1	Gear_1	Order_Q	SEND 1	
Gear_1	Order_Q	Load 1					
			1	Gear_1	Lathe_Q	FIRST 1	
Gear_1	Lathe_Q	wip1=wip1+1	1	Gear_1	Lathe	FIRST 1	
Gear_1	Lathe	WAIT E(3)	1	Gear_1	Mill_Q	FIRST 1	
Gear_1	Mill_Q		1	Gear_1	Mill	FIRST 1	
Gear_1	Mill	WAIT U(3,1)					
		Unload 1					
		wip1=wip1-1					
			1	Gear_1	Orders_Arrival	JOIN 1	
kanban	Mill		1	kanban	kanban_square	FIRST 1	
Gear_2	Brookfield_Forgings		1	Gear_2	Order_Q	SEND 1	
Gear_2	Order_Q	Load 1					
			1	Gear_2	Mill_Q	FIRST 1	
Gear_2	Mill_Q	wip2=wip2+1	1	Gear_2	Mill	FIRST 1	
Gear_2	Mill	WAIT U(4,1)					
		UNLOAD 1					
		wip2=wip2-1					
			1	Gear_2	Orders_Arrival	JOIN 1	
Orders_1	Orders_Arrival	SEND 1 GEAR_1 TO Order_Q					
		JOIN 1 GEAR_1					
		prod1=prod1+1	1	Orders_1	EXIT		FIRST 1
Orders_2	Orders_Arrival	SEND 1 GEAR_2 TO Order_Q					
		JOIN 1 GEAR_2					
		prod2=prod2+1	1	Orders_2	EXIT		FIRST 1

Figure L12.47 shows the layout of the Milwaukee Machine Shop. The processes and the routings are shown in Figure L12.48. The arrival of a customer order (type 1 or 2) at the orders arrival location sends a signal to Brookfield Forgings in the form of a production kanban. The kanban is temporarily attached (LOADed) to a gear forging of the right type at the Order_Q. The gear forgings are sent to the Milwaukee Machine Shop for processing. After they are fully processed, the kanban is separated (UNLOADed). The kanban goes back to the kanban square. The finished gear is united (JOINed) with the appropriate customer order at the orders arrival location.

L12.10 Tracking Cost

ProModel 6.0 includes a cost-tracking feature. The following costs can be monitored:

1. Location cost
2. Resource cost
3. Entity cost

The cost dialog can be accessed from the Build menu (Figure L12.49).

FIGURE 12.49

The Cost option in the Build menu.

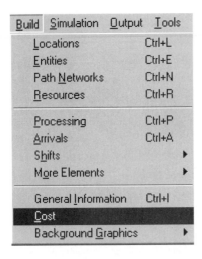

FIGURE L12.50

The Cost dialog box—Locations option.

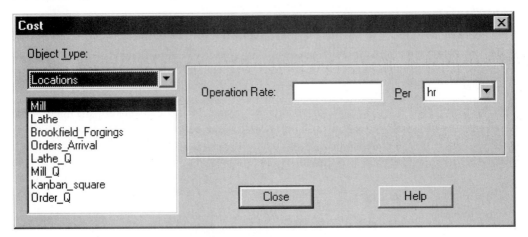

Locations

The Locations Cost dialog box (Figure L12.50) has two fields: Operation Rate and Per Operation Rate specifies the cost per unit of time to process at the selected location. Costs accrue when an entity waits at the location or uses the location. Per is a pull-down menu to set the time unit for the operation rate as second, minute, hour, or day.

Resources

The Resources Cost dialog box (Figure L12.51) has three fields: Regular Rate, Per, and Cost Per Use. Regular Rate specifies the cost per unit of time for a resource used in the model. This rate can also be set or changed during run time

FIGURE L12.51

The Cost dialog box—Resources option.

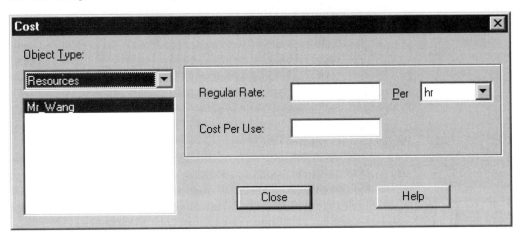

FIGURE L12.52

The Cost dialog box—Entities option.

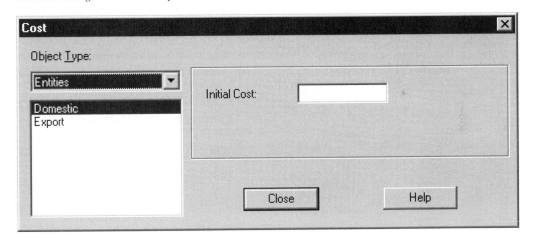

using the SETRATE operation statement. Per is a pull-down menu, defined before. Cost Per Use is a field that allows you to define the actual dollar cost accrued every time the resource is obtained and used.

Entities

The Entities Cost dialog box (Figure L12.52) has only one field: Initial Cost. Initial Cost is the cost of the entity when it arrives to the system through a scheduled arrival.

Increment Cost

The costs of a location, resource, or entity can be incremented by a positive or negative amount using the following operation statements:

- IncLocCost—Enables you to increment the cost of a location.
- IncResCost—Enables you to increment the cost of a resource.
- IncEntCost—Enables you to increment the cost of an entity.

For more information on the cost-tracking feature of ProModel, refer to Lab 6 of the ProModel manual.

Problem Statement

Raja owns a manufacturing cell consisting of two mills and a lathe. All jobs are processed in the same sequence, consisting of an arriving station, a lathe, mill 1, mill 2, and an exit station. The processing time on each machine is normally distributed with a mean of 60 seconds and standard deviation of 5. The arrival rate of jobs is exponentially distributed with a mean of 120 seconds.

Raja, the material handler, transports the jobs between the machines and the arriving and exit stations. Job pickup and release times are uniformly distributed between six and eight seconds. The distances between the stations are given in Table L12.9. Raja can walk at the rate of 150 feet/minute when carrying no load. However, he can walk only at the rate of 80 feet/minute when carrying a load.

The operation costs for the machines are given in Table L12.10. Raja gets paid at the rate of $60 per hour plus $5 per use. The initial cost of the jobs when they enter the system is $100 per piece. Track the number of jobs produced, the total cost of production, and the cost per piece of production. Simulate for 80 hours.

TABLE L12.9   Distances between Stations

From	To	Distance (feet)
Arriving	Lathe	40
Lathe	Mill 1	80
Mill 1	Mill 2	60
Mill 2	Exit	50
Exit	Arrive	80

TABLE L12.10   Operation Costs at Raja's Manufacturing Cell

Machine	Operation Costs
Lathe	$10/minute
Mill 1	$18/minute
Mill 2	$22/minute

FIGURE L12.53

Processes and routings at Raja's manufacturing cell.

		Process			Routing			
Entity	Location	Operation	Blk	Output	Destination	Rule	Move Logic	
Jobs	Arrive_Station		1	Jobs	Lathe	FIRST 1	MOVE WITH Raja THEN FREE	
Jobs	Lathe	wait n(60,5) sec	1	Jobs	Mill1	FIRST 1	MOVE WITH Raja THEN FREE	
Jobs	Mill1	wait n(60,5) sec	1	Jobs	Mill2	FIRST 1	MOVE WITH Raja THEN FREE	
Jobs	Mill2	wait n(60,5) sec	1	Jobs	Exit_Station	FIRST 1	MOVE WITH Raja THEN FREE	
Jobs	Exit_Station	Inc Total_Cost, GetCost()	1	Jobs	EXIT	FIRST 1	Inc Total_Prod Cost_Per_Part=Total_Cost/Total_prod	

FIGURE L12.54

Simulation model of Raja's manufacturing cell.

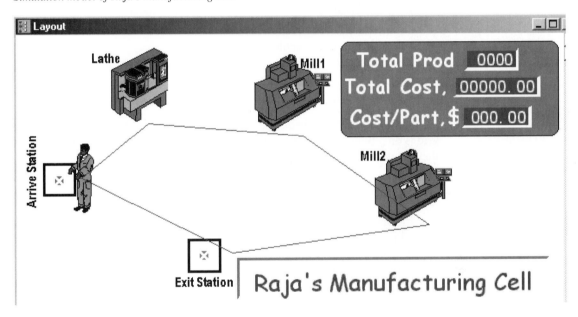

Five locations (Lathe, Mill1, Mill2, Arrive_Station, and Exit_Station) and three variables (Total_Prod, Cost_Per_Part, and Total_Cost) are defined. The processes are shown in Figure L12.53. The simulation model is shown in Figure L12.54.

L12.11 Importing a Background

Background graphics work as a static wallpaper and enhance the look of a simulation model. They make the model realistic and provide credibility during presentations. Many different graphic formats can be imported such as .BMP, .WMF, .GIF, and .PCX. Drawings in CAD formats like .DWG must be saved in one of these file formats before they can be imported. .BMPs and .WMFs can be copied to the clipboard from other applications and passed directly into the background.

AutoCAD drawings can also be copied to the clipboard and pasted into the background. The procedure is as follows:

1. With the graphic on the screen, press <Ctrl> and <C> together. Alternatively, choose Copy from the Edit menu. This will copy the graphic into the Windows clipboard.
2. Open an existing or new model file in ProModel.
3. Press <Ctrl> and <V> together. Alternatively, choose Paste from the Edit menu.

This action will paste the graphic as a background on the layout of the model. Another way to import backgrounds is to use the Edit menu in ProModel:

1. Choose Background Graphics from the Build menu.
2. Select Front of or Behind grid.
3. Choose Import Graphic from the Edit menu.
4. Select the desired file and file type. The image will be imported into the layout of the model.
5. Left-click on the graphic to reposition and resize it, if necessary.

"Front of grid" means the graphic will not be covered by grid lines when the grid is on. "Behind grid" means the graphic will be covered with grid lines when the grid is on.

L12.12 Defining and Displaying Views

Specific areas of the model layout can be predefined and then quickly and easily viewed in ProModel. Each view can be given a unique name and can have a suitable magnification. These views can be accessed during editing of the model by selecting Views from the View menu (Figure L12.55) or by using the keyboard shortcut key. Views can also be accessed in real time during the running of the model with a `VIEW` `"mill_machine"` statement.

Problem Statement

For the **Shipping Boxes Unlimited** example in Lab 7, Section L7.7.2, define the following six views: Monitor Queue, Box Queue, Inspect Queue, Shipping Queue, Pallet Queue, and Full View. Each view must be at 300 percent magnification. Show Full View at start-up. Show the Pallet Queue when an empty pallet arrives at the pallet queue location. Go back to showing the Full View when the box is at the shipping dock. Here are the steps in defining and naming views:

1. Select the View menu after the model layout is finished.
2. Select Views from the View menu.

The Views dialog box is shown in Figure L12.56. Figure L12.57 shows the Add View dialog box. The Full View of the Shipping Boxes Inc. model is shown in Figure L12.58.

FIGURE L12.55

Views command in the View menu.

FIGURE L12.56

The Views dialog box.

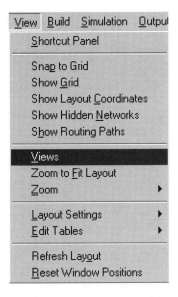

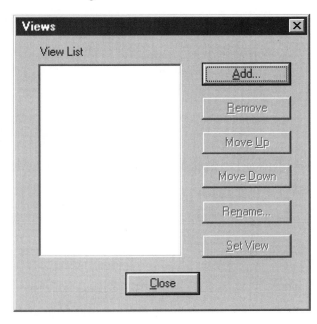

FIGURE L12.57

The Add View dialog box.

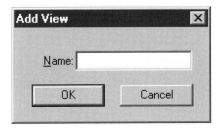

Referencing a View in Model Logic

1. Select General Information (Figure L12.59) from the Build menu. Select Initialization Logic and use the VIEW statement:

```
View "Full View"
```

2. Select Processing from the Build menu. Use the following statement in the Operation field when the Pallet Empty arrives at the Pallet Queue location (Figure L12.60):

```
View "Pallet Queue"
```

Also, use the following statement in the Operation field when the box arrives at the shipping dock:

```
View "Full View"
```

FIGURE L12.58

Full View of the Shipping Boxes Inc. model.

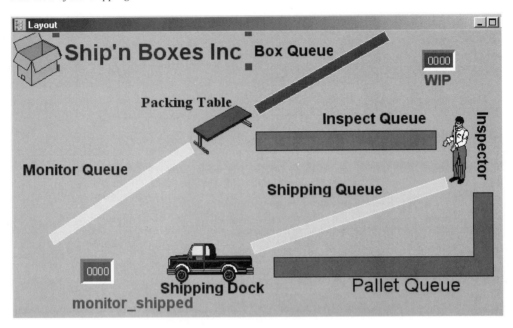

FIGURE L12.59

*General Information
for the Shipping Boxes
Inc. model.*

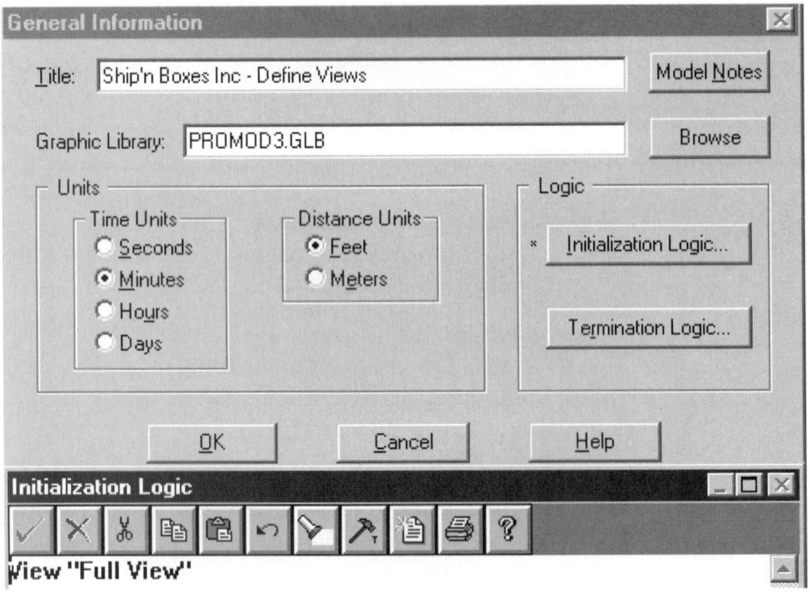

FIGURE L12.60

Processes and routings at Shipping Boxes Inc. incorporating the change of views.

		Process				Routing	
Entity	Location	Operation	Blk	Output	Destination	Rule	
Monitor	Monitor_Queue	wip = wip + 1					
			1	Monitor	Packing_Table	JOIN 1	
Empty_Box	Box_Queue		1	Empty_Box	Packing_Table	FIRST 1	
Empty_Box	Packing_Table	JOIN 1 Monitor					
		WAIT N(5,1)					
			1	Box	Inspect_Queue	FIRST 1	
Box	Inspect_Queue	WHILE Contents(Inspect_Queue) <= 3 DO					
		Wait 60 min					
		wait n(3,1) min	1	Box	Inspector	LOAD 1	
Pallet_empty	Inspector	Load 1					
		RENAME AS Pallet_Full					
Pallet_Full	Inspector		1	Pallet_Full	Shipping_Queue	FIRST 1	
Pallet_Full	Shipping_Queue	unload 1					
		wait U(3,1) min					
			1	Pallet_empty	Pallet_Queue	FIRST 1	
Box	Shipping_Queue		1	Box	Shipping_Dock	FIRST 1	
Box	Shipping_Dock	View "Full View"					
		monitor_shipped = monitor_shipped + 1					
		wip = wip -1	1	Box	EXIT	FIRST 1	
Pallet_empty	Pallet_Queue	View "Pallet Queue"					
			1	Pallet_empty	Inspector	FIRST 1	

L12.13 Creating a Model Package

ProModel has an innovative feature of creating a package of all files associated with the model file. This package file is titled <model name>.PKG—for example, ATM.pkg. The package file can then be archived or distributed to others. This file includes the model file (*.MOD), the graphic library (unless you check the Exclude Graphics Library option), and any external files you defined (such as read files, arrivals files, and shift files); the model package automatically includes bitmaps imported into the background graphics. The package file can be subsequently unpackaged by the receiver to run.

Use the following steps to create a model package:

1. Select Create Model Package from the File menu.
2. Enter the name you wish to use for the model package. ProModel uses the name of the model file as the default name of the package. So ATM.mod will be packaged as ATM.pkg. You can also Browse . . . to select the model name and directory path.
3. Check/uncheck the Exclude Graphics Library box, if you want to include/exclude the graphics library.
4. Check the Protect Model Data box if you want to protect the data in your model and prevent other users from changing and even viewing the model data files.
5. Click OK.

Example

For the **Widgets-R-Us** example in Section L12.6, make a model package that includes the model file and the shift file for operator Joe. Save the model package in a floppy disk. Figure L12.61 shows the Create Model Package dialog.

Unpack

To unpack and install the model package, double-click on the package file. In the Unpack Model Package dialog select the appropriate drive and directory path to install the model file and its associated files (Figure L12.62). Then click Install. After the package file has been installed, ProModel prompts you (Figure L12.63) for loading the model. Click Yes.

FIGURE L12.61

Create Model Package dialog.

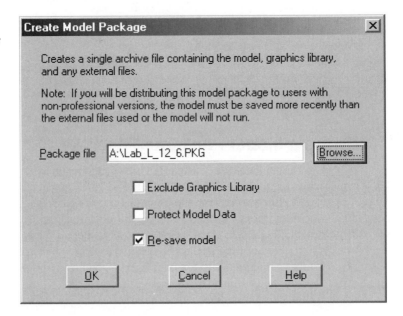

FIGURE L12.62

Unpack Model Package dialog.

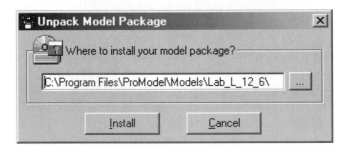

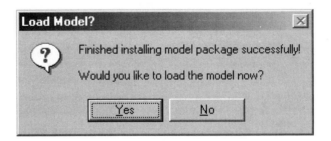

L12.14 Exercises

1. Five different types of equipment are available for processing a special type of part for one day (six hours) of each week. Equipment 1 is available on Monday, equipment 2 on Tuesday, and so forth. The processing time data follow:

Equipment	Time to Process One Part (minutes)
1	5 ± 2
2	4 ± 2
3	3 ± 1.5
4	6 ± 1
5	5 ± 1

Assume that parts arrive at a rate of one every 4 ± 1 hours, including weekends. How many parts are produced each week? How large a storage area is needed for parts waiting for a machine? Is there a bottleneck at any particular time? Why?

2. Customers visit the neighborhood barbershop **Fantastic Dan** for a haircut. Among the customers there are 30 percent children, 50 percent women, and 20 percent men. The customer interarrival time is triangularly distributed with a minimum, mode, and maximum of 8, 11, and 14 minutes respectively. The haircut time (in minutes) depends on the type of customer, as shown in this table:

| | **Haircut Time (minutes)** | |
Customers	Mean	Half-Width
Children	8	2
Women	12	3
Men	10	2

The initial greetings and signing in take Normal (2, · 2) minutes, and the transaction of money at the end of the haircut takes Normal (3, · 3) minutes. Run the simulation model for 100 working days (480 minutes each).

    a. About how many customers of each type does Dan process per day?

    b. What is the average number of customers of each type waiting to get a haircut? What is the maximum?

    c. What is the average time spent by a customer of each type in the salon? What is the maximum?

3. **Poly Castings Inc.** receives castings from its suppliers in batches of one every eleven minutes exponentially distributed. All castings arrive at the raw material store. Of these castings, 70 percent are used to make widget A, and the rest are used to make widget B. Widget A goes from the raw material store to the mill, and then on to the grinder. Widget B goes directly to the grinder. After grinding, all widgets go to degrease for cleaning. Finally, all widgets are sent to the finished parts store. Simulate for 1000 hours.

Widget	*Process 1*	*Process 2*	*Process 3*
Widget A	Mill [N(5,2) min.]	Grinder [U(11,1) min.]	Degrease [7 min.]
Widget B	Grinder [U(9,1) min.]	Degrease [7 min.]	

Track the work-in-process inventory of both types of widgets separately. Also, track the production of finished widgets (both types).

4. The maintenance mechanic in the problem in Section L12.5.1 is an independent contractor and works four hours each day from 10 A.M. until 2 P.M., Monday through Friday, with no lunch break. The rest of the shop works from 8 A.M. until 4 P.M. What will be the impact on the shop (average cycle time and number of widgets made per day) if we hire him full-time and have him work from 8 A.M. until 4 P.M. instead of working part-time?

5. Consider the **United Electronics** Exercise 7, in Section L7.12, with the following enhancements. The soldering machine breaks down on average after every 2000 ± 200 minutes of operation. The repair time is normally distributed (100, 50) minutes. Create a simulation model, with animation, of this system. Simulate this manufacturing system for 100 days, eight hours each day. Collect and print statistics on the utilization of each station, associated queues, and the total number of jobs manufactured during each eight-hour shift (average).

6. Consider the **Poly Casting Inc.** example in Lab 7, Section L7.4, and answer the following questions:

    a. What is the average time a casting spends in the system?

    b. What is the average time a casting waits before being loaded on a mill? After a mill processes 25 castings, it is shut down for a uniformly distributed time between 10 and 20 minutes for cleaning and tool change.

    c. What percentage of the time does each mill spend in cleaning and tool change operations?

    d. What is the average time a casting spends in the system?

    e. What is the average work-in-process of castings in the system?

7. Consider the **NoWaitBurger** stand in Exercise 13, in Section L7.12, and answer the following questions.

    a. What is the average amount of time spent by a customer at the hamburger stand?

    b. Run 10 replications and compute a 90 percent confidence interval for the average amount of time spent by a customer at the stand.

    c. Develop a 90 percent confidence interval for the average number of customers waiting at the burger stand.

8. Sharukh, Amir, and Salman wash cars at the **Bollywood Car Wash.** Cars arrive every 10 ± 6 minutes. They service customers at the rate of one every 20 ± 10 minutes. However, the customers prefer Sharukh to Amir, and Amir over Salman. If the attendant of choice is busy, the customers choose the first available attendant. Simulate the car wash system for 1000 service completions (car washes). Answer the following questions:

    a. Estimate Sharukh's, Amir's, and Salman's utilization.

    b. On average, how long does a customer spend at the car wash?

    c. What is the longest time any customer spent at the car wash?

    d. What is the average number of customers at the car wash?

    Embellishment: The customers are forced to choose the first available attendant; no individual preferences are allowed. Will this make a significant enough difference in the performance of the system to justify this change? Answer questions *a* through *d* to support your argument.

9. Cindy is a pharmacist and works at the **Save-Here Drugstore.** Walk-in customers arrive at a rate of one every 10 ± 3 minutes. Drive-in customers arrive at a rate of one every 20 ± 10 minutes. Drive-in customers are given higher priority than walk-in customers. The number of items in a prescription varies from 1 to 5 (3 ± 2). Cindy can fill one item in 6 ± 1 minutes. She works from 8 A.M. until 5 P.M. Her lunch break is from 12 noon until 1 P.M. She also takes two 15-minute breaks: at 10 A.M. and at 3 P.M. Define a shift file for Cindy named Cindy.sft. Model the pharmacy for a year (250 days) and answer the following questions:

    a. Estimate the average time a customer (of each type) spends at the drugstore.

    b. What is the average number of customers (of each type) waiting for service at the drugstore?

    c. What is the utilization of Cindy (percentage of time busy)?

    d. Do you suggest that we add another pharmacist to assist Cindy? How many pharmacists should we add?

     e. Is it better to have a dedicated pharmacist for drive-in customers and another for walk-in customers?

     f. Create a package file called SaveHere.pkg that includes the simulation model file and the shift file Cindy.sft.

10. A production line with five workstations is used to assemble a product. Parts arrive to the first workstation at random exponentially distributed intervals with a mean of seven minutes. Service times at individual stations are exponentially distributed with a mean of three minutes.

     a. Develop a simulation model that collects data for time in the system for each part.

     b. Additional analysis shows that products from three different priority classes are assembled on the line. Products from different classes appear in random order; products from classes 1, 2, and 3 have highest, medium, and lowest priorities respectively. Modify the simulation program to reflect this situation. Determine the mean time in the system for the three different priority classes.

Class	1	2	3
Probability	0.3	0.5	0.2

     c. The assembly process is modified so that each product to be assembled is placed on a pallet. The production supervisor wants to know how many pallets to place in operation at one time. It takes Uniform(8 ± 2) minutes to move the pallet from station 5 to station 1. Pallet loading and unloading times are Uniform(3 ± 1) minutes each. Modify the simulation model to reflect this situation. Is it better to use 5 or 20 pallets?

11. Two types of parts arrive at a single machine where they are processed one part at a time. The first type of part has a higher priority and arrives exponentially with a mean interarrival time of 30 minutes. The second lower-priority type also arrives exponentially with a mean interarrival time of 30 minutes. Within priority classes, the parts are serviced in FIFO order, but between priority classes the higher-priority type will always be processed first. However, a lower-priority part cannot be interrupted once it has begun processing. The machine processing times for higher- and lower-priority types of parts are uniformly distributed (20 ± 5) and (8 ± 2) minutes, respectively. Simulate the system for 1000 hours.

     a. What are the average time in queue and average time in system for each type of part?

     b. What are the average production rates of each type of part per hour?

12. The manufacture of a certain line of composite aerospace subassembly involves a relatively lengthy assembly process, followed by a short

firing time on an oven, which holds only one sub-assembly at a time. An assembler cannot begin assembling a new sub-assembly until he or she has removed the old one from the oven. The following is the pattern of processes followed by each assembler:

a. Assemble next subassembly.

b. Wait, first-come, first-served, to use the oven.

c. Use the oven.

d. Return to step *a.*

Here are the operating times and relevant financial data:

Operation	Time Required (minutes)
Assemble	30 ± 5
Fire	8 ± 2

Item	Cost Information ($)
Assembler's salary	$35/hour
Oven cost	$180 per 8-hour workday (independent of utilization)
Raw material	$8 per sub-assembly
Sale price of finished sub-assembly	$40 per sub-assembly

Build a simulation model of this manufacturing process. Use the model to determine the optimal number of assemblers to be assigned to an oven. The optimal number is understood in this context to be the one maximizing profit. Base the determination on simulations equivalent to 2000 hours of simulated time. Assume there are no discontinuities within a working day, or in moving between consecutive eight-hour working days. (Adapted from T. Schriber, *Simulation using GPSS,* John Wiley, 1974.)

13. For the following office work, the likelihood of needing rework after audit is 15 percent. Interarrival times are exponentially distributed with a mean of 5 minutes, rework times are exponentially distributed

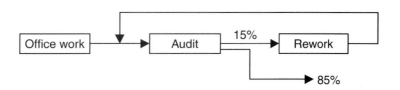

with a mean of 4 minutes, audit time is uniformly distributed between 2 ± 1 minutes, and work times are normally distributed with an average time of 4.5 minutes and standard deviation of 1.0 minute.

a. What is the mean time in the system for the first 200 reworked office files?

b. Develop a histogram of the mean time in the system for the first 1000 departures.

c. Modify the simulation program so that office work is rejected if it fails a second audit. How many office files are inspected twice? How many of these fail the audit a second time?

d. Modify the program to count the number of office files staying in the system longer than a specified length of time. What percentage of the office work stays in the system for more than 12 minutes?

e. Modify the part *a* simulation program so that it takes uniform (3 ± 1) minutes to move the office file from one station to another.

f. Modify the part *a* simulation program so that files to be reworked for second or more times have a lower priority than files needing rework for the first time. Does it seem to make a difference in the overall mean time in the system?

13   MATERIAL HANDLING CONCEPTS

With malice toward none, with charity for all, with firmness in the right, as God gives us to see the right, let us strive on to finish the work we are in . . .
—Abraham Lincoln (1809–1865)

In this lab we introduce you to various material handling concepts in ProModel. First we introduce you to the concepts of conveyors. In the next section we introduce manual and other discrete material handling systems, such as the forklift truck. Section L13.3 introduces the concepts of a crane system. For a more detailed discussion of modeling various material handling systems, please refer to Chapter 13.

L13.1 Conveyors

Conveyors are continuous material handling devices for transferring or moving objects along a fixed path having fixed, predefined loading and unloading points. Some examples of conveyors are belt, chain, overhead, power-and-free, roller, and bucket conveyors.

In ProModel, conveyors are locations represented by a conveyor graphic. A conveyor is defined graphically by a conveyor path on the layout. Once the path has been laid out, the length, speed, and visual representation of the conveyor can be edited by double-clicking on the path and opening the Conveyor/Queue dialog box (Figure L13.1). The various conveyor options are specified in the Conveyor Options dialog box, which is accessed by clicking on the Conveyor Options button in the Conveyor/Queue dialog box (Figure L13.2).

In an accumulating conveyor, if the lead entity comes to a stop, the trailing entities queue behind it. In a nonaccumulating conveyor, if the lead entity is unable to exit the conveyor, then the conveyor and all other entities stop.

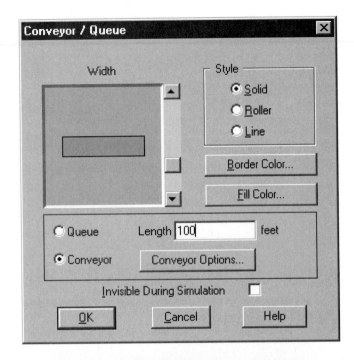

L13.1.1 Multiple Conveyors

Problem Statement

At **Ship'n Boxes Inc.,** boxes of type A arrive at Conv1 at the rate of one every five minutes (exponentially distributed). They are sent to the shipping conveyor. Boxes of type B arrive at Conv2 at the rate of one every two minutes (exponential) and are also sent to the shipping conveyor. The shipping conveyor takes boxes of both type A and B to the truck waiting at the shipping dock. The speed and length of the conveyors are given in Table L13.1. Develop a simulation model and run it for 10 hours. Figure out how many boxes of each type are shipped in 10 hours.

Four locations (Conv1, Conv2, Shipping, and Shipping_Dock) and two entities (Box_A and Box_B) are defined. The processes and layout are shown in Figures L13.3 and L13.4 respectively.

FIGURE L13.3

Processes and routings for the Ship'n Boxes Inc. model.

		Process			Routing		
Entity	Location	Operation	Blk	Output	Destination	Rule	Move Logic
Box_A	CONV1		1	Box_A	SHIPPING	FIRST 1	
Box_B	CONV2		1	Box_B	SHIPPING	FIRST 1	
Box_A	SHIPPING		1	Box_A	SHIPPING_DOCK	FIRST 1	
Box_A	SHIPPING_DOCK	Inc Count_A	1	Box_A	EXIT	FIRST 1	
Box_B	SHIPPING		1	Box_B	SHIPPING_DOCK	FIRST 1	
Box_B	SHIPPING_DOCK	Inc Count_B	1	Box_B	EXIT	FIRST 1	

FIGURE L13.4

Simulation model layout for Ship'n Boxes Inc.

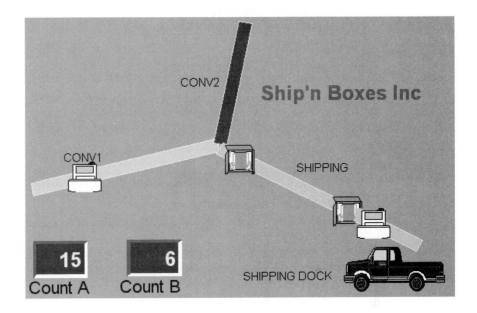

TABLE L13.1 Conveyor Lengths and Speeds

Name	Length	Speed
Conv1	100	30 feet/minute
Conv2	100	50 feet/minute
Shipping	200	20 feet/minute

L13.2 Resources, Path Networks, and Interfaces

A *resource* is a person, piece of equipment, or some other device that is used for one or more of the following functions: transporting entities, assisting in performing operations on entities at locations, performing maintenance on locations, or performing maintenance on other resources. In the following example, an operator is used as an example of a resource.

A *path network* defines the way a resource travels between locations. The specifications of path networks allow you to define the nodes at which the resource parks, the motion of the resource, and the path on which the resource travels. Path networks consist of nodes that are connected by path segments. A beginning node and an ending node define a path segment. Path segments may be unidirectional or bidirectional. Multiple path segments, which may be straight or joined, are connected at path nodes. To create path networks:

1. Select the Path button and then left-click in the layout where you want the path to start.
2. Subsequently, left-click to put joints in the path and right-click to end the path.

Interfaces are where the resource interacts with the location when it is on the path network. To create an interface between a node and a location:

1. Left-click and release on the node (a dashed line appears).
2. Then left-click and release on the location.

Multiple interfaces from a single node to locations can be created, but only one interface may be created from the same path network to a particular location.

L13.2.1 Manual Material Handling Systems

Problem Statement

At **Ghosh's Gear Shop,** a small automotive OEM facility, there are two NC lathes, a degreaser, and inspection equipment. Blanks are received at the receiving area. The blanks are machined on either NC lathe 1 or NC lathe 2 (machines are assigned by turn for successive blanks). After machining, the cogs go to the degreaser for washing. Then the cogs are inspected. An operator moves the parts from one area to another. The operator is also required to run the machines and perform the inspection procedure.

Blanks arrive at the rate of 10 per hour (exponentially distributed). The machining, washing, and inspection processes take Uniform(15 ± 5), Uniform(5 ± 1), and Normal(8, 4) minutes respectively. It takes Exponential(2) minutes to move the material between the processes. The operator is paid $20 per hour. It costs the company $0.10 for each hour the parts spend in processing in the manufacturing shop. The profit per item is $20.00 after deducting direct material costs and all other overhead. Build a simulation model, run it for 100 hours, and determine how many operators the shop should hire (one, two, or three).

Five locations (Receive, NC_Lathe_1, NC_Lathe_2, Inspect, and Degrease), two entities (Blank and Cog), and a resource (Operator) are defined. The processes and routings are shown in Figure L13.5. The blanks arrive with an interarrival time that is exponentially distributed with a mean of 20 minutes. The simulation run hours are set at 100 in the Simulation Option menu. The model is titled Ghosh's Gear Shop in the General Information menu. The layout of Ghosh's Gear Shop is shown in Figure L13.6.

FIGURE L13.5

Process and routing logic at Ghosh's Gear Shop.

		Process			Routing			
Entity	Location	Operation	Blk	Output	Destination	Rule		Move Logic
Blank	Receive		1	Blank	NC_Lathe_1	TURN	1	Get Operator Move with Operator for E(2) min Free Operator
				Blank	NC_Lathe_2	TURN		Get Operator Move with Operator for E(2) min Free Operator
Blank	NC_Lathe_1	Get Operator Wait U(15,5) min Free Operator	1	Cog	Degrease	FIRST	1	Get Operator Move with Operator for E(2) min Free Operator
Blank	NC_Lathe_2	Get Operator Wait U(15,5) min Free Operator	1	Cog	Degrease	FIRST	1	Get Operator Move with Operator for E(2) min Free Operator
Cog	Degrease	Get Operator Wait U(5,1) min Free Operator	1	Cog	Inspect	FIRST	1	Get Operator Move with Operator for E(2) min Free Operator
Cog	Inspect	Get Operator Wait N(8,4) min Free Operator	1	Cog	EXIT	FIRST	1	

FIGURE L13.6

Layout of Ghosh's Gear Shop.

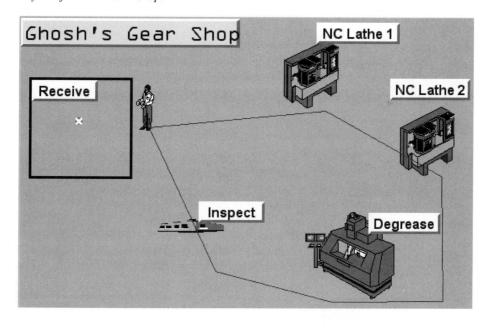

We ran this model with one, two, or three operators working together. The results are summarized in Tables L13.2 and L13.3. The production quantities and the average time in system (minutes) are obtained from the output of the simulation analysis. The profit per hour, the expected delay cost per piece, and the

TABLE L13.2   **Production Performance at Ghosh's Gear Shop**

Number of Operators	Production Quantity Per 100 Hours	Production Rate Per Hour	Gross Profit Per Hour	Average Time in System (minutes)
1	199	1.99	39.80	2423
2	404	4.04	80.80	1782
3	598	5.98	119.60	1354

TABLE L13.3   **Net Profit at Ghosh's Gear Shop**

Number of Operators	Expected Delay Cost ($/piece)	Expected Delay Cost ($/hour)	Expected Service Cost ($/hour)	Total Net Profit ($/hour)
1	$4.04	$8.04	$20	$11.76
2	$2.97	$12.00	$40	$28.80
3	$2.25	$13.45	$60	$46.15

expected delay cost per hour are calculated as follows:

Profit per hour = Production rate per hour × Profit per piece

Expected delay cost ($/piece) = [Average time in system (min.)/60]
× $0.1/piece/hr.

Expected delay cost ($/hr.) = Expected delay cost ($/piece)
× Production rate/hr.

The total net profit per hour after deducting the expected delay and the expected service costs is

Total net profit/hr. = Gross profit/hr. − Expected delay cost/hr.
− Expected service cost/hr.

The net profit per hour is maximized with three operators working in the shop. However, it must be noted that by hiring two operators instead of one, the increase in net profit per hour is about 145 percent, but the increase is only about 60 percent when three operators are hired instead of one.

L13.2.2 Manual versus Automated Material Handling Systems

We will now consider the impact of installing a conveyor system in the manufacturing facility described in the previous section. The conveyor will move all the material from one location to another and will replace the cell operator. The conveyor costs $50 per foot to install and $30,000 per year to maintain. The additional throughput can be sold for $20 per item profit. Should we replace one operator in the previous manufacturing system with a conveyor system? Build a simulation model and run it for 100 hours to help in making this decision.

The layout, locations, and processes and routings are defined as shown in Figures L13.7, L13.8, and L13.9 respectively. The length of each conveyor is 40 feet (Figure L13.10). The speeds of all three conveyors are assumed to be 50 feet/minute (Figure L13.11).

The results of this model are summarized in Tables L13.4 and L13.5. The production quantities and the average time in system (minutes) are obtained from the output of the simulation analysis. The profit per hour, the expected delay cost

FIGURE L13.7

Layout of Ghosh's Gear Shop with conveyors.

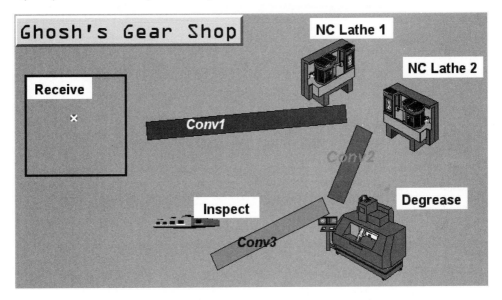

FIGURE L13.8

Locations at Ghosh's Gear Shop with conveyors.

```
*******************************************************
                                          Locations
*******************************************************

Name          Cap        Units  Stats        Rules
----------    --------   -----  -----------  --------------
Receive       inf        1      Time Series  Oldest, ,
NC_Lathe_1    1          1      Time Series  Oldest, ,
NC_Lathe_2    1          1      Time Series  Oldest, ,
Inspect       1          1      Time Series  Oldest, ,
Degrease      1          1      Time Series  Oldest, ,
Conv1         INFINITE   1      Time Series  Oldest, FIFO,
Conv2         INFINITE   1      Time Series  Oldest, FIFO,
Conv3         INFINITE   1      Time Series  Oldest, FIFO,
```

FIGURE L13.9

Process and routing at Ghosh's Gear Shop with conveyors.

Entity	Location	Operation	Blk	Output	Destination	Rule	Move Logic
Blank	Receive		1	Blank	Conv1	FIRST 1	
Blank	Conv1		1	Blank	NC_Lathe_1	TURN 1	
				Blank	NC_Lathe_2	TURN	
Blank	NC_Lathe_1	Wait U(15,5) min	1	Cog	Conv2	FIRST 1	
Blank	NC_Lathe_2	Wait U(15,5) min	1	Cog	Conv2	FIRST 1	
Cog	Conv2		1	Cog	Degrease	FIRST 1	
Cog	Degrease	Wait U(5,1) min					
Cog	Conv3		1	Cog	Conv3	FIRST 1	
Cog	Inspect	Wait N(8,4) min	1	Cog	Inspect	FIRST 1	
			1	Cog	EXIT	FIRST 1	

FIGURE L13.10

Conveyors at Ghosh's Gear Shop.

FIGURE L13.11

Conveyor options at Ghosh's Gear Shop.

TABLE L13.4   **Production and Gross Profit at Ghosh's Gear Shop**

Mode of Transportation	Production Quantity Per 100 Hours	Production Rate Per Hour	Gross Profit Per Hour	Average Time in System (minutes)
Conveyor	747	7.47	$149.40	891

TABLE L13.5   **Net Profit at Ghosh's Gear Shop**

Mode of Transportation	Expected Delay Cost ($/piece)	Expected Delay Cost ($/hour)	Expected Service Cost ($/hour)	Total Net Profit ($/hour)
Conveyor	$1.485	$11.093	$16	$122.307

per piece, and the expected delay cost per hour are calculated as follows:

$$\text{Profit per hour} = \text{Production rate/hour} \times \text{Profit per piece}$$

$$\text{Expected delay cost (\$/piece)} = (\text{Average time in system (min.)}/60) \times \$0.1/\text{piece/hour}$$

$$\text{Expected delay cost (\$/hour)} = \text{Expected delay cost (\$/piece)} \times \text{Production rate/hour}$$

To calculate the service cost of the conveyor, we assume that it is used about 2000 hours per year (8 hrs/day × 250 days/year). Also, for depreciation purposes, we assume straight-line depreciation over three years.

$$\text{Total cost of installation} = \$50/\text{ft.} \times 40 \text{ ft./conveyor segment} \times 3 \text{ conveyor segments}$$
$$= \$6000$$

$$\text{Depreciation per year} = \$6000/3 = \$2000/\text{year}$$

$$\text{Maintenance cost} = \$30{,}000/\text{year}$$

$$\text{Total service cost/year} = \text{Depreciation cost} + \text{Maintenance cost}$$
$$= \$32{,}000/\text{year}$$

$$\text{Total service cost/hour} = \$32{,}000/2000 = \$16/\text{hour}$$

The total net profit per hour after deducting the expected delay and the expected service costs is

$$\text{Total net profit/hour} = \text{Gross profit/hour} - \text{Expected delay cost/hour} - \text{Expected service cost/hour}$$

Comparing the net profit per hour between manual material handling and conveyorized material handling, it is evident that conveyors should be installed to maximize profit.

L13.2.3 Using Operator for Processing

In the following example we use an operator for material handling as well as some of the processing done in the shop.

Problem Statement

Modify the example in Section L7.7.1 by having a material handler (let us call him Joe) return the pallets from the shipping dock to the packing workbench. Eliminate the pallet return conveyor. Also, let's have Joe load the boxes onto pallets [Uniform(3,1) min] and the loaded pallets onto the shipping conveyor. Assume one pallet is in the system. Simulate for 10 hours and determine the following:

1. The number of monitors shipped.
2. The utilization of the material handler.

Seven locations (Conv1, Conv2, Shipping_Q, Shipping_Conv, Shipping_Dock, Packing_Table, and Load_Zone) are defined. The resource (Joe) is defined as shown in Figure L13.12. The path network for the resource and the processes and routings are shown in Figures L13.13 and L13.14 respectively. The arrivals of the entities are shown in Figure L13.15. The layout of Ship'n Boxes Inc. with conveyors is shown in Figure L13.16.

L13.2.4 Automated Manufacturing Cell

Problem Statement

Raja owns a manufacturing cell consisting of two mills and a lathe. All jobs are processed in the same sequence, consisting of arrival station, lathe, mill 1, mill 2,

FIGURE L13.12

Material handler as a resource.

Icon	Name	Units	DTs...	Stats...	Specs...	Search...	Logic...	Pts...	Notes...
	JOE	1	None	By Unit	Net1, N1	None	0	1	

FIGURE L13.13

Path network for the material handler.

Graphic	Name	Type	T/S	Paths...	Interfaces...	Mapping...	Nodes
	Net1	Passing	Speed & Distance 1	2	0	2

From	To	Bi	Distance
N1	N2	Bi	31.30

FIGURE L13.14

Processes and routings for the example in Section L13.2.3.

	Process			Routing			
Entity	Location	Operation	Blk	Output	Destination	Rule	Move Logic
Monitor	CONV1		1	Monitor	Packing_Table	JOIN 1	
Empty_Box	CONV2		1	Empty_Box	Packing_Table	FIRST 1	
Empty_Box	Packing_Table	JOIN 1 Monitor					
		WAIT N(5,1)					
			1	Box	Shipping_Q	FIRST 1	
Box	Shipping_Q	WAIT 0	1	Box	Load_zone	LOAD 1	
Pallet_	Load_zone	load 1					
		USE Joe FOR U(3,1)					
			1	Pallet_	SHIPPING_CONV	FIRST 1	
Pallet_	SHIPPING_CONV		1	Pallet_	SHIPPING_DOCK	FIRST 1	
Pallet_	SHIPPING_DOCK	UNLOAD 1	1	Pallet_	Load_zone	FIRST 1	MOVE WITH JOE THEN FREE
Box	SHIPPING_DOCK		1	Box	EXIT	FIRST 1	

FIGURE L13.15

Arrivals for the example in Section L13.2.3.

```
********************************************************************************
                                  Arrivals
********************************************************************************
```

Entity	Location	Qty each	First Time	Occurrences	Frequency	Logic
Monitor	CONV1	1	0	inf	E(15)	
Empty_Box	CONV2	1	0	inf	E(15)	
Pallet_	Load_zone	1	0	1	5	

FIGURE L13.16

The layout for Shipping Boxes Inc.

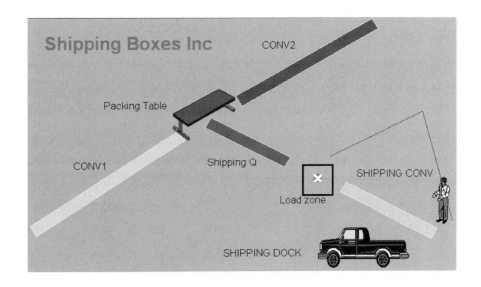

and exit station. The processing time on each machine is normally distributed with a mean of 60 seconds and a standard deviation of 5 seconds. The arrival rate of jobs is exponentially distributed with a mean of 120 seconds.

Raja also transports the jobs between the machines and the arrival and exit stations. Job pickup and release times are uniformly distributed between six and eight seconds. The distances between the stations are given in Table L13.6. Raja can travel at the rate of 150 feet/minute when carrying no load. However, he can walk at the rate of only 80 feet/minute when carrying a load. Simulate for 80 hours.

The layout, locations, path networks, resource specification, and processes and routings are shown in Figures L13.17, L13.18, L13.19, L13.20, and L13.21, respectively.

FIGURE L13.17

Layout of Raja's manufacturing cell.

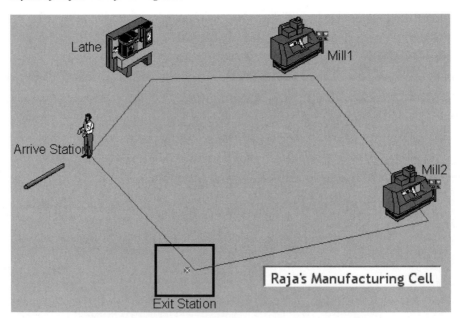

TABLE L13.6   Distances between Stations at Raja's Manufacturing Cell

From	To	Distance (feet)
Arrival station	Lathe	40
Lathe	Mill 1	80
Mill 1	Mill 2	60
Mill 2	Exit	50
Exit	Arrive	80

FIGURE **L13.18**

Locations at Raja's manufacturing cell.

Icon	Name	Cap.	Units	DTs...	Stats...	Rules...
	Lathe	1	1	None	Time Series	Oldest
	Mill1	1	1	None	Time Series	Oldest
	Mill2	1	1	None	Time Series	Oldest
	Arrive_Station	inf	1	None	Time Series	Oldest
	Exit_Station	inf	1	None	Time Series	Oldest

FIGURE **L13.19**

Path networks at Raja's manufacturing cell.

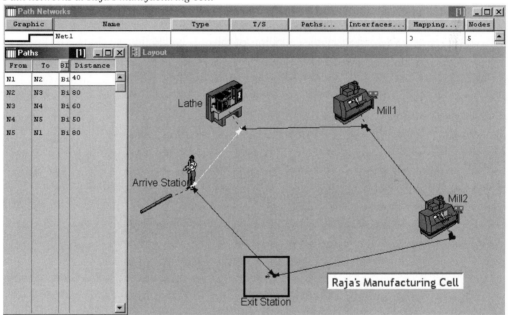

FIGURE **L13.20**

Resource specification at Raja's manufacturing cell.

```
<*******************************************************************************>
                                   Resources
<*******************************************************************************>

                           Res      Ent
Name      Units  Stats    Search   Search  Path        Motion          Cost
------    -----  -------  -------   ------  -----       -------------   -----
Raja      1      By Unit  Closest  Oldest  Net1        Empty: 150 fpm
                                           Home: N1    Full: 80 fpm
```

Process and routing tables at Raja's manufacturing cell.

		Process			Routing			
Entity	Location	Operation	Blk	Output	Destination	Rule	Move Logic	
JOBS	Arrive_Station		1	JOBS	Lathe	FIRST 1	MOVE WITH RAJA THEN FREE	
JOBS	Lathe	WAIT N(60,5) SEC	1	JOBS	Mill1	FIRST 1	MOVE WITH RAJA THEN FREE	
JOBS	Mill1	WAIT N(60,5) SEC	1	JOBS	Mill2	FIRST 1	MOVE WITH RAJA THEN FREE	
JOBS	Mill2	WAIT N(60,5) SEC	1	JOBS	Exit_Station	FIRST 1	MOVE WITH RAJA THEN FREE	
JOBS	Exit_Station		1	JOBS	EXIT	FIRST 1		

L13.3 Crane Systems

ProModel provides the constructs for modeling crane systems. The path networks module and the resources module provide modeling features for the crane systems. Modeling one or multiple cranes operating in the same bay is easily possible. Users can define the following regarding crane systems:

Crane envelope: The crane envelope is a parallelogram-shaped area represented by a crane-type path network. Two rails and two lines connecting the endpoints of the rails bound this envelope. The lines connecting the endpoints of the rails are the two extreme positions of the centerlines of the crane bridge.

Bridge separation distance: This is the minimum distance between the centerlines of any two neighboring cranes. Make sure to define the crane envelope as wide enough to allow sufficient space beyond any serviceable nodes.

Priorities: Two types of priorities are associated with the crane construct in ProModel: resource request priority and crane move priority.

1. *Resource request priorities* are used to resolve any conflicts between multiple entities requesting the same crane at the same time. Lower-priority tasks are preempted to serve higher-priority tasks.

2. *Crane move priorities* are used when multiple cranes are operating in the same work envelope to decide which crane has priority over another to move. Crane move priorities can be subdivided into three categories:

   a. *Task move priority* is used as the base move priority for travel to pick up and travel to drop a load.

   b. *Claim priority* is the same as the task move priority if the crane is moving to its ultimate (task) destination. Otherwise, if it is moving under the influence of another crane, claim priority is equal to the claim priority of the claim inducer bridge.

   c. *Effective claim priority* applies only when there are three or more cranes operating in the same envelope. If multiple cranes are moving in the same direction with overlapping claims, the effective claim priority in the overlapping zone is the maximum of all the claim priorities of those cranes.

For more details on crane operations, and in particular to understand how zone claims are handled, please refer to the ProModel Reference Guide.

Problem Statement

Pritha takes over the manufacturing operation (Section L13.2.4) from Raja and re-names it **Pritha's manufacturing cell.** After taking over, she installs an overhead crane system to handle all the material movements between the stations. Job pickup and release times by the crane are uniformly distributed between six and eight seconds. The coordinates of all the locations in the cell are given in Table L13.7. The crane can travel at the rate of 150 feet/minute with or without a load. Simulate for 80 hours.

Five locations (Lathe, Mill 1, Mill 2, Exit_Station, and Arrive_Station) are de-fined for Pritha's manufacturing cell. The path networks, crane system resource, and processes and routings are shown in Figures L13.22, L13.23, and L13.24.

FIGURE L13.22

Path networks for the crane system at Pritha's manufacturing cell.

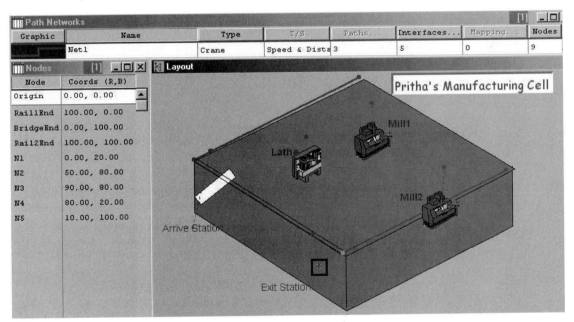

TABLE L13.7   Coordinates of All the Locations in Pritha's Manufacturing Cell (in feet)

Location	X (Rail)	Y (Bridge)
Arriving	10	100
Lathe	50	80
Mill 1	90	80
Mill 2	80	20
Exit	0	20

FIGURE L13.23

The crane system resource defined.

```
(×××××××××××××××××××××××××××××××××××××××××××××××××××××××××××××××××××××
                               Resources                              *
(×××××××××××××××××××××××××××××××××××××××××××××××××××××××××××××××××××××

                          Res      Ent
Name            Units Stats  Search   Search Path       Motion                 Cost
-------------   ----- -------- -------- ------- --------- ----------------------- ----
Crane_System 1        By Unit  Closest  Oldest  Net1      Empty: 150,150 fpm
                                                 Home: Origin Full: 150,150 fpm
                                                          Pickup: u(3,1) Seconds
                                                          Deposit: u(3,1)  Seconds
```

FIGURE L13.24

Process and routing tables defined for Pritha's manufacturing cell.

		Process			Routing		
Entity	Location	Operation	Blk	Output	Destination	Rule	Move Logic
Jobs	Arrive_Station		1	Jobs	Lathe	FIRST 1	MOVE WITH Crane_System THEN FREE
Jobs	Lathe	wait n(60,5) sec	1	Jobs	Mill1	FIRST 1	MOVE WITH Crane_System THEN FREE
Jobs	Mill1	wait n(60,5) sec	1	Jobs	Mill2	FIRST 1	MOVE WITH Crane_System THEN FREE
Jobs	Mill2	wait n(60,5) sec	1	Jobs	Exit_Station	FIRST 1	MOVE WITH Crane_System THEN FREE
Jobs	Exit_Station		1	Jobs	EXIT	FIRST 1	

Select Path Network from the Build menu. From the Type menu, select Crane. The following four nodes are automatically created when we select the crane type of path network: Origin, Rail1 End, Rail2 End, and Bridge End. Define the five nodes N1 through N5 to represent the three machines, the arrival station, and the exit station. Click on the Interface button in the Path Network menu and define all the interfaces for these five nodes.

Select Resource from the Build menu. Name the crane resource. Enter 1 in the Units column (one crane unit). Click on the Specs. button. The Specifications menu opens up. Select Net1 for the path network. Enter the empty and full speed of the crane as 150 ft/min. Also, enter Uniform(3 ± 1) seconds as the pickup and deposit time.

L13.4 Exercises

1. Consider the **DumpOnMe** facility in Exercise 11 of Section L7.12 with the following enhancements. Consider the dump trucks as material handling resources. Assume that 10 loads of coal arrive to the loaders every hour (randomly; the interarrival time is exponentially distributed). Create a simulation model, with animation, of this system. Simulate for 100 days, eight hours each day. Collect statistics to estimate the loader and scale utilization (percentage of time busy). About how many trucks are loaded each day on average?

2. For the **Widgets-R-Us Manufacturing Inc.** example in Section L12.5.1, consider that a maintenance mechanic (a resource) will be hired to do the

repair work on the lathe and the mill. Modify the simulation model and run it for 2000 hours. What is the utilization of the maintenance mechanic?

Hint: Define maintenance_mech as a resource. In the Logic field of the Clock downtimes for Lathe enter the following code:

```
GET maintenance_mech
DISPLAY "The Lathe is Down for Maintenance"
Wait N(10,2) min
FREE maintenance_mech
```

Enter similar code for the Mill.

3. At **Forge Inc.** raw forgings arrive at a circular conveyor system (Figure L13.25). They await a loader in front of the conveyor belt. The conveyor delivers the forgings (one foot long) to three machines located one after the other. A forging is off-loaded to a machine only if the machine is not in use. Otherwise, the forging moves on to the next machine. If the forging cannot gain access to any machine in a given pass, it is recirculated. The conveyor moves at 30 feet per minute. The distance from the loading station to the first machine is 30 feet. The distance between each machine is 10 feet. The distance from the last machine back to the loader is 30 feet. Loading and unloading take 30 seconds each. Forgings arrive to the loader with an exponential interarrival time and a mean of 10 seconds. The machining times are also exponentially distributed with a mean of 20 seconds. Simulate the conveyor system for 200 hours of operation.

a. Collect statistics to estimate the loader and machine utilizations.

b. What is the average time a forging spends in the system?

c. What fraction of forgings cannot gain access to the machines in the first pass and need to be recirculated?

d. What is the production rate of forgings per hour at Forge Inc.?

FIGURE L13.25

Layout of Forge Inc.

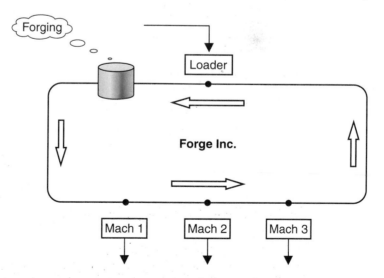

4. **U.S. Construction Company** has one bulldozer, four trucks, and two loaders. The bulldozer stockpiles material for the loaders. Two piles of material must be stocked prior to the initiation of any load operation. The time for the bulldozer to stockpile material is Erlang distributed and consists of the sum of two exponential variables, each with a mean of 4 (this corresponds to an Erlang variable with a mean of 8 and a variance of 32). In addition to this material, a loader and an unloaded truck must be available before the loading operation can begin. Loading time is exponentially distributed with a mean time of 14 minutes for server 1 and 12 minutes for server 2.

   After a truck is loaded, it is hauled and then dumped; it must be returned before it is available for further loading. Hauling time is normally distributed. When loaded, the average hauling time is 22 minutes. When unloaded, the average time is 18 minutes. In both cases, the standard deviation is three minutes. Dumping time is uniformly distributed between two and eight minutes. Following a loading operation, the loader must rest for five minutes before it is available to begin loading again. Simulate this system at the U.S. Construction Co. for a period of one year (2000 working hours) and analyze it.

5. At **Walnut Automotive,** machined castings arrive randomly (exponential, mean of six minutes) from the supplier to be assembled at one of five identical engine assembly stations. A forklift truck delivers the castings from the shipping dock to the engine assembly department. A loop conveyor connects the assembly stations.

   The forklift truck moves at a velocity of five feet per second. The distance from the shipping dock to the assembly department is 1000 feet. The conveyor is 5000 feet long and moves at a velocity of five feet per second.

   At each assembly station, no more than three castings can be waiting for assembly. If a casting arrives at an assembly station and there is no room for the casting (there are already three castings waiting), it goes around for another try. The assembly time is normally distributed with a mean of five minutes and standard deviation of two minutes. The assembly stations are equally distributed around the belt. The load/unload station is located halfway between stations 5 and 1. The forklift truck delivers the castings to the load/unload station. It also picks up the completed assemblies from the load/unload station and delivers them back to the shipping dock.

   Create a simulation model, with animation, of this system. Run the simulation model until 1000 engines have been assembled.

   a. What is the average throughput time for the engines in the manufacturing system?

   b. What are the utilization figures for the forklift truck and the conveyor?

   c. What is the maximum number of engines in the manufacturing system?

   d. What are the maximum and average numbers of castings on the conveyor?

6. At the **Grocery Warehouse,** a sorting system consists of one incoming conveyor and three sorting conveyors, as shown in Figure L13.26. Cases enter the system from the left at a rate of 100 per minute at random times. The incoming conveyor is 150 feet long. The sorting conveyors are 100 feet long. They are numbered 1 to 3 from left to right and are 10 feet apart. The incoming conveyor runs at 10 feet per minute and all the sorting conveyors at 15 feet per minute. All conveyors are accumulating type. Incoming cases are distributed to the three lanes in the following proportions: Lane 1—30 percent; Lane 2—50 percent; and Lane 3—20 percent.

At the end of each sorting conveyor, a sorter (from a group of available sorters) scans each case with a bar code scanner, applies a label, and then places the case on a pallet. One sorter can handle 10 cases per minute on the average (normally distributed with a standard deviation of 0.5 minute).

When a pallet is full (40 cases), a forklift arrives from the shipping dock to take it away, unload the cases, and bring back the empty pallet to an empty pallet queue near the end of the sorting conveyor. A total of five pallets are in circulation. The data shown in Table 13.8 are available for the forklift operation.

FIGURE L13.26

The conveyor sorting system at the Grocery Warehouse.

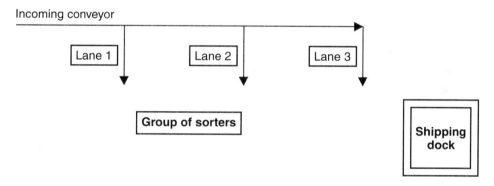

TABLE 13.8   Forklift Operation Data Collected at the Grocery Warehouse

Load time	1 min.
Unload time	3 min.
Travel speed (with and without load)	10 ft./min
Distance from shipping dock to sorting conveyors	250 ft.

Simulate for one year (250 working days, eight hours each day). Answer the following questions:

    a. How many sorters are required? The objective is to have the minimum number of sorters but also avoid overflowing the conveyors.

    b. How many forklifts do we need?

    c. Report on the sorter utilization, total number of cases shipped, and the number of cases palletized by lane.

7. Repeat Exercise 6 with a dedicated sorter in each sorting lane. Address all the same issues.

8. Printed circuit boards arrive randomly from the preparation department. The boards are moved in sets of five by a hand truck to the component assembly department, where the board components are manually assembled. Five identical assembly stations are connected by a loop conveyor.

    When boards are placed onto the conveyor, they are directed to the assembly station with the fewest boards waiting to be processed. After the components are assembled onto the board, they are set aside and removed for inspection at the end of the shift. The time between boards arriving from preparation is exponentially distributed with a mean of five seconds. The hand truck moves at a velocity of five feet per second and the conveyor moves at a velocity of two feet per second. The conveyor is 100 feet long. No more than 20 boards can be placed on the belt at any one time.

    At each assembly station, no more than two boards can be waiting for assembly. If a board arrives at an assembly station and there is no room for the board (there are already two boards waiting), the board goes around the conveyor another time and again tries to enter the station. The assembly time is normally distributed with a mean of 35 seconds and standard deviation of 8 seconds. The assembly stations are uniformly distributed around the belt, and boards are placed onto the belt four feet before the first station. After all five boards are placed onto the belt, the hand truck waits until five boards have arrived from the preparation area before returning for another set of boards.

    Simulate until 100 boards have been assembled. Report on the utilization of the hand truck, conveyor, and the five operators. How many assemblies are produced at each of the five stations? (Adapted from Hoover and Perry, 1989.)

9. In this example we will model the assembly of circuit boards at four identical assembly stations located along a closed loop conveyor (100 feet long) that moves at 15 feet per minute. The boards are assembled from kits, which are sent in totes from a load/unload station to the assembly stations via the top loop of the conveyor. Each kit can be assembled at any one of the four assembly stations. Completed boards are returned in their totes from the assembly stations back to the loading /unloading station. The loading/unloading station (located at the left end of the conveyor) and the four assembly stations (identified by the letters A through D) are equally spaced along the conveyor, 20 feet

apart. The time to assemble each kit is $N(7,2)$ minutes. If an assembly station is busy when a tote arrives, it moves to the next station. Loading and unloading a tote to/from the conveyor takes $N(2,0.2)$ minutes each. Since each tote is one foot long, at most 100 totes will fit on the conveyor at one time. Kits arrive at the load/unload station with an average interarrival time of five minutes, exponentially distributed. Simulate for 1000 hours. Answer the following questions:

a. What are the average and maximum numbers of kits in the system?

b. What is the number of assemblies produced and shipped per hour?

c. What are the average and maximum times spent by a kit in the system?

d. What are the utilizations of the four assembly stations?

e. What percentage of kits go around the conveyor more than once before being assembled?

f. Should we consider installing more assembly stations? Fewer?

10. At the **Assam Plywood Mill,** logs arrive to the mill in truckloads of 15 each, according to a Poisson process distributed with a mean of 1.3 feet and a standard deviation of 0.3 foot. The logs are precut to a standard length of 8 feet, 16 feet, or 24 feet before their arrival to the mill, with a distribution by length of 20 percent, 50 percent, and 30 percent respectively. The arriving logs are unloaded into a bin, from which they are loaded one at a time onto a conveyor. Sensor devices automatically control this loading operation so that the next log is kicked onto the conveyor as soon as sufficient space is available.

The conveyor moves the logs at a speed of 40 feet/minute to a peeling machine, which debarks the logs and peels them into wood strips. These strips are dried and glued together in subsequent phases of the operation to produce plywood. The peeling machine accepts only logs that are 8 feet long. Therefore, the 16-foot and the 24-foot logs must be cut into 8-foot sections before arriving to the peeling machine. This sectioning is done at an automatic cutting station located 32 feet from the start of the conveyor. Whenever the leading end of a 16-foot or 24-foot log reaches the 40-foot mark, the conveyor is automatically stopped and a vertical cut is made at the 32-foot mark. This operation requires 0.5 minute, after which the conveyor can be restarted.

The peeling machine is located beginning at the 48-foot mark along the conveyor. Whenever the trailing end of a log reaches this point, the conveyor is stopped and the log is ready to be moved from the conveyor and loaded into the peeling machine. The loading operation requires 0.8 minute. At the end of this time the conveyor can be restarted, and the debarking/peeling operation begins. The peeling machine processes one 8-foot log at a time at the rate of 20 cubic feet of log per minute.

Simulate the system for 2400 minutes to determine the throughput of wood in cubic feet, the utilization of the peeler, and the number of feet of conveyor occupied.

11. At **Orem Casting Company,** rough castings arrive for machining according to a normal distribution with a mean of 60 seconds and a

standard deviation of 15 seconds. An inspector examines the castings for cracks. Approximately 15 percent of the castings are rejected and scrapped. Three automatic CNC milling machines are used to machine the remaining castings. Two loaders (material handlers) load the mills in a time that is uniformly distributed from 35 to 95 seconds. The loaders will always try to load a casting into an empty machine before waiting for a mill to become available. The milling operation takes a time that varies according to a triangular distribution with a minimum, mode, and maximum of 50, 100, and 130 seconds. The loaders also remove the milled castings in a time that is uniformly distributed from 30 to 60 seconds. Simulate the operation of the company until 1000 castings are machined.

  a. What is the average number of castings in the shop?
  b. What is the maximum number of castings in the shop?
  c. How many castings did the inspector reject?
  d. What is the utilization of the loaders and the machines?
  e. Should we increase the number of loaders? Should we decrease them?

12. At the **Washington Logging Company,** a logging operation involves cutting a tree, which takes 25 minutes on the average. This time is normally distributed with a standard deviation of 10 minutes. The weight of a tree varies uniformly from 1000 to 1500 pounds, based on the density of fir. The logging company has five trucks with a capacity of 30 tons each. The truck will not move the trees to the sawmill until at least 15 tons of trees are loaded. The trip to the sawmill takes two hours, and the return trip also requires two hours. The truck takes 20 minutes to unload at the sawmill. Develop a simulation model of this facility. Determine whether the logging operation has sufficient capability to deliver all the trees cut in one month. The logging operation runs eight hours each day, 20 days a month.

    Embellishments:
    a. Include in the model the logging of alder trees in addition to fir trees. The alder trees are also cut in 25 minutes and segregated at the cutting site. The alder trees have weights ranging from 1500 to 2000 pounds, uniformly distributed. The trucks transport the alder and fir trees in separate trips.
    b. Vary the minimum requirement of trees that are loaded on a truck from 15 to 20 to 25 tons to determine the effect on throughput of this decision variable.

13. **Rancho Cucamonga Coil Company** is considering implementing a crane system as shown in Figure L13.27. Two cranes operate on separate tracks. Crane 1 serves only input station A and is used to transport steel cables to either output station C or D. Crane 2 transports copper cables from input station B to either output station D or E. The distances between the input and output stations are shown in Table 13.9.

    The arrival rate of steel cables to input station A is four rolls per hour (exponential), with 50 percent of the rolls sent to output station C

FIGURE L13.27

Crane system at the Rancho Cucamonga Coil Company.

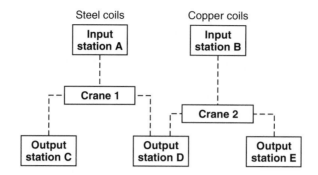

TABLE 13.9   **Distances Between Stations at the Rancho Cucamonga Coil Company**

From	To	Distance (feet)
Input A	Output C	100
Input A	Output D	150
Input A	Input B	300
Input B	Output E	100

and 50 percent to output station D. The arrival rate of copper cables to input station B is two rolls per hour (exponential), with 40 percent of the rolls sent to output station D and 60 percent to output station E. The travel speeds of both the cranes are 20 feet/minute without load and 10 feet/minute with load. The pickup and drop-off times are two minutes each. Simulate the crane system and collect statistics on the throughput from each input station to each output station.

Embellishments:

a. What will be the impact on throughput if the cranes run on a single track with output station D being on the common path of both the cranes? Model the interference in reaching output station D.

b. Establish a cost structure to evaluate and compare the single track (with interference) versus the separate track crane systems.

c. Evaluate the impact of giving priority to jobs that do not require the common track—that is, to jobs destined for output stations C and E.

Reference

S. V. Hoover and R. F. Perry, *Simulation: A Problem Solving Approach,* Addison Wesley, 1989, pp. B93–B95.

14 ADDITIONAL MODELING CONCEPTS

All things good to know are difficult to learn.
—Greek Proverb

In this lab we discuss some of the advanced but very useful concepts in ProModel. In Section L14.1 we model a situation where customers balk (leave) when there is congestion in the system. In Section L14.2 we introduce the concepts of macros and runtime interfaces. In Section L14.3 we show how to generate multiple scenarios for the same model. In Section L14.4 we discuss how to run multiple replications. In Section L14.5 we show how to set up and import data from external files. In Section L14.6 we discuss arrays of data. Table functions are introduced in Section L14.7. Subroutines are explained with examples in Section L14.8. Section L14.9 introduces the concept of arrival cycles. Section L14.10 shows the use of user distributions. Section L14.11 introduces the concepts and use of random number streams.

L14.1 Balking of Customers

Balking occurs when a customer looks at the size of a queue and decides to leave instead of entering the queue because it is too long. Balking is modeled by limiting the capacity or the number of entities that are allowed in the queue. Any customer that arrives when the queue is full must go away. This means loss of customers, loss of business, and lost opportunities.

Problem Statement
All American Car Wash is a five-stage operation that takes 2 ± 1 minutes for each stage of wash (Figure L14.1). The queue for the car wash facility can hold up to three cars. The cars move through the washing stages in order, one car not being

647

Figure L14.1

Locations and layout of All American Car Wash.

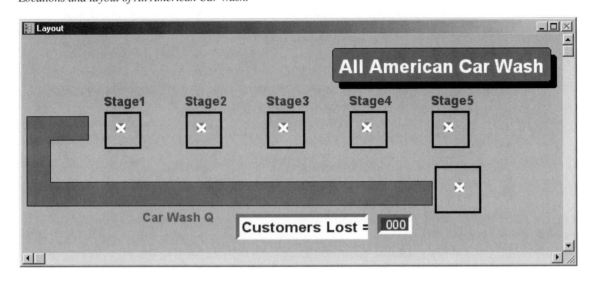

Figure L14.2

Customer arrivals at All American Car Wash.

```
 ********************************************************************************
                                    Arrivals
 ********************************************************************************

 Entity    Location  Qty each    First Time  Occurrences  Frequency      Logic
 --------  --------  ----------  ----------   ------------  ------------  ------
 Car       Arrive    1           0            inf          u (2.5,2) min
```

able to move until the car ahead of it moves. Cars arrive every 2.5 ± 2 minutes for a wash. If a car cannot get into the car wash facility, it drives across the street to Better Car Wash. Simulate for 100 hours.

- a. How many customers does All American Car Wash lose to its competition per hour (balking rate per hour)?
- b. How many cars are served per hour?
- c. What is the average time spent by a customer at the car wash facility?

The customer arrivals and processes/routings are defined as shown in Figures L14.2 and L14.3. The simulation run hours are set at 100. The total number of cars that balked and went away to the competition across the street is 104 in 100 hours (Figure L14.4). That is about 10.4 cars per hour. The total number of customers served is 2377 in 100 hours (Figure L14.5). That is about 23.77 cars per hour. We can also see that, on average, the customers spent 14 minutes in the car wash facility.

FIGURE L14.3

Process and routing tables at All American Car Wash.

```
                              Process                     Routing

Entity    Location    Operation              Blk  Output  Destination Rule
_____  _____   _____     ___  _____ _____ _____

Car       Arrive      If Contents(Car_Wash_Q) <3 Then
                          Route 1
                      Else
                         Begin
                            LOST=LOST+1
                            Route 2
                      End                      1   Car     Car_Wash_Q  FIRST 1
                                               2   Car     EXIT        FIRST 1
Car       Car_Wash_Q                           1   Car     Stage1      FIRST 1
Car       Stage1      wait u(2.1) min          1   Car     Stage2      FIRST 1
Car       Stage2      wait u(2.1) min          1   Car     Stage3      FIRST 1
Car       Stage3      wait u(2.1) min          1   Car     Stage4      FIRST 1
Car       Stage4      wait u(2.1) min          1   Car     Stage5      FIRST 1
Car       Stage5      wait u(2.1) min          1   Car     EXIT        FIRST 1
```

FIGURE L14.4

Cars that balked.

		Variables for lab_I_13_1, Normal Run		
Name	Total Changes	Avg Time Per Change (MIN)	Minimum Value	Maximum Value	Current Value
LOST	104	57	0	104	104

FIGURE L14.5

Customers served.

Entity Activity	Entity States	Variables	Location Costing	Resource Costing	Entity Costing	Logs

			Entity Activity for lab_I_13_1, Normal Run				
Name	Total Exits	Current Qty In System	Avg Time In System (MIN)	Avg Time In Move Logic (MIN)	Avg Time Wait For Res (MIN)	Avg Time In Operation (MIN)	Avg Time Blocked (MIN)
Car	2377	3	14	0	0	11	3

L14.2 Macros and Runtime Interface

If text, a set of statements, or a block of code is used many times in the model, it is convenient to substitute a macro and use the macro as many times as needed. Macros are created and edited in the following manner:

1. Select More Elements from the Build menu.

2. Select Macros.

The runtime interface (RTI) is a useful feature through which the user can interact with and supply parameters to the model without having to rewrite it. Every time the simulation is run, the RTI allows the user to change model parameters defined in the RTI. The RTI provides a user-friendly menu to change only the macros that the modeler wants the user to change. An RTI is a custom interface defined by a modeler that allows others to modify the model or conduct multiple-scenario experiments without altering the actual model data. All changes are saved along with the model so they are preserved from run to run. RTI parameters are based on macros, so they may be used to change any model parameter that can be defined using a macro (that is, any field that allows an expression or any logic definition).

An RTI is created and used in the following manner:

1. Select Macros from the Build menu and type in a macro ID.
2. Click the RTI button and choose Define from the submenu. This opens the RTI Definition dialog box.
3. Enter the Parameter Name that you want the macro to represent.
4. Enter an optional Prompt to appear for editing this model parameter.
5. Select the parameter type, either Unrestricted Text or Numeric Range.
6. For numeric parameters:
    a. Enter the lower value in the From box.
    b. Enter the upper value in the To box.
7. Click OK.
8. Enter the default text or numeric value in the Macro Text field.
9. Use the macro ID in the model to refer to the runtime parameter (such as operation time or resource usage time) in the model.
10. Before running the model, use the Model Parameters dialog box or the Scenarios dialog box to edit the RTI parameter.

Problem Statement

Widgets-R-Us Inc. receives various kinds of widget orders. Raw castings of widgets arrive in batches of one every five minutes. Some widgets go from the raw material store to the mill, and then on to the grinder. Other widgets go directly to the grinder. After grinding, all widgets go to degrease for cleaning. Finally, all widgets are sent to the finished parts store. The milling and grinding times vary depending on the widget design. However, the degrease time is seven minutes per widget. The layout of Widgets-R-Us is shown in Figure L14.6.

Define a runtime interface to allow the user to change the milling and grinding times every time the simulation is run. Also, allow the user to change the total quantity of widgets being processed. Track the work-in-process inventory (WIP) of widgets. In addition, define another variable to track the production (PROD_QTY) of finished widgets.

The macros are defined as shown in Figure L14.7. The runtime interface for the milling time is shown in Figure L14.8. Figure L14.9 shows the use of the parameters Mill_Time and Grind_Time in the process and routing tables. To

FIGURE L14.6

*Layout of
Widgets-R-Us.*

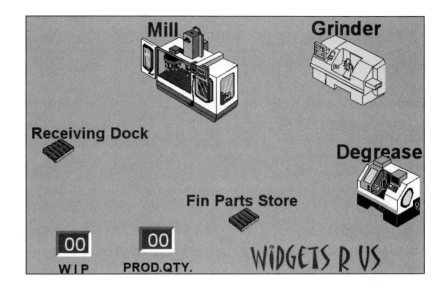

FIGURE L14.7

Macros created for Widgets-R-Us simulation model.

ID	Text...	Options
Mill_Time_Avg	5	RTI
Mill_Time_sd	2	RTI
Order_Qty	100	RTI
Grind_Time_Avg	12	RTI
Grind_Time_Halfwidth	3	RTI

FIGURE L14.8

*The runtime interface
defined for the milling
operation.*

RTI Definition for Mill_Time_Avg

Parameter
Name: Mill_Time_Avg

Prompt: What is the Average Milling Time?
Milling time is a normal distribution.
Enter the Avg. time in minutes.

○ Unrestricted Text

○ Record Range

● Numeric Range

From 5   To 20

[OK]   [Cancel]   [Help]

Figure L14.9

The process and routing tables showing the Mill_Time and Grind_Time parameters.

		Process			Routing		
Entity	Location	Operation	Blk	Output	Destination	Rule	Move Logic
Casting	Receiving_Dock	WIP=WIP+1 WAIT 0	1	Widget	Mill	FIRST 1	MOVE FOR 1
Widget	Mill	Wait N(Mill_Time_avg,Mill_Time_sd)	1	Widget	Grinder	FIRST 1	MOVE FOR 1
Widget	Grinder	Wait U(Grind_Time_Avg,Grind_Time_Halfwidth)	1	Widget	Degrease	FIRST 1	MOVE FOR 1
Widget	Degrease	wait 7 WIP = WIP - 1 PROD_QTY = PROD_QTY + 1	1	Widget	EXIT	FIRST 1	move for 1

Figure L14.10

Model Parameters in the Simulation menu.

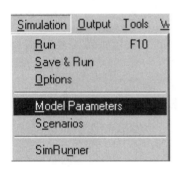

Figure L14.11

The Model Parameters dialog box.

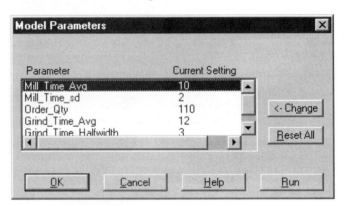

Figure L14.12

The Grind_Time_Halfwidth parameter dialog box.

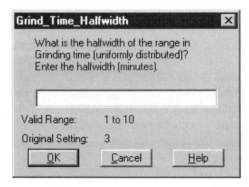

view or change any of the model parameters, select Model Parameters from the Simulation menu (Figure L14.10). The model parameters dialog box is shown in Figure L14.11. To change the Grind_Time_Halfwidth, first select it from the model parameters list, and then press the Change button. The Grind_Time_Halfwidth dialog box is shown in Figure L14.12.

L14.3 Generating Scenarios

When a modeler alters one or more input parameters or elements of the model, he or she creates model scenarios. For example, running a model first with 5 units of pallets and then with 10 units to test the effect of increased resource availability is the same thing as creating two scenarios.

A model builder has the option of defining multiple scenarios for a model using the runtime interface. Using scenarios allows the model user to alter various model parameters to run a series of "what-if" analyses without changing the model directly. This technique makes it easy for analysts who have no interest in dealing with the nuts and bolts of a model to make changes safely and conveniently to the model to learn what would happen under a variety of circumstances.

The Scenarios dialog box is accessed from the Simulation menu (Figure L14.13). Scenarios are saved with the model for future use. They are created using the Scenarios dialog box (Figure L14.14).

For the **Shipping Boxes Unlimited** example in Lab 7, Section L7.7.2, generate two scenarios (Figure L14.14):

Scenario 1: Five pallets in circulation in the system.

Scenario 2: Ten pallets in circulation in the system.

Define a model parameter for the number of pallets in the system (no_of_pallets) in the runtime interface. The RTI dialog box is obtained from the Build/More Elements/Macros menu. Edit the scenarios (Figure L14.15) and change the parameter no_of_pallets to the values of 5 and 10, respectively, for the two scenarios. In the Arrivals menu, change the occurrences for Pallet to the parameter name no_of_pallets. This parameter will take on the values of 5 and 10 in scenarios 1 and 2, respectively (Figure L14.16).

To run the scenarios, click on the Run Scenarios button in the Scenarios dialog box. Both scenarios will be run in a sequential manner, and reports for both scenarios will be created (Figure L14.17).

FIGURE L14.13

The Scenarios menu.

FIGURE L14.14

Scenarios defined for Shipping Boxes Unlimited in Section L7.7.2.

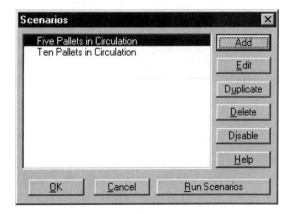

FIGURE L14.15

Editing the scenario parameter.

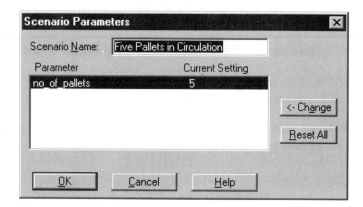

FIGURE L14.16

Arrivals table for Shipping Boxes Unlimited in Section L7.7.2.

```
***********************************************************************************
                                  Arrivals                                      *
***********************************************************************************

Entity      Location          Qty each   First Time  Occurrences   Frequency   Logic
----------  ----------------  ---------  ----------  ------------  ----------  -------
Monitor     CONU1             1          0           inf           E(5)
Empty_Box   CONU2             1          0           inf           E(4)
Pallet_     Pallet_Conveyor   1          0           no_of_pallets  1
```

FIGURE L14.17

Reports created for both the scenarios.

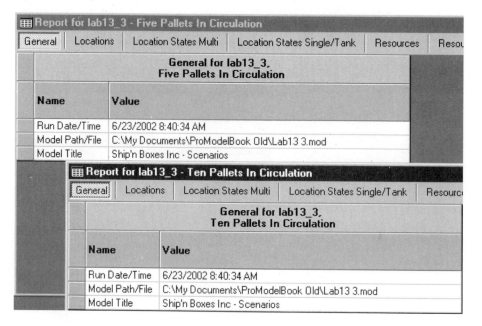

L14.4 External Files

External files are used to read data directly into the simulation model or to write output data from the simulation model. External files are also used to store and access process sheets, arrival schedules, shift schedules, bills of material, delivery schedules, and so forth. All external files used with a model must be listed in the External Files Editor and are accessed from the Build menu. All external files must be generally in the .WK1 format only and can be of the following six types.

General Read File.   These files contain numeric values read into a simulation model using a READ statement. A space, comma, or end of line can separate the data values. Any nonnumeric data are skipped automatically. The syntax of the READ statement is as follows:

READ <file ID>, <variable name>

If the same file is to be read more than once in the model, it may be necessary to reset the file between each reading. This can be achieved by adding an arbitrary end-of-file marker 99999 and the following two lines of code:

```
Read MydataFile1, Value1
If Value1D99999 Then Reset MydataFile1
```

The data stored in a general read file must be in ASCII format. Most spreadsheet programs (Lotus 1-2-3, Excel, and others) can convert spreadsheets to ASCII files (MydataFile1.TXT).

General Write.   These files are used to write text strings and numeric values using the WRITE and WRITELINE statements. Text strings are enclosed in quotes when written to the file, with commas automatically appended to strings. This enables the files to be read into spreadsheet programs like Excel or Lotus 1-2-3 for viewing. Write files can also be written using the XWRITE statement, which gives the modeler full control over the output and formatting. If you write to the same file more than once, either during multiple replications or within a single replication, the new data are appended to the previous data.

The WRITE statement writes to a general write file. The next item is written to the file immediately after the previous item. Any file that is written to with the WRITE statement becomes an ASCII text file and ProModel attaches an end-of-file marker automatically when it closes the file.

WRITE <file ID>, <string or numeric expression>

The WRITELINE statement writes information to a general write file and starts a new line. It always appends to the file unless you have RESET the file. Any file that is written to with the WRITELINE statement becomes an ASCII text file, and ProModel attaches an end-of-file marker automatically when it closes the file.

The syntax of the WRITE, WRITELINE, and the XWRITE statements is as follows:

```
WRITE MyReport, "Customer Service Completed At:"
WRITELINE MyReport, CLOCK(min)
```

The XWRITE statement allows the user to write in any format he or she chooses.

```
XWRITE <file ID>, <string or numeric expression>
XWRITE MyReport2, "Customer Service Completed At:" $FORMAT(Var1,5,2)
```

Entity Location.   An entity–location file contains numeric expressions listed by entity and location names. Entity names should appear across the top row, beginning in column 2, while location names should be entered down the first column beginning in row 2. A numeric expression for each entity–location combination is entered in the cell where the names intersect. These files are always in .WK1 format. Both Excel and Lotus 1-2-3 files can be saved in .WK1 format. When saving an entity–location file, make sure to choose Save As from the File menu (Excel or Lotus 1-2-3) and change the file format to .WK1. Other .TXT files also can be opened in Excel or Lotus 1-2-3 and, after minor editing, saved in .WK1 format.

Arrival.   An arrivals file is a spreadsheet file (.WK1 format only) containing arrival information normally specified in the Arrival Editor. One or more arrival files may be defined and referenced in the External Files Editor. Arrival files are automatically read in following the reading of the Arrival Editor data. The column entries must be as in Table L14.1.

Shift.   A shift file record is automatically created in the External Files Editor when a shift is assigned to a location or resource. If shifts have been assigned, the name(s) of the shift file(s) will be created automatically in the External Files Editor. Do not attempt to create a shift file record in the External Files Editor yourself.

.DLL.   A .DLL file is needed when using external subroutines through the XSUB() function.

Problem Statement

At the **Pomona Castings, Inc.,** castings arrive for processing at the rate of 12 per hour (average interarrival time assumed to be five minutes). Seventy percent

Table L14.1   Entries in an External Arrivals File

Column	Data
A	Entity name
B	Location name
C	Quantity per arrival
D	Time of first arrival
E	Number of arrivals
F	Frequency of arrivals
G	Attribute assignments

FIGURE L14.18

An external entity-location file in .WK1 format.

P14_5.wk1

	A	B	C
1		Casting_A	Casting_B
2	Mill	N(5,2)min	
3	Grinder	U(11,1) min	U(9,1) min
4	Degrease	7 min	7 min

FIGURE L14.19

File ID and file name created for the external file.

External Files

ID	Type...	File Name...
SvcTms	Entity Location	A:\P14_5.wkl

FIGURE L14.20

The process table referring to the file ID of the external file.

Process			Routing				
Entity	Location	Operation	Blk	Output	Destination	Rule	Move Logic
Casting	Receiving_Dock	WIP=WIP+1					
			1	Casting_A	Receiving_Dock	0.700000 1	
				Casting_B	Receiving_Dock	0.300000 1	
Casting_A	Receiving_Dock		1	Casting_A	Mill	FIRST 1	MOVE FOR 1
Casting_A	Mill	WAIT SvcTms()	1	Casting_A	Grinder	FIRST 1	MOVE FOR 1
Casting_A	Grinder	WAIT SvcTms()	1	Casting_A	Degrease	FIRST 1	MOVE FOR 1
Casting_B	Receiving_Dock		1	Casting_B	Grinder	FIRST 1	MOVE FOR 1
Casting_B	Grinder	WAIT SvcTms()	1	Casting_B	Degrease	FIRST 1	MOVE FOR 1
ALL	Degrease	WAIT SvcTms()	1	ALL	Fin_Parts_Store	FIRST 1	MOVE FOR 1
ALL	Fin_Parts_Store	wait 0 WIP = WIP - 1 PROD_QTY = PROD_QTY + 1					
			1	ALL	EXIT	FIRST 1	move for 1

of the castings are processed as casting type A, while the rest are processed as casting type B.

For Pomona Castings, Inc., create an entity–location file named P14_5 .WK1 to store the process routing and process time information (Figure L14.18). In the simulation model, read from this external file to obtain all the process information. Build a simulation model and run it for 100 hours. Keep track of the work-in-process inventory and the production quantity.

Choose Build/More Elements/External Files. Define the ID as SvcTms. The Type of file is Entity Location. The file name (and the correct path) is also provided (Figure L14.19). In the Process definition, use the file ID (Figure L14.20) instead of the actual process time—for example, WAIT SvcTms(). Change the file path to point to the appropriate directory and drive where the external file is located. A snapshot of the simulation model is shown in Figure L14.21.

FIGURE L14.21

A snapshot of the simulation model for Pomona Castings, Inc.

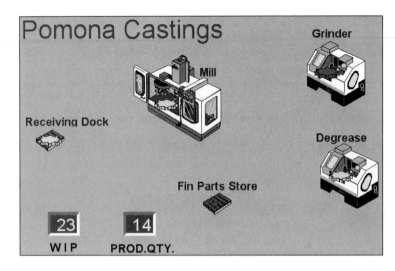

L14.5 Arrays

An *array* is a collection of values that are related in some way such as a list of test scores, a collection of measurements from some experiment, or a sales tax table. An array is a structured way of representing such data.

An array can have one or more dimensions. A two-dimensional array is useful when the data can be arranged in rows and columns. Similarly, a three-dimensional array is appropriate when the data can be arranged in rows, columns, and ranks. When several characteristics are associated with the data, still higher dimensions may be appropriate, with each dimension corresponding to one of these characteristics.

Each cell in an array works much like a variable. A reference to a cell in an array can be used anywhere a variable can be used. Cells in arrays are usually initialized to zero, although initializing cells to some other value can be done in the initialization logic. A WHILE-DO loop can be used for initializing array cell values.

Suppose that electroplating bath temperatures are recorded four times a day at each of three locations in the tank. These temperature readings can be arranged in an array having four rows and three columns (Table L14.2). These 12 data items can be conveniently stored in a two-dimensional array named Temp[4,3] with four rows and three columns.

An external Excel file (BathTemperature.xls) contains these bath temperature data (Figure L14.22). The information from this file can be imported into an array in ProModel (Figure L14.23) using the Array Editor. When you import data from an external Excel spreadsheet into an array, ProModel loads the data from left to right, top to bottom. Although there is no limit to the quantity of values you may use, ProModel supports only two-dimensional arrays. Figure L14.24 shows the Import File dialog in the Array Editor.

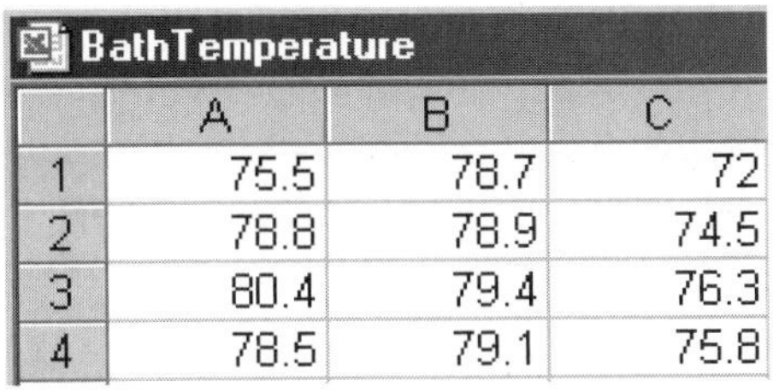

FIGURE L14.22

An external file containing bath temperatures.

FIGURE L14.23

The Array Editor in ProModel.

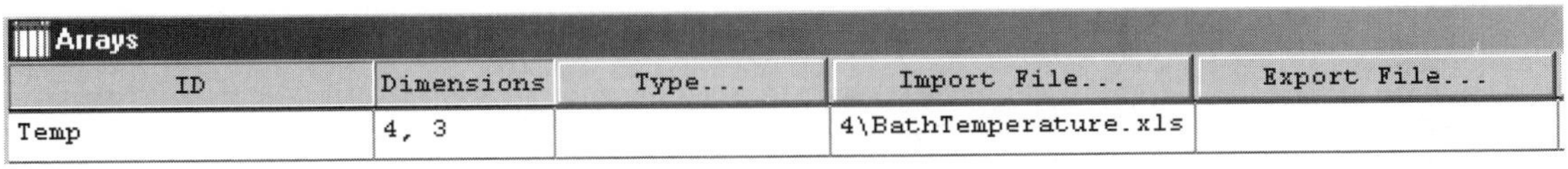

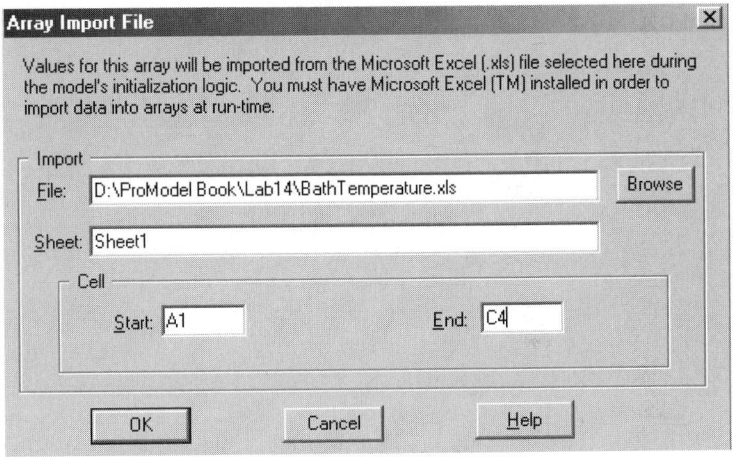

FIGURE L14.24

The Import File dialog in the Array Editor.

TABLE L14.2   Electroplating Bath Temperatures

Time	Location 1	Location 2	Location 3
1	75.5	78.7	72.0
2	78.8	78.9	74.5
3	80.4	79.4	76.3
4	78.5	79.1	75.8

Value of an array can also be exported to an Excel spreadsheet (*.xls) during the termination logic. Microsoft Excel must have been installed in your computer in order to export data from an array at runtime. Array exporting can be turned off in the Simulations Options dialog.

Problem Statement

Table L14.3 shows the status of orders at the beginning of the month at **Joe's Job-shop.** In his shop, Joe has three machines through which three types of jobs are routed. All jobs go to all machines, but with different routings. The data for job routings and processing times (exponential) are given in Table L14.4. The processing times are given in minutes. Use a one-dimensional array to hold the information on the order status. Simulate and find out how long it will take for Joe to finish all his pending orders.

The layout, locations, and arrival of jobs at Joe's Jobshop are shown in Figures L14.25, L14.26, and L14.27. The pending order array is shown in

FIGURE L14.25

Layout of Joe's Jobshop.

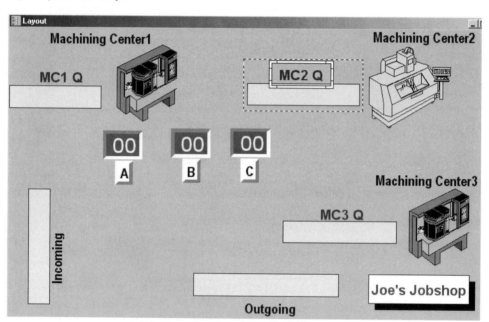

TABLE L14.3	Data for Pending Orders at Joe's Jobshop
Jobs	*Number of Pending Orders*
A	25
B	27
C	17

TABLE L14.4	Process Routings and Average Process Times	
Jobs	*Machines*	*Process Times*
A	2–3–1	45–60–75
B	1–2–3	70–70–50
C	3–1–2	50–60–60

Figure L14.28. The initialization logic is shown in Figure L14.29. The processes and routings are shown in Figure L14.30.

It took about 73 hours to complete all the pending work orders. At eight hours per day, it took Joe a little over nine days to complete the backlog of orders.

FIGURE L14.26

Locations at Joe's Jobshop.

Icon	Name	Cap.	Units	DTs...	Stats...	Rules...
	Machining_Center1	1	1	None	Time Series	Oldest
	Machining_Center2	1	1	None	Time Series	Oldest
	Machining_Center3	1	1	None	Time Series	Oldest
	Incoming	INFINITE	1	None	Time Series	Oldest, FIFO
	Outgoing	INFINITE	1	None	Time Series	Oldest, FIFO
	MC1_Q	inf	1	None	Time Series	Oldest, FIFO
	MC2_Q	inf	1	None	Time Series	Oldest, FIFO
	MC3_Q	inf	1	None	Time Series	Oldest, FIFO

FIGURE L14.27

Arrival of jobs at Joe's Jobshop.

```
(XXXXXXXXXXXXXXXXXXXXXXXXXXXXXXXXXXXXXXXXXXXXXXXXXXXXXXXXXXXXXXXXXXXXXXXXXXXXXXXXXXXX)
                              Arrivals
(XXXXXXXXXXXXXXXXXXXXXXXXXXXXXXXXXXXXXXXXXXXXXXXXXXXXXXXXXXXXXXXXXXXXXXXXXXXXXXXXXXXX)


Entity     Location  Qty each    First Time  Occurrences         Frequency   Logic
--------   --------  ---------   ----------  ---------------     ---------   -------
Job_A      Incoming  1           0           Order_Pending[1]    e(60)
Job_B      Incoming  1           0           Order_Pending[2]    e(60)
Job_C      Incoming  1           0           Order_Pending[3]    e(60)
```

FIGURE L14.28

Pending order array for Joe's Jobshop.

ID	Dimensions	Type...	Import File...	Export File...	Notes...
Order_Pending	3	Integer			

FIGURE L14.29

Initialization logic in the General Information menu.

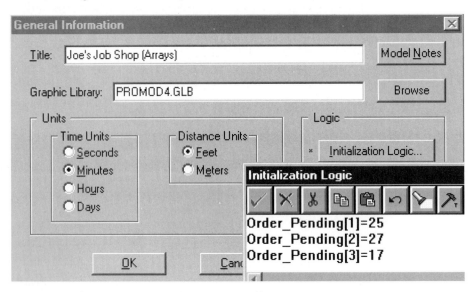

FIGURE L14.30

Process and routing tables for Joe's Jobshop.

	Process				Routing	
Entity	Location	Operation	Blk	Output	Destination	Rule
Job_A	Incoming		1	Job_A	MC2_Q	FIRST 1
Job_A	MC2_Q		1	Job_A	Machining_Center2	FIRST 1
Job_A	Machining_Center2	wait e(45) min	1	Job_A	MC3_Q	FIRST 1
Job_A	MC3_Q		1	Job_A	Machining_Center3	FIRST 1
Job_A	Machining_Center3	wait e(60) min	1	Job_A	MC1_Q	FIRST 1
Job_A	MC1_Q		1	Job_A	Machining_Center1	FIRST 1
Job_A	Machining_Center1	wait e(75) min				
		qty_A = qty_A + 1	1	Job_A	EXIT	FIRST 1
Job_B	Incoming		1	Job_B	MC1_Q	FIRST 1
Job_B	MC1_Q		1	Job_B	Machining_Center1	FIRST 1
Job_B	Machining_Center1	wait e(70) min	1	Job_B	MC2_Q	FIRST 1
Job_B	MC2_Q		1	Job_B	Machining_Center2	FIRST 1
Job_B	Machining_Center2	wait e(70) min	1	Job_B	MC3_Q	FIRST 1
Job_B	MC3_Q		1	Job_B	Machining_Center3	FIRST 1
Job_B	Machining_Center3	wait e(50) min				
		qty_B = qty_B + 1	1	Job_B	EXIT	FIRST 1
Job_C	Incoming		1	Job_C	MC3_Q	FIRST 1
Job_C	MC3_Q		1	Job_C	Machining_Center3	FIRST 1
Job_C	Machining_Center3	wait e(50) min	1	Job_C	MC1_Q	FIRST 1
Job_C	MC1_Q		1	Job_C	Machining_Center1	FIRST 1
Job_C	Machining_Center1	wait e(60) min	1	Job_C	MC2_Q	FIRST 1
Job_C	MC2_Q		1	Job_C	Machining_Center2	FIRST 1
Job_C	Machining_Center2	wait e(60) min				
		qty_C = qty_C + 1	1	Job_C	EXIT	FIRST 1

L14.6 Table Functions

Table functions are a useful way to specify and set up relationships between independent and dependent variables. A table function editor is provided in Pro-Model to build tables. This feature is especially useful when building empirical distributions or functions into a model. Table functions are defined by the user and return a dependent value based on the independent value passed as the function argument. Independent values are entered in ascending order. The dependent value is calculated by linear interpolation. For a linear function, only two endpoints are needed. For a nonlinear function, more than two reference values are specified.

Problem Statement

Customers arrive at the **Save Here Grocery** store with a mean time between arrival (exponential) that is a function of the time of the day, (the number of hours elapsed since the store opening), as shown in Table L14.5. The grocery store consists of two aisles and a cashier. Once inside the store, customers may choose to shop in one, two, or none of the aisles. The probability of shopping aisle one is 0.75, and for aisle two it is 0.5. The number of items selected in each aisle is described by a normal distribution with a mean of 15 and a standard deviation of 5. The time these shoppers require to select an item is five minutes. When all desired items have been selected, customers queue up at the cashier to pay for them. The time to check out is a uniformly distributed random variable that depends on the number of items purchased. The checkout time per item is 0.6 ± 0.5 minutes (uniformly distributed). Simulate for 10 days (16 hours/day).

The locations, the arrival of entities, the processes and routings, and the layout of the grocery store are shown in Figures L14.31, L14.32, L14.33, and L14.34, respectively. The arrival frequency is derived from the table function `arrival time(HR)`:

```
e(arrival_time(CLOCK(HR) - 16*TRUNC(CLOCK(HR)/16))) min
```

The table function `arrival_time` is shown in Figure L14.35.

TABLE L14.5 **Time between Arrivals at Save Here Grocery**

Time of Day	Time between Arrivals (minutes)
0	50
2	40
4	25
6	20
8	20
10	20
12	25
14	30
16	30

FIGURE L14.31

Locations in Save Here Grocery.

```
**********************************************************************
                              Locations
**********************************************************************

Name           Cap        Units Stats        Rules            Cost
-----------    --------   ----- -----------  ---------------  -----
Aisle1         INFINITE   1     Time Series  Oldest, FIFO,
Aisle2         INFINITE   1     Time Series  Oldest, FIFO,
Incoming       1          1     Time Series  Oldest, ,
Cashier        1          1     Time Series  Oldest, ,
Cash_Q         INFINITE   1     Time Series  Oldest, FIFO,
check          1          1     Time Series  Oldest, ,
```

FIGURE L14.32

Arrivals at Save Here Grocery.

```
*********************************************************************************
                              Arrivals                                *
*********************************************************************************

Entity    Location  Qty each   First Time  Occurrences  Frequency                                              Logic
--------  --------  --------   ----------  -----------  -------------------------------------------------------
Customer  Incoming  1          0           inf          e(arrival_time(CLOCK(HR)-16*TRUNC(CLOCK(HR)/16))) min
```

FIGURE L14.33

Process and routing tables for Save Here Grocery.

```
                         Process                                    Routing

Entity     Location  Operation               Blk   Output     Destination  Rule
--------   --------  -------------------      ----  ---------  -----------  --------
Customer   Incoming                          1     Customer   Aisle1       0.750000 1
                                                   Customer   check        0.250000
Customer   Aisle1    INT item1
                     item1 = N(15,5)
                     wait item1*5 min
                     item = item+item1
                                             1     Customer   check        FIRST 1
Customer   check                             1     Customer   Aisle2       0.500000 1
                                                   Customer   Cash_Q       0.500000
Customer   Aisle2    INT item2
                     item2=N(15,5)
                     wait  item2*5 min
                     item = item+item2
                                             1     Customer   Cash_Q       FIRST 1
Customer   Cash_Q                            1     Customer   Cashier      FIRST 1
Customer   Cashier   wait item*U(.6,.5) min
                                             1     Customer   EXIT         FIRST 1
```

FIGURE **L14.34**

The layout of Save Here Grocery.

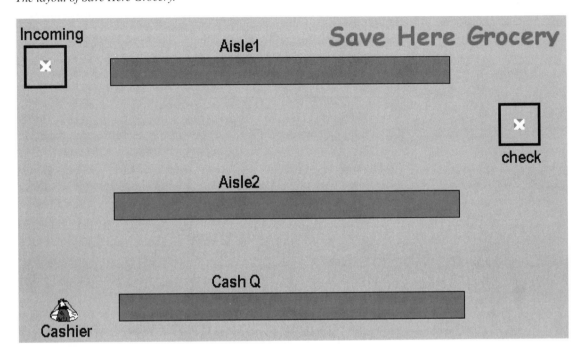

FIGURE **L14.35**

The Table Functions dialog box.

ID	Table...
arrival_time	Defined
0	50
2	40
4	25
6	20
8	20
10	20
12	25
14	30
16	30

L14.7 Subroutines

A subroutine is a separate section of code intended to accomplish a specific task. It is a user-defined command that can be called upon to perform a block of logic and optionally return a value. Subroutines may have parameters or local variables (local to the subroutine) that take on the values of arguments passed to the subroutine. There are three variations for the use of subroutines in ProModel:

1. A subroutine is called by its name from the main block of code.
2. A subroutine is processed independently of the calling logic so that the calling logic continues without waiting for the subroutine to finish. An ACTIVATE statement followed by the name of the subroutine is needed.
3. Subroutines written in an external programming language can be called using the XSUB() function.

Subroutines are defined in the Subroutines Editor in the More Elements section of the Build menu. For more information on ProModel subroutines, please refer to the ProModel Users Guide.

Problem Statement

At **California Gears,** gear blanks are routed from a fabrication center to one of three manual inspection stations. The operation logic for the gears is identical at each station except for the processing times, which are a function of the individual inspectors. Each gear is assigned two attributes during fabrication. The first attribute, OuterDia, is the dimension of the outer diameter of the gear. The second attribute, InnerDia, is the dimension of the inner diameter of the gear. During the fabrication process, the outer and inner diameters are machined with an average of 4.015 ± 0.01 and 2.015 ± 0.01 (uniformly distributed). These dimensions are tested at the inspection stations and the values entered into a text file (Quality.doc) for quality tracking. After inspection, gears are routed to a shipping location if they pass inspection, or to a scrap location if they fail inspection.

Gear blanks arrive at the rate of 12 per hour (interarrival time exponentially distributed with a mean of five minutes). The fabrication and inspection times (normal) are shown in Table L14.6. The specification limits for the outer and inner

TABLE L14.6   Fabrication and Inspection Times at California Gears

	Mean	*Standard Deviation*
Fabrication time	3.2	0.1
Inspector 1	4	0.3
Inspector 2	5	0.2
Inspector 3	4	0.1

diameters are given in Table L14.7. The layout, locations, entities, and arrival of raw material are shown in Figures L14.36, L14.37, L14.38, and L14.39 respectively. The subroutine defining routing logic is shown in Figure L14.40. Figure L14.41 shows the processes and routing logic. The external file in which quality data will be written is defined in Figure L14.42. Figure L14.43 shows a portion of the actual gear rejection report.

FIGURE L14.36

Layout of California Gears.

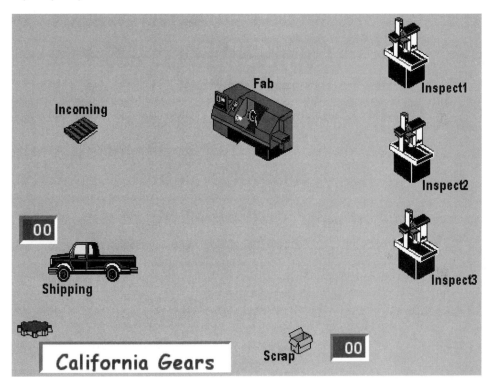

TABLE L14.7 The Specification Limits for Gears

	Lower Specification Limit	Upper Specification Limit
Outer diameter	4.01"	4.02"
Inner diameter	2.01"	2.02"

FIGURE L14.37

Locations at California Gears.

```
(×××××××××××××××××××××××××××××××××××××××××××××
                                    Locations
(×××××××××××××××××××××××××××××××××××××××××××××

Name        Cap  Units  Stats        Rules      Cost
_____    ___  _____  _____  _____  _____

Fab         1    1      Time Series  Oldest,  ,
Inspect1    1    1      Time Series  Oldest,  ,
Inspect2    1    1      Time Series  Oldest,  ,
Inspect3    1    1      Time Series  Oldest,  ,
Incoming    inf  1      Time Series  Oldest,  ,
Shipping    inf  1      Time Series  Oldest,  ,
Scrap       inf  1      Time Series  Oldest,  ,
```

FIGURE L14.38

Entities at California Gears.

```
(×××××××××××××××××××××××××××××××××××××××××××××>
                                    Entities
(×××××××××××××××××××××××××××××××××××××××××××××>

Name        Speed (fpm)  Stats        Cost
_____  _____  _____  _____

Gear_Blank  150          Time Series
Gear        150          Time Series
Gear_Scrap  150          Time Series
```

FIGURE L14.39

Arrival of raw material at California Gears.

```
××××××××××××××××××××××××××××××××××××××××××××××××××××××××××××××××××××××××××
                                    Arrivals
××××××××××××××××××××××××××××××××××××××××××××××××××××××××××××××××××××××××××

Entity      Location  Qty each  First Time  Occurrences  Frequency  Logic
_____  _____  _____  _____  _____  _____  _____

Gear_Blank  Incoming  1         0           inf          e(5) min
```

FIGURE L14.40

Subroutine defining routing logic.

```
×××××××××××××××××××××××××××××××××××××××××××××××××××××××××××××××××××××××××××
                                    Subroutines                              ×
×××××××××××××××××××××××××××××××××××××××××××××××××××××××××××××××××××××××××××××

ID          Type      Parameter  Type      Logic
_____    _____   _____  _____   _____

InspProc    None      M          Integer   WAIT N(M,SD)
                      SD         Real      If OuterDia > 4.01 and
                                                OuterDia < 4.02 and
                                                    InnerDia > 2.01 and
                                                        InnerDia < 2.02 THEN
                                               ROUTE 1
                                           ELSE
                                               ROUTE 2
```

FIGURE L14.41

Process and routing tables for California Gears.

```
*******************************************************************************
*                              Processing                                    *
*******************************************************************************

                          Process                        Routing

Entity      Location  Operation           Blk  Output      Destination Rule     Move Logic
----------  --------  ------------------   ---  ----------  ----------- -------  ----------
Gear_Blank  Incoming                       1    Gear_Blank  Fab         FIRST 1
Gear_Blank  Fab       wait N(3.2,.1) min
                      OuterDia =U(4.015,.01)
                      InnerDia = U(2.015,.01)
                      WRITE QUALITYINFO, "Gear Rejected At Time="
                      WRITE QUALITYINFO, CLOCK(min),3,2
                      WRITE QUALITYINFO, "Outer Dia="
                      WRITE QUALITYINFO, OuterDia,1,3
                      WRITE QUALITYINFO, "Inner Dia="
                      WRITE QUALITYINFO, InnerDia,1,3

                                           1    Gear        Inspect1    TURN 1
                                                Gear        Inspect2    TURN
                                                Gear        Inspect3    TURN
Gear        Inspect1  InspProc(4,.3)       1    Gear        Shipping    FIRST 1
                                           2    Gear_Scrap  Scrap       FIRST 1
Gear        Inspect2  InspProc(5,.2)       1    Gear        Shipping    FIRST 1
                                           2    Gear_Scrap  Scrap       FIRST 1
Gear        Inspect3  InspProc(4,.1)       1    Gear        Shipping    FIRST 1
                                           2    Gear_Scrap  Scrap       FIRST 1
Gear        Shipping  ship=ship+1          1    Gear        EXIT        FIRST 1
Gear_Scrap  Scrap     reject=reject+1      1    Gear_Scrap  EXIT        FIRST 1
```

FIGURE L14.42

External file for California Gears.

```
*******************************************************************
                                              External Files
*******************************************************************

ID            Type            File Name        Prompt
-----------   -------------   -------------    -------
qualityinfo   General Write   quality.doc
```

FIGURE L14.43

Gear rejection report (partial) for California Gears.

```
"Gear Rejected At Time=",    3.40, "Outer Dia=", 4.013, "Inner Dia=", 2.015,
"Gear Rejected At Time=",    7.74, "Outer Dia=", 4.013, "Inner Dia=", 2.012,
"Gear Rejected At Time=",   11.09, "Outer Dia=", 4.014, "Inner Dia=", 2.022,
"Gear Rejected At Time=",   14.38, "Outer Dia=", 4.011, "Inner Dia=", 2.008,
"Gear Rejected At Time=",   17.64, "Outer Dia=", 4.005, "Inner Dia=", 2.017,
"Gear Rejected At Time=",   20.95, "Outer Dia=", 4.009, "Inner Dia=", 2.015,
"Gear Rejected At Time=",   23.98, "Outer Dia=", 4.005, "Inner Dia=", 2.018,
"Gear Rejected At Time=",   27.18, "Outer Dia=", 4.006, "Inner Dia=", 2.024,
"Gear Rejected At Time=",   30.48, "Outer Dia=", 4.013, "Inner Dia=", 2.015,
"Gear Rejected At Time=",   33.50, "Outer Dia=", 4.008, "Inner Dia=", 2.023,
"Gear Rejected At Time=",   36.78, "Outer Dia=", 4.016, "Inner Dia=", 2.017,
"Gear Rejected At Time=",   39.97, "Outer Dia=", 4.007, "Inner Dia=", 2.009,
"Gear Rejected At Time=",   45.44, "Outer Dia=", 4.017, "Inner Dia=", 2.009,
"Gear Rejected At Time=",   48.53, "Outer Dia=", 4.006, "Inner Dia=", 2.024,
```

L14.8 Arrival Cycles

Sometimes the arrival of customers to a system follows a definite cyclic pattern. Examples are arrival of flights to an airport, arrival of buses to a bus stand or depot, traffic on the freeway, and customers coming to a takeout restaurant. Arrival cycles are defined in the Arrival Cycles edit table (Figure L14.44), which is part of More Elements in the Build menu.

Arrivals can be expressed either as a percentage or in quantity. The arrival percentage or quantity can be in either a cumulative or a noncumulative format. The time is always specified in a cumulative format. Once the arrival cycles are defined, they can be assigned to an arrivals record in the Arrivals edit table.

Problem Statement
At the **Newport Beach Burger** stand, there are three peak periods of customer arrivals: breakfast, lunch, and dinner (Table L14.8). The customer arrivals taper out

FIGURE L14.44

Arrival Cycles edit menu.

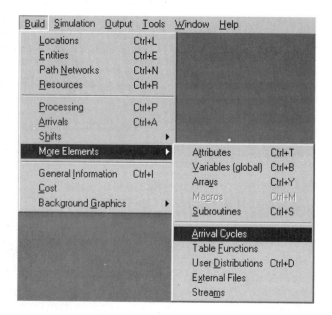

TABLE L14.8   **Customer Arrival Pattern at Newport Beach Burger**

From	*To*	*Percent*
6:00 A.M.	6:30 A.M.	5
6:30 A.M.	8:00 A.M.	20
8:00 A.M.	11:00 A.M.	5
11:00 A.M.	1:00 P.M.	35
1:00 P.M.	5:00 P.M.	10
5:00 P.M.	7:00 P.M.	20
7:00 P.M.	9:00 P.M.	5

before and after these peak periods. The same cycle of arrivals repeats every day. A total of 100 customers visit the store on an average day (normal distribution with a standard deviation of five). Upon arrival, the customers take Uniform(5 ± 2) minutes to order and receive food and Normal(15, 3) minutes to eat, finish business discussions, gossip, read a newspaper, and so on. The restaurant currently has only one employee (who takes the order, prepares the food, serves, and takes the money). Simulate for 100 days.

The locations, processes and routings, arrivals, and arrival quantities are shown in Figures L14.45, L14.46, L14.47, and L14.48, respectively. The arrival cycles and the probability density function of arrivals at Newport Beach Burger are shown in Figure L14.49. A snapshot of the simulation model is shown in Figure L14.50.

FIGURE L14.45

Locations at Newport Beach Burger.

```
********************************************************************
                              Locations
********************************************************************

Name        Cap       Units Stats        Rules          Cost
----------- --------- ----- ------------ -------------- ------
Incoming    inf       1     Time Series  Oldest,  ,
Server      1         1     Time Series  Oldest,  ,
Server_Q    INFINITE  1     Time Series  Oldest,  FIFO,
dining      1         6     Time Series  Oldest,  , First
dining.1    1         1     Time Series  Oldest,  ,
dining.2    1         1     Time Series  Oldest,  ,
dining.3    1         1     Time Series  Oldest,  ,
dining.4    1         1     Time Series  Oldest,  ,
dining.5    1         1     Time Series  Oldest,  ,
dining.6    1         1     Time Series  Oldest,  ,
```

FIGURE L14.46

Process and routing tables at Newport Beach Burger.

		Process			Routing		
Entity	Location	Operation	Blk	Output	Destination	Rule	Move Logic
Customer	Incoming		1	Customer	Server_Q	FIRST 1	
Customer	Server_Q		1	Customer	Server	FIRST 1	
Customer	Server	wait U(5,2) min	1	Customer	dining	FIRST 1	
Customer	dining	wait N(15,3) min	1	Customer	EXIT	FIRST 1	

FIGURE L14.47

Arrivals at Newport Beach Burger.

```
**********************************************************************
                              Arrivals                             *
**********************************************************************

Entity    Location  Qty each                First Time Occurrences Frequency Logic
--------- --------- ----------------------- ---------- ----------- --------- -----
Customer  Incoming  N(100,5); Burger_Arrival 0            100         24
```

FIGURE L14.48

Arrival quantity defined for the Arrivals table.

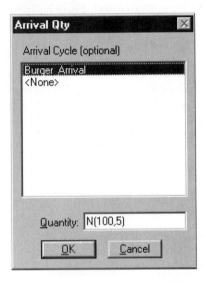

FIGURE L14.49

Arrival cycles and probability density function of arrivals defined at Newport Beach Burger.

ID	Qty / %	Cumulative...	Table...
Burger_Arrival	Percent	No	Defined

Table for Burger_Arrival [1]

Time (Hours)	Qty / %
0.5	5
2	20
5	5
7	35
11	10
13	20
15	5

FIGURE L14.50

A snapshot of Newport Beach Burger.

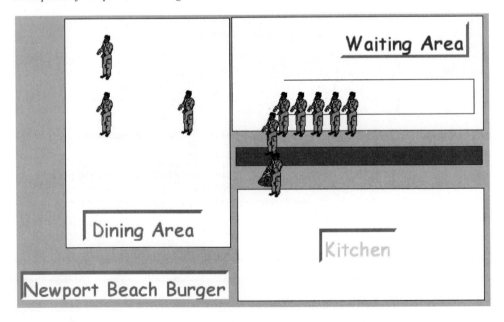

L14.9 User Distributions

Sometimes it becomes necessary to use the data collected by the user directly instead of using one of the built-in distributions. User distribution tables can be accessed from Build/More Elements (Figure L14.51). These user-defined or empirical distributions can be either discrete or continuous. The user distributions can be expressed in either a cumulative (cumulative density function or cdf) or a non-cumulative (probability density function or pdf) fashion.

Problem Statement

The customers at the **Newport Beach Burger** stand arrive in group sizes of one, two, three, or four with the probabilities shown in Table L14.9. The mean time between arrivals is 15 minutes (exponential). Ordering times have the probabilities

FIGURE L14.51

User Distributions menu.

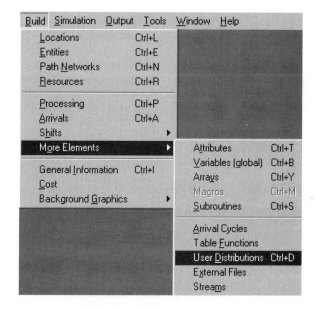

TABLE L14.9   **Probability Density Function of Customer Group Sizes**

Group Size	Probability
1	.4
2	.3
3	.1
4	.2

shown in Table L14.10. The probability density function of eating times are shown in Table L14.11. Simulate for 100 hours.

The user distributions, group size distribution, order time distribution, and eating time distribution are defined as shown in Figures L14.52, L14.53, L14.54, and L14.55, respectively.

FIGURE L14.52

User distributions defined for Newport Beach Burger.

User Distributions			
ID	Type...	Cumulative...	Table...
group_size	Discrete	No	Defined
order_time	Continuous	No	Defined
eat_time	Continuous	No	Defined

FIGURE L14.53

Group size distribution for Newport Beach Burger.

Table for group_size	
Percentage	Value
40	1
30	2
10	3
20	4

FIGURE L14.54

Order time distribution for Newport Beach Burger.

Table for order_time	
Percentage	Value
0	3
35	4
35	5
30	8

FIGURE L14.55

Eating time distribution for Newport Beach Burger.

Table for eat_time	
Percentage	Value
0	10
30	12
35	14
35	16

TABLE L14.10   Probability Density Function of Ordering Times

Ordering Time	Probability
$P(0 <= X < 3)$	.0
$P(3 <= X < 4)$	.35
$P(4 <= X < 5)$	.35
$P(5 <= X < 8)$	.3
$P(8 <= X)$	.0

TABLE L14.11   Probability Density Function of Eating Times

Eating Time	Probability
$P(0 <= X < 10)$	.0
$P(10 <= X < 12)$	.3
$P(12 <= X < 14)$	.35
$P(14 <= X < 16)$	.35
$P(16 <= X)$	.0

L14.10 Random Number Streams

The sequence of random numbers generated by a given seed is referred to as a *random number stream.* A random number stream is a sequence of an independently cycling series of random numbers. The number of distinct values generated before the stream repeats is called the stream's *cycle length.* Ideally, the cycle length should be long enough so that values do not repeat within the simulation. For a more thorough discussion on generating random numbers, please refer to Section 3.4.1 in Chapter 3.

In ProModel, up to 100 streams may be used in the model. A random number stream always generates numbers between 0 and 1, which are then used to sample from various hypothetical distributions by a method called *inverse transformation,* which is discussed in Section 3.4.2 in Chapter 3.

By default, all streams use a seed value of 1. Also, the random number streams are not reset between each replication if multiple replications are run. The Streams editor is used to assign a different seed to any stream and also to reset the seed value to the initial seed value between replications.

Problem Statement

Salt Lake Machine Shop has two machines: Mach A and Mach B. A maintenance mechanic, mechanic Dan, is hired for preventive maintenance on these machines. The mean time between preventive maintenance is 120 ± 10 minutes. Both machines are shut down at exactly the same time. The actual maintenance takes 10 ± 5 minutes. Jobs arrive at the rate of 7.5 per hour (exponential mean time between arrival). Jobs go to either Mach A or Mach B selected on a random basis. Simulate for 100 hours.

The layout, resources, path network of the mechanic, and process and routing tables are shown in Figures L14.56, L14.57, L14.58, and L14.59. The arrival of customers is defined by an exponential interarrival time with a mean of three

FIGURE L14.56

Layout of Salt Lake Machine Shop.

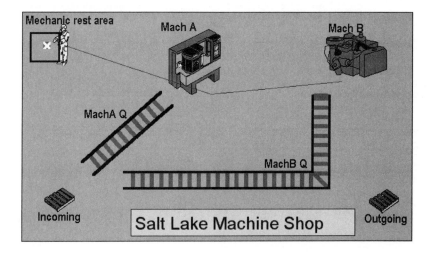

minutes. The downtimes and the definition of random number streams are shown in Figures L14.60 and L14.61, respectively. Note that the same seed value (Figure L14.61) is used in both the random number streams to ensure that both machines are shut down at exactly the same time.

FIGURE L14.57

Resources at Salt Lake Machine Shop.

```
**************************************************************************
                              Resources
**************************************************************************

                             Res      Ent
Name            Units Stats   Search   Search Path        Motion           Cost
-----------     ----- -----   ------   ------ --------     ----------       ----
Mechanic_Dan 1        By Unit Closest  Oldest Net1         Empty: 150 fpm
                                              Home: N1     Full: 150 fpm
                                              (Return)
```

FIGURE L14.58

Path network of the mechanic at Salt Lake Machine Shop.

```
**************************************************************************
                         Path Networks                                 *
**************************************************************************

Name   Type      T/S               From   To     BI   Dist/Time  Speed Factor
-----  --------  ----------------  -----  -----  ----  ---------  ------------
Net1   Passing   Speed & Distance  N1     N2     Bi    24.07      1
                                   N2     N3     Bi    20.24      1
```

FIGURE L14.59

Process and routing tables at Salt Lake Machine Shop.

```
                              Process                        Routing

Entity        Location  Operation          Blk  Output        Destination Rule
-----------   --------  -----------------  ---  -----------   ----------- ---------
Raw_Material  Incoming                      1   Raw_Material  MachA_Q     RANDOM 1
                                                Raw_Material  MachB_Q     RANDOM
Raw_Material  MachA_Q                        1   Raw_Material  Mach_A      FIRST 1
Raw_Material  Mach_A    wait U(10,5) min     1   Raw_Material  Outgoing    FIRST 1
Raw_Material  Outgoing                       1   Raw_Material  EXIT        FIRST 1
Raw_Material  MachB_Q                        1   Raw_Material  Mach_B      FIRST 1
Raw_Material  Mach_B    wait U(10,5) min     1   Raw_Material  Outgoing    FIRST 1
Raw_Material  Outgoing                       1   Raw_Material  EXIT        FIRST 1
```

FIGURE L14.60

Clock downtimes for machines A and B at Salt Lake Machine Shop.

```
********************************************************************  *******
                  Clock downtimes for Locations                        *
********************************************************************  *******

Loc      Frequency         First Time  Priority   Scheduled Disable   Logic
------   ---------------   ----------  --------   --------- -------   -------------------------
Mach_A   U(120,10,10) min              99         Yes       No        USE 1 Mechanic_Dan FOR U(10, 5)
                                                                      DISPLAY "Machine A down !!!!"

Mach_B   U(120,10,11) min              99         Yes       No        USE 1 Mechanic_Dan FOR U(10, 5)
                                                                      DISPLAY "Machine B down !!!!"
```

FIGURE L14.61

Definition of random number streams.

```
(**************************************************:
                                          Streams
(**************************************************:

Stream #          Seed #            Reset
_____       _____       _____
10                27                No
11                27                No
```

L14.11 Exercises

1. Differentiate between the following:
   a. Table functions versus arrays.
   b. Subroutines versus macros.
   c. Arrivals versus arrival cycles.
   d. Scenarios versus replications.
   e. Scenarios versus views.
   f. Built-in distribution versus user distribution.

2. What are some of the advantages of using an external file in ProModel?

3. **HiTek Molding,** a small mold shop, produces three types of parts: Jobs A, B, and C. The ratio of each part and the processing times (minimum, mean, and maximum of a triangular distribution) are as follows:

Job Type	Ratio	Minimum	Mean	Maximum
Job A	0.4	30 sec.	60 sec.	90 sec.
Job B	0.5	20 sec.	30 sec.	40 sec.
Job C	0.1	90 sec.	120 sec.	200 sec.

Create an array (mold-time) and an external file (L14_11_3.xls) for the molding processing time data.

Orders for jobs arrive at an average rate of four every minute (interarrival time exponentially distributed). Each machine can mold any type of job, one job at a time. Develop scenarios with one, two, three, or four molding machines to compare with each other. Simulate each scenario for 100 hours and compare them with each other. What would be the appropriate criteria for such a comparison?

4. For the **Southern California Airline** (Exercise 4, Section L7.12), use an external file to read interarrival and service time distribution data directly into the simulation model.

5. For the **Bengal Toy Company** (Exercise 9, Section L7.12), use an external file to store all the data. Read this file directly into the simulation model.

6. In Exercise 1, of Section L12.14, use an array to store the process time information for the five pieces of equipment. Read this information directly into the simulation model.

7. For the **Detroit ToolNDie** plant (Exercise 14, Section L7.12), generate the following three scenarios:
   a. Scenario I: One tool crib clerk.
   b. Scenario II: Two tool crib clerks.
   c. Scenario III: Three tool crib clerks.

   Run 10 replications of each scenario. Analyze and compare the results. How many clerks would you recommend hiring?

8. In **United Electronics** (Exercise 7 in Section L7.12), use an array to store the process time information. Read this information from an external Excel spreadsheet into the simulation model.

9. For **Salt Lake City Electronics** (Exercise 10 in Section L7.12), use external files (arrivals and entity_location files) to store all the data. Read this file directly into the simulation model.

10. For **Ghosh's Gear Shop** example in Section L13.2.1, create macros and suitable runtime interfaces for the following processing time parameters:

Average machining time	half width of machining time
Average washing time	half width of washing time
Average inspection time	std. dev. of inspection time

11. **West Coast Federal** a drive-in bank, has one teller and space for five waiting cars. If a customer arrives when the line is full, he or she drives around the block and tries again. Time between arrivals is exponential with mean of 10 minutes. Time to drive around the block is normally distributed with mean 3 min and standard deviation 0.6 min. Service time is uniform at 9 ± 3 minutes. Build a simulation model and run it for 2000 hours (approximately one year of operation).
    a. Collect statistics on time in queue, time in system, teller utilization, number of customers served per hour, and number of customers balked per hour.
    b. Modify the model to allow two cars to wait after they are served to get onto the street. Waiting time for traffic is exponential with a mean of four minutes. Collect all the statistics from part *a*.
    c. Modify the model to reflect balking customers leaving the system and not driving around the block. Collect all the same statistics. How many customers are lost per hour?
    d. The bank's operating hours are 9 A.M. till 3 P.M. The drive-in facility is closed at 2:30 P.M. Customers remaining in line are served until the last customer has left the bank. Modify the model to reflect these changes. Run for 200 days of operation.

12. **San Dimas Mutual Bank** has two drive-in ATM kiosks in tandem but only one access lane (Figure L14.62). In addition, there is one indoor

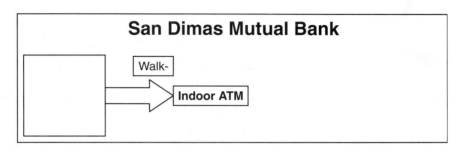

Parking Lot

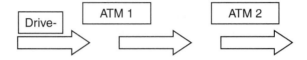

ATM for customers who decide to park (30 percent of all customers) and walk in. Customers arrive at intervals that are spaced on average five minutes apart (exponentially distributed). ATM customers are of three types—save money (deposit cash or check), spend money (withdraw cash), or count money (check balance). If both ATMs are free when a customer arrives, the customer will use the "downstream" ATM 2. A car at ATM 1 cannot pass a car at the ATM in front of it even if it has finished.

Type of Customer	Fraction of All Customers	Time at ATM
Save Money	1/3	Normal(7,2) minutes
Spend Money	1/2	Normal(5,2) minutes
Count Money	1/6	Normal(3,1) minutes

Create a simulation model of San Dimas Mutual Bank. The ATM kiosks operate 24/7. Run the model until 2000 cars have been served. Analyze the following:

a. The average and maximum queue size.

b. The average and maximum time spent by a customer waiting in queue.

c. The average and maximum time in the system.

d. Utilization of the three ATMs.

e. Number of customers served each hour.

Embellishment: Modify the model so that if four or more cars are in the ATM drive-in system at the bank when the customer arrives, the customer decides to park and walk in to use the ATM in the lobby. The time at this ATM is same as the drive-in ATMs. Run the model until 2000 cars have been served. Analyze the following:

a. Average and maximum number of customers deciding to park and walk in. How big should the parking lot be?

b. The average and maximum drive-in queue size.

c. The average and maximum time spent by a customer waiting in the drive-in queue.

d. The average and maximum walk-in queue size.

e. The average and maximum time spent by a customer waiting in the walk-in queue.

f. The average and maximum time in the system.

g. Utilization of the three ATMs.

h. Number of customers served each hour.

III CASE STUDY ASSIGNMENTS

These case studies have been used in senior- or graduate-level simulation classes. Each of these case studies can be analyzed over a three- to five-week period. A single student or a group of two to three students can work together on these case studies. If you are using the student version of the software, you may need to make some simplifying assumptions to limit the size of the model. You will also need to fill in (research or assume) some of the information and data missing from the case descriptions.

TOY AIRPLANE MANUFACTURING

A toy company produces three types (A, B, and C) of toy aluminum airplanes in the following daily volumes: A = 1000, B = 1500 and C = 1800. The company expects demand to increase for its products by 30 percent over the next six months and needs to know the total machines and operators that will be required. All planes go through five operations (10 through 50) except for plane A, which skips operation 40. Following is a list of operation times, move times, and resources used:

Opn	Description	Operation Time	Resource	Move Time to Next Operation	Movement Resource
10	Die casting	3 min. (outputs 6 parts)	Automated die caster	.3 min.	Mover
20	Cutting	Triangular (.25, .28, .35)	Cutter	none	
30	Grinding	Sample times: .23, .22, .26, .22, .25, .23, .24, .22, .21, .23, .20, .23, .22, .25, .23, .24, .23, .25, .47, .23, .25, .21, .24, .22, .26, .23, .25, .24, .21, .24, .26	Grinder	.2 min.	Mover
40	Coating	12 min. per batch of 24	Coater	.2 min	Mover
50	Inspection and packaging	Triangular (.27, .30, .40)	Packager	To exit with 88% yield	

After die casting, planes are moved to each operation in batch sizes of 24. Input buffers exist at each operation. The factory operates eight hours a day, five days per week. The factory starts out empty at the beginning of each day and ships all parts produced at the end of the day. The die caster experiences downtimes every 30 minutes exponentially distributed and takes 8 minutes normally distributed with a standard deviation of 2 minutes to repair. One maintenance person is always on duty to make repairs.

Find the total number of machines and personnel needed to meet daily production requirements. Document the assumptions and experimental procedure you went through to conduct the study.

MI CAZUELA—MEXICAN RESTAURANT

Maria opened her authentic Mexican restaurant Mi Cazuela (a *cazuela* is a clay cooking bowl with a small handle on each side) in Pasadena, California, in the 1980s. It quickly became popular for the tasty food and use of fresh organic produce and all-natural meats. As her oldest child, you have been asked to run the restaurant. If you are able to gain her confidence, she will eventually hand over the restaurant to you.

You have definite ideas about increasing the profitability at Mi Cazuela. Lately, you have observed a troubling trend in the restaurant. An increasing number of customers are expressing dissatisfaction with the long wait, and you have also observed that some people leave without being served.

Your initial analysis of the situation at Mi Cazuela indicates that one way to improve customer service is to reduce the waiting time in the restaurant. You also realize that by optimizing the process for the peak time in the restaurant, you will be able to increase the profit.

Customers arrive in groups that vary in size from one to four (uniformly distributed). Currently, there are four tables for four and three tables for two patrons in the dining area. One table for four can be replaced with two tables for two, or vice versa. Groups of one or two customers wait in one queue while groups of three or four customers wait in another queue. Each of these waiting lines can accommodate up to two groups only. One- or two-customer groups are directed to tables for two. Three- or four-customer groups are directed to tables for four.

There are two cooks in the kitchen and two waiters. The cooks are paid $100/day, and the waiters get $60/day. The cost of raw material (vegetables, meat, spices, and other food material) is $1 per customer. The overhead cost of the restaurant (rent, insurance, utilities, and so on) is $300/day. The bill for each customer varies uniformly from $10 to $16 or $U(13,3)$.

The restaurant remains open seven days a week from 5 P.M. till 11 P.M. The customer arrival pattern is as follows. The total number of customer groups visiting the restaurant each day varies uniformly between 30 and 50 or $U(40,10)$:

Customer Arrival Pattern

From	To	Percent
5 P.M.	6 P.M.	10
6 P.M.	7 P.M.	20
7 P.M.	9 P.M.	55
9 P.M.	10 P.M.	10
10 P.M.	11 P.M.	5

Processes at the Restaurant

When a table of the right size becomes available and a waiter is free, he or she seats the customer, writes down the order, and delivers the order to the kitchen. Cooks prepare the food in the kitchen and bring it out. Any available waiter delivers the food to the customer. Customers enjoy the dinner. A waiter cleans the table and collects payment from the customers. The customers leave the restaurant. The various activity times are as follows:

Activity #	Activity	Activity Time Distributions
1	Waiter seats the customer group.	$N(2, 0.5)$ min
2	Waiter writes down the order.	$N(3, 0.7)$ min
3	Waiter delivers the order to the kitchen.	$N(2, 0.5)$ min
4	Cook prepares food.	$N(5, 1)$ min
5	Cook brings out the food.	$N(2, 0.5)$ min
6	Waiter delivers food to customer group.	$N(2, 0.5)$ min
7	Customers eat.	$N(10, 2)$ min
8	Waiter cleans table and collects payment + tips.	$N(3, 0.8)$ min

Part A

Analyze and answer the following questions:

1. What is the range of profit (develop a $\pm3\sigma$ confidence interval) per day at Mi Cazuela?
2. On average, how many customers leave the restaurant (per day) without eating?
3. What is the range of time (develop a $\pm3\sigma$ confidence interval) a customer group spends at the restaurant?
4. How much time (develop a $\pm3\sigma$ confidence interval) does a customer group wait in line?

Part B

You would like to change the mix of four-seat tables and two-seat tables in the dining area to increase profit and reduce the number of balking customers. You would also like to investigate if hiring additional waiters and/or cooks will improve the bottom line (profit).

Part C

You are thinking of using an automated handheld device for the waiters to take the customer orders and transmit the information (wireless) to the kitchen. The order entry and transmission (activities #2 and 3) is estimated to take $N(1.5, 0.2)$ minutes. The rent for each of these devices is $2/hour. Will using these devices improve profit? Reduce customer time in the system? Should you invest in these handheld devices?

Part D

The area surrounding the mall is going through a construction boom. It is expected that Mi Cazuela (and the mall) will soon see an increase in the number of patrons per day. Soon the number of customer groups visiting the restaurant is expected to grow to 50–70 per day, or $U(60,10)$. You have been debating whether to take over the adjoining coffeeshop and expand the Mi Cazuela restaurant. The additional area will allow you to add four more tables of four and three tables of two customers each. The overhead cost of the additional area will be $200 per day. Should you expand your restaurant? Will it increase profit?

How is your performance in managing Mi Cazuela? Do you think Mama Maria will be proud and hand over the reins of the business to you?

CASE STUDY 3

JAI HIND CYCLES INC. PLANS NEW PRODUCTION FACILITY

Mr. Singh is the industrial engineering manager at Jai Hind Cycles, a producer of bicycles. As part of the growth plan for the company, the management is planning to introduce a new model of mountain bike strictly for the export market. Presently, JHC assembles regular bikes for the domestic market. The company runs one shift every day. The present facility has a process layout. Mr. Singh is considering replacing the existing layout with a group technology cell layout. As JHC's IE manager, Mr. Singh has been asked to report on the impact that will be made by the addition of the mountain bike to JHC's current production capabilities.

Mr. Singh has collected the following data from the existing plant:

1. The present production rate is 200 regular bikes per day in one 480-minute shift.
2. The following is the list of all the existing equipment in JHC's production facility:

Equipment Type	Process Time	Quantity
Forging	60 sec/large sprocket	2
	30 sec/small sprocket	
Molding	2 parts/90 sec	2
Welding	1 weld/60 sec	8
Tube bender	1 bend/30 sec	2
Die casting	1 part/minute	1
Drill press	20 sec/part	1
Punch press	30 sec/part	1
Electric saw	1 cut/15 sec	2
Assembly	30–60 minutes	

Table 1 shows a detailed bill of materials of all the parts manufactured by JHC and the machining requirements for both models of bikes. Only parts of the regular and the mountain bikes that appear in this table are manufactured within the plant. The rest of the parts either are purchased from the market or are subcontracted to the vendors.

A job-shop floor plan of the existing facility is shown in Figure 1. The whole facility is 500,000 square feet in covered area.

The figures for the last five years of the combined total market demand are as follows:

Year	Demand
1998	75,000
1999	82,000
2000	80,000
2001	77,000
2002	79,000

At present, the shortages are met by importing the balance of the demand. However, this is a costly option, and management thinks indigenously manufactured bikes of good quality would be in great demand.

Tasks

1. Design a cellular layout for the manufacturing facility, incorporating group technology principles.
2. Determine the amount of resources needed to satisfy the increased demand.
3. Suggest a possible material handling system for the new facility—conveyor(s), forklift truck(s), AGV(s).

TABLE 1   Detailed Bill of Materials for Jai Hind Cycles
Bicycle Parts and Process List

Assembly Name	Subassembly Name	Part Name	Operations			
1 Regular bike			Assembly			
	1.1 Bike frame		Assembly			
		1.1.1 Top tube	Cutting			
		1.1.2 Seat tube	Cutting			
		1.1.3 Down tube	Cutting			
		1.1.4 Head tube	Cutting			
		1.1.5 Fork blade	Cutting	Bending		
		1.1.6 Chainstay	Cutting			
		1.1.7 Seatstay	Cutting	Bending		
		1.1.8 Rear fork tip	Welding			
		1.1.9 Front fork tip	Welding			
		1.1.10 Top tube lug	Casting	Welding		
		1.1.11 Down tube lug	Casting	Welding		
		1.1.12 Seat lug	Casting	Welding		
		1.1.13 Bottom bracket	Casting	Welding		
	1.2 Handlebar and stem assembly		Assembly			
		1.2.1 Handlebars	Cutting	Bending		
		1.2.2 Handlebar plugs	Molding			
		1.2.3 Handlebar stem	Casting	Cutting		
	1.3 Saddle post assembly		Assembly			
		1.3.1 Saddle	Molding			
		1.3.2 Seat post	Cutting			
	1.4 Drive chain assembly		Assembly			
		1.4.1 Crank spider	Forging			
		1.4.2 Large sprocket	Forging			
		1.4.3 Small sprocket	Forging			
2 Mountain bike			Assembly			
	2.1 Frame and handle bar		Assembly			
		2.1.1 Hub	Cutting			
		2.1.2 Frame legs	Cutting	Bending	Bending	
		2.1.3 Handlebar tube	Cutting	Welding	Welding	Welding	Welding
		2.1.4 Saddle post tube	Cutting			
		2.1.5 Handlebar	Cutting			
		2.1.6 Balance bar	Cutting	Welding	Welding	
	2.2 Saddle and seat post		Assembly			
		2.2.1 Handlebar post	Cutting			
		2.2.2 Saddle post	Cutting	Drill press		
		2.2.3 Mount brackets	Cutting	Drill press	Welding	Welding
		2.2.4 Axle mount	Cutting	Punch press	Welding	Welding
		2.2.5 Chain guard	Molding			

4. How many shifts per day does JHC need to work?

5. Develop a staffing plan for the present situation and for the new situation.

6. Develop a cost model and economic justification for the growth plan. Is the increased production plan justified from an economic standpoint?

FIGURE 1

Floor plan for Jai Hind Cycles.

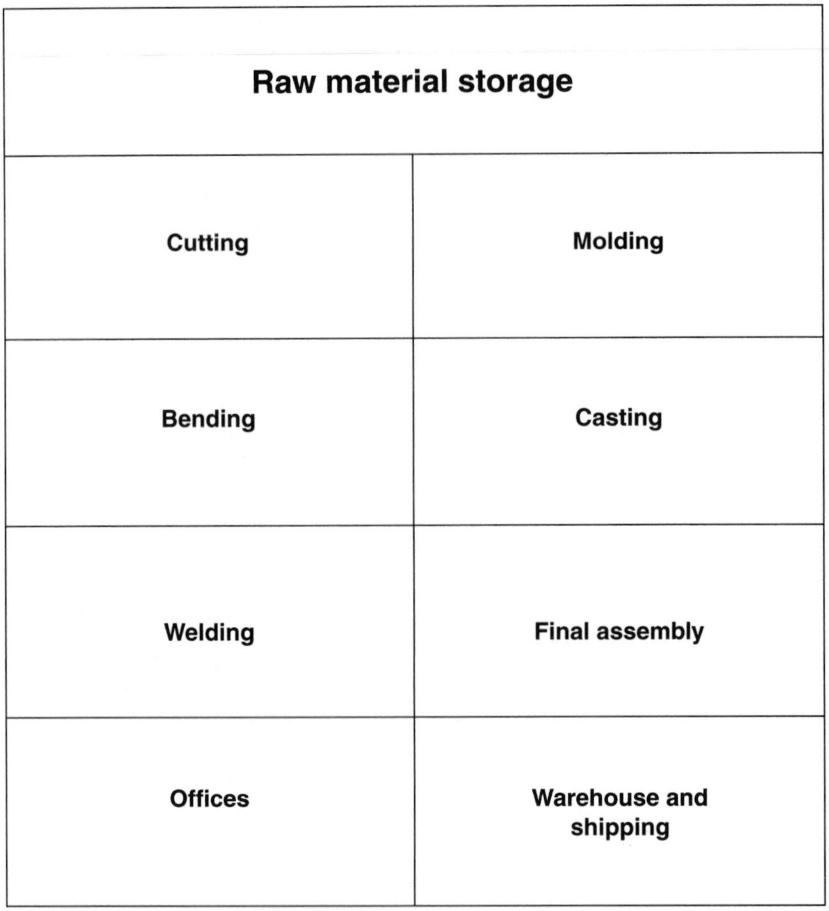

Raw material storage
Cutting / Molding
Bending / Casting
Welding / Final assembly
Offices / Warehouse and shipping

CASE STUDY 4

THE FSB COIN SYSTEM

George A. Johnson
Idaho State University

Todd Cooper
First Security Bank

Todd had a problem. First Security Bank had developed a consumer lending software package to increase the capacity and speed with which auto loan applications could be processed. The system consisted of faxed applications combined with online processing. The goal had been to provide a 30-minute turnaround of an application from the time the bank received the faxed application from the dealer to the time the loan was either approved or disapproved. The system had recently been installed and the results had not been satisfactory. The question now was what to do next.

First Security Bank of Idaho is the second largest bank in the state of Idaho with branches throughout the state. The bank is a full-service bank providing a broad range of banking services. Consumer loans and, in particular, auto loans make up an important part of these services. The bank is part of a larger system covering most of the intermountain states, and its headquarters are in Salt Lake City.

The auto loan business is a highly competitive field with a number of players including full-line banks, credit unions, and consumer finance companies. Because of the highly competitive nature, interest rates tend to be similar and competition is based on other factors. An important factor for the dealer is the time it takes to obtain loan approval. The quicker the loan approval, the quicker a sale can be closed and merchandise moved. A 30-minute turnaround of loan applications would be an important factor to a dealer, who has a significant impact on the consumer's decision on where to seek a loan.

The loan application process begins at the automobile dealership. It is there that an application is completed for the purpose of borrowing money to purchase a car. The application is then sent to the bank via a fax machine. Most fax transmissions are less than two minutes in length, and there is a bank of eight receiving fax machines. All machines are tied to the same 800 number. The plan is that eight machines should provide sufficient capacity that there should never be the problem of a busy signal received by the sending machine.

Once the fax transmission is complete, the application is taken from the machine by a runner and distributed to one of eight data entry clerks. The goal is that data entry should take no longer than six minutes. The goal was also set that there should be no greater than 5 percent errors.

Once the data input is complete, the input clerk assigns the application to one of six regions around the state. Each region has a group of specific dealers determined by geographic distribution. The application, now electronic in form, is distributed to the regions via the wide area network. The loan officer in the respective region will then process the loan, make a decision, and fax that decision back to the dealer. The goal is that the loan officer should complete this function within 20 minutes. This allows about another two minutes to fax the application back to the dealer.

The system has been operating approximately six months and has failed to meet the goal of 30 minutes. In addition, the error rate is running approximately 10 percent. Summary data are provided here:

Region	Applications	Average Time	Number of Loan Officers
1	6150	58.76	6
2	1485	37.22	2
3	2655	37.00	4
4	1680	51.07	2
5	1440	37.00	2
6	1590	37.01	3

A weighted average processing time for all regions is 46.07 minutes.

Information on data input indicates that this part of the process is taking almost twice as long as originally planned. The time from when the runner delivers the document to when it is entered is currently averaging 9.5 minutes. Also, it has been found that the time to process an error averages six minutes. Errors are corrected at the region and add to the region's processing time.

Todd needed to come up with some recommendations on how to solve the problem. Staffing seemed to be an issue in some regions, and the performance of the data input clerks was below expectations. The higher processing times and error rates needed to be corrected. He thought that if he solved these two problems and increased the staff, he could get the averages in all regions down to 30 minutes.

CASE STUDY 5

AUTOMATED WAREHOUSING AT ATHLETIC SHOE COMPANY

The centralized storage and distribution operation at Athletic Shoe Company (ASC) is considering replacement of its conventional manual storage racking systems with an elaborate automated storage and retrieval system (AS/RS). The objective of this case study is to come up with the preliminary design of the storage and material handling systems for ASC that will meet the needs of the company in timely distribution of its products.

On average, between 100,000 and 150,000 pairs of shoes are shipped per day to between 8000 and 10,000 shipping destinations. In order to support this level of operations, it is estimated that rack storage space of up to 3,000,000 pairs of shoes, consisting of 30,000 stock-keeping units (SKUs), is required.

The area available for storage, as shown in Figure 1, is 500,000 square feet. The height of the ceiling is 40 feet. A first-in, first-out (FIFO) inventory policy is adopted in the

FIGURE 1

Layout of the Athletic Shoe Company warehouse.

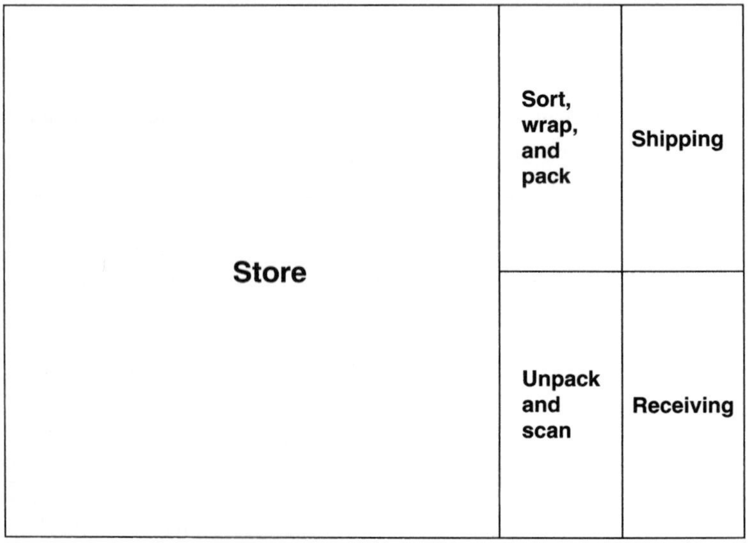

warehouse. For storage and retrieval, consider the following options:

   a. A dedicated picker for each aisle.
   b. A picker that is shared between the storage aisles.

For material handling in the shipping and receiving areas, and also between the storage areas and the shipping/receiving docks, consider one or more of the following options:

   a. Forklift trucks.
   b. Automated guided vehicles.
   c. Conveyors.

You as the warehouse manager would like to discourage many different types of material handling devices. The number of types of devices should be kept to a minimum, thus avoiding complicated interface problems.

   The weight of the shoeboxes varies from one to six pounds with a mode of four pounds. All the boxes measure 18" long 12" wide 5" high. Because of the construction of the boxes and the weight of the shoes, no more than eight boxes can be stacked up on each other.

   The general process flow for receiving and shipping of shoes is as follows:

Receiving

   1. Unload from truck.
   2. Scan the incoming boxes/pallets.
   3. Send to storage racks.
   4. Store.

Shipping

   1. Batch pick shipping orders.
   2. Send to sortation system.
   3. Wrap and pack.
   4. Load in outgoing truck.

Tasks

   1. Construct a simulation model of the warehouse and perform experiments using the model to judge the effectiveness and efficiency of the design with respect to parameters such as flows, capacity, operation, interfacing, and so on.
   2. Write a detailed specification of the storage plan: the amount of rack storage space included in the design (capacity), rack types, dimensions, rack configurations, and aisles within the layout.
   3. Design and specify the material handling equipment for all of the functions listed, including the interfaces required to change handling methods between functions.
   4. Design and specify the AS/R system. Compare a dedicated versus a shared picker system.

5. Plan the staffing requirements.

6. How many shifts should this warehouse be working?

7. Estimate the throughput for the warehouse per shift, per day, per year. What are the design parameters necessary for each function to attain the indicated level of throughput?

8. Develop a detailed facilities layout of the final design, including aisles and material handling equipment.

9. Develop a cost estimate for the proposed design.

CASE STUDY 6

CONCENTRATE LINE AT FLORIDA CITRUS COMPANY

Wai Seto, Suhandi Samsudin, Shi Lau, and Samson Chen
California State Polytechnic University–Pomona

Florida Citrus Company (FCC), located in Tampa, Florida, is a subdivision of Healthy Foods Inc., which currently has 12,000 employees in 50 food production facilities around the world.

FCC has specialized in producing a wide range of juice products for the past 50 years. FCC employs about 350 employees. Its juice products are primarily divided in the following four categories: aseptic products, juice concentrate, jug products, and cup products.

Based on the product categories, the manufacturing facility is divided into four cells. Each cell has a different kind of machine setting and configuration. The machines and equipment are mostly automatic. The aseptic cell is comprised of 44£125 machine, 36£125 machine, J.J. Var., and J.J. Rainbows. Depending on the demands of these sizes, the equipment is flexible to interchange based on the customer orders. The concentrate line generally produces concentrated juices for different private labels such as Vons, Stater Bros., Ralph's, Kroger, and Lucky's. The concentrate line produces fruit concentrates such as kiwi–strawberry, apple, orange, lemonade, and grape. The jug line seldom operates as the demand is poor. The cup line is the fastest-growing product in the business. It produces mainly apple, orange, grape, and lemonade flavors.

The concentrate line is currently not able to meet the budgeted case rate standard. FCC is seeking to find the real causes that contribute to the current production problem. The company also wants to improve the facility layout and reduce inventory levels in the concentrate line.

The concentrate line is divided into five stations:

1. Depalletizer: Tri-Can
2. Filler: Pfaudler
3. Seamer: Angelus
4. Palletizer: Currie
5. Packer: Diablo

The current concentrate line production is as follows:

Equipment Description	Rated Speed	Operating Speed
Filler	1750 cases/hr	600 cans/min
Seamer	1500 cases/hr	600 cans/min
Packer	1800 cases/hr	28 cases/min
Palletizer	1800 cases/hr	28 cases/min
Depalletizer	1800 cases/hr	600 cans/min
Bundler	1500 cases/hr	550 cans/min

The concentrate line stations and the flow of production are shown in Figure 1. The concentrate line starts from the receiving area. Full pallet loads of 3600 empty cans in 10 layers arrive at the receiving area. The arrival conveyor transports these pallets to the depalletizer (1). The cans are loaded onto the depalletizer, which is operated by Don.

FIGURE 1

Concentrate line stations for Florida Citrus Company.

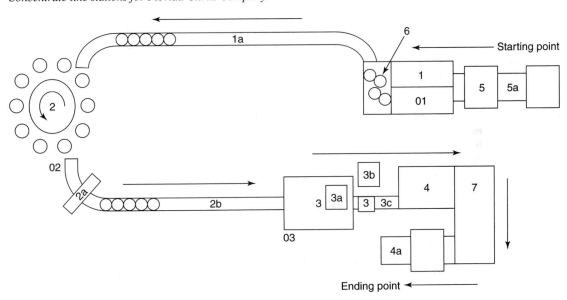

1. Depalletizer
2. Pfaudler bowl
3. Packmaster
4. Palletizer
5. The pallet
6. The concentrate cans
7. The box organizer
8. The box

1a. Depalletizer conveyor
2a. Lid stamping mechanism
2b. Filler bowl conveyor
3a. Glue mechanism
3b. Cardboard feeding machine
3c. Packmaster conveyor
4a. Exit conveyor
5a. Conveyor to the depalletizer

01 Operator 1
02 Operator 2
03 Operator 3

The depalletizer pushes out one layer of 360 cans at a time from the pallet and then raises up one layer of empty cans onto the depalletizer conveyor belt (1a). Conveyor 1a transports the layer of cans to the depalletizer dispenser. The dispenser separates each can from the layer of cans. Individual empty cans travel on the empty can conveyor to the Pfaudler bowl.

The Pfaudler bowl is a big circular container that stores the concentrate. Its 36 filling devices are used to fill the cans with concentrate. Pamela operates the Pfaudler bowl. Empty cans travel on the filler bowl conveyor (2b) and are filled with the appropriate juice concentrate. Filled cans are sent to the lid stamping mechanism (2a) on the filler bowl conveyor. The lid stamping closes the filled cans. As the closed cans come through the lid stamping mechanism, they are transported by the prewash conveyor to the washing machine to be flushed with water to wash away any leftover concentrate on the can. Four closed cans are combined as a group. The group of cans is then transported by the accumulate conveyor to the accumulator.

The accumulator combines six such groups (24 cans in all). The accumulated group of 24 cans is then transported by the prepack conveyor to the Packmaster (3), operated

FIGURE 2

Process flow for Florida Citrus Company.

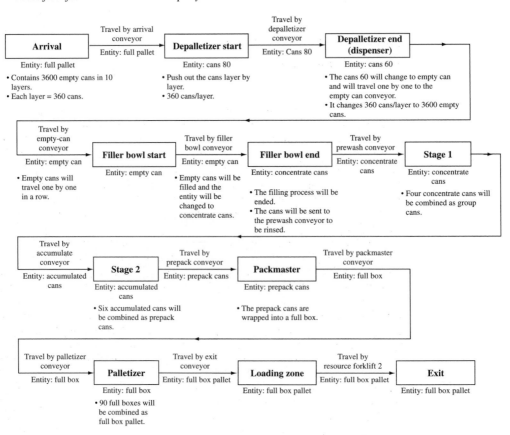

by Pat. Pat loads cardboard boxes onto the cardboard feeding machine (3b) next to the Packmaster. Then the 24 cans are wrapped and packed into each cardboard box. The glue mechanism inside the Packmaster glues all six sides of the box. The boxes are then raised up to the palletizer conveyor (3c), which transports the boxes to the palletizer (4).

The box organizer (7) mechanism loads three boxes at a time onto the pallet. A total of 90 boxes are loaded onto each pallet (10 levels, 9 boxes per level). The palletizer then lowers the pallet onto the exit conveyor (4a) to be transported to the loading zone. From the loading zone a forklift truck carries the pallets to the shipping dock. Figure 2 describes the process flow.

A study conducted by a group of Cal Poly students revealed the cause of most downtime to be located at the Packmaster. The Packmaster is supposed to pack a group of cans into a cardboard box. However, if the cardboard is warped, the mechanism will stop the operation. Another problem with the Packmaster is its glue operation. The glue heads sometimes are clotted.

All these machines operate in an automatic manner. However, there are frequent machine stoppages caused by the following factors: change of flavor, poor maintenance, lack of communication between workers, lack of attention by the workers, inefficient layout of the concentrate line, and bad machine design.

All the stations are arranged in the sequence of the manufacturing process. As such, the production line cannot operate in a flexible or parallel manner. Also, the machines depend on product being fed from upstream processes. An upstream machine stoppage will cause eventual downstream machine stoppages.

Work Measurement

A detailed production study was conducted that brought out the following facts:

Packmaster

Juice Flavors	Working Time (%)	Down Time (%)
Albertson's Pink Lemonade	68.75	31.25
Albertson's Pink Lemonade	77.84	22.16
Best Yet Orange Juice	71.73	28.27
Crisp Lemonade	65.75	34.25
Flav-R-Pac Lemonade	76.35	23.65
Fry's Lemonade	78.76	21.24
Hy-Top Pink Lemonade	68.83	31.17
IGA Grape Juice	83.04	16.96
Ladylee Grape Juice	93.32	6.68
Rosauer's Orange Juice	51.40	48.60
Rosauer's Pink Lemonade	61.59	38.41
Smith's Kiwi Raspberry	75.16	24.84
Smith's Kiwi Strawberry	85.05	14.95
Stater Bros. Lemonade	21.62	78.38
Stater Bros. Pink Lemonade	86.21	13.79
Western Family Pink Lemonade	64.07	35.93

The production study also showed the label change time on the Packmaster as follows:

Flavor from	Flavor to	Label Change Time (sec)
Albertson's Pink Lemonade	Western Family Pink Lemonade	824
Fry's Lemonade	Flav-R-Pac Lemonade	189
IGA Grape Juice	Ladylee Grape Juice	177
Rosauer's Pink Lemonade	Albertson's Pink Lemonade	41
Smith's Kiwi Raspberry	IGA Grape Juice	641
Smith's Kiwi Strawberry	Smith's Kiwi Raspberry	66
Stater Bros. Lemonade	Stater Bros. Pink Lemonade	160

The Packmaster was observed for a total of 45,983 sec. Out of this time, the Packmaster was working for a total of 24,027 sec, down for 13,108 sec, and being set up for change of flavor for 8848 sec. The average flavor change time for the Pfaudler bowl is 19.24 percent of the total observed time. The number of cases produced during this observed time was 11,590. The production rate is calculated to be (11,590/46,384)3600, or about 907 cases per hour.

It was also observed that the Packmaster was down because of flipped cans (8.6 percent), sensor failure (43.9 percent), and miscellaneous other reasons (47.5 percent).

The following information on the conveyors was obtained:

Name of Conveyor	Length (ft.)	Speed (ft/min)
Arrival conveyor		
Depalletizer conveyor	28.75	12.6
Empty-cans conveyor	120	130
Filler bowl conveyor	10	126
Prewash conveyor	23.6	255
Accumulate conveyor	38	48
Prepack conveyor	12	35
Palletizer conveyor	54.4	76
Exit conveyor		

The Pfaudler bowl was observed for a total of 46,384 sec. Out of this time, the bowl was working for 27,258 sec, down for 10,278 sec, and being set up for change of flavor for 8848 sec. The average flavor change time for the Pfaudler bowl is 19.08 percent of the total observed time. The number of cases produced in this observed time was 11,590. The production rate is calculated to be (11,590/46,384)3600, or about 900 cases per hour.

Pfaudler Bowl

Fruit Juice Flavors	Working Time (%)	Down Time (%)
Albertson's Pink Lemonade	74.81	25.19
Albertson's Wild Berry Punch	88.20	11.80
Best Yet Grape Juice	68.91	31.09
Best Yet Orange Juice	86.08	13.92
Crisp Lemonade	53.21	46.79

(continued)

Fruit Juice Flavors	Working Time (%)	Down Time (%)
Flav-R-Pac Lemonade	79.62	20.38
Flavorite Lemonade	69.07	30.93
Fry's Lemonade	80.54	19.46
Hy-Top Pink Lemonade	81.85	18.15
IGA Grape Juice	89.93	10.07
IGA Pink Lemonade	45.54	54.46
Ladylee Grape Juice	94.36	5.64
Ladylee Lemonade	91.86	8.14
Rosauer's Orange Juice	64.20	35.80
Rosauer's Pink Lemonade	100.00	0.00
Smith's Kiwi Raspberry	92.71	7.29
Smith's Kiwi Strawberry	96.49	3.51
Special Value Wild Berry Punch	80.09	19.91
Stater Bros. Lemonade	26.36	73.64
Stater Bros. Pink Lemonade	90.18	9.82
Western Family Pink Lemonade	66.30	33.70

The flavor change time was observed as given in the following table:

Flavor from	Flavor to	Flavor Change Time (sec)
Albertson's Lemonade	Rosauer's Pink Lemonade	537
Albertson's Lemonade	Albertson's Pink Lemonade	702
Albertson's Limeade	Fry's Lemonade	992
Albertson's Pink Lemonade	Western Family Pink Lemonade	400
Albertson's Pink Lemonade	IGA Pink Lemonade	69
Albertson's Wild Berry Punch	Special Value Apple Melon	1292
Best Yet Grape Juice	Special Value Wild Berry Punch	627
Flav-R-Pac Lemonade	Flavorite Lemonade	303
Flav-R-Pac Orange Juice	Rosauer's Orange Juice	42
Flavorite Lemonade	Ladylee Lemonade	41
Fry's Lemonade	Flav-R-Pac Lemonade	183
Furr's Orange Juice	Best Yet Orange Juice	684
Hy-Top Grape Juice	Best Yet Grape Juice	155
Hy-Top Pink Lemonade	Flav-R-Pac Lemonade	49
IGA Grape Juice	Ladylee Grape Juice	67
IGA Pink Lemonade	Best Yet Pink Lemonade	0
Ladylee Grape Juice	Albertson's Grape Juice	100
Ladylee Lemonade	Crisp Lemonade	49
Ladylee Pink Lemonade	Hy-Top Pink Lemonade	0
Rosauer's Orange Juice	Flavorite Orange Juice	0
Rosauer's Pink Lemonade	Albertson's Pink Lemonade	98
Smith's Apple Melon	Smith's Kiwi Strawberry	382
Smith's Kiwi Raspberry	IGA Grape Juice	580
Smith's Kiwi Strawberry	Smith's Kiwi Raspberry	53
Special Value Wild Berry Punch	Albertson's Wild Berry Punch	62
Stater Bros. Lemonade	Stater Bros. Pink Lemonade	50
Western Family Pink Lemonade	Safeway Pink Lemonade	1153

Tasks

1. Build simulation models and figure out the production capacity of the concentrate line at FCC (without considering any downtime).
2. What would be the capacity after considering the historical downtimes in the line?
3. What are the bottleneck operations in the whole process?
4. How can we reduce the level of inventory in the concentrate line? What would be the magnitude of reduction in the levels of inventory?
5. If we address the bottleneck operations as found in task 3, what would be the increase in capacity levels?

CASE STUDY 7

BALANCING THE PRODUCTION LINE AT SOUTHERN CALIFORNIA DOOR COMPANY

Suryadi Santoso
California State Polytechnic University–Pomona

Southern California Door Company produces solid wooden doors of various designs for new and existing homes. A layout of the production facility is shown in Figure 1. The current production facility is not balanced well. This leads to frequent congestion and stockouts on the production floor. The overall inventory (both raw material and work in process) is also fairly high. Mr. Santoso, the industrial engineering manager for the company, has been asked by management to smooth out the flow of production as well as reduce the levels of inventory. The company is also expecting a growth in the volume of sales. The production manager is asking Mr. Santoso to find the staffing level and equipment resources needed for the current level of sales as well as 10, 25, 50, and 100 percent growth in sales volume.

A preliminary process flow study by Mr. Santoso reveals the production flow shown in Figure 2.

Process Flow

Raw wood material is taken from the raw material storage to carriage 1. The raw material is inspected for correct sizes and defects. Material that does not meet the specifications is moved to carriage 1B. Raw wood from carriage 1 is fed into the rip saw machine.

In the rip saw machine, the raw wood is cut into rectangular cross sections. Cut wood material coming out of the rip saw machine is placed on carriage 3. Waste material from the cutting operation (rip saw) is placed in carriage 2.

Cut wood from carriage 3 is brought to the moulding shaper and grooved on one side. Out of the moulding shaper, grooved wood material is placed on carriage 4. From carriage 4, the grooved wood is stored in carriage 5 (if carriage 5 is full, carriage 6 or 7 is used). Grooved wood is transported from carriages 5, 6, and 7 to the chop saw working table.

One by one, the grooved wood material from the chop saw working table is fed into the chop saw machine. The grooved wood material to be fed is inspected by the operator to see if

FIGURE 1

Layout of production facility at Southern California Door Company.

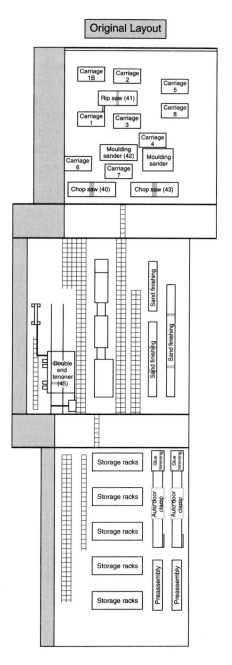

there are any defects in the wood. Usable chopped parts from the chop saw machine are stored in the chop saw storage shelves. Wood material that has defects is chopped into small blocks to cut out the defective surfaces using the chop saw and thrown away to carriage 8.

The chopped parts in the chop saw storage shelves are stacked into batches of a certain number and then packed with tape. From the chop saw storage shelves, some of the batches

FIGURE 2

Process sequences and present input/output flow for Southern California Door Company.

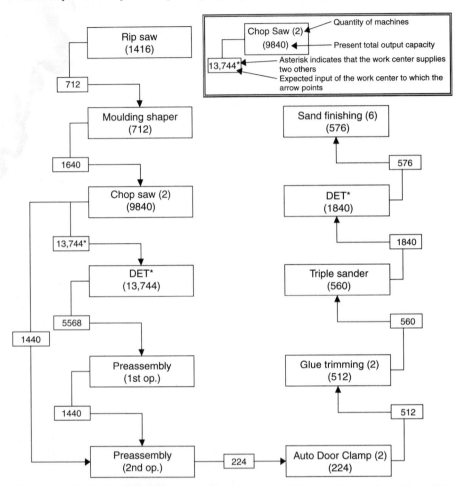

The present total output capacity of DET represents the number of units of a single product manufactured in an eight-hour shift. The DET machine supplies parts for two other work centers, preassembly (1st op.) and sand finishing. In reality, the DET machine has to balance the output between those two work centers mentioned; in other words, the DET machine is shared by two different parts for two different work centers during an eight-hour shift.

are transported to the double end tenoner (DET) storage, while the rest of the batches are kept in the chop saw storage shelves.

The transported batches are unpacked in the DET storage and then fed into the DET machine to be grooved on both sides. The parts coming out of the DET machine are placed on a roller next to the machine.

The parts are rebatched. From the DET machine, the batches are transported to storage racks and stored there until further processing. The batches stored in the chop saw storage shelves are picked up and placed on the preassembly table, as are the batches stored in the storage racks. The operator inspects to see if there is any defect in the wood. Defective parts are then taken back from the preassembly table to the storage racks.

The rest of the parts are given to the second operator in the same workstation. The second operator tries to match the color pattern of all the parts needed to assemble the door (four frames and a center panel). The operator puts glue on both ends of all four frame parts and preassembles the frame parts and center panel together.

The frame–panel preassembly is moved from the preassembly table to the auto door clamp conveyor and pressed into the auto door clamp machine. The pressed assembly is taken out of the auto door clamp machine and carried out by the auto door clamp conveyor.

Next, the preassembly is picked up and placed on the glue trimming table. Under a black light, the inspector looks for any excess glue coming out of the assembly parting lines. Excess glue is trimmed using a specially designed cutter.

From the glue trimming table, the assembly is brought to a roller next to the triple sanding machine (the auto cross grain sander and the auto drum sander). The operator feeds the assembly into the triple sander. The assembly undergoes three sanding processes: one through the auto cross grain sander and two through the auto drum sander. After coming out of the triple sander machine, the sanded assembly is picked up and placed on a roller between the DET and the triple sander machine. The sanded assembly waits there for further processing. The operator feeds the sanded assembly into the DET machine, where it is grooved on two of the sides.

Out of the DET machine, the assembly is taken by the second operator and placed temporarily on a roller next to the DET machine. After finishing with all the assembly, the first operator gets the grooved assembly and feeds it to the DET machine, where the assembly is grooved again on the other two sides. Going out of the machine, the grooved assembly is then placed on a roller between the DET machine and the triple sander machine.

The assembly is stored for further processing. From the roller conveyor, the grooved assembly is picked up by the operators from the sand finishing station and placed on the table. The operators finish the sanding process on the table using a handheld power sander. After finishing the sanding, the assembly is placed on the table for temporary storage. Finally, the sanded assembly is moved to a roller next to the storage racks to wait for further processes.

Work Measurement

A detailed work measurement effort was undertaken by Santoso to collect data on various manufacturing processes involved. Table 1 summarizes the results of all the time studies.

The current number of machines and/or workstations and their output capacities are as follows:

		Output Capacities	
Machine	*Number of Machines*	*Units/Hour*	*Units/Shift*
Rip saw	1	177	1416
Moulding shaper	1	89	712
Chop saw	2	615	4920
DET	1	3426	13,744
Preassembly 1	1	696	5568
Preassembly 2	1	90	720
Auto door clamp	2	14	224
Glue trimming	2	32	512
Triple sander	1	70	560
DET	1	460	1840
Sand finishing	6	12	576

TABLE 1   Time Studies Results for Southern California Door Company

Machine Name	Machine Number	Operation	Task Number	Task Description	Task Observations (seconds)
Rip saw	41	Cut raw material into correct cross-sectional dimensions	1	Inspect size and make adjustments	5.7,6.6,5.05,6.99,5.93,7.52, 5.37,7.21,8.96,6.68
			2	Grasp raw wood and feed into rip saw machine	6.79,6.3,7.52,6.15,6.53,6.03, 6.09,7.31,7,5.78,
			3	Cut wood with rip saw	12.4,11.53,11.26,12.88,11.56, 10.38,11.31,11.85,12.78,11.88
			4	Remove cut pieces from rip saw and place in carriage 3; throw waste into carriage 4	10.56,9.94,9.78,11.9,11.44, 8.87,7.35,10.93,12.47,10.34
Moulding shaper	42	Grooving one side of the raw material	1	Get cut material from carriage 3	11.52,12.83,14.64,8.25,12.58, 13.81,13.68,12.21,6.17,15.06, 11.93
			2	Place cut pieces onto moulding shaper	16.61,16.58,14.43,21.16, 18.17,25.14,26.15,30.06,35.16, 25.06,24.37
			3	Groove one side of material with moulding shaper	28.13,29.41,29.07,31.41,30.75, 38.95,39.83,42.27,39.32, 40.12,36.3
Chop saw	40, 43	Chop material into proper lengths	1	Inspect grooved material for size and defects	2.68,2.08,2.24,1.61,2.3,2.99, 3.02,3.11,3.21,3.02,3.06,2.79, 2.51,2.96,3.23,2.37,2.64
			2	Feed material and chop to smaller pieces	16.81,14.56,18.81,20.13, 23.25,18.53,16.53,25.56,25.3, 24.78,15.42,13.92,15.48,20.51, 17.79,23.54,17.01
			3	Throw away defective parts to carriage 8 and stack chopped parts	9.19,9.03,13.16,10.78,5.69, 4.1,6.9,9.16,3.6,3.22,8.83, 12.63,14.94,12.86,10.25,0.76, 10.63
Double end tenoner (DET)	45	Grooving frames	1	Get 2 to 4 frames from stack and feed into DET	9.33,14.7,10.18,14.47,13.12, 12.49,13.12,12.76,32.15, 33.94,13.23,11.97,9.21,24.86, 18.29,29.74,16.53,14.24, 12.78,15.38
			2	DET grooves both sides of frames	60.21,58.77,57.23,59.81, 61.64,60.29,59.85,61.43, 63.59,62.71,61.2,59.19,58.47, 60.27,59.73,60.21,61.82, 62.85,58.94,57.23
			3	Remove frames and stack	10.4,11.57,19.15,16.94,12.68, 31.47,36.97,13,14.5,14.62, 15.76,26.82,32.14,30.67,22.43, 29.61,34.92,18.27,20.31,24.88
Preassembly 1		Inspecting and matching frame parts	1	Get frame parts from storage racks	49.82,50.08,19.35,32.54, 35.31,33.43,37.84,42.17, 49.04,55.09

(continued)

Machine Name	Machine Number	Operation	Task Number	Task Description	Task Observations (seconds)
			2	Inspect for defects and match frame parts by color	36.52,35.99,29.09,57.43,53.6, 42.45,57.77,61.21,63.96, 56.41
Preassembly 2		Preassembling and gluing frames	1	Match center panel and four frame parts by color	8.13,8.32,7.43,10.63,6.28, 6.48,7.29,7.34,4.82,5.24
			2	Glue and preassemble frame parts and center panel	19.5,24.1,23.84,22.94,21.75, 22.47,23.66,25.63,29.59,30.09
			3	Place assembly in auto door clamp conveyor	4.38,2.71,4.35,3.69,3.04,2.62,3, 3.78,3,3.23
Auto door clamps	52, 54	Clamping preassembly	1	Conveyor feeds preassembled parts (preassy) into machine	6.55,5.05,6.86,4.77,7.68,5.33, 5.24,7.3,5.71,6.55
			2	Press the preassy	221.28,222,220.35,224.91, 194.4,231.82,213.34,206.75, 223.62,227.44
			3	Assy comes out of machine	4.22,5.69,7.15,5.78,5.1,4.75, 5.53,5.1,4.24,4.84
Glue trimming		Trimming excess glue out of the assembly	1	Remove assy from auto door clamp machine and inspect for excess glue	35.74,17.96,30.59,17.39, 21.48,10.15,16.89,10.87, 10.59,10.26,14.23,11.92, 24.87,10.91,11.77,15.48, 29.71,10.86,19.64
			2	Trim excess glue	58.53,90.87,67.93,70.78, 70.53,77.9,85.88,86.84,78.9, 95.6,78.5,72.65,72.44,91.01, 86.12,84.9,72.56,79.09,77.75
Triple sander	46, 47, 48	Sanding the assembly through three different sanding machines	1	Get assy from stack and feed into sander	2.45,3.56,3.18,3.16,3.32,3.58, 4.22,2.27,4.76,3.9
			2	Sand the assembly	30.72,32.75,34.13,35.66,37, 36.31,36.84,37.03,37.44,38.54
			3	Remove sanded assy and stack	3.31,6.54,5.03,5.51,5.22,5.84, 5.38,6.69,4.22,6.44
Double end tenoner (DET)	45	Grooving sanded assembly	1	Feed assy into DET	5.99,6.14,6.49,6.46,6.42,6.64, 3.21,4.11,3.71,4.2
			2	Groove assy	31.97,32.93,35.11,33.67, 34.06,33.21,33.43,35.23, 33.87,33.72
			3	Remove assy and stack	3.84,3,3.06,2.93,3.06,2.85, 2.88,3.22,1.87,2.41
Sand finishing		Sand finishing the assembly	1	Get part and place on table	3.49,3.42,3.47,3.29,3.36,3.2, 5.73,3.02,3.39,3.54,3.71,3.48
			2	Sand finish the part	215.8,207.57,244.17,254.28, 238.36,218.76,341.77,247.59, 252.63,308.06,221.27,233.66
			3	Stack parts	2.26,2.95,2,1.41,3.79,2.74,4.7, 3.35,3.09,2.75,2.59,2.71

FIGURE 3

Groups of operators for Southern California Door Company.

Work Center/ Machine	Minimum Quantity Required	Number of Operators Working	Utilization (Shift)	Group of Operators	Notes
Sand finishing	6	6	1.00	*	
Triple sander	2	4	0.51	****	
Glue trimming	3	3	0.75	***	
Auto door clamp	6	0	0.86		
Preassembly (2nd op.)	1	1	0.80	**	
Preassembly (1st op.)	1	1	0.21	*****	
DET	1	2	0.31	****	Grooving assembled doors
	1	2	0.08	****	Grooving frames
Chop saw	1	1	0.47	****	
Moulding shaper	1	1	0.54	*****	
Rip saw	1	1	0.27	****	

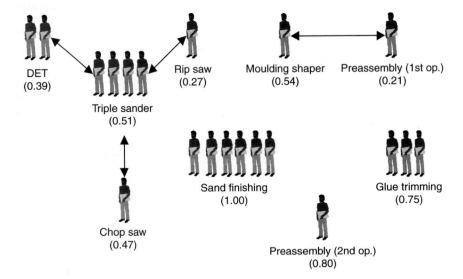

DET (0.39)   Triple sander (0.51)   Rip saw (0.27)   Moulding shaper (0.54)   Preassembly (1st op.) (0.21)   Chop saw (0.47)   Sand finishing (1.00)   Preassembly (2nd op.) (0.80)   Glue trimming (0.75)

Additional data are shown in Figures 2 and 3.

Tasks

Build simulation models to analyze the following:

1. Find the manufacturing capacity of the overall facility. What are the current bottlenecks of production?
2. How would you balance the flow of production? What improvements in capacity will that make?
3. What would you suggest to reduce inventory?
4. How could you reduce the manufacturing flow time?

5. The production manager is asking Mr. Santoso to find out the staffing and equipment resources needed for the current level of sales as well as 10, 25, 50, and 100 percent growth in sales volume.

6. Develop layouts for the facility for various levels of production.

7. What kind of material handling equipment would you recommend? Develop the specifications, amount, and cost.

CASE STUDY 8

MATERIAL HANDLING AT CALIFORNIA STEEL INDUSTRIES, INC.

Hary Herho, David Hong, Genghis Kuo, and Ka Hsing Loi
California State Polytechnic University–Pomona

California Steel Industries (CSI) is located in Fontana, California, approximately 50 miles east of Los Angeles. The company's facility, which occupies the space of the old Kaiser steel plant, covers about 400 acres. The facility is connected by 8.5 miles of roads and a 22-mile railroad system. CSI owns and operates seven diesel locomotives and 140 flat and gondola cars.

Founded in 1984, CSI is a fairly new company. CSI produces and ships over 1 million tons of steel annually. Future projections are for increased production. Therefore, CSI has invested hundreds of millions of dollars in modernizing its facilities.

The basic boundary of our defined system runs around the three main buildings: the tin mill (TM), the #1 continuous galvanizing line, and the cold sheet mill (M). Within the tin mill and cold sheet mill are several important production units that will be examined (see Figure 1).

The tin mill contains a 62" continuous pickling line, the 5-stand tandem cold mill (5-stand), the box annealing furnaces, and the #2 continuous galvanizing line. The cold sheet mill contains the cleaning line and additional box annealing furnaces. The #1 continuous galvanizing line is contained by itself in its own building.

CSI produces three main steel coil products: galvanized, cold rolled, and full hard. Roughly, galvanized coils make up about 60 percent of the total coils produced. The cold rolled coils are 35 percent of the coils and the full hard are the remaining 5 percent. Coils are also categorized into heavy gauge and light gauge. Assume that heavy-gauge coils are produced 60 percent of the time and light-gauge coils 40 percent of the time.

The study will begin with the coils arriving off the 5-stand. The coils weigh from 5 to 32 tons, with a mode of 16 tons. Most of the coils, 70 percent, that will be cold rolled coils will be processed first through the cleaning line at the cold sheet mill in order to remove the grime and residue left from the cold reduction process at the 5-stand. The other 30 percent are a "mill clean" product, which will not need to undergo the cleaning process since the coils are treated additionally at the 5-stand. After exiting the 5-stand, the coils are moved by a crane to the 5-stand bay to await transportation to the next stage by coil haulers. The coils that are to be annealed need to be upended, or rotated so that the coils' core is vertical. The only upender is located at the cold sheet mill. Since about 66 percent of the coils are annealed at the tin mill, they need to be transported to the tin mill and brought back to the cold sheet mill to be upended again.

Currently, coils are transported to and from different mill buildings by human-driven diesel coil haulers that have a 60-ton payload capability. A smaller 40-ton coil hauler

FIGURE 1

*Layout for California
Steel Industries.*

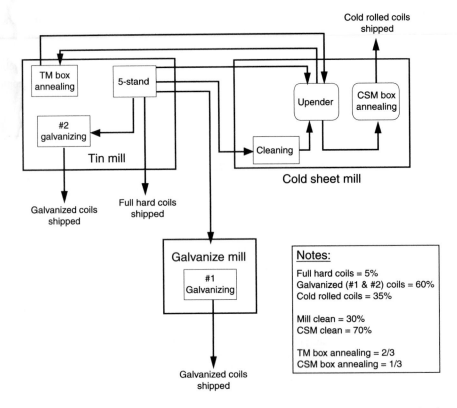

transports coils that are to be moved around within a building. There are two 60-ton haulers and two 40-ton haulers. Assume that one hauler will be down for maintenance at all times.

The following are the process times at each of the production units:

5-stand	Normal(8,2) min
#1 galvanizing line	Normal(30,8) min
#2 galvanizing line	Normal(25,4) min
Cleaning	Normal(15,3) min
Annealing	5 hr/ton

Annealing is a batched process in which groups of coils are treated at one time. The annealing bases at the cold sheet mill allow for coils to be batched three at a time. Coils can be batched 12 at a time at the tin mill annealing bases.

Assume that each storage bay after a coil has been processed has infinite capacity. Coils that are slated to be galvanized will go to either of the two galvanizing lines. The #1 continuous galvanizing line handles heavy-gauge coils, while the #2 galvanizing line processes the light-gauge coils.

The proposed layout (see Figure 2) will be very much like the original layout. The proposed material handling system that we are evaluating will utilize the railroads that connect the three main buildings. The two rails will allow coils to be moved from the tin mill to the cold sheet mill and the #1 galvanizing line. The top rail is the in-process rail, which will

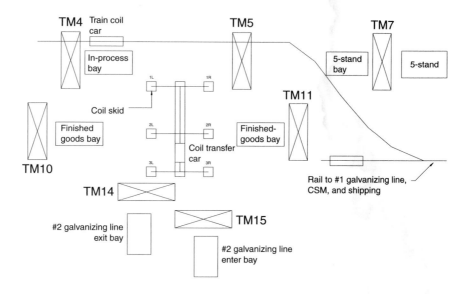

FIGURE 2

Proposed coil handling layout for California Steel Industries.

move coils that need to be processed at the cold sheet mill or the #1 galvanizing line. The bottom rail will ship out full hard coils and coils from the #2 galvanizing line. The train coil cars will be able to carry 100 tons of coils.

In addition, a coil transfer car system will be installed near the #2 galvanizing line. The car will consist of a smaller "baby" car that will be held inside the belly of a larger "mother" car. The "mother" car will travel north–south and position itself at a coil skid. The "baby" car, traveling east–west, will detach from the "mother" car, move underneath the skid, lift the coil, and travel back to the belly of the "mother" car.

Crane TM 7 will move coils from the 5-stand to the 5-stand bay, as in the current layout. The proposed system, however, will move coils to processing in the #2 galvanizing line with the assistance of four main cranes, namely TM 5, TM 11, TM 14, and TM 15. Crane TM 5 will carry coils to the coil skid at the north end of the rail. From there, the car will carry coils to the south end of the rail and place them on the right coil skid to wait to be picked up by TM 15 and stored in the #2 galvanizing line entry bay. This crane will also assist the line operator to move coils into position to be processed. After a coil is galvanized, crane TM 14 will move the coil to the #2 galvanizing line delivery bay. Galvanized coils that are to be shipped will be put on the southernmost coil skid to be transported by the coil car to the middle skids, where crane TM 11 will place them in either the rail or truck shipping areas.

One facility change that will take place is the movement of all the box annealing furnaces to the cold sheet mill. This change will prevent the back and forth movement of coils between the tin mill and cold sheet mill.

Tasks

1. Build simulation models of the current and proposed systems.
2. Compare the two material handling systems in terms of throughput time of coils and work-in-process inventory.
3. Experiment with the modernized model. Determine what will be the optimal number of train coil cars on the in-process and finished-goods rails.

Common Continuous and Discrete Distributions*

A.1 Continuous Distributions

Beta Distribution (min, max, *p*, *q*)

$$f(x) = \frac{1}{B(p,q)} \frac{(x - \min)^{p-1}(\max - x)^{q-1}}{(\max - \min)^{p+q-1}}$$

$\min \leq x \leq \max$

$\min$ = minimum value of x

$\max$ = maximum value of x

p = lower shape parameter > 0

q = upper shape parameter > 0

$B(p, q)$ = beta function

$$\text{mean} = (\max - \min)\left(\frac{p}{p+q}\right) + \min$$

$$\text{variance} = (\max - \min)^2 \left(\frac{pq}{(p+q)^2(p+q+1)}\right)$$

$$\text{mode} = \begin{cases} (\max - \min)\left(\dfrac{p-1}{p+q-2}\right) + \min & \text{if } p > 1, q > 1 \\ \min \text{ and } \max & \text{if } p < 1, q < 1 \\ \min & \text{if } (p < 1, q \geq 1) \text{ or if } (p = 1, q > 1) \\ \max & \text{if } (p \geq 1, q < 1) \text{ or if } (p > 1, q = 1) \\ \text{does not uniquely exist} & \text{if } p = q = 1 \end{cases}$$

The beta distribution is a continuous distribution that has both upper and lower finite bounds. Because many real situations can be bounded in this way, the beta distribution can be used empirically to estimate the actual distribution before many data are available. Even when data are available, the beta distribution should fit most data in a reasonable fashion, although it may not be the best fit. The uniform distribution is a special case of the beta distribution with $p, q = 1$.

As can be seen in the examples that follow, the beta distribution can approach zero or infinity at either of its bounds, with p controlling the lower bound and q controlling the upper bound. Values of $p, q < 1$ cause the beta distribution to approach infinity at that bound. Values of $p, q > 1$ cause the beta distribution to be finite at that bound.

Beta distributions have been used to model distributions of activity time in PERT analysis, porosity/void ratio of soil, phase derivatives in communication theory, size of progeny in *Escherichia coli*, dissipation rate in breakage models, proportions in gas mixtures, steady-state reflectivity, clutter and power of radar signals, construction duration, particle size, tool wear, and others. Many of these uses occur because of the doubly bounded nature of the beta distribution.

---

*Adapted by permission from *Stat::Fit Users Guide* (South Kent, Connecticut: Geer Mountain Software Corporation, 1997).

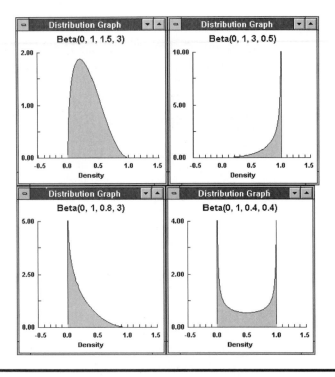

Erlang Distribution (min, *m*, *β*)

$$f(x) = \frac{(x - \min)^{m-1}}{\beta^m \Gamma(m)} \exp\left(-\frac{[x - \min]}{\beta}\right)$$

min = minimum x
m = shape factor = positive integer
β = scale factor > 0
mean = min + $m\beta$
variance = $m\beta^2$
mode = min + $\beta(m - 1)$

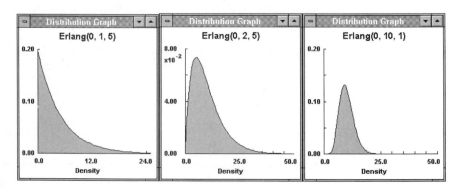

The Erlang distribution is a continuous distribution bounded on the lower side. It is a special case of the gamma distribution where the parameter m is restricted to a positive integer. As such, the Erlang distribution has no region where $f(x)$ tends to infinity at the minimum value of x [$m < 1$] but does have a special case at $m = 1$, where it reduces to the exponential distribution.

The Erlang distribution has been used extensively in reliability and in queuing theory, and thus in discrete event simulation, because it can be viewed as the sum of m exponentially distributed random variables, each with mean beta. It can be further generalized (see Banks and Carson 1984; Johnson et al. 1994).

As can be seen in the examples, the Erlang distribution follows the exponential distribution at $m = 1$, has a positive skewness with a peak near 0 for m between 2 and 9, and tends to a symmetrical distribution offset from the minimum at larger m.

Exponential Distribution (min, β)

$$f(x) = \frac{1}{\beta} \exp\left(-\frac{[x - \text{min}]}{\beta}\right)$$

$\text{min} = \text{minimum } x \text{ value}$
$\beta = \text{scale parameter}$
$\text{mean} = \text{min} + \beta$
$\text{variance} = \beta^2$
$\text{mode} = \text{min}$

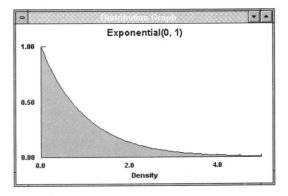

The exponential distribution is a continuous distribution bounded on the lower side. Its shape is always the same, starting at a finite value at the minimum and continuously decreasing at larger x. As shown in the example, the exponential distribution decreases rapidly for increasing x.

The exponential distribution is frequently used to represent the time between random occurrences, such as the time between arrivals at a specific location in a queuing model or the time between failures in reliability models. It has also been used to represent the service times of a specific operation. Further, it serves as an explicit manner in which the time dependence on noise may be treated. As such, these models are making explicit use of the lack of history dependence of the exponential distribution; it has the same set of probabilities when shifted in time. Even when exponential models are known to be inadequate to describe the situation, their mathematical tractability provides a good starting point. A more complex distribution such as Erlang or Weibull may be investigated (see Johnson et al. 1994, p. 499; Law and Kelton 1991, p. 330).

Gamma Distribution (min, α, β)

$$f(x) = \frac{(x - \min)^{\alpha-1}}{\beta^\alpha \Gamma(\alpha)} \exp\left(-\frac{[x - \min]}{\beta}\right)$$

$\min$ = minimum x

α = shape parameter > 0

β = scale parameter > 0

mean = $\min + \alpha\beta$

variance = $\alpha\beta^2$

$$\text{mode} = \begin{cases} \min + \beta(\alpha - 1) & \text{if } \alpha \geq 1 \\ \min & \text{if } \alpha < 1 \end{cases}$$

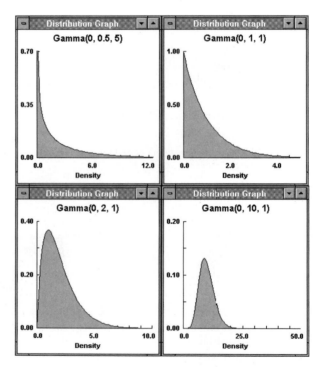

The gamma distribution is a continuous distribution bounded at the lower side. It has three distinct regions. For $\alpha = 1$, the gamma distribution reduces to the exponential distribution, starting at a finite value at minimum x and decreasing monotonically thereafter. For $\alpha < 1$, the gamma distribution tends to infinity at minimum x and decreases monotonically for increasing x. For $\alpha > 1$, the gamma distribution is 0 at minimum x, peaks at a value that depends on both alpha and beta, and decreases monotonically thereafter. If α is restricted to positive integers, the gamma distribution is reduced to the Erlang distribution.

Note that the gamma distribution also reduces to the chi-square distribution for $\min = 0$, $\beta = 2$, and $\alpha = n\mu/2$. It can then be viewed as the distribution of the sum of squares of independent unit normal variables, with $n\mu$ degrees of freedom, and is used in many statistical tests.

The gamma distribution can also be used to approximate the normal distribution, for large α, while maintaining its strictly positive values of x [actually $(x\text{-min})$].

The gamma distribution has been used to represent lifetimes, lead times, personal income data, a population about a stable equilibrium, interarrival times, and service times. In particular, it can represent lifetime with redundancy (see Johnson et al. 1994, p. 343; Shooman 1990).

Examples of each of the regions of the gamma distribution are shown here. Note the peak of the distribution moving away from the minimum value for increasing α, but with a much broader distribution.

Lognormal Distribution (min, μ, σ)

$$f(x) = \frac{1}{(x - \min)\sqrt{2\pi\sigma^2}} \exp\left(-\frac{[\ln(x - \min) - \mu]^2}{2\sigma^2}\right)$$

$\min$ = minimum x
μ = mean of the included normal distribution
σ = standard deviation of the included normal distribution
mean = $\min + \exp(\mu + (\sigma^2/2))$
variance = $\exp(2\mu + \sigma^2)(\exp(\sigma^2) - 1)$
mode = $\min + \exp(\mu - \sigma^2)$

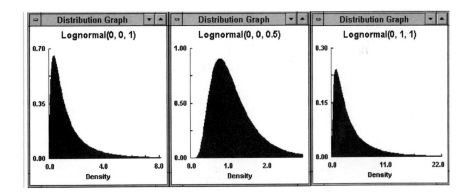

The lognormal distribution is a continuous distribution bounded on the lower side. It is always 0 at minimum x, rising to a peak that depends on both μ and σ, then decreasing monotonically for increasing x.

By definition, the natural logarithm of a lognormal random variable is a normal random variable. Its parameters are usually given in terms of this included normal distribution.

The lognormal distribution can also be used to approximate the normal distribution, for small σ, while maintaining its strictly positive values of x [actually $(x\text{-min})$].

The lognormal distribution is used in many different areas including the distribution of particle size in naturally occurring aggregates, dust concentration in industrial atmospheres, the distribution of minerals present in low concentrations, the duration of sickness absence, physicians' consulting time, lifetime distributions in reliability, distribution of income, employee retention, and many applications modeling weight, height, and so forth (see Johnson et al. 1994, p. 207).

The lognormal distribution can provide very peaked distributions for increasing σ—indeed, far more peaked than can be easily represented in graphical form.

Normal Distribution (μ, σ)

$$f(x) = \frac{1}{\sqrt{2\pi\sigma^2}} \exp\left(-\frac{[x-\mu]^2}{2\sigma^2}\right)$$

μ = shift parameter
σ = scale parameter = standard deviation
mean = μ
variance = σ^2
mode = μ

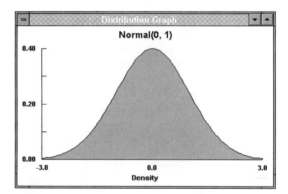

The normal distribution is an unbounded continuous distribution. It is sometimes called a Gaussian distribution or the bell curve. Because of its property of representing an increasing sum of small, independent errors, the normal distribution finds many, many uses in statistics; however, it is wrongly used in many situations. Possibly, the most important test in the fitting of analytical distributions is the elimination of the normal distribution as a possible candidate (see Johnson et al. 1994, p. 80).

The normal distribution is used as an approximation for the binomial distribution when the values of n and p are in the appropriate range. The normal distribution is frequently used to represent symmetrical data but suffers from being unbounded in both directions. If the data are known to have a lower bound, they may be better represented by suitable parameterization of the lognormal, Weibull, or gamma distribution. If the data are known to have both upper and lower bounds, the beta distribution can be used, although much work has been done on truncated normal distributions (not supported in Stat::Fit).

The normal distribution, shown here, has the familiar bell shape. It is unchanged in shape with changes in μ or σ.

Pearson 5 Distribution (min, α, β)

$$f(x) = \frac{\beta^\alpha}{\Gamma(\alpha)(x - \min)^{\alpha+1}} \exp\left(-\frac{\beta}{[x - \min]}\right)$$

$\min = \text{minimum } x$
$\alpha = \text{shape parameter} > 0$
$\beta = \text{scale parameter} > 0$

$$\text{mean} = \begin{cases} \min + \dfrac{\beta}{\alpha - 1} & \text{for } \alpha > 1 \\ \text{does not exist} & \text{for } 0 < \alpha \leq 1 \end{cases}$$

$$\text{variance} = \begin{cases} \dfrac{\beta^2}{(\alpha - 1)^2(\alpha - 2)} & \text{for } \alpha > 2 \\ \text{does not exist} & \text{for } 0 < \alpha \leq 2 \end{cases}$$

$$\text{mode} = \min + \frac{\beta}{\alpha + 1}$$

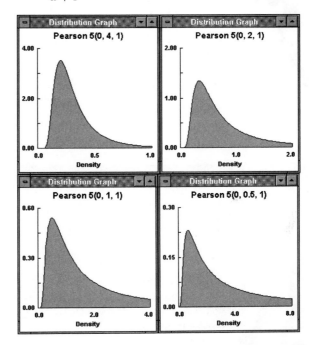

The Pearson 5 distribution is a continuous distribution with a bound on the lower side and is sometimes called the inverse gamma distribution due to the reciprocal relationship between a Pearson 5 random variable and a gamma random variable.

The Pearson 5 distribution is useful for modeling time delays where some minimum delay value is almost assured and the maximum time is unbounded and variably long, such as time to complete a difficult task, time to respond to an emergency, time to repair a tool, and so forth.

The Pearson 5 distribution starts slowly near its minimum and has a peak slightly removed from it, as shown here. With decreasing α, the peak gets flatter (see the vertical scale) and the tail gets much broader.

Pearson 6 Distribution (min, *β*, *p*, *q*)

$$f(x) = \frac{\left(\dfrac{x - \min}{\beta}\right)^{p-1}}{\beta\left[1 + \left(\dfrac{x - \min}{\beta}\right)\right]^{p+q} B(p, q)}$$

$x > \min$

$\min \in (-\infty, \infty)$

$\beta > 0$

$p > 0$

$q > 0$

$\beta(p, q) = $ beta function

$$\text{mean} = \begin{cases} \min + \dfrac{\beta p}{q - 1} & \text{for } q > 1 \\ \text{does not exist} & \text{for } 0 < q \le 1 \end{cases}$$

$$\text{variance} = \begin{cases} \dfrac{\beta^2 q(p + q - 1)}{(q - 1)^2(q - 2)} & \text{for } q > 2 \\ \text{does not exist} & \text{for } 0 < q \le 2 \end{cases}$$

$$\text{mode} = \begin{cases} \min + \dfrac{\beta(p - 1)}{q + 1} & \text{for } p \ge 1 \\ \min & \text{otherwise} \end{cases}$$

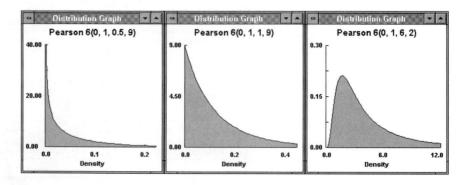

The Pearson 6 distribution is a continuous distribution bounded on the low side. The Pearson 6 distribution is sometimes called the beta distribution of the second kind due to the relationship of a Pearson 6 random variable to a beta random variable.

Like the gamma distribution, the Pearson 6 distribution has three distinct regions. For $p = 1$, the Pearson 6 distribution resembles the exponential distribution, starting at a finite value at minimum x and decreasing monotonically thereafter. For $p < 1$, the Pearson 6 distribution tends to infinity at minimum x and decreases monotonically for increasing x. For $p > 1$, the Pearson 6 distribution is 0 at minimum x, peaks at a value that depends on both p and q, and decreases monotonically thereafter.

The Pearson 6 distribution appears to have found little direct use, except in its reduced form as the F distribution, where it serves as the distribution of the ratio of independent estimators of variance and provides the final test for the analysis of variance.

The three regions of the Pearson 6 distribution are shown here. Also note that the distribution becomes sharply peaked just off the minimum for increasing q.

Triangular Distribution (min, max, mode)

$$f(x) = \begin{cases} \dfrac{2(x - \text{min})}{(\text{max} - \text{min})(\text{mode} - \text{min})} & \text{if min} \leq x \leq \text{mode} \\[2ex] \dfrac{2(\text{max} - x)}{(\text{max} - \text{min})(\text{max} - \text{mode})} & \text{if mode} < x \leq \text{max} \end{cases}$$

$$\text{min} = \text{minimum } x$$
$$\text{max} = \text{maximum } x$$
$$\text{mode} = \text{most likely } x$$
$$\text{mean} = \frac{\text{min} + \text{max} + \text{mode}}{3}$$
$$\text{variance} = \frac{\text{min}^2 + \text{max}^2 + \text{mode}^2 - (\text{min})(\text{max}) - (\text{min})(\text{mode}) - (\text{max})(\text{mode})}{18}$$

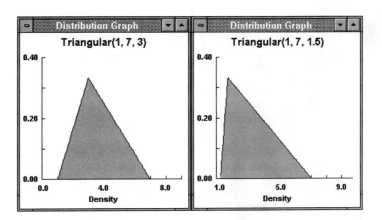

The triangular distribution is a continuous distribution bounded on both sides. The triangular distribution is often used when no or few data are available; it is rarely an accurate representation of a data set (see Law and Kelton 1991, p. 341). However, it is employed as the functional form of regions for fuzzy logic due to its ease of use.

The triangular distribution can take on very skewed forms, as shown here, including negative skewness. For the exceptional cases where the mode is either the min or max, the triangular distribution becomes a right triangle.

Uniform Distribution (min, max)

$$f(x) = \frac{1}{\text{max} - \text{min}}$$

$\text{min} = \text{minimum } x$

$\text{max} = \text{maximum } x$

$$\text{mean} = \frac{\text{min} + \text{max}}{2}$$

$$\text{variance} = \frac{(\text{max} - \text{min})^2}{12}$$

mode does not exist

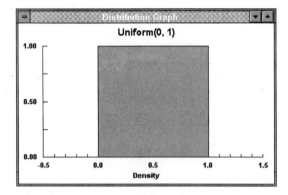

The uniform distribution is a continuous distribution bounded on both sides. Its density does not depend on the value of x. It is a special case of the beta distribution. It is frequently called the rectangular distribution (see Johnson et al. 1995, p. 276). Most random number generators provide samples from the uniform distribution on (0,1) and then convert these samples to random variates from other distributions.

The uniform distribution is used to represent a random variable with constant likelihood of being in any small interval between min and max. Note that the probability of either the min or max value is 0; the end points do *not* occur. If the end points are necessary, try the sum of two opposing right triangular distributions.

Weibull Distribution (min, α, β)

$$f(x) = \frac{\alpha}{\beta} \left(\frac{x - \text{min}}{\beta} \right)^{\alpha - 1} \exp \left(-\left(\frac{[x - \text{min}]}{\beta} \right)^{\alpha} \right)$$

$\text{min} = \text{minimum } x$

$\alpha = \text{shape parameter} > 0$

$\beta = \text{scale parameter} > 0$

$\text{mean} = \text{min} + \alpha\beta$

$\text{variance} = \alpha\beta^2$

$$\text{mode} = \begin{cases} \text{min} + \beta(\alpha - 1) & \text{if } \alpha \geq 1 \\ \text{min} & \text{if } \alpha < 1 \end{cases}$$

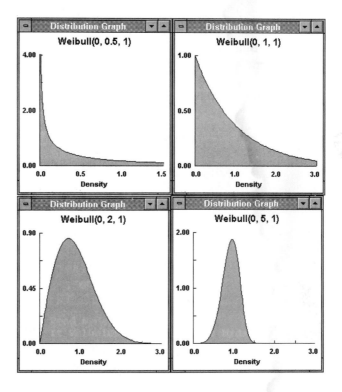

The Weibull distribution is a continuous distribution bounded on the lower side. Because it provides one of the limiting distributions for extreme values, it is also referred to as the Frechet distribution and the Weibull–Gnedenko distribution. Unfortunately, the Weibull distribution has been given various functional forms in the many engineering references; the form here is the standard form given in Johnson et al. 1994, p. 628.

Like the gamma distribution, the Weibull distribution has three distinct regions. For $\alpha = 1$, the Weibull distribution is reduced to the exponential distribution, starting at a finite value at minimum x and decreasing monotonically thereafter. For $\alpha < 1$, the Weibull distribution tends to infinity at minimum x and decreases monotonically for increasing x. For $\alpha > 1$, the Weibull distribution is 0 at minimum x, peaks at a value that depends on both α and β, and decreases monotonically thereafter. Uniquely, the Weibull distribution has negative skewness for $\alpha > 3.6$.

The Weibull distribution can also be used to approximate the normal distribution for $\alpha = 3.6$, while maintaining its strictly positive values of x [actually $(x$-min$)$], although the kurtosis is slightly smaller than 3, the normal value.

The Weibull distribution derived its popularity from its use to model the strength of materials, and has since been used to model just about everything. In particular, the Weibull distribution is used to represent wear-out lifetimes in reliability, wind speed, rainfall intensity, health-related issues, germination, duration of industrial stoppages, migratory systems, and thunderstorm data (see Johnson et al. 1994, p. 628; Shooman 1990, p. 190).

A.2 Discrete Distributions

Binomial Distribution (*n*, *p*)

$$p(x) = \binom{n}{x} p^x (1 - p)^{n-x}$$

$x = 0, 1, \ldots, n$
$n = $ number of trials
$p = $ probability of the event occurring
$$\binom{n}{x} = \frac{n!}{x!(n - x)!}$$
mean $= np$
variance $= np(1 - p)$
$$\text{mode} = \begin{cases} p(n + 1) - 1 \text{ and } p(n + 1) & \text{if } p(n + 1) \text{ is an integer} \\ p(n + 1) & \text{otherwise} \end{cases}$$

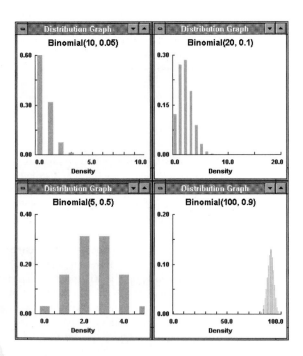

The binomial distribution is a discrete distribution bounded by [0, *n*]. Typically, it is used where a single trial is repeated over and over, such as the tossing of a coin. The parameter *p* is the probability of the event, either heads or tails, either occurring or not occurring. Each single trial is assumed to be independent of all others. For large *n*, the binomial distribution may be approximated by the normal distribution—for example, when $np > 9$ and $p < 0.5$ or when $np(1 - p) > 9$.

As shown in the examples, low values of p give high probabilities for low values of x and vice versa, so that the peak in the distribution may approach either bound. Note that the probabilities are actually weights at each integer but are represented by broader bars for visibility.

The binomial distribution can be used to describe

- The number of defective items in a batch.
- The number of people in a group of a particular type.
- Out of a group of employees, the number of employees who call in sick on a given day.

It is also useful in other event sampling tests where the probability of the event is known to be constant or nearly so. See Johnson et al. (1992, p. 134).

Discrete Uniform Distribution (min, max)

$$p(x) = \frac{1}{\max - \min + 1}$$

$x = \min, \min + 1, \ldots, \max$

$\min = \text{minimum } x$

$\max = \text{maximum } x$

$\text{mean} = \dfrac{\min + \max}{2}$

$\text{variance} = \dfrac{(\max - \min + 1)^2 - 1}{12}$

mode does not uniquely exist

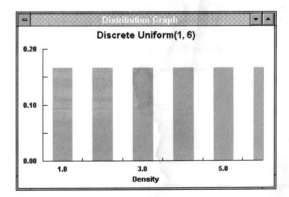

The discrete uniform distribution is a discrete distribution bounded on [min, max] with constant probability at every value on or between the bounds. Sometimes called the discrete rectangular distribution, it arises when an event can have a finite and equally probable number of outcomes (see Johnson et al. 1992, p. 272). Note that the probabilities are actually weights at each integer but are represented by broader bars for visibility.

Geometric Distribution (*p*)

$$p(x) = p(1 - p)^x$$

p = probability of occurrence

$$\text{mean} = \frac{1 - p}{p}$$

$$\text{variance} = \frac{1 - p}{p^2}$$

$$\text{mode} = 0$$

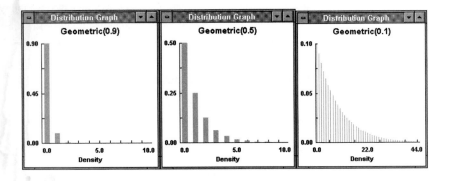

The geometric distribution is a discrete distribution bounded at 0 and unbounded on the high side. It is a special case of the negative binomial distribution. In particular, it is the direct discrete analog for the continuous exponential distribution. The geometric distribution has no history dependence, its probability at any value being independent of a shift along the axis.

The geometric distribution has been used for inventory demand, marketing survey returns, a ticket control problem, and meteorological models (see Johnson et al. 1992, p. 201; Law and Kelton 1991; p. 366.)

Several examples with decreasing probability are shown here. Note that the probabilities are actually weights at each integer but are represented by broader bars for visibility.

Poisson Distribution (λ)

$$p(x) = \frac{e^{-\lambda} \lambda^x}{x!}$$

λ = rate of occurrence

$\text{mean} = \lambda$

$\text{variance} = \lambda$

$$\text{mode} = \begin{cases} \lambda - 1 \text{ and } \lambda & \text{if } \lambda \text{ is an integer} \\ \lfloor \lambda \rfloor & \text{otherwise} \end{cases}$$

The Poisson distribution is a discrete distribution bounded at 0 on the low side and unbounded on the high side. The Poisson distribution is a limiting form of the hypergeometric distribution.

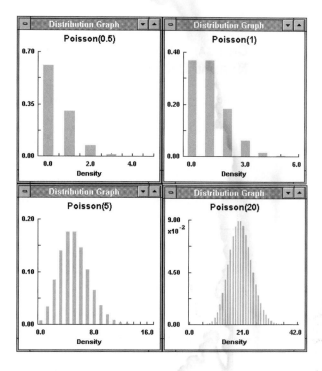

The Poisson distribution finds frequent use because it represents the infrequent occurrence of events whose rate is constant. This includes many types of events in time and space such as arrivals of telephone calls, defects in semiconductor manufacturing, defects in all aspects of quality control, molecular distributions, stellar distributions, geographical distributions of plants, shot noise, and so on. It is an important starting point in queuing theory and reliability theory (see Johnson et al. 1992, p. 151). Note that the time between arrivals (defects) is exponentially distributed, which makes this distribution a particularly convenient starting point even when the process is more complex.

The Poisson distribution peaks near λ and falls off rapidly on either side. Note that the probabilities are actually weights at each integer but are represented by broader bars for visibility.

References

Banks, Jerry, and John S. Carson II. *Discrete-Event System Simulation.* Englewood Cliffs, NJ: Prentice Hall, 1984.

Johnson, Norman L.; Samuel Kotz; and N. Balakrishnan. *Continuous Univariate Distributions.* Vol. 1. New York: John Wiley & Sons, 1994.

Johnson, Norman L.; Samuel Kotz; and N. Balakrishnan. *Continuous Univariate Distributions.* Vol. 2. New York: John Wiley & Sons, 1995.

Johnson, Norman L.; Samuel Kotz; and Adrienne W. Kemp. *Univariate Discrete Distributions.* New York: John Wiley & Sons, 1992.

Law, Averill M., and W. David Kelton. *Simulation Modeling and Analysis.* New York: McGraw-Hill, 1991.

Shooman, Martin L. *Probabilistic Reliability: An Engineering Approach.* Melbourne, Florida: Robert E. Krieger, 1990.

CRITICAL VALUES FOR STUDENT'S t DISTRIBUTION AND STANDARD NORMAL DISTRIBUTION

(Critical values for the standard normal distribution (z_α) appear in the last row with $df = \infty$. $z_\alpha = t_{\infty,\alpha}$.)

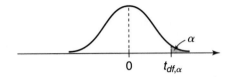

Degrees of Freedom (df)	$\alpha = 0.40$	$\alpha = 0.30$	$\alpha = 0.20$	$\alpha = 0.10$	$\alpha = 0.05$	$\alpha = 0.025$	$\alpha = 0.01$	$\alpha = 0.005$
1	0.325	0.727	1.367	3.078	6.314	12.706	31.827	63.657
2	0.289	0.617	1.061	1.886	2.920	4.303	6.965	9.925
3	0.277	0.584	0.978	1.638	2.353	3.182	4.541	5.841
4	0.271	0.569	0.941	1.533	2.132	2.776	3.747	4.604
5	0.267	0.559	0.920	1.476	2.015	2.571	3.365	4.032
6	0.265	0.553	0.906	1.440	1.943	2.447	3.143	3.707
7	0.263	0.549	0.896	1.415	1.895	2.365	2.998	3.499
8	0.262	0.546	0.889	1.397	1.860	2.306	2.896	3.355
9	0.261	0.543	0.883	1.383	1.833	2.262	2.821	3.250
10	0.260	0.542	0.879	1.372	1.812	2.228	2.764	3.169
11	0.260	0.540	0.876	1.363	1.796	2.201	2.718	3.106
12	0.259	0.539	0.873	1.356	1.782	2.179	2.681	3.055
13	0.259	0.538	0.870	1.350	1.771	2.160	2.650	3.012
14	0.258	0.537	0.868	1.345	1.761	2.145	2.624	2.977
15	0.258	0.536	0.866	1.341	1.753	2.131	2.602	2.947
16	0.258	0.535	0.865	1.337	1.746	2.120	2.583	2.921
17	0.257	0.534	0.863	1.333	1.740	2.110	2.567	2.898
18	0.257	0.534	0.862	1.330	1.734	2.101	2.552	2.878
19	0.257	0.533	0.861	1.328	1.729	2.093	2.539	2.861
20	0.257	0.533	0.860	1.325	1.725	2.086	2.528	2.845
21	0.257	0.532	0.859	1.323	1.721	2.080	2.518	2.831
22	0.256	0.532	0.858	1.321	1.717	2.074	2.508	2.819
23	0.256	0.532	0.858	1.319	1.714	2.069	2.500	2.807
24	0.256	0.531	0.857	1.316	1.708	2.060	2.485	2.797
25	0.256	0.531	0.856	1.316	1.708	2.060	2.485	2.787
26	0.256	0.531	0.856	1.315	1.706	2.056	2.479	2.779
27	0.256	0.531	0.855	1.314	1.703	2.052	2.473	2.771
28	0.256	0.530	0.855	1.313	1.701	2.048	2.467	2.763
29	0.256	0.530	0.854	1.310	1.697	2.042	2.457	2.756
30	0.256	0.530	0.854	1.310	1.697	2.042	2.457	2.750
40	0.255	0.529	0.851	1.303	1.684	2.021	2.423	2.704
60	0.254	0.527	0.848	1.296	1.671	2.000	2.390	2.660
120	0.254	0.526	0.845	1.289	1.658	1.980	2.358	2.617
∞	0.253	0.524	0.842	1.282	1.645	1.960	2.326	2.576

F Distribution for $\alpha = 0.05$

	Numerator Degrees of Freedom [df(Treatment)]																				
	1	*2*	*3*	*4*	*5*	*6*	*7*	*8*	*9*	*10*	*12*	*14*	*16*	*20*	*24*	*30*	*40*	*50*	*100*	*200*	*∞*
3	10.13	9.55	9.28	9.12	9.01	8.94	8.89	8.85	8.81	8.79	8.74	8.71	8.69	8.66	8.64	8.62	8.59	8.58	8.55	8.54	8.54
4	7.71	6.94	6.59	6.39	6.26	6.16	6.09	6.04	6.00	5.96	5.91	5.87	5.84	5.80	5.77	5.75	5.72	5.70	5.66	5.65	5.63
5	6.61	5.79	5.41	5.19	5.05	4.95	4.88	4.82	4.77	4.74	4.68	4.64	4.60	4.56	4.53	4.50	4.46	4.44	4.41	4.39	4.36
6	5.99	5.14	4.76	4.53	4.39	4.28	4.21	4.15	4.10	4.06	4.00	3.96	3.92	3.87	3.84	3.81	3.77	3.75	3.71	3.69	3.67
7	5.59	4.74	4.35	4.12	3.97	3.87	3.79	3.73	3.68	3.64	3.57	3.53	3.49	3.44	3.41	3.38	3.34	3.32	3.27	3.25	3.23
8	5.32	4.46	4.07	3.84	3.69	3.58	3.50	3.44	3.39	3.35	3.28	3.24	3.20	3.15	3.12	3.08	3.04	3.02	2.97	2.95	2.93
9	5.12	4.26	3.86	3.63	3.48	3.37	3.29	3.23	3.18	3.14	3.07	3.03	2.99	2.94	2.90	2.86	2.83	2.80	2.76	2.73	2.71
10	4.96	4.10	3.71	3.48	3.33	3.22	3.14	3.07	3.02	2.98	2.91	2.86	2.83	2.77	2.74	2.70	2.66	2.64	2.59	2.56	2.54
11	4.84	3.98	3.59	3.36	3.20	3.09	3.01	2.95	2.90	2.85	2.79	2.74	2.70	2.65	2.61	2.57	2.53	2.51	2.46	2.43	2.41
12	4.75	3.89	3.49	3.26	3.11	3.00	2.91	2.85	2.80	2.75	2.69	2.64	2.60	2.54	2.51	2.47	2.43	2.40	2.35	2.32	2.30
13	4.67	3.81	3.41	3.18	3.03	2.92	2.83	2.77	2.71	2.67	2.60	2.55	2.51	2.46	2.42	2.38	2.34	2.31	2.26	2.23	2.21
14	4.60	3.74	3.34	3.11	2.96	2.85	2.76	2.70	2.65	2.60	2.53	2.48	2.44	2.39	2.35	2.31	2.27	2.24	2.19	2.16	2.13
15	4.54	3.68	3.29	3.06	2.90	2.79	2.71	2.64	2.59	2.54	2.48	2.42	2.38	2.33	2.29	2.25	2.20	2.18	2.12	2.10	2.07
16	4.49	3.63	3.24	3.01	2.85	2.74	2.66	2.59	2.54	2.49	2.42	2.37	2.33	2.28	2.24	2.19	2.15	2.12	2.07	2.04	2.01
17	4.45	3.59	3.20	2.96	2.81	2.70	2.61	2.55	2.49	2.45	2.38	2.33	2.29	2.23	2.19	2.15	2.10	2.08	2.02	1.99	1.96
18	4.41	3.55	3.16	2.93	2.77	2.66	2.58	2.51	2.46	2.41	2.34	2.29	2.25	2.19	2.15	2.11	2.06	2.04	1.98	1.95	1.92
19	4.38	3.52	3.13	2.90	2.74	2.63	2.54	2.48	2.42	2.38	2.31	2.26	2.21	2.16	2.11	2.07	2.03	2.00	1.94	1.91	1.88
20	4.35	3.49	3.10	2.87	2.71	2.60	2.51	2.45	2.39	2.35	2.28	2.23	2.18	2.12	2.08	2.04	1.99	1.97	1.91	1.88	1.84
22	4.30	3.44	3.05	2.82	2.66	2.55	2.46	2.40	2.34	2.30	2.23	2.17	2.13	2.07	2.03	1.98	1.94	1.91	1.85	1.82	1.78
24	4.26	3.40	3.01	2.78	2.62	2.51	2.42	2.36	2.30	2.25	2.18	2.13	2.09	2.03	1.98	1.94	1.89	1.86	1.80	1.77	1.73
26	4.23	3.37	2.98	2.74	2.59	2.47	2.39	2.32	2.27	2.22	2.15	2.09	2.05	1.99	1.95	1.90	1.85	1.82	1.76	1.73	1.69
28	4.20	3.34	2.95	2.71	2.56	2.45	2.36	2.29	2.24	2.19	2.12	2.06	2.02	1.96	1.91	1.87	1.82	1.79	1.73	1.69	1.66
30	4.17	3.32	2.92	2.69	2.53	2.42	2.33	2.27	2.21	2.16	2.09	2.04	1.99	1.93	1.89	1.84	1.79	1.76	1.70	1.66	1.62
35	4.12	3.27	2.87	2.64	2.49	2.37	2.29	2.22	2.16	2.11	2.04	1.99	1.94	1.88	1.83	1.79	1.74	1.70	1.63	1.60	1.56
40	4.08	3.23	2.84	2.61	2.45	2.34	2.25	2.18	2.12	2.08	2.00	1.95	1.90	1.84	1.79	1.74	1.69	1.66	1.59	1.55	1.51
45	4.06	3.20	2.81	2.58	2.42	2.31	2.22	2.15	2.10	2.05	1.97	1.92	1.87	1.81	1.76	1.71	1.66	1.63	1.55	1.51	1.47
50	4.03	3.18	2.79	2.56	2.40	2.29	2.20	2.13	2.07	2.03	1.95	1.89	1.85	1.78	1.74	1.69	1.63	1.60	1.52	1.48	1.44
60	4.00	3.15	2.76	2.53	2.37	2.25	2.17	2.10	2.04	1.99	1.92	1.86	1.82	1.75	1.70	1.65	1.59	1.56	1.48	1.44	1.39
70	3.98	3.13	2.74	2.50	2.35	2.23	2.14	2.07	2.02	1.97	1.89	1.84	1.79	1.72	1.67	1.62	1.57	1.53	1.45	1.40	1.35
80	3.96	3.11	2.72	2.49	2.33	2.21	2.13	2.06	2.00	1.95	1.88	1.82	1.77	1.70	1.65	1.60	1.54	1.51	1.43	1.38	1.33
100	3.94	3.09	2.70	2.46	2.31	2.19	2.10	2.03	1.97	1.93	1.85	1.79	1.75	1.68	1.63	1.57	1.52	1.48	1.39	1.34	1.28
200	3.89	3.04	2.65	2.42	2.26	2.14	2.06	1.98	1.93	1.88	1.80	1.74	1.69	1.62	1.57	1.52	1.46	1.41	1.32	1.26	1.19
∞	1.04	3.00	2.61	2.37	2.21	2.10	2.01	1.94	1.88	1.83	1.75	1.69	1.64	1.57	1.52	1.46	1.40	1.35	1.25	1.17	1.03

Denominator Degrees of Freedom [df(Error)]

CRITICAL VALUES FOR CHI-SQUARE DISTRIBUTION

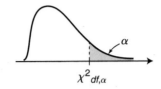

$$\chi^2_{df,\alpha}$$

Degrees of Freedom (df)	$\alpha = 0.4$	$\alpha = 0.3$	$\alpha = 0.2$	$\alpha = 0.1$	$\alpha = 0.05$	$\alpha = 0.025$	$\alpha = 0.01$	$\alpha = 0.005$
1	0.708	1.074	1.642	2.706	3.841	5.024	6.635	7.879
2	1.833	2.408	3.219	4.605	5.991	7.378	9.210	10.597
3	2.946	3.665	4.642	6.251	7.815	9.348	11.345	12.838
4	4.045	4.878	5.989	7.779	9.488	11.143	13.277	14.860
5	5.132	6.064	7.289	9.236	11.070	12.832	15.086	16.750
6	6.211	7.231	8.558	10.645	12.592	14.449	16.812	18.548
7	7.283	8.383	9.803	12.017	14.067	16.013	18.475	20.278
8	8.351	9.524	11.030	13.362	15.507	17.535	20.090	21.955
9	9.414	10.656	12.242	14.684	16.919	19.023	21.666	23.589
10	10.473	11.781	13.442	15.987	18.307	20.483	23.209	25.188
11	11.530	12.899	14.631	17.275	19.675	21.920	24.725	26.757
12	12.584	14.011	15.812	18.549	21.026	23.337	26.217	28.300
13	13.636	15.119	16.985	19.812	22.362	24.736	27.688	29.819
14	14.685	16.222	18.151	21.064	23.685	26.119	29.141	31.319
15	15.733	17.322	19.311	22.307	24.996	27.488	30.578	32.801
16	16.780	18.418	20.465	23.542	26.296	28.845	32.000	34.267
17	17.824	19.511	21.615	24.769	27.587	30.191	33.409	35.718
18	18.868	20.601	22.760	25.989	28.869	31.526	34.805	37.156
19	19.910	21.689	23.900	27.204	30.144	32.852	36.191	38.582
20	20.951	22.775	25.038	28.412	31.410	34.170	37.566	39.997
21	21.992	23.858	26.171	29.615	32.671	35.479	38.932	41.401
22	23.031	24.939	27.301	30.813	33.924	36.781	40.289	42.796
23	24.069	26.018	28.429	32.007	35.172	38.076	41.638	44.181
24	25.106	27.096	29.553	33.196	36.415	39.364	42.980	45.558
25	26.143	28.172	30.675	34.382	37.652	40.646	44.314	46.928
26	27.179	29.246	31.795	35.563	38.885	41.923	45.642	48.290
27	28.214	30.319	32.912	36.741	40.113	43.195	46.963	49.645
28	29.249	31.391	34.027	37.916	41.337	44.461	48.278	50.994
29	30.283	32.461	35.139	39.087	42.557	45.722	49.588	52.335
30	31.316	33.530	36.250	40.256	43.773	46.979	50.892	53.672
40	41.622	44.165	47.269	51.805	55.758	59.342	63.691	66.766
60	62.135	65.226	68.972	74.397	79.082	83.298	88.379	91.952
120	123.289	127.616	132.806	140.233	146.567	152.211	158.950	163.648